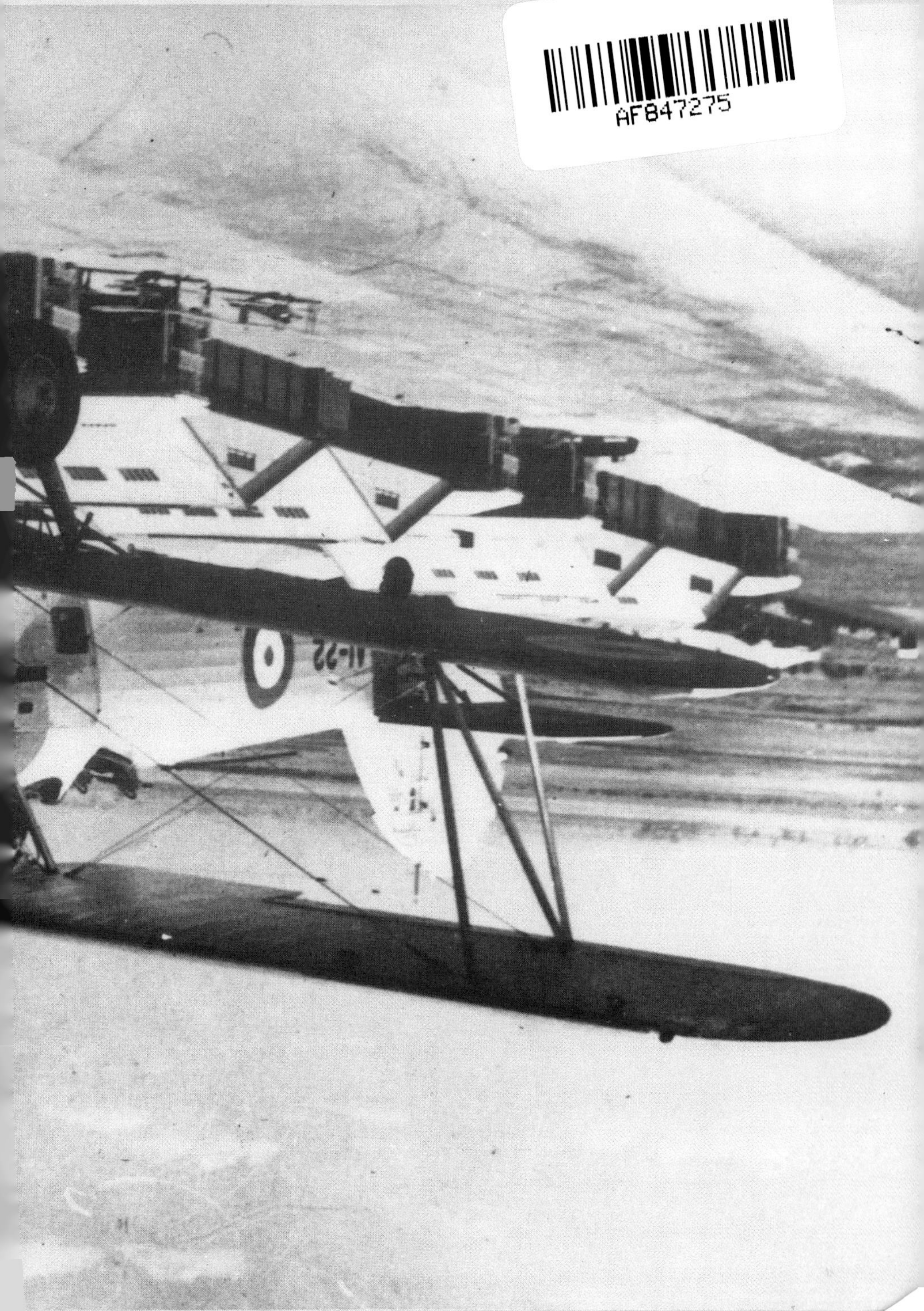

The Third Brother

The Third Brother

The Royal Australian Air Force 1921–39

C.D. COULTHARD-CLARK

ALLEN & UNWIN
in association with
The Royal Australian Air Force

First published in 1991
Allen & Unwin Australia Pty Ltd
8 Napier Street, North Sydney, NSW 2059 Australia

National Library of Australia
Cataloguing-in-publication entry:

Coulthard-Clark, C. D. (Christopher David), 1951– .
The third brother : the Royal Australian Air Force 1921–39.

Bibliography.
Includes index.
ISBN 0 04 442307 1.
ISBN 0 04 442308 x (leatherbound edn.)

1. Australia. Royal Australian Air Force — History.
I. Title.

358.400994

Set in 10/11pt Times New Roman by Excel Imaging, St Leonards
Printed by South Wind Productions, Singapore

Dedicated to members of the Royal Australian Air Force between the World Wars, with particular acknowledgement to Air Commodore 'Paddy' Heffernan who pressed so long for the story to be written.

The army and navy must recognise in the air force the birth of a third brother—younger but none the less important in the great military family.

Giulio Douhet *(1909)*

Contents

Foreword

I well recall that, when I was a young cadet at RAAF College in the mid-1950s, RAAF history was the stuff of lore and legend. Almost nothing had been written on the development, operations and activities of our nation's air force. F.M. Cutback's history of the Australian Flying Corps covered the history of Australian military aviation up to the formation of the RAAF. Events from 1939 were well known to me and my fellow cadets—even if they were not well documented—for they had occurred within our lifetimes and many of our instructors and educational staff had been participants, and they passed much information to us. The period from 1921 to 1939 was, however, a veritable black hole. When I graduated from RAAF College in 1956 I knew almost nothing of the period within which the fledgling RAAF was nurtured and developed.

In the years since then I learned a lot more but, even then, my knowledge was not great and there were many gaps within it. That was mainly because the documentation of the period has been so thin. For example, George Odgers in his 1965 *Illustrated History* of the RAAF devotes less than four pages of text to the period. *The Golden Years*, which was commissioned by the Minister for Air to commemorate the Service's 50th anniversary, devotes just 10 illustrated pages of its 124 to the first 18 years of the RAAF. Douglas Gillison in his volume of the history of the RAAF in the Second World War provided some broad background of the RAAF's development; John McCarthy in his 1976 work, *Australia and Imperial Defence 1918–39*, hinted at a real story behind the bare facts of Gillison's account; and soon thereafter Sir Richard Williams in his autobiography confirmed that there was indeed a considerable story to be told of how our air force was formed and developed.

That story has now been told. Chris Coulthard-Clark in his aptly titled history has illuminated what has been an obscured period in our Air Force's history. With the professional historian's eye for the significant and unusual and with an equally professional commitment to objectivity and balance, he has opened for us the rich and varied history of the RAAF's early years.

The Third Brother is a work of both interest and power. As I have read it I have found that I have been urging myself on, trying to read more quickly as yet another corner of our history is illuminated and one more gap in my knowledge is filled. For me, so many of the beliefs and attitudes of members of the RAAF of that period with whom I came in contact in my own career are now more explicable—and many of the effects of those beliefs and attitudes are with us to this day.

We should all be grateful that my predecessor, Air Marshal 'Jake' Newham, had the deep appreciation of RAAF heritage which was, and remains, a characteristic of his professional life for it was he who commissioned Chris Coulthard-Clark to document the history of the RAAF from its foundation until the outbreak of the Second World War. The decision to commission the history is one for which all of us who value our heritage can be grateful, and the selection of Chris Coulthard-Clark has ensured that the history is not only diligently researched but is expressed lucidly and with true regard for the humanity of the many people involved.

As the date of the RAAF's 70th birthday approaches I am delighted that the first 18 years of its existence have now been properly illuminated for our education and contemplation.

R.G. FUNNELL
Air Marshal
Chief of the Air Staff

October 1990

Acknowledgements

A great many people have been involved with getting this project off the ground and administering it while in full flight. Here, I must begin with the current Chief of the Air Staff, Air Marshal Ray Funnell, and his predecessor Air Marshal John Newham. Although the latter had retired before my appointment as 'CAS Historian', I am in his debt for the work done during his time in office to initiate the project. Air Marshal Funnell shared this keen interest, and the discussions generated by our occasional meetings in the corridor were invariably encouraging. Others in the Office of the CAS to whom I am grateful are: the three officers who have filled the appointment of director during the life of the project, Group Captains Peter Nicholson, Nick Ford and Brian Kavanagh; the three Special Projects officers, Wing Commanders Phil Morrall, Roger Parsons and Ruth Taylor (to whom fell all the negotiations for publication); the staff of the Office of the CAS, especially Warrant Officer Jeff Bromhead and Mr John Siviour; and members of the RAAF Historical Section, especially Mr Richard Fisher, Mr Bob Piper and Mr David Wilson.

Among those just mentioned, I must pay a special tribute to Dave Wilson for the many hours he devoted to privately transcribing taped interviews, and the many administrative tasks which he undertook outside his normal duties — all of which helped lift some of the burden off me and for which he has my sincerest gratitude.

Completion of this book in two and a half years has been possible solely because of the remarkable level of support which the project received from everyone associated with it. From my wife and children who put up with my periodic absences and everything else that writing a work on this scale entailed, to the many people who agreed to be interviewed or in other ways provided material of relevance to my research, the assistance given has been, at all times, unstinting. Most of those with material are listed in the Bibliography, but separate mention needs to be made here of a number of individuals whose help would otherwise go unrecognised.

Firstly, I should like to express my appreciation to Associate Professor John McCarthy and Dr Jeff Grey at the Australian Defence Force Academy, for their advice, guidance and encouragement while I completed this work as a preliminary to submitting a thesis on Australian air power in the same period. Also valuable was the assistance of Mr Arthur Skimin, of the Defence Department's Archives and Historical Section, whose staff identified relevant record holdings and made available taped interviews with prominent Air Force personalities no longer living. Early in my research Group Captain Keith Isaacs very kindly gave me the benefit of his broad knowledge of the field as a steer towards the major sources and publications which were relevant. Mr Colin Owers generously provided me with notes and photocopies of material derived from his own research, hence saving me an enormous amount of time in the Archives.

The copying of photographs was a task expertly undertaken by several agencies: in Canberra, Mr Gerry Sebastian and his staff at the Defence Public Relations Photographic Section handled a number of requests involving material on short-term loan, and Squadron Leader Norm Cooper also assisted in this way using the resources of the RAAF Base, Fairbairn; the bulk of this task, however, was handled by Squadron Leader Graham Walton and the staff of the RAAF School of Photography, East Sale, and the Central Photographic Establishment, Laverton. Again, all have my thanks for their efficient and ready cooperation.

At the RAAF Museum at Point Cook, I was always made welcome and afforded every assistance by the successive officers-in-charge, Wing Commander Peter Allen, Squadron Leader (now Wing Commander) John Strickland and Squadron Leader Jeremy Clark. Special mention must be made of the Curator, Warrant Officer David Gardner, who handled all arrangements that enabled me to get the most out of my visits and was a most hospitable host besides. Mr John McKenzie and LAC Robyn Hirvonen also have my thanks.

Among the many groups and institutions whose assistance deserves special mention are the RAAF Association and the Pre-War RAAF Association — both of which put me in contact with valuable sources. Much of my research was conducted at the National Library of Australia and the Australian War Memorial at Canberra, where the staff were — as always most helpful. During a visit to London in 1989, valuable assistance was provided by the Royal Air Force Museum, and especially the Keeper of Aviation Records there, Mr Peter Murton; the Public Record Office; and the Imperial War Museum.

While all those mentioned have helped me the better to do my job, responsibility for any errors which appear, and for the opinions or conclusions expressed, is mine alone.

Introduction

In the years before the First World War, Australia—alone of Britain's Dominions—followed the mother country's lead in laying the foundations of military air power. In April 1912, when Britain created the Royal Flying Corps (comprising naval and military wings, and a central flying school to train pilots for both), the Australian Army was already recruiting competent 'mechanists and aviators'[1] for a Central Flying School (CFS) of its own, planned at the site of the new national capital, Canberra. The proposed location of the school was changed in 1913, to an area of 734 acres (295 hectares) at Point Cook on the western side of Port Phillip Bay in Victoria, on the basis that this was a safer and less isolated site.[2] Placing the CFS by the sea also facilitated proposals that it 'should be available to the Naval as well as the Military Forces'.[3]

Early in 1914 Australia's aviation arm was almost ready to take to the air. Five aircraft had been purchased from Britain to serve training needs, as well as equipping the single flight of an air squadron which would comprise the initial element of Australia's flying corps.[4] An Aviation Instructional Staff had also been formed. This comprised two officer pilots — 28-year-old Henry Petre (pronounced 'Peter') and 26-year-old Eric Harrison — and four air mechanics: Richard Chester, Ted Shorland, Cyril Heath, and George Fonteneau.[5] All had been engaged in England where they received their training, but Chester and Harrison were actually Australian — the latter being a former motor mechanic from Castlemaine, Victoria.[6]

Test-flying began at Point Cook from 1 March 1914, and by July, when the peacetime Army establishment was amended to acknowledge the existence of No.1 Flight of the Australian Flying Corps (AFC),[7] plans were underway for the commencement of flying training. Applications were called for in July, but war had begun in Europe before the first four students — Captain Thomas White and Lieutenants David Manwell, George Merz and Richard Williams — arrived on 17 August. It was Williams, the only one of the four who was a member of the permanent staff (the others held commissions in the citizen forces), who became the first to qualify as a pilot on 11 November.[8]

The staff of the CFS was also augmented at this time with the addition of Sergeant S.J. Hendy, transferred from the garrison artillery to become squadron quartermaster-sergeant in August, and six airmen — Privates Les Carter, Norman Dyer, George Mackinolty, H.A. McIntosh, R.P. Mason, and Arthur Murphy. The latter were all soon promoted corporal following the promotion of the original four mechanics to sergeant rank, and Chester was made warrant officer from 1 July 1914.[9]

The new flying school's existence at the outbreak of war made it possible for actual air operations to be contemplated within Australia's region. When information was received in late November about an armed enemy base reported to be located on the Sepik River, in the German territory of New Guinea which had been recently brought under Australian control, a military operation was planned to deal with it. An infantry battalion being assembled at Sydney for garrison duties in German Pacific territories north of New Guinea was detailed for this mission, taking with it two aircraft from Point Cook which were hastily sent by train from Werribee and loaded in crates aboard the troopship. With the machines went two pilots and four mechanics.[10]

With Army authorities anticipating a 'decent scrap',[11] the air party busied themselves during the voyage north. As no prepared landing fields would be available, the aircraft were expected to operate as floatplanes. Only one of the embarked machines was a seaplane however — having been a gift from a public-spirited businessman — and plans were consequently carried out to fit the second machine with floats *en route*. Two of the mechanics also filled in their time improvising a supply of aerial bombs by adding small propellers to the nose-caps of 16-kg artillery shells.[12] In the event, no sooner had the vessel arrived off Madang than the alleged enemy base was found to be a fiction, a fraud perpetrated by a paid informer on gullible intelligence authorities in Sydney.[13] The force sent to deal with it went on to Rabaul, where the aircraft — still in their crates — were stored in the Customs House before being returned to Melbourne in January 1915.

Though this opportunity to lay the basis of a tradition for Australia's fledgling air corps proved abortive, a second chance immediately arose. In February a request was received from the Indian government for aerial assistance during a planned campaign in Mesopotamia (modern Iraq). The Australian government agreed to help, despatching what became known as the Half-Flight, AFC, from Melbourne on 20 April. Four officers and 41 other ranks served in the disastrous Mesopotamian campaign of 1915–16, the pilots becoming flying officers in the RFC.[14]

At the end of 1915 the Army Council in London suggested that Dominion governments might prefer to raise complete squadrons for service with the military wing of the British flying corps, instead of encouraging enlistment in those services by individuals. According to the official history of the AFC: 'The Australian Government, perceiving the value of an air force for future training as well as for the present emergency, promptly adopted the suggestion'.[15] By 1916 the AFC had expanded to four front-line squadrons (three on the Western

The entire personnel of the Central Flying School, Point Cook, in late 1914. *Rear row:* A.W. Murphy (*), W.J. Lord (caretaker), L.G. Carter (*), N.M. Dyer (*), R.P. Mason; *centre row:* G.J.W. Mackinolty (*), V. Dearman (clerk), G.A. Fonteneau (*), A.E. Shorland (*), C.V. Heath, — Pivot; *front row:* S.J. Hendy, G. Merz, H.A. Petre, E. Harrison (*), T.W. White, R.H. Chester; *seated at feet:* R. Williams (*), D.T. Manwell. Manwell, Merz, White and Williams were students on the first flying course; Petre and Harrison were the instructors. Those indicated with an asterisk (*) served in the RAAF after the First World War. (*RAAF Historical*)

front and one in the Middle East), and before the following year was out had established, too, a training wing of four squadrons based in Gloucestershire, England.[16]

Providing aviators on this scale for overseas service naturally entailed a substantial commitment of resources back in Australia. Eight courses totalling 120 trainees entered the CFS during the war years.[17] In addition to the output of trained pilots from this source, a second flying school with two instructors and biplanes bought from the United States was opened in 1916 by the New South Wales government on a site at Richmond, near Windsor, outside Sydney. Six courses averaging 13 trainees passed through this State Aviation School before it ceased to operate in late 1918,[18] although its graduates went mainly to the British air services rather than the AFC.[19]

By the war's end, a great many Australians had gained experience in air operations through their service. According to one account, 460 officers and 2234 other ranks had served with the AFC, and about 200 other pilots and observers flew with other flying services during the 1914–18 War.[20] As a statement of Australia's total contribution to winning the air war, this estimate seems too low, however, as although it appears to include the 200 RFC commissions offered to Australians by the British War Office in July 1916,[21] there were many more men besides who gained entry to either the naval or military wings of the British air service. The actual total of Australians who were members of the various air services must be 3–4000.

Though Australia's part in the air war was far smaller than that of Canada, which provided some 21 000 airmen,[22] the fact remains that until the war's last days there were no formations that specifically had a Canadian identity. The AFC, on the other hand, existed as an embryo national air force, quite distinct even while it operated within a much larger British structure. Though British authorities persisted in giving Australian units RFC designations — the combat squadrons (Nos.1–4, AFC) were known until January 1918 as Nos.67, 68, 69 and 71 (Australian) Squadrons, RFC[23] — their identity remained unmistakable and their contribution readily recognisable.

Sixty-five pilots who were members of the AFC, or were Australians serving in the British naval or military flying corps, became accredited aces by shooting down the five enemy machines needed to qualify; between them they accounted for over 650 aircraft.[24] One Australian — Lieutenant Frank McNamara of No.1 Squadron, AFC — also achieved the distinction of winning the Victoria Cross, and many, many more received lesser decorations. While four Australians serving with the British naval or military flying corps rose to the command of a wing (that is, a formation of several squadrons), one AFC officer also attained this level. This was Lieutenant-Colonel Richard Williams, who succeeded from the command of his own No.1 Squadron in 1918 to that of 40th Wing, or the 'Palestine Brigade, Royal Air Force', comprising the four squadrons which played the leading part in the air war against the Turks.

In France and Belgium, Mesopotamia, and in Palestine, Australians had their eyes opened to the transformation of the battlefield which came with the application of air power. In the great variety of aerial operations undertaken on the Western front, and in the night defence of England against Zeppelin raids, they witnessed the new ways in which war would be fought. Even in Palestine, something of a sideshow in relation to the European theatre, AFC members were afforded the most graphic demonstrations of the effectiveness of air power, as in the case of a column of the Turkish 7th Army — 7000 strong — which was caught retreating along a narrow wady in September 1918 and annihilated in four hours' solely by air attacks.[25]

Australians had also been participants in some of the significant technical innovations which revolutionised the conduct of battlefield operations, whether it was dropping food and supplies (including a millstone delivered by parachute) to the garrison at Kut beseiged by Turks in Mesopotamia;[26] or No.1 Squadron's attempts using local resources to match the German achievement of synchronising a machinegun to fire forward through an aircraft's revolving propeller[27]; or the hasty devising of an aerial system of supplying ammunition to forward troops during the Australian Corps' attack at Hamel.[28]

As the war came to an end in the last months of 1918, the thoughts of

Lieutenant-Colonel Williams, a figure of major importance in the story of Australian air power, were turning to the shape that the future air defence of his country might, or should, have. Writing on 21 December 1918 from 'In the Field', Williams reminded the General Officer Commanding the AIF in Egypt (Lieutenant-General Sir Harry Chauvel) that he had been 'detailed by you to put in writing, at the conclusion of the War, the results of the experiences of the Australian Flying Corps on this front from all points of view in order that lessons learnt shall not be lost'.

Now, he said, 'owing to unofficial information that it appears to be the intention of the Commonwealth to have its Squadrons, at present in England, France and Egypt, returned to Australia with complete equipment' to form the basis of a peacetime air force, he was taking the opportunity to present his views on the types of aircraft and other equipment best suited to Australian needs and conditions.[29] Though Williams put his opinions at length, there is no evidence that any action was taken to communicate them to Melbourne or London — and it was here that the post-war future of Australian air forces was being determined.

Map of places referred to in the book

1

Decision for an air force

The birth of the Royal Australian Air Force in the years immediately following the end of the First World War was not accomplished amid unanimity and general confidence that this was a necessary or even an appropriate step. Instead, it was a compromise solution arrived at after three years of constant and often futile debate which was to leave a lasting legacy of bitterness within the Australian defence community. As this chapter describes, the decision to form the nation's air defence resources as a single new service, separate from either the Army or the Navy but serving their needs, was both the product and the further cause of inter-service rivalry.

Competing Army and Navy schemes

Even before the First World War ended, thinking in Australia had begun to focus on establishing aviation in the nation's defence services on more than a special wartime basis. The result was that in the first half of 1918 both the Army and Navy evolved schemes to establish separate air services meeting their particular requirements. The Royal Australian Navy (RAN) had begun developing proposals for an Australian naval air service in June 1917, but got nowhere before obtaining an experienced airman from the British Admiralty some six months later.[1] This was 32-year-old Commander Hugh Maguire of the RNAS, who arrived in Melbourne in May 1918 as the Naval Board's air adviser with the rank of lieutenant-colonel in the newly inaugurated RAF.

Pre-empting Navy's move, on 29 April 1918 the CGS (Major-General Gordon Legge) put forward an Army bid for an air corps of 300 officers and 3000 other ranks and called for an immediate start to the construction of 200 aircraft and 12 balloons. Behind Legge's request at this time was the German offensive launched in France in March, which was enjoying considerable success. The CGS feared that the Allies might be forced to a disadvantageous settlement of the war in Europe, leaving Australia exposed to Japan's ambitions in the Pacific, hence Legge attached great urgency to authorisation of his plan. Cabinet was sufficiently impressed by Legge's arguments that on 1 May the

Minister for Defence, Senator Pearce, instructed him to initiate the steps necessary to get local manufacture underway and to produce estimates of the cost involved.[2]

A fortnight later, on 16 May, the Naval Board had Maguire's first report before it. The scheme envisaged a naval air service of 2000 officers and men operating three airship stations at Sydney, Melbourne and Fremantle; two schools; and two seaplane stations.[3] Anticipating the Board's agreement with the broad lines of his plan, Maguire actually began inspecting sites for a seaplane school, recommending an area on Corio Bay at Limeburners Creek (now the northern outskirts of Geelong) which included a portion of the Avalon Estate owned by the Austin family. Maguire's scheme was put to the government on 24 June with a request for £500 000 to allow a start to be made in implementing the program. Faced now with two schemes, Cabinet decided to refer both to the Council of Defence for detailed consideration.

By this time, the CGS had reworked his initial plan prepared back in April, and the day after the Naval Board revealed its scheme Legge submitted a revised program for a military air service which included fifteen air squadrons, schools, factories and repair shops, aircraft parks and depots, and a balloon wing. To man this service, 654 officers and 7209 other ranks were required, most of them being citizen force personnel. This whole scheme was expected to cost £1 million in initial capital outlays, and £475 000 in annual recurring expenses.[4]

The closeness of the dates associated with these events suggests that the presentation of competing schemes was probably not entirely coincidental but reflected an underlying concern by Navy and Army not to be outdone by each other. In any case, the basis for a fierce battle over funds had now been laid. Understandably, a committee appointed by the Council of Defence on 26 June to assess the schemes was unwilling to make a choice between the two bids, given that both CNS and CGS insisted that their respective programs were essential as they stood and no changes could be contemplated.[5] The committee accordingly interpreted its function as merely to ensure that the schemes were presented in a form which allowed the government to judge the costs involved.[6]

Presumably alert to the emerging inter-service wrangling involved, Cabinet also shrank from making the hard decision when it considered the recommendations put to it in mid-August. Instead, a figure of £3 million was decided upon as an allocation for defence aviation over three years to June 1921 and the Defence Council was directed to arrange how the sum was to be allocated.[7] While large, this funding was not sufficiently generous to allow both schemes to be adopted without modification, so that both parties were faced with the need to compromise.

This thorny problem having returned to it unresolved, the Defence Council appointed another sub-committee to consider the options. This established only that neither service was willing to budge, and from its first meeting in October the sub-committee became the forum for vehement disagreement between the senior Army and Navy representatives.[8] Navy argued for it to be allowed to start its own program with half the money available. Legge—realising that this would leave insufficient funds for Army to do much at all—flatly refused to modify his scheme, arguing that 'without an air service of 400 planes, 600 officers and 7000 other ranks it would be murder to send our men to meet a properly equipped enemy'.[9]

An important modifying influence in these exchanges now proved to be Maguire, whose views had changed from those he expressed soon after arriving in Australia. On 17 September he proposed to the Naval Board that Australia follow the British course and establish a single air service with its own minister and staff. The Naval Board allegedly did not endorse the idea but failed to clearly reject it either, with the result that Maguire thereafter acted as if the proposal had Board approval.[10]

If the Naval Board was not actually keen on Maguire's idea, the CGS strongly objected when it was suggested in the sub-committee. As a result of this deadlock, the report presented on 18 October by the chairman, George Swinburne (chairman of the Defence Board of Business Administration), was a mish-mash of half measures which held little prospect of advancing the project very far at all. Importantly though, Maguire's suggested approach was taken up by Swinburne and included as a recommendation that 'the whole air service should be under one administration and authority', a clear recognition that a single service meeting the needs of both the Army and Navy represented the only immediate and economical way forward.[11] Legge still refused to countenance this solution and submitted a minority report warning that any attempt to establish a combined service would be inadvisable.[12]

After yet another sub-committee of more junior officers went over exactly the same ground again, concluding once more that 'the most efficient and economical way to establish Aviation in Australia was to inaugurate a separate Air Force under its own Minister and staff', Swinburne produced a new report on 20 January 1919 which plumped firmly for this course. He recommended the immediate appointment of a temporary body to work out complete details of a new 'Australian Flying Corps' and charged with gradually implementing these. With Legge still dissenting, the Defence Council accepted the Swinburne proposals and recommended them for government approval.[13]

The temporary body established to organise the new air corps was the Air Service Committee (ASC), which met for the first time on 31 January 1919. The members of this body were Swinburne (chairman), Maguire and Legge, the latter now apparently resigned to seeing the course he had hitherto resisted being adopted. As proposed by Swinburne, the ASC was also given administrative powers, including authority to spend £500 000 for buildings and plant.[14] At long last the bitterly disputed question of whether aviation was to be embodied in separate corps of the existing naval and military forces, or constituted as a new, joint entity, appeared to have been settled.

The achievement of this outcome had not been without considerable cost to those involved. On 20 November 1918, Maguire applied to the Naval Board for permission to return to England, citing as his reason the tortuous and (till then) inconclusive passage of the air service proposals: 'I was sent out here to advise on the subject of Naval Aviation and having given that advice I submit I can do no more under present circumstances . . . I would strongly submit that without wasting more time and delaying indefinitely I may be permitted to proceed as suggested'.

Notwithstanding the resolution reached two months later, Maguire still saw no reason to stay. Arguing that someone needed to deal with the British Air Ministry about the staff and personnel required to start the Australian air service, he continued to plan for his return to England. Taking advantage of the knowledge that he had acquired of Australian conditions, it was decided to have Maguire assume the post of Australian liaison officer when he arrived back in the Air Ministry in May 1919. To assist Maguire in selecting personnel, the Minister for Defence also decided to retain in London the senior AFC officer then filling the post of Staff Officer for Aviation at AIF Headquarters, Lieutenant-Colonel Richard Williams.[15]

Work of the Air Service Committee

In the months following the formation of the ASC, a myriad of details entailed in bringing the new flying corps into being needed resolution. Among the important questions requiring consideration was that of acquiring aircraft. The commander of the AIF overseas (General Sir William Birdwood) had already cabled on 29 January 1919 that, after consulting the Air Staff in London, he understood 'Australia is entitled to receive equipment for four Service Squadrons'. What he meant by this is unclear, as there was no suggestion at this stage of Australia receiving a free gift of air equipment and any aircraft acquired had presumably to be paid for. Birdwood nonetheless recommended that units should each be equipped with a different aircraft type, and proposed these types comprise single-seat Sopwith Snipe fighters, two-seat reconnaissance Bristol fighters, DH9A day bombers and Vickers Vimy long-range night bombers. These squadrons, he suggested, should operate as cadre training units, holding half service aircraft types and half training machines. In addition, therefore, Australia would need to obtain a trainer type, which he suggested should be the Avro 504.[16]

Accepting most of this advice, the ASC recommended the purchase of 15 of each of the types listed by Birdwood; with 20 Avro 504Ks and 12 Sopwith 'Pup' Scouts which were already on order for the CFS at Point Cook, these numbers were considered sufficient for initial purposes.[17] On 1 March 1919 the ASC reduced the order for Vimy aircraft to just four machines. In June these arrangements underwent further adjustment after the British government announced that it proposed to make a gift of 100 aircraft to any of the Dominions wishing to establish their own air forces, an offer which Australia readily accepted.[18]

In its first report to the Council of Defence dated 8 February, the ASC stated its initial plans for the new corps. In addition to the existing CFS, it was planned to have air stations located at Corio Bay and near Sydney. There would now be only one seaplane or flying-boat squadron, four landplane units, and single airship and balloon sections. Strength of the corps was expected to grow to 1350 officers and men within two years, it being estimated a force this size would cost about £500 000 in capital expenditure and a roughly equivalent sum in annual expenses. To staff the Air Board which it was proposed would administer the service, the committee proposed an organisation totalling 35 personnel: one brigadier, two colonels, four majors, four captains, three military clerks and 21 civilian clerks. This report was adopted by the Council of Defence on 14 February.[19]

A fortnight later Maguire, due now to sail for England on 20 March, suggested that applications for appointment to the new corps should be called for from among AFC members presently preparing for repatriation to Australia. Steps were initiated in April to fill the first posts required—a Director of Air Services (brigadier), a Director of Equipment (colonel) and two station commanders (lieutenant-colonels). Officers and men with experience of seaplanes, and airships and kite balloons, were also wanted. By June the Air Council was ready to submit names for the two senior posts mentioned, but this matter went no further when Prime Minister Hughes, then overseas, cabled back to the acting prime minister (W.A. Watt): 'I earnestly hope that you will not appoint any but Australians to these or any other positions. In my opinion such a policy would be not only quite wrong, but exceedingly foolish.'[20] A decision was accordingly made not to proceed with these appointments for the time being, because they were considered unnecessary until the scheme was 'more matured'.[21]

Although arrangements were proceeding, there remained many difficult areas to be sorted out. In the matter of bases, for example, there was agreement about the proposed seaplane station on Corio Bay and detailed plans for laying this out were begun by the Works and Railways Department.[22] However, a site had still to be located for the air station near Sydney. Several sites were considered, including the Randwick Rifle Range, an area at the confluence of Shea's Creek Canal and Cook's River, and an area near Liverpool. During the deliberations taking place at this time, it first emerged that the New South Wales government had in mind offering to the Commonwealth the site of the aviation school it had been operating at Richmond, outside Sydney, since 1916 and which was about to be closed down. Nonetheless, as late as August the Army still favoured the inland Liverpool site, a choice which scarcely seemed designed to serve naval as well as military needs.[23] As Swinburne admitted, 'The requirements of the Navy on a site for seaplanes . . . will need separate treatment'.

In May, the process underwent a delay occasioned by the visit of Admiral of the Fleet Lord Jellicoe of Scapa to report on Australia's maritime defences. The government was understandably unwilling to take action until the findings of Jellicoe's report, submitted in August, were known. In the event the program recommended by Jellicoe, involving twelve flying-boat squadrons, with additional units of torpedo-bombers, ships aircraft, and harbour defence fighters, appeared so lavish as to have no bearing on arrangements at all.[24] The most important consequence of this visit was the taking of a decision (on the advice of the Admiral's air adviser, Lieutenant-Colonel E.D.M. Robertson) to defer, 'pending further developments', the formation of airship and balloon sections as part of the new air service.[25]

The Jellicoe visit was not the only pressure delaying the planning process being carried out by the ASC. Private calls were also reaching senior government figures to defer matters pending the return of AIF men with the most up-to-date experience of aviation in the war. One of those urging this course was Australian war correspondent H.S. (later Sir Henry) Gullett, who wrote to Prime Minister Hughes from London on 3 May 1919 to comment on the fact that the ASC had been formed to function as 'something in the nature of an Air Board'. This was likely, he warned, 'to cause deep dissatisfaction in the AIF and probably result in much trouble with the Government'. Gullett particularly questioned the credentials of both Legge and Maguire for the task entrusted to the ASC, and suggested that the prime minister should ensure matters were placed on hold at least until the return to Australia of Major-General Sir Brudenell White.[26]

Whether it was calls such as these which finally decided the government, or the more general need to delay all non-essential expenditure to reduce the financial burden imposed by the war, matters now came to an abrupt pause. The hard-won progress of more than eighteen months of bitter debate and seemingly endless committees was set aside when Cabinet abruptly suspended the whole air service scheme on 15 September, leaving only a limited provision on the 1919–20 estimates to keep the CFS functioning as little more than a stores section.[27]

This was too much for Maguire. He had gone to London with definite instructions to negotiate with the Air Ministry for the key personnel and equipment needed for the Australian air service, then been forced to watch as later instructions cancelled virtually all the arrangements he had made except in regard to the flying-boat squadron. Clearly frustrated at seeing all his efforts thrown away, he resigned his RAF commission and asked to be released from duty as Australian liaison officer from 15 October 1919. The day before he made his departure, Lieutenant-Colonel Williams also embarked to return to Melbourne for duty back in Australia.[28]

Understandably, Swinburne was also upset at the government's decision, especially after all that it had entailed for him personally to patiently take matters to the stage which had been reached. Although the scheme was not actually cancelled, the ASC was now wound up and held its final meeting on 8 October. By this time Swinburne was the sole remaining member of the original Board of Business Administration, the other two members having severed their connection with the Defence Department in July 1919. Although he continued to have a role in major issues of defence policy, both by virtue of his having a seat on the Council of Defence and as Civil and Finance Member of the Military Board, this evidently did not ease his disappointment over the air service question. According to his biographers:

> This enforced impotence of a highly skilled body of men irked him and no doubt made him anxious to resign. In September 1919 he asked to be relieved, and though it was then impossible, he continued to press his request, and in March 1920, he left the department.[29]

Although suspension was a serious setback, the air service scheme was not entirely dead but lay waiting for further events to revive it.

The question of control

Among the matters receiving the attention of the ASC when that body first began considering the detailed outline of the proposed single air service, was the potentially difficult question of control. In fact, this issue was to emerge as the new focus for bitter inter-service competition which was to further delay the birth of the new air force and bedevil its early years. The first round in this debate took place within the ASC in February 1919, when the question was raised whether legislation needed to be put through Parliament covering the air service. The committee initially took the view that any matters that arose could be adequately dealt with by regulation, but early the next month Legge presented the committee with draft regulations he had prepared, under which the air service was to be raised and maintained as part of the military forces though placed under the dual control of the ministers for Defence and the Navy, and an Air Board. The minutes of the ASC record that: 'After discussion . . . Colonel Maguire expressed his strong desire that the Air Administration should be given complete control and that neither the Military Board nor Naval Board should have any jurisdiction in the matter'.

After further considering this aspect, the committee recommended to the Council of Defence that a 'Joint Air Force' should be created 'separate from both the Naval and Military Forces of the Commonwealth and under legislation to be provided in that behalf'. The rest of Legge's proposals were mainly adopted, including the intention that:

> the organisation, administration and control of the Australian Air Force should be subject to the joint control of the Ministers for the Navy and Defence exercised through a Directorate to be known as the Air Board and consisting of, say, three persons appointed by them, consisting of a Director of Air Service, a Director of Equipment and a Civil Member. Orders for the Air Force should be issued by the Air Board.[30]

Unfortunately this was not the end of the issue, as in June 1919 a special committee had been formed under the chairmanship of Swinburne to consider 'Certain Matters of Defence Policy', among which was the 'proposed policy and programme for the establishment of an Air Service'. The other members of the committee, which sat for five days and completed its report on 30 June, were Generals White, Legge, and Sir James McCay, and Claude Reading (a well-known company director who was then Swinburne's deputy on the Board of Business Administration). The most important aspect of the committee's findings on the air question concerned the system of control for the new service.[31]

The course recommended for adoption was the creation of an 'Australian Air Corps' formed into two wings meeting the needs of both the other two services—that is, essentially a return to the pre-war British model of the RFC. The Australian Air Corps, it was proposed, would be under the control of one Minister (the Minister for Defence *not* the Navy) and administered by an Air Board composed of members of the naval and military forces, with the Board having a 'permanent Administrative Committee to control the corps'. The wings allotted to Navy and Army were to be controlled by the naval and military boards respectively, while the CFS training pilots for both the wings would be under the control of the Air Board.

The conclusions of this body actually took matters no closer to resolution, because it failed to take account of naval opinion. Not only had the committee lacked any naval representation, but the new CNS (Rear-Admiral Sir Percy Grant) was unaware of its existence until he received a copy of its report. In the circumstances he was not at all inclined to endorse findings which, in effect, would see Navy lose any control, via its Minister, over air service administration. CNS fired off an angry letter to the prime minister strongly opposing the report's findings and suggesting that the entire air service scheme be held over for the time being, at least until Lord Jellicoe's report on Australian naval defence was available.[32] It was at this point that Cabinet suspended the air service scheme, this particular dispute possibly having been an additional factor in the government's decision. Not until December did the aviation question resurface, when CNS—perhaps worried that Navy Office continued to be excluded from planning—asked the Minister for Defence for any new information on the air service proposals.[33]

The result of CNS's enquiry was yet another meeting on 14 January 1920 involving Swinburne, Legge, White, and Grant, this time with the ministers for the Navy (Cook) and Defence (Pearce) present. Also in attendance were Lieutenant-Colonels Williams, fresh from London, and Stanley James ('Jimmy') Goble of the RAF, an Australian who had served in the RNAS before taking over from Maguire as the Naval Board's air adviser in March 1919. The meeting was told that, although the air service scheme had been suspended, Navy 'desired meanwhile that an Air Board, which it had been proposed to establish, should take over the administration of present air defence activities, including the Central Flying School, in order to provide better facilities for the development and study of the Naval aspects of Air Defence'.[34]

In the discussion that followed, Legge attempted to argue that the facilities requested by the Navy could be provided by the temporary Australian Air Corps (AAC) which the Army had just formed by reorganising the CFS. This did not satisfy Grant, however, who insisted that if the AAC was to meet Navy needs too then it needed to be removed from Army control. He wanted to see an air board formed, both to administer the AAC and to advise on organising the permanent air service which would replace it.[35] Having heard both sides of the argument, Cook and Pearce agreed to recommend to Cabinet that an air board be constituted as urged by Navy.[36]

Over the next fortnight the two departments nominated the officers who were to be their representatives on the Air Board—Brigadier-General Thomas Blamey and Williams for Army, Captain Wilfred Nunn and Goble for Navy. On 16 January Pearce also wrote to Cook indicating that he was prepared to make available a public servant, Charles Auty, to temporarily act as secretary to the Air Board if this was acceptable to Navy. At the first meeting of the new body on 29 January the Minister for Defence outlined the work he expected the Board to perform, which was principally the preparation of a scheme for a single air service which Pearce could submit to the Council of Defence the following week. While leaving Board members free to make recommendations on matters of control and administration if they chose, Pearce cautioned them about financial aspects and the fact that the government did not propose to establish a separate ministry.[37]

Finally, Pearce indicated to the meeting that, pending the adoption of a permanent system of control, Williams (as Acting Director of Military Air Services) would be responsible to the Air Board for all matters connected with the existing AAC. This action later drew protests from Navy that there was no need for an appointment which effectively made Williams head of the existing air force and that, if such a position was necessary, then it should be held jointly by representatives of Navy and Army. Only on the understanding that the post was for administering the existing small unit at Point Cook did Navy acquiesce.[38]

By 7 February the new Air Board had produced the report Pearce required. This called for a six-year program which, on completion, would see Australia with an air force of 36 squadrons having 644 aircraft on strength and a further 322 machines in reserve. Costs of this organisation were admitted to be only best estimates, but were expected to total over £4 million in the first two years alone. As the report made clear, the role of aircraft in the defence of Australia was not viewed as involving anything more than 'auxiliary work for the Army and Navy', a fact which was clearly reflected in the basis for calculating the numbers of squadrons and aircraft. All that had been done was for an Army wish-list of 21 squadrons to be added to the Navy wish-list of 15 squadrons embodied in the Jellicoe report. The report's remarks on the issue of control also revealed the extent to which the other services saw their vested interests to be at risk:

> . . . the Australian Air Corps for the present will have no independent function other than in its relation to the Naval and Military Forces . . . It is vital, therefore, that the controlling body of the Australian Air Forces should be strongly representative of the Naval and Military forces, in order to insure proper co-ordination of the Naval and Military requirements. At the same time, it is obvious that a body of this nature could not control the internal organization and administration of the Air Corps.
>
> The administration and discipline of any corps can be best effected by a Commanding Officer. There are, however, quite definite objections to this as regards the Air Corps. If it were placed under a definite command, it would be essential that the Commander should have direct access to the Minister in charge of the corps, and it would be trusting to personalities too much to insure that this officer did not tend to eliminate the Naval and Military controlling body. In fact, it might be forecasted with certainty that should a strong personality obtain the command this result would be inevitable. This would affect the value of the corps in the defence of Australia in a very adverse manner.[39]

The report went on to state that Board members were unable to come to agreement over how control of the air force was to be effected, with Navy claims to the right to appoint the first director being rejected by Army. Alternatively, the naval representatives proposed a system of joint control which envisaged having one naval and one military officer of equal rank heading their own wing of the corps. This solution did not meet with the approval of Army either, on the grounds that it would 'tend to divide the Air Corps into two opposing corps, and bring about similar conditions to those which existed in England prior to the organization of the Royal Air Force under one control. It is considered desirable to aim at unity within the corps from the outset.'[40] Unable to agree, the Board referred this vexed question back to the Council of Defence.

With the Air Board deadlocked over a crucial point of policy, and having presented a supremely naive and quite inappropriate bid for resources, the air service question palpably languished for almost five months. In his autobiography written many years later, Williams distanced himself from the Air Board's report, claiming that it had been the work of Blamey and that although he had signed it there were aspects with which he had disagreed:

> I was, of course, still an army officer. I had not previously worked with General Blamey and found him quite different to other senior officers in that he preferred to state his own opinions quite definitely and was disinclined to discuss them with juniors.[41]

The government reportedly accepted the Air Board scheme on the recommendation of the Council of Defence, but there was virtually no prospect that the funding needed for implementation would be forthcoming. Moreover, despite the apparent success of the 14 January meeting, Navy was still smarting over some aspects of the Army approach to the interim air corps at Point Cook. A naval historian records that when draft regulations were prepared under the Defence Act to formalise the reorganisation already effected at the CFS from 1 January, CNS complained that this was 'the thin edge of the wedge to obtain military control over all aviation'. Even the name of the AAC smacked to him of a 'military, not joint, organisation', prompting him to propose the early establishment of an air council to ensure joint control, as well as an air board to provide for the administration of the force.[42] These were part of the scheme put to government by the Council of Defence in the wake of the Air Board report.

At this point there was a change in the nomenclature of the various bodies proposed for the control of the air service. The permanent administrative committee which had been discussed previously was now planned to assume the title of Air Board when formed, while the superior coordinating body initially having that name would be known as the Air Council. On 17 June Pearce sought the Navy Department's nomination of representatives on the Air Council. More than a month passed before Cook replied to this request, the reason being a rear-guard action by the Naval Board to convince their Minister of the need to limit the influence that Army might have over the air force. The Board discovered to its frustration, however, that Cook refused to sign the letter it had drafted taking up the issue.[43]

Not dissuaded by their Minister's attitude, the Board pressed the point by preparing an even more strongly worded draft:

> After careful consideration, and from experience of the Air Board as at present constituted, the Naval Board wish to point out that they are of opinion that a radical change will have to be made in former proposals as to its constitution, as they feel sure that the proposed system will prove unworkable.[44]

Mention here of the Navy's experience of the Air Board 'as at present constituted' was, as one naval historian points out, probably a reference to an incident that month in which 'Williams antagonised the Naval Board by interfering in the naval chain of command'.[45] What had happened was that an aircraft from Point Cook, an Avro 504 floatplane, had been embarked in the Navy's flagship, HMAS *Australia*, to take part in seaplane trials. Three members of the AAC—Captain H.F. De La Rue and two mechanics, Sergeants R. Grange and C.C. Collins—were to accompany and operate the machine; in accordance with the Defence Act, they were required to volunteer for service outside Australian waters and to submit to the Naval Discipline Act.

Prior to embarkation, Williams had told the commander of the CFS to give De La Rue instructions about rendering reports on the trials, as he wanted information 'as will enable us to decide whether the carrying of Float Seaplanes on ships . . . is a practicable and desirable proposition'. Notwithstanding Williams' natural interest as Director in a matter of this kind, the Naval Board had bristled at the very idea of members of a ship's company reporting to any authority outside naval channels and insisted on 14 July that the orders issued to De La Rue be cancelled. All reports were to be made to the captain of *Australia*, who would send them on to the Naval Board for eventual transmission to Air authorities.[46]

As Cook refused to sign the revised letter drafted by the Naval Board, on 20 July the Board produced an alternative scheme for his consideration. Under this proposed organisation, the Air Council was to be presided over by a Minister other than Defence or the Navy and would have two representatives each from the Army and Navy as its members. Two separate naval and Army wings would each be under their own directors, while a third director controlled an equipment pool, with all of these officers being responsible to the Air Council. In accordance with this scheme, the Naval Board argued that 'the Navy Department should be voted the Naval proportion of the air Vote to inaugurate its own flying wing, and carry out its own system of training and administration, and directly control the operation'. Not surprisingly, Cook refused to allow this scheme to go to the Council of Defence, despite Grant's request to be told what the Minister objected to in it 'so that the Board could reconsider their proposal'.[47]

By the time the Navy nominees for the Air Council—Captain Charles Hardy (Second Naval Member of the Naval Board) and Goble—had been conveyed to Pearce on 21 July, the latter Minister had already approved, on 14 July, of the Army representatives being the CGS (White) and Deputy CGS (Blamey). With all nominations now in, on 23 July he directed that the Air Council was to be constituted on a temporary basis until the Estimates were passed by Parliament. The first meeting of the new Air Council was held the very next day, attended only by Hardy, Goble and Blamey, and with Williams in attendance as a consultant. Following this meeting, Pearce decided that—without prejudice to appointments subsequently made to the Administrative Air Board, whenever that was constituted—Williams was to take a seat on the Air Council until the First Air Member of the Air Board was available to do so. Williams was to be responsible to the Air Council for all matters presently in his charge.[48]

Not until 9 November 1920 were regulations bringing both the Air Council and the Air Board into existence approved by the Governor-General. These reduced Navy and Army representation on the Council to one member each (CNS and CGS), while other members—apart from the Minister as president—now comprised two members of the Air Board (Williams and Goble), and the Controller of Civil Aviation (CCA), Lieutenant-Colonel Horace Brinsmead. The regulations, however, also spelt out that in relation to administration and command of the air force, the Air Board 'shall be responsible to, and subject to the control of, the Air Council'. This was an arrangement which made clear that the Australian Air Force soon to be officially inaugurated would be anything but an independent and co-equal third service, and held serious implications for the good administration of the RAAF, as later recounted (see chapter 12). On 18 November, Pearce told Cook that he had instructed the Military Board that responsibility for the AAC would, 'as far as present legislation will permit', devolve on the Air Board and Air Council from 22 November.[49]

Stagnation at the Central Flying School

While all the foregoing debate over the shape of the future permanent air force had been taking place, the nation's existing military aviation establishment at Point Cook had undergone a prolonged period of stagnation and uncertainty following the end of the First World War. The armistice in November 1918 came barely two months after the CFS had undergone an extensive shake-up. Under the reorganisation which became effective on 9 September, the school became a unit of the Australian Imperial Force (AIF) and consisted of No.1 Home Training Squadron and an Aeroplane Repair Section. The Australian Flying Corps (AFC) reinforcements camp at Point Cook, now redesignated No.1 Home Training Depot, also came under the umbrella of the CFS.

The establishment given for the CFS in the reorganisation (but apparently meaning only the training squadron) provided for an AIF staff of 12 officers and 57 other ranks, formed into a headquarters of three officers and 18 other ranks and three flights each of three officers and 13 other ranks, supported by 32 home service personnel filling positions as clerks, cooks, labourers, etc.; the repair section was separately mentioned as having three officers and 52 other ranks, with four home service personnel employed as clerks and storemen. Commanding this remodelled establishment was Major William Sheldon, a 29-year-old officer of the permanent forces who had left Australia with No.1 Squadron, AFC, in 1916 and later held the command of No.4 Squadron (1916–17) and No.2 Squadron (1918).[50]

As part of these changed arrangements, a separate training establishment, the School of Military Aeronautics, was also formed to carry out 'the elementary training of Flight Cadets'.[51] Courses of up to 28 trainees were to undergo instruction in the theory of flight, aircraft types, construction, rigging, engine testing, meteorology, aerial photography, mapping, and air gunnery, along with training in drill, discipline and service customs—'everything, in fact, that can be taught "on the ground" '—before transferring to the CFS for flying training and practical rigging.[52] The school was to be staffed by three officers (a chief instructor, an assistant instructor and an adjutant), nine AIF non-commissioned instructors, and 21 home service personnel, and was expected to open its doors to the first course before Christmas that year.[53]

Lieutenant Haneil ('Neil') Kilby, an RAF officer on exchange in Australia, was appointed as OC and chief instructor from 26 August 1918 and given temporary rank of captain.[54] Kilby was actually an Australian who had obtained a commission in the 21st (Empress of India) Lancers and served with the RFC in France as an observer. A technical officer, he had been an instructor at the RFC's wartime School of Aeronautics at Queen's College, Oxford, engaged in training many Australians in aeronautical navigation prior to their undergoing flying training.[55] The first staff member to take up duty at the new school was actually the adjutant and quartermaster, Lieutenant Eric Roberts, who began on 1 November—just ten days before the armistice. Roberts had served in Egypt and Palestine with No.1 Squadron, AFC, until returned to Australia in 1917 to convalesce from injuries.

In view of the limited accommodation available at Point Cook, and to save on the cost of erecting special buildings for the purpose, the new school was to be located at Point Franklin, near Portsea, on the eastern side of Port Phillip Bay.[56] Here it would occupy disused artillery barracks in the old fort which had been built in 1889 on a cliff commanding shipping lanes at the entrance to the Bay. To Roberts and a small staff fell the task of beginning the conversion of the Fort Franklin barracks for their new purpose.[57] The school had still not formally opened, however, when the Air Service Committee decided on 17 March to close it down.[58] A caretaker was left in charge and the other staff transferred back to Point Cook, Kilby to become adjutant at the CFS.

The demise of the short-lived School of Military Aeronautics was already being matched by a wind-down in activity at the CFS itself, with the CGS (Legge) advising the ASC on 5 February that he was putting men off wherever possible.[59] The officer strength of No.1 Home Training Squadron was certainly under a dozen; in addition to Major Sheldon (CO), Captain Kilby (adjutant), and Lieutenant G.F. Thomas (quartermaster), the pilots were Captains Frank ('Dilb') McNamara, VC, and P.G. (later Sir Gordon) Taylor, and Lieutenants J.V. Tunbridge, Horatio ('Horrie') Miller, F.E. Tregilles, Frederick Sheppard, F.G. Huxley, and Roland ('Steve') Oakes. Apart from McNamara, whose post-nominal initials were unique in the AFC, most of the other pilots also had active service experience. Taylor, Tunbridge, and Huxley all wore the ribbon of the Military Cross won in combat, and Tunbridge additionally bore physical evidence of his experience in the form of burn disfigurements suffered in May 1917, when a bullet from a German Fokker aircraft which he had engaged exploded a Very-light cartridge in his machine and sent him down with his clothing on fire.[60] Oakes, too, was no novice in air fighting; a permanent officer (a 1916 graduate of the Royal Military College at Duntroon), he had served with No.37 Squadron, RFC, flying against Zeppelins raiding England, and later with No.102 Squadron, RFC, in night bombing in northern France.[61]

But, notwithstanding that there was available an ample and fully experienced staff, there was little military purpose to much of what was occurring at Point Cook. Oakes later recalled that, during his time at the CFS up to the end of 1919, the staff 'never got down to any real training programme, but filled in time with all sorts of unimportant odd jobs'.[62] Among these was a task given to Oakes and Miller in December 1918 to survey from the ground an air route from Melbourne to Sydney in a T Model Ford.[63] Within a few months Oakes tested out the suitability of the grounds he had helped select, when, following the sale of surplus wartime CFS aircraft (see chapter 6), he and Gordon Taylor flew two Airco DH6 machines to Sydney for delivery to their new civil owners.[64] Another extraneous task which fell to the CFS in the immediate post-war period was propaganda flights, such as the Peace Loan of August-September 1919 intended to promote sales of government bonds.[65]

Only occasionally were training activities conducted, and when undertaken they were not always successful. One reconnaissance flight to Ballarat by a formation of Sopwith Pups and Avros in June 1919 encountered such a strong north wind that all but one of the machines were forced to turn back, prompting critical comment in the Melbourne press.[66] A similar flight a few days later to welcome members of the AFC returning to Melbourne on the liner *Kaiser-I-Hind* was also curtailed due to adverse weather, again provoking adverse comment in the press.[67] The response to critical judgements of the CFS's ability or utility was, according to Oakes, to carry out 'strafing' flights over the central business district of Melbourne: 'flying at building height over Bourke, Collins and Flinders Streets, giving the pedestrians and traffic quite a fright. We could get away with that sort of thing in those days, with no disciplinary action being taken.'[68] Another notable flight was the CFS performance given at the Melbourne showground during the annual Royal Show, recalled by Oakes as having been staged with lamentable disregard for public safety.[69]

CFS resources were put to more creditable and purposeful use after the Australian government announced on 20 March 1919 that it was sponsoring a competition carrying a prize of £10 000 for the first Australians to fly from England to Australia. An important requirement arose to survey the route that competitors would follow once they reached the northern coast of the Australian mainland and made their way south. Two survey parties comprising former AFC pilots were despatched overland to select suitable landing places along the route from Darwin to Melbourne. Lieutenants Paul McGinness and Hudson Fysh were tasked with surveying the stretch between Darwin and Longreach in western Queensland, while Major Rolf Brown and Lieutenant A.R. ('Roly') McComb covered the section from Cootamundra to Longreach. At the same time it was decided that the proposed route from Cootamundra to Cloncurry (Qld) would be traversed by air, as a separate check on the suitability of the places selected as landing grounds.

Captain Henry Wrigley, recently returned from AIF service overseas, was recalled from leave and detailed to make this special flight south-to-north across Australia—in fact, the first transcontinental air crossing in any direction. Accompanying him as mechanic would be Sergeant Arthur ('Spud') Murphy, a choice which, in Wrigley's words, was 'a particularly happy one, for Murphy was an old school-mate of mine and was a first-rate mechanic who later was to be one of the most outstanding practical engineers in the Royal Australian Air Force'.[70] Murphy was actually a pilot, too, having served with No.1 Squadron,

During August–September 1919, aircraft from Point Cook were used to publicise and raise subscriptions to the First Peace Loan. Sopwith Pup C528 was flown around country New South Wales by Captain Gordon Wilson. Travelling from Richmond to Bathurst on September 17, he returned eight days later having visited Blayney, Carcoar, Orange, Molong, Cumnock, Wellington, Mudgee, Gulgong, Dubbo, Narromine, Peak Hill, Parkes, Grenfell, Cowra and Young. C528 was later transferred to the RAAF as A4-8. (*Mrs S. Clerehan*)

AFC, in Palestine as a temporary second lieutenant. After flying missions in the desert in support of Arab forces under Colonel T.E. Lawrence, he was awarded the DFC and the Hejaz Order of Nahda before returning to Australia and reverting to non-commissioned rank.

The machine to be used on the flight was hardly the ideal type to ensure success, being a two-seat BE2E advanced trainer with a 90hp air-cooled engine and a top speed of 72 miles per hour. As Wrigley noted, 'It was, in fact, an obsolete type, even for training purposes, and B6183 had seen a very considerable amount of hard wear and tear in training activities since it had left the works of the maker . . . However, it was structurally sound and airworthy.'[71] In preparation for the flight, the plane was modified by the addition of two fuel tanks under the centre section of the top wing, which increased capacity to 52 gallons (enough to cover 400 miles between refuelling stops), and an extra oil tank to the side of the engine.[72]

After several false starts followed by repairs and further modifications, Wrigley and Murphy finally took off from Point Cook on 16 November 1919, a Sunday morning, escorted by a formation of the remaining aircraft of the station. On 12 December—27 days later—they touched down at Darwin, at a grassed landing ground prepared at Fanny Bay, north of the town, and situated between the Vesteys meatworks and the Northern Territory gaol, with its western edge only 360 metres from the sea. The successful competitors in the race for the £10 000 prize—the brothers Ross and Keith Smith flying a Vickers Vimy bomber—had already arrived two days earlier, and Wrigley brought the BE2E in for a graceful landing alongside the parked Vickers machine.[73]

Captain H.N. Wrigley (*right*) and Sergeant A.W. Murphy pictured at Darwin on 12 December 1919, after completing a survey of the route to be followed by competitors in the England-to-Australia air race. Theirs was the first aerial crossing of the Australian continent and was hailed as almost as significant an achievement as that of the Smith brothers who won the race. The distance involved in the flight (4500 kilometres) was covered in just under 47 hours' flying time but took 27 days to accomplish. (*NLA*)

Despite the length of the interval since they had left Point Cook, Wrigley and Murphy had been in the air a total of only 46 hours and 50 minutes. In that time they had traversed 2800 miles (4500 kilometres), including several reconnaissances to both sides of their route, often over country which was lacking in any communications or was even unexplored, and for much of which they lacked accurate maps.[74] As Hudson Fysh, shortly to become a founder of the QANTAS airline, reported to Melbourne in a letter dated 6 January 1920:

> Captain Wrigley's trip was a remarkable one, and considering the type of machine used, distance & country covered quite as great as that of Capt. Ross Smith. Unfortunately here he was not accorded a very enthusiastic welcome as far as shouting went; but people recognised his trip for its full worth. I sincerely hope he & Sgt Murphy will be recognised in the South. I never had any hopes of seeing him here, and I am sure another trip like it will never be undertaken again or ever has been.[75]

In recognition of their remarkable feat, Wrigley and Murphy were both awarded the Air Force Cross. So far as the epic nature of the trip was concerned, of course, time would prove Fysh wrong.

While at Darwin, the two military aviators found that criticism in Parliament of what was held by some to have been a foolhardy undertaking had caused a change of plan. It had been originally intended that, after briefing the Smith brothers on the route, Wrigley and Murphy would fly themselves back to Melbourne—if possible by the alternative route down the overland telegraph line to Adelaide. A telegram was now received from the Defence Department, however, instructing that they should on no account attempt this, but were to arrange for the aircraft to be shipped back south. Accordingly the next fortnight was spent dismantling the aircraft and packing it for shipment on the steamer *Mataram*, on which they travelled to Sydney. Arriving there on 10 January 1920, they left for Melbourne on the express six days later.[76]

The return of Wrigley and Murphy to Point Cook was in time for them to be involved in preparations to welcome the Smiths' Vickers Vimy to Melbourne the next month. Arrangements were made for the brothers (now knighted) and their mechanics, Sergeants William Shiers and James Bennett, to be given an official reception at Flemington racecourse on 24 February 1920 at which they would receive their prize. As part of these celebrations, aircraft from the CFS (including the BE2E) were to meet up with the Vimy over Essendon as it came in from the north and escort it to a landing at Flemington. There it was intended that, while the presentation of the winners' cheque was being made, the CFS machines would provide a display of aerobatics for the assembled crowd.[77]

Unfortunately for these plans, engine trouble at Henty prevented the Smiths arriving that day and a crowd of more than 50 000 disappointed people returned home. Not until the next day was the big Vimy machine sighted over Melbourne, before landing at Point Cook.[78] Presentation of the prize money was made on 27 February at a lunch at Parliament House—a setting vastly different to that originally planned.

The Australian Air Corps

By the time the Smith brothers were making their triumphant progress to Melbourne, events at Point Cook had already taken a new turn. From 31 December 1919 the Aviation Instructional Staff which had been in existence since 1912, and the units created at the CFS by the reorganisation of September 1918, were all disbanded.[79] Taking their place from 1 January 1920 was a new body called the Australian Air Corps (AAC). As the CGS (Legge) made clear when initiating the change on 8 November 1919, the AAC was intended purely as a temporary expedient to provide the 55 other ranks personnel needed for 'the maintenance of existing equipment' pending the establishment of a new and permanent organisation.[80]

By 1 December Legge had altered his ideas slightly, and announced to the secretary of the Defence Department that:

> In addition to the officers and men required for the sole duty of maintaining equipment in good order, I propose to add a few others for flying duties, in order

> that the equipment may not remain without doing useful work . . . It is proposed to continue the preparation of air route maps, at first by means of motor transit, and then verifying by aeroplane visits, which will also tend to make the Air Service known in country districts.

Under the revised provisions he prepared there were now to be nine officers—a captain as commanding officer, a lieutenant performing duties as adjutant, a captain in charge of the workshops, a lieutenant to serve as 'Test Pilot' and to perform additional duties in the workshops, a captain and three lieutenants as 'Flying Officers', and a medical officer holding captain's rank. The number of other ranks required had risen to 70.[81]

On 23 December, Legge sought the approval of the Minister for Defence to make the first officer appointments. He explained that it was intended to appoint an ex-AFC officer, Major William Anderson, as commander at the CFS, but as he was currently on sick leave it was proposed that Major Rolf Brown should assume the post in a temporary capacity. The other names submitted were: Captain Kilby, to continue as adjutant; Captain William Stutt, formerly of the NSW State Aviation School, to take charge of the repair section's workshops; Captain Adrian Cole as flight commander; and Captain Wrigley (then at Darwin) as one of the flying officers. 'Other names', Legge said, 'will be submitted when replies have been received from those asked whether they are willing to take temporary employment', but the proposed appointments were needed at once to take over from the present CFS staff by 1 January. Gaining the approval sought, Legge instructed Brown to visit Point Cook and take in hand the selection of the other ranks who were to be offered employment in the AAC. 'The men selected should be such as could be appointed to at least the same, or a higher position on the creation of an Air

The BE2E used by Wrigley and Murphy on their transcontinental flight is dismantled at Darwin in December 1919, in preparation for shipment back to Point Cook. (*NLA*)

Pilots of the Australian Air Corps await the arrival of the Smith brothers at Flemington racecourse, Melbourne, on 24 February 1920. The officer at right is Captain F.W.F. Lukis. The official reception planned for the winners of the England-to-Australia air race had to be cancelled when their Vickers Vimy was delayed by engine trouble. (*Mrs S. Clerehan*)

Force', and preference was to be given to returned soldiers. By 2 January 1920 Brown had made 50 selections for appointment, leaving 20 positions to be filled by public advertisement.[82]

The creation of the new corps and the making of these appointments produced an enormous upheaval at the CFS. The fact that not all existing staff were offered employment meant that there was a departure by many of those associated with the school throughout the previous year. Major Sheldon, for example, received a curt note from headquarters only on 29 December telling him that he was out of a job and to hand over to Brown.[83] Oakes, being a regular, found himself reposted for duty with the garrison artillery at Fremantle (WA),[84] and Miller departed to become a mechanic with a small aviation company in South Australia.[85]

In fact, the matter of appointments in the new corps became the cause of considerable acrimony. By one account, the process of selecting officers and men who were to be offered appointment as members of the AAC was entrusted to a committee specially set up for this purpose. According to

Edward Hart, the honorary secretary of the New South Wales section of the Australian Aero Club and also the managing editor of *Sea, Land and Air* magazine, he had held discussions with Legge in July 1919 about 'the selection of returned airmen for the proposed Australian Air Force'. During these talks he pressed an aero club proposal that wartime commanders of AFC wings and squadrons should be consulted regarding the qualifications and suitability of individuals seeking to be appointed. As a consequence, a selection committee comprising Lieutenant-Colonel Oswald Watt, and Majors William Anderson, Alan Murray Jones, Rolf Brown, A.W.L. Ellis, S.W. Addison, R.C. Phillips and William Sheldon was constituted at Victoria Barracks.[86]

This committee's recommendations evidently did not carry a lot of weight with the Army hierarchy, given that Sheldon appears to have been a casualty of the purge which followed. Nor indeed, does it appear that the committee continued to be consulted as a body. In early December 1919, however, Legge did telegram Colonel Watt seeking his views on the officer proposed for appointment as the AAC's first flight commander and the other candidates being considered for subordinate posts as lieutenants. It was seemingly as a result of Watt's response on 9 December that Cole was selected for the senior job.[87]

Not all those approached about accepting appointments wanted them. Captain E.F. Pflaum declined, and when Captain Roy King (now working with the Larkin-Sopwith Aviation Company in Melbourne) was contacted, he wrote back on 30 January 1920 to say:

> In view of the fact that public opinion is so very bitter against the discharge of several officers who held appointments at Point Cook for a considerable period, I feel I cannot accept the appointment offered to me. There is a particularly bitter feeling around owing to Captain McNamara, VC, not being given an appointment in the new Corps, and I feel I must forfeit my place in favor of this very good and gallant officer.[88]

McNamara was belatedly given an appointment on 1 April, by which time Captains Lawrence Wackett, Ernest Jones and Frank Lukis had also joined the AAC.[89]

On 19 February Major Brown relinquished the command at Point Cook to Major Anderson, who now became the corps' senior officer.[90] In February, too, Captain Wrigley took over as the adjutant of CFS, *vice* Kilby who departed to become aide-de-camp to the Governor-General.[91] Further appointments in the AAC appear to have been made purely as the need arose. For example, when it was decided to embark an Avro 504 seaplane in HMAS *Australia* for trials in July 1920, the AAC had no experienced naval pilots or observers on strength. It was due to the requirement arising from the naval trials that Captain Ferdinand Hippolyte ('Kanga') De La Rue and Lieutenant William Walne were appointed to the AAC. De La Rue was an Australian-born ex-member of the RNAS who had been demobilised but applied to Goble at Navy Office for employment.[92] Initially the trials were conducted without an observer for the seaplane aboard, until Walne was located and offered employment (see chapter 7).[93]

Further pressure for additional staff came with the arrival of the imperial gift equipment. Early in March the Director of Air Services (Williams) drew to the Minister's attention the fact that the first shipload of this materiel was expected in Melbourne at the end of the month, and was understood to comprise 600 packing cases (550 of these being over 96 cubic feet in capacity), eight lorries, two trailers, eleven side-cars, seven tons of miscellaneous stores and 44 000 gallons of oil! He pointed out that the AAC had no equipment staff to take charge of this mass of stores, which was only the first instalment of what was on the way. He therefore urged that the existing AAC establishment of nine officers and 70 other ranks, all located at the CFS, should be increased by a further seven officers and 36 other ranks. Five of the officers (a captain and four lieutenants), and all the additional other ranks were for the purpose of managing the gift equipment, while two officers were to be concerned with buildings to accommodate the arriving stores.[94]

Senator Pearce was initially unconvinced of the need for more staff, objecting on 13 March that: 'there is no necessity at this juncture to submit proposals for any appointments, transfers, rates of pay or rearrangements of duties that will in any way compromise action to be taken when the Cabinet has finally decided what the future air policy is to be'. Assured that the proposed engagements were purely temporary and involved no right to subsequent permanent employment, he was persuaded that the request for more manpower was reasonable and gave his approval the following day.[95] In the event this proved to be the least of Williams' worries at this time, as his attempts to obtain sufficient staff to handle the gift equipment suddenly became caught up in an inter-service intrigue.

At a meeting of the Air Board on 16 March, Williams' recommendations on this matter came under the scrutiny of the Navy representatives, Captain C. Round-Turner and Goble. It was their suggestion that Major H.J. Berryman RAF, an Australian ex-RNAS aviator whom the Naval Board had in mind to command the planned seaplane squadron, should be placed in charge of the AAC's technical staff to handle the incoming stores and equipment. Williams objected that this officer was not qualified, arguing that he needed an equipment specialist who would be responsible to him for work carried out in this area. Under pressure from the naval representatives, Williams agreed to Berryman being placed under his authority within the Defence Department for duty in connection with the gift equipment.[96]

Williams' reluctance to take Berryman appears to have derived from suspicion of Navy motives in this matter—not without cause, as subsequent events showed. A fortnight after this Board meeting, Round-Turner and Goble produced a memorandum which argued that the branch concerned with technical stores, equipment, and buildings was of such importance that it needed to be kept apart from the administration of personnel and training. As Williams was responsible—by ministerial direction—for existing air services (that is, the aerodrome at Point Cook), not the future air force, they argued that 'it would seem that such [arriving stores for the AAC] should not be placed under the direction of an officer who is already saddled with other responsibilities'. Round-Turner and Goble went on to say that, since the gift equipment needed to be under an engineer with flying and administrative experience, 'and knowing of no other officer who is an Australian, with the necessary qualifications, [we] suggest that Major Berryman would be a suitable officer for the appointment'.

Williams was quick to respond to this suggestion, which attempted to limit his authority and presented him with a naval rival additional to Goble. It can only be speculated upon whether he recognised the deeper threat contained in the Navy proposal: that is, if Williams accepted that he was solely in charge of the CFS and allowed a separate equipment empire to be established under Berryman, there was a case for appointing a coordinating authority over both establishments—and the most likely appointee to that command would be Goble. In any event, Williams dealt with the Navy proposal on 7 April by recommending the appointment of three ex-AFC officers—Captain Percy McBain, and Lieutenants Charles Harman and Hector Johnston—as equipment officers instead of Berryman. The two separate proposals were debated by the Air Board at a meeting on 16 April, of which the minutes record only that, after 'lengthy consideration of the various aspects of the questions involved', it was agreed to submit Williams' nominations to the Minister for appointment as a temporary measure; Williams conceded nothing more than to place Berryman in charge of one of the equipment depots which it would be necessary to establish. Pearce approved these arrangements three days later.[97]

Even this was not the end of manpower problems for the AAC. On 23 April Williams reported to the secretary of the Defence Department that the number of clerks, drivers and storemen authorised by the Minister for temporary employment in connexion with the gift equipment had now been found to be insufficient for the purpose. The arrival of the first shipment had revealed more work to be involved than was anticipated. The motor vehicles, being pre-used, required overhaul, in addition to which they had suffered damage during handling on the voyage. To cope with this additional work, he sought approval for 42 technical and 12 non-technical personnel over and above the previously authorised AAC strength of 106. Not all were needed immediately, he emphasised, as this depended on the rate of arrival of further consignments of gift equipment, though he stated that 'the majority could now be employed'. The Minister approved this submission on 27 April, with the result that advertisements were placed in newspapers in all States calling for applicants.[98]

Although the AAC had now grown to a strength of 160 other ranks, it remained very much a temporary body whose identity harkened back to the now defunct AFC rather than the forthcoming Australian air force. Emphasising the corps' role as a purely transitional entity was the order regarding uniform issued by the CFS adjutant (Captain Kilby) less than three weeks after the AAC's formation:

> Owing to the temporary nature of the Australian Air Corps, it has been approved that no distinct uniform shall be designed—that uniform at present in possession of members may be worn out, and that should members find it necessary to purchase new uniform while the Corps is constituted at present, AIF pattern, as worn by AFC, is to be worn ... [99]

Despite its lack of permanence, the AAC nonetheless made its presence felt with some notable achievements using the newly arrived gift machines which, regardless of being old war birds, were still very modern machines judged by the standard of aircraft hitherto flown in Australia. On 17 June 1920 Captain Cole, with Captain De La Rue as passenger, took up a DH9A (F2779) on a height test during which they reached 27 000 feet before being forced down by lack of oxygen.[100] Apart from inspiring the medical officer at Point Cook, Captain Arthur Lawrence, to begin 'rather haphazard experiments [into the problems of high-altitude flying which] can be said to mark the beginning of work in aviation medicine in Australia',[101] the altitude attained by Cole was still the Australian record more than a decade later.[102]

Williams also used a DH9A to achieve the first non-stop flight between Sydney and Melbourne on 22 July.[103] With Warrant Officer Les Carter (one of the members of the original CFS staff of 1913) as his passenger, he covered the distance in five hours and 29 minutes. Williams records in his autobiography that this was actually the return leg of a flight undertaken in company with another DH9A flown by Captain Lawrence Wackett. The two machines flew from Point Cook and landed at Canberra before going on to Richmond, to inspect the building and machinery at the site of the former State aviation school being offered for sale to the Commonwealth by the New South Wales government.[104]

Another notable achievement for the AAC followed soon afterwards during the tour of Australia by the Prince of Wales, when a requirement arose at short notice to take over the carriage of mail for the royal visitor. An arrangement had been made to bring the prince's mail, which had arrived in Fremantle (WA) by ship from England, across to Sydney before his scheduled departure on board HMS *Renown* on 21 August. A prominent Mildura financier and publicist, C.J. ('Jack') De Garis, had offered to pick up the mail when the transcontinental train brought it to Port Augusta (SA) and then fly it to Sydney. Unfortunately for this plan, the plane De Garis intended to use became unserviceable, with the result that the AAC was called upon to take over the task. Two DH9As, one crewed by Williams and Carter, and the other by Captain Jones and Sergeant Herbert Chester, ensured the success of this undertaking. As one publication noted at the time, 'what was promised was accomplished and the *Renown* passed out of Sydney Heads shortly after 10.30 pm, while the Prince was reading his letters from Home'.[105]

In fact, the visit of the Prince of Wales revealed how thinly stretched the AAC was. When, back in May, Williams was instructed by the visit organiser (Major-General White) to arrange an aerial escort for the arrival of *Renown* at Melbourne, he soon appreciated that he had more aeroplanes available than pilots. He planned to send four aircraft to intercept the prince's cruiser outside the Port Phillip Heads and escort it to Queenscliff before returning to the CFS. These would be followed by another four planes which were to resume the escort from near Williamstown to the Port Melbourne pier before also returning to Point Cook. For an official procession up St. Kilda Road, however, it was hoped to have 'all available machines'—between 15 and 20 aircraft—in formations at 1000–1500 feet above the route, with another two planes carrying photographers.

Aviation 'firsts' came easy in the 1920s. According to the *Australasian* on 26 June 1920, an Avro seaplane flown from Point Cook by Captain H.F. De La Rue was the first to land on the Yarra River. The machine was dismantled and taken to the Exhibition Building where it joined a public display of enemy and allied aircraft and equipment. Avro H3042 later carried the RAAF number A3-47. (*Australasian*)

As Williams explained to the secretary of the Defence Department on 13 May, the AAC had insufficient pilots for this last effort 'but there are several volunteers from AFC, AIF pilots to fly on the day'. These were not seeking to be paid, but just wanted to take part. Williams recommended accepting their services, 'as apart altogether from HRH visit this flying will serve to keep a few additional pilots in training'. The Minister (Pearce) approved this suggestion, though stressing that in the event of accidents volunteer pilots would receive medical treatment 'as though still members of AIF' but would not be entitled to pensions or compensation.[106]

A reliance on ex-AFC pilots was also a feature of the Second Peace Loan undertaken Australia-wide in August 1920, which involved seventeen Avro 504s from the AAC. In a letter dated 25 August, Williams pointed out to Treasury that in order to have the required number of machines ready for use in the loan campaign 'certain persons had to be employed from civilian sources'.[107] In fact, it appears that no more than fifteen AAC pilots and mechanics were available for this work,[108] so that the rest were apparently all returned servicemen temporarily engaged. In addition, as Williams went on to note, it was necessary to send AAC personnel to Point Cook after their normal work was completed in Melbourne and to work all weekends to get the aircraft ready in the time allowed.

The actual conduct of the Loan campaign was only a mixed success, due primarily to the bad weather encountered in August which caused many pilots to fail to complete their program. By the end of the Loan, four of the nineteen AAC planes used had been so badly damaged in accidents as to warrant being written off. Despite this aspect of the Peace Loan, it has been concluded by one analyst of the flights that they 'were a success from the AAC's point of view':

> Aircraft were made serviceable in a very short time and operated in areas which had never seen an aircraft before, under the most primitive of conditions, and in the most appalling weather. The pilots gained valuable practical experience in operating under Australian conditions. The AAC did not lose one man in carrying out the tour.[109]

Apart from this, the tours had created a high profile for the air service. While refusing to accede in a suggestion by the Victorian organising committee that a defence plane should land in Collins Street in Melbourne's central business district, on the grounds that 'such a procedure will be detrimental to the forthcoming air policy in Australia',[110] there were other special stunts undertaken. These included flying exhibitions and leaflet drops over Collins Street each Saturday between 12 noon and 1 pm, and a special 'Aeroplane Drive' arranged for the afternoon of 28 August which saw four defence machines flying over every race meeting and football match in Melbourne and surrounding districts, dropping leaflets and stunting. Defence aircraft also participated in the welcome at Flemington racecourse on 31 August for Lieutenants Raymond Parer and John McIntosh—the second competitors to finish the England-to-Australia race (even though the Smiths had already won the prize). Four Avros also took part in the aerial derby at Serpentine (Vic) on 27 August—believed to have been the first conducted in Australia.[111]

One task which came the way of the AAC in this period was to have tragic consequences. On 22 September 1920 a ministerial instruction was received at Point Cook to prepare two of the new gift DH9As to undertake an air search for the *Amelia J.*,[112] a 350-ton three-masted schooner which had gone missing in August while sailing from Newcastle to Hobart laden with coal. Also taking part in the search was HMAS *Platypus*.[113] Taking off next day, the aircraft were to cross Bass Strait and proceed down the Tasmanian east coast looking for any sign of wreckage. One machine was crewed by Major Anderson and Sergeant Chester, and the other by Captain Stutt and Sergeant Abner Dalzell.

When Anderson's machine landed at Brighton, north of Hobart, on the evening of 23 September, its occupants had sighted nothing.[114] There was, however, now no sign of the other DH9A, which had last been seen by Anderson and Chester entering cloud at about 11 000 feet near Cape Barren Island.[115] Subsequent reports were received that residents on Vansittart Island in Bass Strait had heard an aircraft passing overhead with a badly missing engine; other reports of a ball of fire seen to drop into the sea pointed to the possibility of a fuel leak having ignited on board the aircraft. Although a combined sea and air search for the missing aircraft was begun on 27 September, this also found no trace of wreckage.[116]

Stutt and Dalzell had been popular members of the small station community at Point Cook, and their loss in such circumstances was felt as a sharp blow. 'Billy' Stutt and his wife Stella lived on the station with their three children aged between seven years and six months. Both Billy and his wife were devotees of classical music, and held regular concerts in their home.[117]

Dalzell had served as a rigger with the AFC in England and France, before returning to Australia and joining the AIS in September 1919.[118] He also was married.

A court of inquiry into the tragedy was convened at Victoria Barracks in November 1920. Headed by Brigadier-General Walter Coxen, this body was critical in its findings about several aspects of the way in which the two aircraft had been hastily despatched, but concluded that these faults had been 'due to the existing strength at the Central Flying School being inadequate for efficient administration and the carrying out of the work allotted to that school'.[119] Regarding the missing aviators, in the absence of any definite information it could only be concluded that they had suffered an accident. In fact, nothing ever did come to light regarding their fate; while wreckage believed to be from the *Amelia J.* was washed up on Flinders Island in June 1921,[120] the disappearance of the aviators remained complete. As Stutt and Dalzell had been engaged on official duty at the time, the court of inquiry recommended that their widows receive the maximum compensation allowable under the finance regulations—£550 to Mrs Stutt and £248 to Mrs Dalzell.

When the court's recommendation about the appropriateness of paying compensation was considered by Cabinet on 2 December 1920, it was promptly accepted. The government also recognised that members of the AAC were generally not able to obtain adequate private insurance—flying was seen as a risky endeavour and premiums were correspondingly high—and decided therefore to treble the amount payable in the instances of both men.[121] But this was, in fact, still a meagre recompense for the loss of a bread-winner, and the issues raised by the disappearance of Stutt and Dalzell were still with the Defence Department when the RAAF came into being in 1921.

The Air Board formed

Soon after the regulations for the Air Council and Air Board were promulgated on 9 November 1920, the first Air Board came into being although it did not hold its first meeting until 22 December. As prescribed by the regulations, this body consisted of just four members: the three service officers filling the positions of Director of Intelligence and Organisation (DIO), Director of Personnel and Training (DPT) and Director of Equipment (DofE), and a civilian who was the Finance Member (FM). The first appointees to the Board accordingly were Lieutenant-Colonel Williams (DIO), Wing Commander Goble (DPT), Captain McBain (DofE), and Mr Albert Joyce (FM).

In the choice of Williams and Goble for the two senior posts on the Board, an obvious attempt was made to balance and hence placate the aspirations and interests of Navy and Army. According to Williams, this meant going to some extraordinary and even unjustified lengths. He claims in his autobiography that the suggestion of the Naval Board that their nominee, Goble, must be the senior member of the new Air Board was regarded by the CGS (White) as insulting to Army. Navy's insistence on this point finally led to Williams being called 'one day' to the Minister's office, where he found Senator Pearce had with him his Navy counterpart, Cook. It was here, Williams stated, that he

The original Air Board formed in November 1920 to run the Air Force, shown in 1928. *At rear, from left:* Major P.E. Coleman (Secretary), Mr A.C. Joyce (Finance Member); *front row:* GpCapt (then WgCdr) S.J. Goble (Director of Personnel & Training), ACdre (then WgCdr) R. Williams (Director of Intelligence & Organisation), WgCdr (then SqnLdr) P.A. McBain (Director of Equipment). (*RAAF Historical*)

first learnt of the compromise worked out by the two ministers. Both Goble and himself were to have the same rank of wing commander from the date of the formation of the new air service, although Williams was to be the senior. 'But the decision of the Ministers did not stop there', he claimed, 'they directed that I should chair the first meeting of the Air Board and that after that Goble and I should alternate in that capacity'.[122]

Understandably, Williams was dissatisfied with and disparaging of this arrangement. He later recounted that the effect of the ministers' agreement was to grant Goble an undeserved promotion from the rank he had achieved in the field and above the rank he held in his present service, the RAF. As he later wrote:

> It was thoroughly inequitable, but the Chief of the Naval Staff had almost achieved his point. Promotion over others was given to an officer of very limited experience for no other reason than that he held a commission in the 'senior service' in war. It was not a recommendation that Goble's own Service, the Royal Air Force, would have made . . . If it had been designed to create the wrong sort of atmosphere to start a new disciplined Service, it could hardly have been better.[123]

Unfortunately for Williams' reliability as a witness, his observations regarding Goble were so wide of the mark as to be discreditable to himself. The fact of the matter was that Goble was in all respects as qualified for the rank he was to receive in the new service as was Williams. Far from having received an undeserved promotion, Goble was in an identical if not superior position to Williams at the time the ministers decided they should both be wing commanders. Whereas Goble was already a substantive squadron leader and honorary wing commander in the RAF, Williams was only a substantive captain holding the brevet rank of major and the honorary rank of lieutenant-colonel until December 1920, when he was promoted major and brevet lieutenant-colonel—curiously backdated to the previous January.[124]

So far as respective service and command experience was concerned, Goble's record in no way suffered by comparison. Williams could claim the distinction of having commanded an RAF wing on operations, though this was only from June 1918 and in the less demanding Middle East theatre where he had served the whole of the war. Goble's experience had been gained in the central crucible of the Western front. As a member of No.8 Squadron, RNAS, he had taken part in fierce fighter combat during the Battle of the Somme in 1916, and was commanding No.5 Squadron, RNAS, a bomber unit, when the 1918 German offensive began. His conspicuous bravery and skill was already well recognised even before his performance during the 1918 offensive became the subject of a special memorandum issued by RNAS Headquarters on 25 March to all wing and squadron commanders throughout his service.

Commending the achievement of No.5 Squadron in maintaining operations while their aerodrome was under shellfire until evacuated the following day, then subsequently shifting twice more but the whole time continuing their attacks, the memorandum declared:

> The greatest credit is due to Squadron Commander S.J. Goble . . . and the Officers and Men of Naval Squadron No.5, and this work shows what can be accomplished by a Squadron with good leadership and a good spirit amongst the officers and men. The whole secret lies in the ability of the Squadron Commander to inspire this spirit into his squadron.[125]

These same deeds drew also the congratulations of RFC Headquarters and the Air Council on 27 March 1918.[126] The fact is inescapable that Goble was what

Williams was not: an air ace with seven confirmed victories in aerial combat. While Williams wore on his chest the ribbons of the Distinguished Service Order and an Officer in the Order of the British Empire, so did Goble—with the Distinguished Service Cross and the French Croix de Guerre besides.

By most accounts there was also a gulf separating the personalities of the two men. Williams was the son of a Cornish miner in the copper town of Moonta, South Australia, and came from a staunch Methodist background. He began his military career in the militia while working as a bank clerk, and by determination and application gained a permanent commission in the military forces in 1912. He was, by most accounts, a somewhat puritanical, stiff-necked individual with an autocratic style. Goble was from a similar social background, his father being a stationmaster with the Victorian Railways. He also entered the railways service, and on the outbreak of war refused to accept the rejection on medical grounds of his two attempts to enlist in the AIF. He paid his own passage to England to achieve acceptance as a trainee pilot in the RNAS in 1915. In contrast to Williams, he was reputedly an amiable and gregarious character with an engaging sense of humour, and a keen punter on the horses.

Recognising the threat that Goble represented to his own position, Williams had been waging a jealous campaign against his rival even before they were thrust together as colleagues on the Air Board. On 4 October 1920, Goble complained that he had gone out to Point Cook that day in accordance with a standing arrangement whereby CFS aircraft would be made available to him for flying practice. Though he had exercised this same arrangement just the previous week, on this occasion he was told that he could not be permitted to fly without the permission of Williams, who had given instructions to the OC to this effect: 'I might say that obstacles have always been placed in my way with regard to obtaining flying practice, and at the same time I understand that considerable capital has been made in certain quarters of the fact that I seldom take the air'.

Goble's complaint led to the Naval Board requesting that the Minister give instructions to ensure that Goble was not denied the opportunity to maintain his piloting skills.[127] That such should have been necessary did not augur well for the working relationship between these two men who would hold the prime responsibility for the development of Australia's air force in their charge.

Final impetus to formation of the AAF

Although the Defence Department was firmly committed along the path towards establishing an Australian air force, a time-frame attached to this intention was still lacking at the start of 1921. Given a government reluctance to spend money on defence in the immediate post-war period, this situation might have continued indefinitely without the emergence of some impelling reason to take the final step towards formally creating the new service. Just such an impetus fortunately arose out of the government's desire to establish air mail services in some of the nation's more remote regions.

The idea of using service aircraft to establish a trial air service had received Cabinet approval on 9 September 1920. The proposal envisaged that a unit of the air force would be formed 'as soon as funds, material and personnel are available' and allotted a route to be decided upon by the Air Council and the Postmaster-General's (PMG) Department. This air force unit would carry mails over the selected route for a definite period, to enable data to be collected on which the government could form a judgement on the advantages (or otherwise) of such a service. If the trial was successful, it was expected that the service would be taken over by a civil aviation company.[128]

Adding to, if not actually prompting, the government's interest in this idea at the time was a formal letter by the Premier of Western Australia, H.P. Colebatch, to Prime Minister Hughes on 6 September 1920, pointing out the use being made of service aircraft in the United States for civil purposes such as forest patrols during fire seasons. The premier suggested that the Commonwealth follow this example by employing some of its airmen to assist in such things as providing air medical services to residents of remote districts, thereby adding knowledge of areas of strategic importance to their training. In responding to this appeal on 26 October, the prime minister was careful not to give much encouragement to Colebatch, even though Cabinet had already endorsed almost exactly the approach urged by the premier. Hughes said that using service aircraft for civil purposes would be investigated more fully when the 'Commonwealth Air Forces' had been sufficiently developed, because the organisation of these forces was not then far enough advanced to make their use possible, however desirable. Progress in developing the air force would have to be gradual for financial and other reasons, and it would be necessary to keep the small force planned concentrated to derive maximum benefits in training and discipline: 'premature dispersion of force would be inimical to efficiency'.

The negative tone of this reply strongly suggests that it was drafted for Hughes' signature by Williams, for it soon emerged that he was not at all in sympathy with the idea of using military aircraft for purposes outside the defence sphere, at least at this very early stage. When, on 25 January 1921, the Controller of Civil Aviation (Brinsmead) met with Charles Bright, acting secretary of the PMG's Department, to discuss the experimental air service, it was only to have the latter lament the fact that the Cabinet direction in this matter had 'not been proceeded with beyond an interview with Colonel Williams and the exchange of a letter or two'. Valuable data, Bright observed, might otherwise have been available for formulating a basis for the sort of contracts employed in the United States.

If Williams had been—as Bright seemed clearly to imply—deliberately dragging his feet over the air service proposal, arguing that nothing of this nature could be attempted until the air force itself had some sort of firm existence, then he must be credited with shrewd tactics. Brinsmead duly reported his conversation with Bright in a note to the secretary of the Defence Department, and this was promptly referred to the Air Council by the Minister for Defence, Senator Pearce, who pointedly appended the observation on 27 January that it was 'the policy of the Cabinet that an experimental aerial mail service should be run by the Air Force by arrangement with PMG's Dept.'.

The Air Council considered the matter at its meeting on 3 February 1921, and decided to instruct the Air Board 'to prepare and submit proposals under existing Acts for the development of the Air Force forthwith, to such a degree as is necessary to enable the immediate establishment of an Air Force, and to initiate an experimental aerial mail service'.[129] Williams now had the authority he needed to get on with bringing the air force into being, even if it was now bound up with the unwanted requirement for the new force to carry the air mails.

In naming a start date for the Australian Air Force (AAF) Williams suggested 31 March 1921 rather than the next day, which then marked the start of the financial year; as he later recounted, he made it that date 'rather than the commencement of the following month to prevent nasty people referring to us as "April Fools" '.[130] Since 1 April was good enough for the birthday of the RAF, he might be considered as being unduly sensitive. But then, judged by the tone of the Australian debate on the air service proposal to this point, perhaps his precaution was not unwarranted.

2

Raising the new service

The agenda for the historic first meeting of the reconstituted Air Board on 22 December 1920 included several matters which laid important foundations for the future air force. Among these were a scheme outlined by Williams as DIO for obtaining junior officers on transfer or attachment from the other two services; another proposal from Williams that agreement be sought from the Navy and the Army to the Air Force using, in the interests of economy, 'certain existing administrative machinery' under their control (such things as joint use of libraries, instructors, hospitals and medical officers, chaplains, supply and storage facilities); and a proposal from Goble as DPT for the adoption of RAF rank titles. It is not clear from Board minutes that all these matters were actually discussed at the first meeting. Goble's proposal was certainly withdrawn for later submission,[1] and Williams' attempt to obtain junior officers from the other services is shown as having been discussed at what appears to have been a further meeting of the Board the next day.[2]

In any event, by far the most important of the matters scheduled for discussion at this time was a submission from Williams covering the 'Proposed immediate programme and general lines of development' of the air force.[3] Under his plan the force was to have seven squadrons, comprising two fighter squadrons for air defence, two corps reconnaissance squadrons for Army cooperation, and three squadrons for naval cooperation (one each of float-seaplanes, flying boats and torpedo-carriers). Controlling these units would be an air force headquarters and two wing headquarters. Completing the organisation would be a liaison office in London, a flying training school, a recruit and non-technical training centre, and a stores depot. Williams' program was definitely discussed at the first Board meeting, and was duly considered by a meeting of the Air Council the next day at which it was 'approved generally'.[4]

In the long memorandum outlining his proposals, Williams acknowledged the magnitude of the task confronting the Board:

> The task is the greater because Australia has not previously attempted to do anything in this direction further than the actual preliminary flying training of a limited number of pilots ... There is no peacetime experience in organisation,

> administration and training of a service such as a modern air force in any part of the Empire on which to rely, for such a service on a peace basis was never attempted until the formation of the permanent RAF after the recent war.[5]

The initial plan of development

The program for the new AAF initially approved by the government was that proposed by Williams, with only slight modifications. It was a not overly ambitious scheme which would have seen Australia provided with an air service of quite reasonable size and capability. Outlining the plan in the Senate on 8 April 1921 during the debate on the Air Defence Bill which he had introduced, Defence Minister Pearce explained that at the end of the first year of the program the air force would comprise: a headquarters, one aircraft depot, a flying training school, a recruit and non-technical training depot, and six aircraft squadrons. A liaison office would also be opened at the Air Ministry in London.[6]

The six service squadrons planned were initially to be split between two States, Victoria and New South Wales: two landplane squadrons and a flying-boat squadron in the former; and two landplane units, with a squadron of seaplanes to operate on ships of the RAN, in the latter. One of the landplane squadrons in each State was to comprise primarily Citizen Air Force personnel. The scheme of development also provided for early acquisition of a site in Western Australia, however, to enable the establishment of a unit there in the second year of the program.

As the Minister went on to state, this plan envisaged a force with a total strength of almost 1500 officers and men. The permanent component was to comprise 150 officers and 1000 other ranks, while there would be 36 officers and 300 other ranks in the CAF. However, he predicted that: 'This proportion of Permanent Forces to Citizen Forces will not be continued in later years, new units in the future being composed mostly of Citizen Forces'.

Speaking again during the resumed Second Reading debate on the Air Defence Bill on 22 April, Senator Pearce made clear that it was anticipated that full use would be made of the gift aircraft which Australia had received from Britain. Of the 151 aircraft the Air Force held, 72 would be brought into commission—24 at No.1 Flying Training School (six DH9s, three SE5As, twelve Avro 504Ks and three Sopwith Pups) and 48 in the four landplane squadrons (sixteen DH9s, eight DH9As and 24 SE5As). This would leave 79 machines (six DH9s, 21 DH9As, eight SE5As, 36 Avro 504s and eight Pups) as spares to replace losses and in reserve. To complete the projections, nine flying boats and twelve ships' seaplanes had yet to be purchased, but Pearce assured the Senate that orders would be placed for these 'without delay'.[7]

The size of this program, though hardly vast, at least made it respectable to seek the prefix 'Royal' for the new service. A proposal to obtain the approval of the king to the granting of this title was made by the Air Council as early as February 1921, even before the force officially existed.[8] The Governor-General was notified that the request had been successful in a despatch from London dated 11 May, but this advice was not received in the Defence Department until 20 June. The news became public a few days later,[9] but despite this it was not until 13 August that the necessary order signed by the Governor-General was gazetted, making the new name official from that date.[10]

Recruitment of personnel

When the Australian Air Force was formally inaugurated on 31 March 1921, the personnel for this new organisation was obtained by taking over the bulk of the strength of the old AAC. Not all AAC members chose to make the transfer, and these were still being discharged as late as 30 March.[11] According to often-repeated figures, the numbers making up the initial strength of the AAF were 21 officers and 130 other ranks.[12] Subsequent research confirms the identity of the 21 officers, but a nominal roll of the first 1013 men enlisted in the RAAF up to September 1926 indicates that just 128 other ranks had 31 March 1921 as their date of joining. No.1 on this latter list was Arthur Murphy, who was commissioned as a flying officer four months later.

The 21 original officers of the force were all granted only temporary Air Force commissions, and included: on the General List, Wing Commanders R. Williams, DSO, OBE, and S.J. Goble, DSO, OBE, DSC; Squadron Leader W.H. Anderson, DFC; Flight Lieutenants E. Harrison (honorary Squadron Leader), R.S. Brown, AFC (honorary Squadron Leader), A.M. Jones, MC, DFC (honorary Squadron Leader), H.J.T. Berryman, AFC, A.P. Lawrence, MC (medical officer), L.J. Wackett, DFC, AFC, A.T. Cole, MC, DFC, H.F. De La Rue, DFC, and H.N. Wrigley, DFC, AFC; Flying Officers F.H. McNamara, VC (honorary Flight Lieutenant), A.H. Cobby, DSO, DFC, F.W.F. Lukis, J.K. Fryer-Smith (honorary Flight Lieutenant) and W.S.J. Walne; on the Quartermasters List were Squadron Leader P.A. McBain, MBE, and Flying Officers C.J. Harman, MSM, H. Johnston, MSM, and J.H. Rogers, MSM. Of these, Williams, Anderson, Brown, Wackett and Harrison were also shown as still on secondment from the Army's Staff Corps.[13]

By 30 June the AAF's strength had risen to 25 officers and allegedly 195 airmen,[14] although the nominal roll previously mentioned suggests enlistments of other ranks still had not exceeded 145 by this stage. Increasing the force to its projected level from this small base was clearly going to be a task of considerable magnitude, and in July the Air Board adjusted its recruitment goal to 108 officers and 791 other ranks—this 'to be attained to gradually'.[15] In fact, recruitment efforts had begun back in March, when advertisements were placed in daily newspapers across Australia seeking applicants in the permanent Air Force. Vacancies existed for officers 'between the ranks of Pilot Officer and Squadron Leader' and for other ranks 'between the gradings of Aircraftsman and Warrant Officers, Class 1' in a wide range of technical, non-technical and administrative fields.[16]

By late September it was reported in *Aircraft* magazine that appointments to the force now totalled 48 officers and 258 airmen. A listing of the officers showed that a number of the original appointees had already been promoted, while newcomers were: on the General List, Flight Lieutenants I.N.C. Clarke, DSC, A.A.N.D. Pentland, MC, DFC, A. Hepburn, DFC, and G.F. Cherry (medical officer); Flying Officers R.J. Brownell, MC, MM, G. Jones, DFC, E.D. Cummings, DFC, J.R. Bell, E.P. Kenny, DFC, E.A. Mustard, DFC, A.W. Murphy, DFC, AFC, E.J. Stephens, DFC, G.F. Thomas, W.D. Bostock, L.J. Balderson, J.H. Summers, H.C. Harrison; Pilot Officers J.D. Jarman and H.A. Wilson; on the Quartermasters List, Flight Lieutenants T.R. Marsden, DSO (honorary Squadron Leader) and F. Knight; Flying Officers E.J. Howells, MC, W.E. Cowper, F.C. Hawley, DFC, and G.J.W. Mackinolty; Pilot Officers T.A. Swinbourne and T.H. Kempson.[17]

Announcement of these appointments immediately became the focus of a blazing controversy when it was reported the next month that there had been 'an outburst of dissatisfaction among rejected applicants'. *Aircraft* claimed to have been inundated with complaints, including allegations that candidates from Victoria had been given preference over applicants from other States. Citing the particular case of Queensland, from which no officer appointments had been made, the editor of *Aircraft,* Edward Hart, remarked that, if the complaints were true and were borne out with other States too, 'then the air force promises to be an expensive toy'.[18]

Williams reacted promptly to this potentially damaging situation. In an interview with Hart on 10 October he firmly rebutted suggestions that selection had been based on any other grounds than taking the best Australian fliers who offered their services, although he agreed that there were other excellent qualified men who might have been selected but were not applicants for the RAAF, and added: 'In support of our contention that we have made a very fair pick I would ask you to peruse the list of officers already appointed, and note the number of decorations possessed by them and awarded for work in the air.' Regarding the absence of any appointees from Queensland, Williams stated that fourteen applications had been received but of these four had not been considered because they lacked flying experience or were not certificated pilots, and the remaining ten held no decorations for work they had done as pilots on war service. In fact, he said, one offer of appointment had been made by the RAAF, but was declined for health reasons.

Commenting on Williams' response to the allegations, Hart noted that the question of decorations as a qualification threatened to become highly controversial. As he pointed out, there was a number of well-known aviators who were undecorated, and it was striking that of the nineteen officers without decorations who had been appointed no fewer than eleven were from Victoria. In all, 30 of the total of 48 officers were Victorians, while the officer corps of the new force lacked any members from either Queensland or South Australia. Hart went on to assert that the apppointment of other ranks showed a similar State bias, with no fewer than 43 of 55 non-commissioned officers being from Victoria, as were 117 of the remaining 203 airmen. Perhaps the most interesting aspect of the figures cited in the course of this (viewed from a distance in time) quite petty dispute was the glimpse afforded of the RAAF's rank structure at this time. According to Hart's figures, the Air Force had two sergeant-majors class 1, four sergeant-majors class 2, seven flight sergeants, sixteen sergeants and two acting sergeants, 24 corporals, 33 leading aircraftsmen (LACs), 161 aircraftsmen 1 and nine aircraftsmen 2.

Williams did not let matters rest here. In the next issue of *Aircraft* he pointed out that a large proportion of the NCOs shown as Victorians in the numbers Hart had been quoting were, in fact, originally from other States but had been serving with the AAC (which was then wholly based in Victoria) at the time they were transferred to the RAAF. He also pointed to the simple fact that the largest proportion of applications for vacancies advertised in newspapers had come from Victoria—719 out of 1678 applicants, or 43 per cent.[19]

While this controversy duly fizzled out, Williams had on his hands another dispute concerning the initial recruitment of officers for the RAAF which held far more importance for the success and well-being of the new force. The problem was in many respects an extension of the bad feeling which had developed between Williams and Goble, reflecting the fact that the service of one had been in Australia's wartime flying corps while the other had served in one of the British air services. Grumblings were now heard that members of the latter group were discriminated against, in favour of those with AFC experience, in both initial appointments and the allocation of ranks. As Air Commodore Ray Brownell, an ex-RFC pilot commissioned in the RAAF's first few months, later wrote:

> I ascertained that to obtain an appointment I would have to agree to be demoted to the lowly rank of Flying Officer and that many pilots who had served in the Australian Flying Corps were to be given precedence over those of us who had served in the Royal Flying Corps, regardless of experience or of active service flying. As I had joined the RFC before units of the Australian Flying Corps were operating on the Western Front, I assumed I would at least be granted equivalent rank to that which I held when I left the Royal Air Force. I also realised that not many of the Australian Flying Corps pilots could have had as much wartime experience as I had had.[20]

In reality, however, it was not just the ex-RFC types who felt they had something to complain about. Air Marshal Sir George Jones, a former AFC pilot, recalls:

> Jealousy was terrible when the RAAF was formed, because it was felt Williams had looked after his friends from No.1 Squadron in the AFC—Murray Jones, Lukis, Cole, and others like McNamara, VC. He did take others in, like Wrigley and Anderson from No.3 Squadron, and Cobby from No.4 because he had such a high score of enemy aircraft in the war. And me, they gave me a short service commission as a Flying Officer (although I'd been a captain in the AFC in France) which I thought was a bit mean even though I accepted it.[21]

The tensions unleashed actually worked in three directions. The primary tension was between AFC members and those who had served with the RAF, but there were also differences between those whose experience had been with the AFC or RFC and those from the RNAS. Jones remembers the jealousy becoming 'quite bad at times':

> For instance, at Point Cook we were acting the goat one night . . . when John Summers got the bright idea that he'd throw me out of the Mess, but he found he wasn't strong enough to do it. Roy Marsden [the CO] evidently saw what had happened and ordered Summers to apologise to me. Summers came down to my workshop the next morning and apologised quite handsomely, we shook hands and

> that was it. But he went to Cole, the station commander, telling him that I did a bad job down in the workshops—I heard of this later from Cole. The feeling was that the AFC blokes didn't know their stuff—the RAF had been the real thing and those who'd been in that service were king-pin ... I got a bellyful of the extreme competition, I can tell you.[22]

The biographer of Squadron Leader 'Jerry' Pentland, an ex-RAF air ace with the eighth highest tally of enemy aircraft downed by an Australian in the war, also remarked on this 'unfortunate sense of discriminatory feeling between the two sets of officers':

> Most of the pilots who served with British units had seen more active service and gained wider experience than their Australian Flying Corps counterparts, but there was a definite impression that they were being overlooked when it came to filling key positions. There was no open hostility, but occasionally tempers flared and once Pentland and De La Rue threw an uppity AFC man into the Mess fireplace, which didn't improve matters.[23]

Conflicts of this sort were also being fought at the RAAF's senior levels, though generally in a more concealed fashion. Almost immediately upon the Air Force's formation arrangements were made to remove Goble from the scene by sending him to England. According to Williams, this move originated with Goble himself: 'The decision regarding the alternating of the chairman at meetings of the Air Board was, I believe, a little embarrassing for Goble and soon after the Service was formed ... he suggested that he be sent for attachment to the RAF for a period of two years'. Williams goes on to remark:

> No objection to this proposal came from the Chief of the Naval Staff ... at the meeting of the Air Council. This was rather surprising after the Navy's efforts to have Goble appointed the senior of the Service for it meant that he would now have no hand in the first years of its development. It also put an end to the 'alternative chairman' of the Board.[24]

Ostensibly the purpose of Goble's trip was attendance at courses to provide him with a knowledge of the latest methods of training in naval cooperation, and also to interview applicants seeking appointment to the flying-boat squadron due to be raised as part of the RAAF.[25] Air Council records make clear, however, that it was intended to hold him at the Air Ministry after he had completed these courses by employing him as the Commonwealth's air liaison officer at the Air Ministry, during which time it was initially planned that an RAF officer, Wing Commander E.D.M. Robertson, would fill his position on the Air Board in Melbourne.[26] In the event, when Goble sailed in October, his place on the Board was temporarily taken by a local officer, Squadron Leader Anderson.

During the period Goble was on course, it was planned that liaison duties at the Air Ministry would be carried out by a RAAF technical officer acting as his deputy. This post was to be filled by Eric Harrison. The ALO already in London, Major William Coates, MC, was 'desirous of relinquishing his position and returning to Australia at an early date' (to accept a commission in the Air Force), while Harrison had officially appealed against the rank to which he had been relegated on the RAAF's formation, evidently considering that his appointment in 1912 to the original Aviation Instructional Staff entitled him to something better than he had received within the RAAF rank structure.

The Air Council therefore approved a proposal that Harrison be appointed as Coates' replacement and promoted to the substantive rank of squadron leader as a technical officer. Such was the haste to settle this friction that when the Director of Equipment told the Air Board on 1 July that a complete stocktake of the stores currently in Harrison's charge would be necessary prior to him sailing for England, and that this was expected to take three months, the Board decided that it was not advisable to delay Harrison's departure and authorised the DofE 'to make such arrangements as he considered satisfactory to himself'.[27]

Use of 'existing defence machinery'

With the doubling of the RAAF's original strength which took place within the first six months of the force's existence, a requirement emerged early for induction training of new recruits. Although 80 per cent of the personnel enlisted had served during the war,[28] all but a small proportion had returned to civilian life since the end of the war and stood to benefit from some refresher training. The program approved for the Air Force provided for the establishment of a recruit and non-technical training centre for precisely this purpose, but—in an unexpected twisting of Williams' policy of using 'existing defence machinery' as an economy measure wherever possible—the Military Board came up with an alternative proposal (as explained more fully in chapter 3). Consequently the first enlistees—both officers and other ranks—received their formal introduction to the RAAF at a course at the Army's Central Training Depot (CTD) at Holsworthy, just outside the Sydney suburb of Liverpool.

The first detachment of 22 officers and 145 airmen went into camp at the CTD on 12 September 1921,[29] forming what was known as 'No.1 Air Force Course'.[30] Course members were organised into four divisions or platoons, with all the officers forming the first of these. It was evidently intended that another intake would follow after this first group had completed their training. On 6 September 1921, Flight Lieutenant Lawrence Wackett told an intending recruit: 'All persons now enlisting go first of all to a course of non-technical training at Liverpool, NSW. The first course starts on Monday next and the second in three months time.'[31] However, for some unexplained reason a second detachment of 10 officers and 74 airmen marched in barely four weeks later, on 7 November; this became 'No.2 Course' and trained concurrently with the first arrivals.[32]

Although administered as 'RAAF Details, Liverpool' with Squadron Leader Rolf Brown as OC, for the next fourteen weeks the RAAF personnel underwent drill and other infantry training from Army instructors under Brigadier-General J.M. ('Bull') Antill, the Chief Instructor of the depot. This training was by no means demanding. One airman, Francis Hakanson, recalls: 'The purpose of the camp seemed to be just to get everyone to know one another. We had the weekends off. I really enjoyed it.'[33] Sir George Jones has recorded that, taking advantage of the weekend leave situation, he brought his wife and baby with him to Sydney and installed them in a boarding house in Elizabeth Bay Road.[34]

While not especially testing, according to one account the course produced some interesting clashes between the military style of doing things and the Air Force ethos left over from the days of the AFC. Warrant Officer Robert Buchanan, a Scot with the inevitable nickname of 'Jock', was a short round-shouldered man not well-made for demonstrating orthodox military deportment. He had been the chief master mechanic of No.5 Squadron, AFC, during the war, and served on with the AAC before finding himself at Holsworthy. When challenged by a major on the CTD staff for riding a bicycle in the camp, he reportedly explained, respectfully, that he was orderly sergeant that day, and Air Force usage was that the orderly sergeant always had a bike to get around an aerodrome. 'Ha!' snorted the major, 'But this is not an aerodrome. Anything else to say?' Buchanan, still respectful, pointed out that a RAAF technical warrant officer was responsible only to the CO of a unit, and rode away from the interview on his bike.[35]

For some, the course provided a taste of military life that was too much for their liking. Under the regulations, enlistees could claim a free discharge if they applied within three months of their attestation. Records reveal that at least a dozen men took advantage of this provision while at camp, while one man was discharged by the Chief Instructor for 'gross misconduct'.[36]

During the course, too, the depot received a number of official visitors who came to see what was going on: the Governor-General, Lord Forster, visited on 24 October,[37] and Sir Walter Davidson, the New South Wales Governor, attended a parade and sports meeting on 12 December, three days before the second course was wound up.[38] With the completion of training, everyone proceeded on leave over Christmas before being required to report back at Point Cook.

For reasons explained in the next section of this chapter, no more Air Force personnel were sent through courses at the CTD. However, this was not the last occasion on which the Air Force found the policy of using 'existing defence machinery' was turned against it. With personnel numbers bolstered by the return of the Liverpool courses, in 1922 the senior medical officer at Point Cook, Squadron Leader Arthur Lawrence, was transferred to Headquarters to fill a new appointment as Director of Air Force Medical Services, his former post at Point Cook being filled by Flight Lieutenant George Cherry. Lawrence was now told by the Director-General of Army Medical Services, Major-General Sir Neville Howse, VC, that 'until the air force expanded further it was proposed, for the sake of economy, to use existing army machinery as far as possible'.[39]

Taking the principle one step further, a conference was held later in 1922 to consider, among other issues, whether to amalgamate the medical services of the Army, Navy and Air Force. Lawrence was appointed to a sub-committee of the medical directors of the three services which, reportedly under strong personal influence of Howse, recommended in July the formation of a single medical corps to provide all services for the defence forces. Lawrence and his naval counterpart were, however, not fully enthusiastic about the plan and

Officers attending the RAAF 'No. 1 Course' held at the army's Central Training Depot, Holsworthy, pictured with the Governor-General, Lord Forster, and depot staff on 24 October 1921. *Rear row:* FOff G.F. Thomas, Lt Hudson, POff H.A. Wilson, FOff E.J. Stephens, FLt I.N.C. Clark, FOff J.H. Summers, Major Christie, FOff W.D. Bostock, FOff J.R. Bell, POff J.D. Jarman, FOff A.W. Murphy, FOff E.A. Mustard; *centre row:* FOff H.C. Harrison, FLt A.A.N.D. Pentland, SqnLdr R.S. Brown, Capt Traill, Brig-Gen Antill, Lord Forster, Brig-Gen Cox, Brig-Gen Cox-Taylor, Major Selby; *front row:* POff T.H. Kempson, FOff G. Jones, FLt H.F. De La Rue, FOff F.C. Hawley, FOff E.D. Cummings, FOff R.J. Brownell, FOff W.E. Cowper, FOff E.P. Kenny, FOff L.J. Balderson. (*Mrs J. Stewart*)

consequently it was still unimplemented when, in September 1925, the Minister for Defence (which happened to be Sir Neville Howse) ordered the amalgamation proposal to be re-examined. With Navy now strongly opposing the idea, Lawrence also advised the Air Board that he was withdrawing his endorsement of the recommendation of the original sub-committee and advocating cooperation among the three services instead of amalgamation.

Although a new committee which met in November 1925 decided that amalgamation was not feasible, it appointed another sub-committee to consider means of improving administration of the three separate medical services. The report submitted by this body favoured cooperation, although the Army representative still pressed for amalgamation. After a Royal Commission of Health in 1925 argued for coordinating the medical services of Army, Navy and Air Force under one director-general with sub-directors in each service, in September 1927 Howse's successor as Minister for Defence (Major-General Sir William Glasgow) decided that the Army should assume control of the medical and dental services of the RAAF. The Director-General of Army Medical Services was to be responsible to the Air Board for the administration of these services from 1 October 1927. Although the Board protested this decision, the Minister stood firm and instructed that his ruling was to be put into effect.

The termination of Lawrence's commission as a wing commander in the PAF on 24 September 1928 and his continuance as Deputy-Director of Medical Services (Air) on the CAF list was the logical outcome of the process set in train by the 'existing defence machinery' policy which Williams initially espoused.

Cutback in the development program

At the time the AAF came into formal existence, its organisation was neither large nor widely spread. Headquarters were established on Melbourne's St. Kilda Road, inside the Victoria Barracks which housed the rest of the central administration of the Defence Department. Occupying a two-storey bluestone structure known as 'J' Block which was built in the 1870s as the original barracks hospital,[40] RAAF Headquarters had just 13 servicemen on its strength. These were the three uniformed members of the Air Board, Williams, Goble and McBain, each with an attached staff officer (A.M. Jones, Wrigley and Rogers respectively), and seven airmen.[41] At Point Cook the old CFS had become No.1 Flying Training School (FTS) manned by twelve officers and 67 airmen.[42] At two locations around Melbourne there was also the basis of what was already being referred to in April 1921 as No.1 Aircraft Depot (AD), in the form of stores sections at Spotswood and North Fitzroy.[43] The former, responsible for handling imperial gift equipment, occupied wheatsheds rented from the Victorian Railways, and was under Squadron Leader Berryman; the latter, located in vacant brewery buildings at 470 Nicholson Street, became the motor transport repair section and was under Squadron Leader Harrison.[44] Between them these two sections employed a total of three officers and 56 airmen.[45]

On 1 July 1921 No.1 AD was formally created as a unit based at Point Cook, with the section at North Fitzroy remaining as a detachment of the depot. The section at Spotswood was closed down at this time, with personnel being posted elsewhere (except for the OC, Squadron Leader Berryman, who relieved Harrison at North Fitzroy when the latter proceeded overseas) and the stores being left under a caretaker. Squadron Leader Anderson was appointed as temporary CO of the unit, in addition to his duties as commander of the whole Point Cook base which was now redesignated as No.1 Station, RAAF. With the expected return of personnel from Liverpool in January 1922, No.1 AD was reorganised into a headquarters, stores, aircraft repair and engine repair sections. The Motor Transport Repair Section initially continued its separate existence at North Fitzroy but in April it was moved to Point Cook to join the rest of the unit.[46]

The completion of the Liverpool courses also enabled a start to be made on forming the rest of the unit organisation envisaged under the approved program, supposedly from 1 January 1922 although the personnel allocated to each unit were certainly not available for some days—and in some cases weeks—after this date. Under economy steps decided upon in the previous July, the squadrons to be formed would not wholly comprise permanent personnel but would initially have a portion of the headquarters and just one of their three flights—the remainder to be formed later from CAF personnel.[47]

According to the disposition approved by the Air Board on 20 December 1921 the initial allocation of men and aircraft was as follows:

No.1 AD
Squadron Leader A.M. Jones; Flight Lieutenants Coates and Cherry; Flying Officers Howells, Johnston, Mackinolty and Fryer-Smith; Pilot Officers Jarman, Swinbourne, Wilson and Kempson; 88 airmen.
No.1 FTS (eight Avro 504Ks, four Sopwith Pups)
Flight Lieutenant Clarke; Flying Officers Brownell, Bell and Stephens; 12 airmen.
No.1 Squadron (Corps reconnaissance—four DH9s)
Flight Lieutenant S.G. Brearley, DFC; Flying Officers Summers and Balderson; 5 airmen.
No.2 Squadron (Fighter—four SE5As)
Flight Lieutenant Cobby; Flying Officer Cummings; 4 airmen.
No.3 Squadron (Corps reconnaissance—four DH9s)
Squadron Leaders Brown and Berryman; Flight Lieutenant Lukis; Flying Officers Cowper, Kenny and Bostock; 27 airmen.
No.4 Squadron (Fighter—four SE5As)
Flight Lieutenant Pentland; Flying Officer G. Jones; 7 airmen.
No.5 Squadron (Naval Cooperation—three Fairey IIIDs)
Flight Lieutenant De La Rue; Flying Officers Murphy, Mustard, Hawley and Walne; 17 airmen.[48]

According to the plan, the first four of these units would be located at Point Cook, while Nos. 3 and 4 Squadrons were the units which would make up No.2 Station in New South Wales, at a locality which had been decided upon in July 1921 as Richmond.[49] No.5 Squadron, the seaplane unit to work with the fleet, was to be based at a site on Sydney Harbour, negotiations for the purchase of which were still proceeding.[50] Not yet formed (due to the lack of suitable planes) was the flying-boat unit which would become No.6 Squadron, although early in 1922 an area of 116 acres at North Geelong, on the foreshore of Corio Bay, was acquired by the government for this purpose.[51]

Unfortunately for the RAAF at this crucial juncture in its growth, powerful external factors now intervened. On 2 March the new Defence Minister (Massy Greene) announced to the Air Council that, following the apparent success of the disarmament conference at Washington the previous December, the government intended to limit its spending on defence purposes during the next financial year. For the Air Force this meant that funding would be held to the same level as the current year. Although the RAAF had been allocated £500 000 in its first year, it had been possible to spend only £135 195 of this amount because the service was formed only three months before the financial year ended. Notwithstanding these special circumstances, the Air Force vote was cut to £382 551 for 1921–22, of which only £245 479 was spent due to various delays. With the government's desire to pin its faith in the Washington Treaties and limit spending further, the Air Force now found itself restricted to just £250 000 as its total vote for 1922–23.[52]

This was a blow which threatened the RAAF's program of development. Indeed, Williams was later moved to express his certainty that 'had the Air Force not been formed when it was . . . the financial depression which followed immediately after might well have resulted in its not having been established at all'.[53] As it was, the service now faced curtailment before even having reached its initial limited goals. A revised policy blueprint dated 6 April 1922 accepted that it was impossible to proceed with the unit organisation originally planned. Instead, the RAAF would aim to create units mainly composed of CAF personnel formed around permanent cadres. The force would also concentrate on preserving a capacity for future expansion by focusing on training personnel and building up the essential ground organisation—that is, land, buildings and workshops. Accordingly the new plan called for an early start to be made on permanent buildings to house No.1 AD at a new site at Laverton, and the formation of a single squadron to be based at Richmond, outside Sydney.[54]

In line with the new reality, the five squadrons formed only months before were disbanded from 1 July 1922, leaving the men and 25 machines of these embryo units grouped as a single mixed squadron of six flights at No.1 FTS. In November an establishment for what was referred to as No.1 Squadron was approved: three flights comprising 39 officers and 155 other ranks.[55] An establishment was also applied to the whole Air Force from 1 July 1922, reducing the number of airmen who could be enlisted to a maximum of 300.[56] In fact, for the next two years to mid-1924 the strength of the RAAF was to remain—in round figures—50 officers and 300 other ranks.[57] There was one other ramification of cut-backs in funding. As Williams pointed out to the Parliamentary Standing Committee on Public Works in September 1922, flying personnel in the RAAF actually numbered no more than about 30 men and after eighteen months in existence the Air Force had yet to begin the training of any new pilots: 'Our present staff is really a nucleus, and quite inadequate alone to effectively serve in the case of a possible emergency'.[58]

For many of those who had entered the service with high expectations of what their new career had to offer, there was now only disappointment. Several officers re-evaluated their future in the changed circumstances and decided not to continue. When Squadron Leader Berryman learnt that he was to become the transport officer at Point Cook, a post he saw as being of 'relatively minor importance' and incompatible with his rank and seniority, he applied to resign, seeking compensation on the grounds that he had become excess to the RAAF establishment. The Board, in considering his request in July, noted that:

> there is no necessity for any Air Force officers to resign and no invitation to do so has been given. On the other hand, if compensation were offered, there are probably several officers and other ranks, whose services could ill be spared, who would avail themselves of the opportunity to resign now, owing to the greatly reduced prospects of promotion &c. brought about by the reduction of the establishments previously approved.[59]

The Board therefore decided it could not recommend the payment of compensation to any personnel who chose to voluntarily resign. This did not deter Berryman, who decided to press ahead in any case. He later became the Australian agent for the German aircraft firm of Junkers. Leaving at the same time and under the same conditions was the RAAF's first signals officer, Flying Officer Hector Johnston.[60] The unrest among Australian officers was also evident from the admission of the Minister for Defence (Bowden) in June 1923,

Following the abandonment of the RAAF's original approved structure in mid-1922, these four SE5As (A2-6, A2-13, A2-2, A2-19) formed 'C' Flight of a composite squadron at Point Cook which was the sole Air Force unit until mid-1925. *(FLt F.C.H. Hakanson)*

speaking in Parliament to the reintroduced Air Defence Bill, that some RAAF officers were leaving to take commissions in the RAF (which had been increased to 34 squadrons) and quoting the British Air Minister as having welcomed Australian applicants.[61]

To avoid retrenching any airmen, the Air Board adopted the approach of encouraging redundant personnel to requalify for remuster to other trade categories where there were unfilled vacancies.[62] Despite this arrangement, a rash of applications were received from other ranks wanting to quit even if they had to purchase their discharge. By October 1922 the numbers seeking release by this means had reached the stage where notice was given that: 'In view of the number of applications lately received, the question of discharges by purchase has been reconsidered, and it has been decided that, for the future, they will be granted only for very urgent reasons'.[63]

Fresh cause for dissatisfaction among the officer ranks arose early in 1923 when the results of examinations to confirm their initial probationary ranks became known. Those achieving a score over 70 per cent were recommended for permanent commissions, and three—Bostock, Stephens and Swinbourne—were noted for promotion. Those with lower scores, however, were recommended only for short service commissions—although some were given the prospect of being reconsidered for permanency later. Four officers scoring less than 60 per cent in the examination were not recommended for confirmation at their existing rank; two were offered temporary appointment at lower ranks, while two—Coates and Cummings—preferred to resign.[64]

Early activities of the RAAF

With the bulk of its personnel at the Central Training Depot during the second half of 1921, it was not until 1922 that the Air Force actually settled down to any sort of normal work. The Melbourne *Argus* reported that the air station at Point Cook had been 'practically deserted' for some time, and the aerodrome guarded only by a small care and maintenance party, before the return of the 'first detachment' of 30 officers and 130 men from Sydney by special train on 10 January 1922.[65] Actually, an advance party had arrived back a week earlier, comprising Squadron Leader Murray Jones in command, Flight Lieutenant Edward Howells as adjutant, Sergeant-Major Hugh Cully as the RAAF's first Disciplinary Sergeant-Major (DSM), and a few airmen.[66] The remainder of the personnel returning from the CTD arrived on 3 February.[67]

One of the first activities was the issuing of new RAAF uniforms. According to Sir George Jones, it had been the intention to provide these to new enlistees during their time at Holsworthy.[68] This did not happen, however, due to the non-completion of contracts,[69] and photographic evidence shows that personnel at the CTD wore Army-style AFC attire throughout their course. Evidence in the Melbourne press indicates that the issue finally proceeded in mid-January 1922,[70] with *Aircraft* magazine reporting: 'Two complete uniforms are now issued to all ranks of the Royal Australian Air Force. The summer uniform is of white drill, the winter dress of 'horizon blue' cloth similar to that worn by officers of the Royal Air Force.[71] In fact, the winter uniform was closer to navy blue than to the colour adopted in the RAF (see chapter 4), and the summer uniform was soon changed, with the impractical white giving way to a pale khaki colour though a white tropical full dress uniform for officers was retained.[72]

Francis Hakanson, one of those who attended the CTD course, later recalled the situation initially experienced following the arrival at Point Cook of the new personnel:

> There was not a lot of work done in the first two or three months. Five or six Bessoneau hangars were half up and the new arrivals had to finish erecting them. After two or three months, men were allotted to jobs. I went to 3 Squadron, which operated the few SE5As that had been erected before everyone arrived from Liverpool camp.[73]

Late in April the RAAF undertook its first exercise in Army cooperation, which involved spotting for the garrison artillery at Queenscliff (see chapter 7). This was followed in May by the participation of six RAAF machines in the New South Wales Aerial Pageant held at Sydney's Victoria Park racecourse. Four SE5As flown by Flight Lieutenant 'Jerry' Pentland (leader) and Flying Officers Ray Brownell, Eric Stephens and John Summers gave the 'first exhibition of formation flying seen in NSW and followed with a priceless display of aerobatics', then engaged in mock combat with two DH9As flown by Flight Lieutenants Stanley Brearley and Frank Lukis.[74]

Mock attacks quickly became a popular means of showing the general public and important figures the new service's combat capability. During a visit to Point Cook in December by the Governor-General, Lord Forster, a show practically identical to that given at the Sydney air pageant was again mounted with four SE5As pitted against two DH9s.[75] Other accounts speak of similar displays being given to visiting school parties and journalists.[76] In March 1923 one mock combat was staged over the centre of Melbourne, with eight DH9As making a bold daylight bombing attack on Victoria Barracks against a defence mounted by four SE5As and two Sopwith Pups. Press accounts claim that the raid had the 'definite object' of showing 'whether our airmen were fitted to make and repel attacks from the air'. That being the case, the results were hardly reassuring. The fighters reportedly failed to intercept the attacking bombers until after these had made their pass over the Barracks, scoring what were claimed to be direct hits, and had to settle for harrying them all the way back to Point Cook.[77]

While these activities were important in bringing the force to the notice of the public, another—unfortunately more sombre occasion—also achieved this result. In April 1922 two national aviation heroes, Sir Ross Smith and his mechanic J.M. Bennett, half the team which won the 1919 England-to-Australia air race, were killed in an aircrash outside London. Their remains were brought back to Australia for burial, Bennett in Melbourne and Smith in Adelaide. Inevitably the RAAF was heavily involved. For the funeral in Adelaide on 15 June, a flight of DH9s was sent to South Australia via Warracknabeal and Nhill (Vic), along with a 50-man firing party which went by rail.[78] The funeral of Bennett in Melbourne two days later involved an equally large commitment of RAAF resources.

A party of airmen marched, in slow time with arms reversed, for most of the distance from Parliament House at the top of Bourke Street to St. Kilda Junction, where a halt was made outside Bennett's garage while a fly-past of RAAF aircraft trailing black streamers went overhead, then from there to the St. Kilda Cemetery.[79] Well might Williams later issue a message of congratulations, declaring:

> This is practically the first time the RAAF has come prominently under the notice of the general Public in any way, and it is believed that it acquitted itself creditably. It will of course be realised that attention to drill is not perhaps the most important feature of an Air Force but the Board sees no reason why just as high a standard cannot be maintained in the branches of Air Force work as was shown in the drill and general conduct of all ranks last week.[80]

However, even after 65 years one participant recalled the Bennett funeral only too well:

> As far as I can recall the whole torture lasted just on four hours, and we were loaded into a Leyland lorry for a stand-up trip back to Point Cook . . . The whole thing was an outrage—a scandalous misuse of an ordinary man's funeral to stage a public relations extravaganza, and it was extreme physical abuse of the personnel concerned.[81]

In view of the prominent public profile so soon achieved by the Air Force, it is not surprising that steps to form a band were quickly taken. In January 1923 Corporal B.D. Thacker, a motor driver mechanic, started up a voluntary station band with himself as acting bandmaster. In August the Air Board approved, at Thacker's suggestion, the engagement of an expert instructor 'in order to train the band up to a standard adequately finished to enable it to take its part in such ceremonial parades as are required outside the Station'. This resulted in the appointment of a civilian conductor, Hugh Niven, as part-time Bandmaster of what became the RAAF Band, initially as warrant officer but from July 1934 with the honorary rank of flying officer.[82] Niven's long association with the band was matched by LAC Hal Harding, who became Drum Major when Thacker left the Air Force in 1927 and who was also commissioned.

In November 1923 practically the whole of the RAAF's activities were suspended during the police strike in Victoria, with 'all personnel and many vehicles employed on Guards, transport, &c., or standing by for emergencies'.[83] Air Force vehicles had to be loaned to the military forces at the height of this testing time, as that service was still entirely dependent on horse-drawn vehicles.[84]

For the RAAF, 1924 turned out to be another year studded with special public occasions, beginning in March when Australia was visited by a British naval squadron led by the giant battlecruiser *Hood.* The squadron's arrival at Melbourne on 17 March was marked by an unprecedented display involving 22 RAAF machines. At first light two Fairey IIIDs met the British ships at the entrance to Port Phillip Bay, and from there the aerial escort provided was virtually constant—first three DH9As, then five DH9s, followed by two formations each of five Avro 504s, in addition to one SE5A and a Sopwith Pup. According to the Melbourne press, these 'kept in close contact with the warships during the journey up the bay, and performed various manoeuvres in the air. When the warships had reached their berths the aeroplanes returned to Point Cook and landed without mishap'. After the ships docked at Prince's Pier, the squadron's commander, Vice-Admiral Sir Frederick Field, was received by an Air Force guard of honour and the RAAF Band playing 'Rule Britannia'. Despite its naval significance, the occasion was very much Air Force in flavour.[85]

When the squadron sailed on up the east coast, six DH9s were dispatched from Point Cook on 4 April to Richmond, from where they provided an air contingent for the welcoming celebrations at Sydney as well. These machines were flown by Flight Lieutenants Ray Brownell (commander), John Summers,

DH9s stop-over at Nhill (Vic) on their way to Adelaide for the funeral of Sir Keith Smith on 15 June 1922. The lettering visible on the cowling of the aircraft denote them as machines presented by the British Government. The middle aircraft (A6-21) was flown by Flight Lieutenant W.D. Bostock. *(Author)*

Following the deaths of Sir Ross Smith and Lieutenant J.M. Bennett in an aircraft accident in England in April 1922, their remains were returned to Australia for burial. The funerals, in Adelaide and Melbourne respectively, were solemn occasions in which the RAAF was heavily involved. For the airmen taking part in Bennett's funerals, however, this event was chiefly notable for entailing a tortuous march from Parliament House in the city to the St. Kilda cemetery, a distance of more than seven kilometres—most of which was covered in slow time. *(RAAF Museum)*

Ernest Mustard and Arthur Hempel, and Flying Officers Edward Leggatt and Vernon Burgess. Press reports of the day provide vivid descriptions of the squadron's entry into Sydney Harbour, reporting that the first two RAAF aircraft were spotted coming up from the south as the ships were about two miles off the Heads:

> One, when about Coogee, swept east to meet the oncoming vessels, the other turned west and performed several stunts over the harbour. Right on the stroke of 10 o'clock the giant Hood came abreast of North Head. Almost simultaneously six more aeroplanes, flying in battle formation, swept up from the south, and flying high over the Hood when between the Heads, gave some realism to actual warfare by dropping a number of smoke bombs and then dropping some rockets, which exploded. It was the airmen's welcome from aloft.[86]

Completing the flight's sojourn in Sydney was its involvement with the ceremonies marking the sinking on 12 April 1924 of HMAS *Australia,* former flagship of the Australian fleet, under the terms of the Washington Treaty. As the scuttling was carried out off Sydney Heads, two RAAF planes trailing farewell pennants photographed the scene for official purposes, and as the ship went down one circled low to drop a wreath over the floating debris which was the only evidence of its passing.[87]

The following August DH9s were once more flying north, this time to take part in the centenary celebrations of the city of Brisbane. Unfortunately for the RAAF's image, all three aircraft originally sent were forced down at various

The RAAF Band and honour guard march off Prince's Pier after ceremonies welcoming the British special service squadron, led by the giant battlecruiser HMS *Hood,* on 17 March 1924. Because the Point Cook base was the largest permanent defence presence in proximity to Melbourne, RAAF personnel became involved in many special occasions of a non-Air Force nature. *(Mrs Y.C. Bonwick)*

places around New South Wales—at Griffith, Marulan, and on the Clarence River.[88] This necessitated the hasty despatch of flight mechanics from Point Cook to get these planes airborne again. So hurried was departure that one of those sent, Warrant Officer Walter Nicholson, left without warm flying gear and suffered freezing to his face which caused later medical problems.[89]

In September RAAF members were again involved in an air display at Essendon. Flight Lieutenant Harry Cobby was one of two pilots involved in giving an exhibition of balloon bursting, while a mock battle was staged by a DH9A flown by Squadron Leader Bill Anderson and two SE5As flown by Flight Lieutenants Summers and Brownell.[90] The popularity of air shows prompted the Air Force to mount the first pageant of its own at Flemington racecourse on 13 December 1924, held in aid of several charities including the RAAF Benevolent Fund. The program featured displays of landing in confined areas, picking up and dropping messages, low-level bombing, formation flying, aerial drill, ammunition dropping, aerobatics, and races.

Attracting a paying crowd estimated at 7000 people (with many more viewing from outside the ground) the pageant was, according to one newspaper, 'one of the most spectacular flying exhibitions yet seen in Australia'. The only hiccough to the program apparently was the unscheduled crash of a DH9 on completion of the aerial combat display.[91] If this was what spectators were hoping to see, the real excitement had already taken place during the rehearsals three days earlier. No fewer than four aircraft—an Avro 504 and three DH9As—were damaged during landings on the racecourse, the Avro extensively after it 'struck a fence, carrying away some of the timber, and came to rest on the steeplechase course'.[92]

In April 1925 the RAAF also came before the public eye while playing host to one of the great early epic flights. This was the visit by Italian air force officer, Major the Marquis Francesco de Pinedo, and his engineer, Ernesto Companelli. Pinedo reached Broome (WA) in a small Savoia-Marchetti flying boat on the last day of May, only the third plane to fly to Australia from Europe. Once the machine arrived in Melbourne on 9 June it was serviced and overhauled by Air Force mechanics at Point Cook. Pinedo left again in August for Japan, arriving back in Rome in November. When Australians serving in the Allied advance through Italy in 1944 came across a derelict aircraft bearing an RAAF crest on its fuselage, there was considerable puzzlement until Pinedo's visit to Australia 20 years earlier was remembered. As Sir George Jones recalls, the emblem had been painted on by his workshop personnel.[93]

Of all the activities which fell to the Air Force in the first years of its existence none was as portentous for future development as the program, publicly mentioned in mid-1921, to survey air routes in all the States.[94] At first this scheme was linked with an earlier proposal to use the RAAF in initiating an experimental airmail service (see previous chapter). It will be recalled that

One of the RAAF's early characters, A.J. Abicair, pictured with a Civil Aviation Branch DH37 during an air display at Essendon Aerodrome in 1925.
(Mr C.H.G. Gardiner)

When the flying boat used by the Marquis de Pinedo on his epic round-the-world flight reached Melbourne in June 1925, it was serviced by RAAF mechanics at Point Cook. Commemorating the connection, the flying boat later carried a RAAF crest painted on its fuselage (visible just forward of the 'Savoia' marking). *(RAAF Museum)*

this idea had figured strongly in the government's reasons for finally inaugurating the Air Force after much delay. Records show that following the Air Council decision on 3 February 1921 instructing Williams to proceed immediately, Colonel Brinsmead (CCA) had plunged headlong into planning for the RAAF's involvement in starting up an airmail service in Western Australia.

On 8 March Brinsmead outlined the scheme he had in mind to Defence Minister Pearce. In his opinion a satisfactory weekly service could be established in the west using a RAAF flight of six DH9s. Some additional transport might be required, he estimated, but 'it is understood that in any event this establishment is to be allocated to Western Australia' under the program approved for the Air Force. He therefore wanted a PMG Department's report advocating consideration of services either between Geraldton and Derby (WA) or Brisbane and Longreach or Winton (Qld) to be brought before the next Air Council meeting, and for DOI (Williams) to be instructed to prepare 'his necessary orders in anticipation of Air Council concurrence with the Postmaster-General's recommendation'. Pearce agreed to this course, and the Council accordingly considered the question on 15 March. At this meeting it was decided to call on the Air Board to submit its estimates of the likely costs involved, and the men and materiel needed, to establish a western service, along with details of how the costs would be divided between the Air Force and Civil Aviation votes, probable time before the service could commence, suggested routes, stages and terminal points, and so on.[95]

Only at this point did Williams directly reveal that he was flatly opposed to the whole scheme. Observing that he personally doubted that Cabinet's expectations regarding the commercial viability of such a service would be realised, he challenged the economics of operating DH9s over the route once operating costs were matched with the likely amount of paying freight. He also calculated that at least eleven or twelve aircraft would be needed, not just six, and at least seven pilots, not to mention the likelihood that twelve landing grounds and at least eight emergency grounds (possibly more, depending on the country) would have to be constructed. All this would inevitably mean that the planned formation of squadrons in Victoria and New South Wales must be deferred.

Brinsmead was understandably disappointed with this response, considering that national economic benefit outweighed all Williams' objections. As he made clear on 19 March, he also believed that Williams had exaggerated the demands likely to fall on the RAAF. Other members of the Air Board, he noted, had not actually had the opportunity to look at Williams' figures, and he suggested that perhaps when they did DOI's appreciation might be modified. If the Board ended up agreeing with Williams then there would be no alternative but to defer the proposal until more efficient aircraft were available or until PMG Department could select another route not presenting the same difficulties.

As a West Australian himself, Senator Pearce was rather irritated at the thought that bureaucratic infighting might cause the benefits of the proposed aerial service to go to another State. He therefore directed the Air Board on 19 March to prepare separate costings of Brinsmead's scheme and its own estimates of what would be involved in instituting a service. Brinsmead's clear intention of circumventing Williams by manipulating the Minister failed to work. When the Board examined the issue on 12 April it endorsed Williams' arguments and concurred in his figures. Observing that the country traversed by part of the proposed route appeared to 'entail undue risk to pilots and machines', Board members supported Williams' contention that a service could certainly be run but not as a commercial success. If the scheme was to proceed anyhow, the Board pointed to the need for a preliminary survey from the ground to be carried out along the whole of the route. When Brinsmead supported this suggestion, the Air Council decided to await a route report before making any recommendations.

As it happened, the Air Force already had an officer in the west during this debate. On 28 March Western Australia had suffered its first fatal air crash when John McIntosh, Parer's partner in the 1919 England-to-Australia air race, was killed while barnstorming in his Avro 504K at Pithara. In response to a request from the coroner conducting the inquest into McIntosh's death for a flying expert to advise him, Flight Lieutenant Adrian Cole (himself the son of a Melbourne coroner for 25 years, Dr R.H. Cole) left on the express for Perth on 30 March. Written instructions from Williams which he took with him stated: 'While in Perth, you are also . . . to obtain as much information as possible regarding the country between Geraldton and Derby over which a proposal has been made for the running of an aerial service'.[96]

Unluckily for Cole, he had only just returned to Melbourne from the McIntosh inquest when the surveying of the Geraldton-Derby route became a burning issue. Given fresh instructions by Williams on 9 May, he was promptly turned around and sent back to perform this new task, apparently also with the intention that he would be placed in charge of the RAAF unit running the service if this was eventually decided upon. Cole's luck was really out, however, because 1921 turned out to be a year of record rain in the western State. Areas through which he was expected to travel to select suitable landing grounds were subjected to as much as twenty times the average rainfall. At Northam he hired a 1920 Dodge touring car recently used during the Prince of Wales' visit, complete with its owner-driver, and the two men set out in mid-May on a journey expected to cover more than 1900 kilometres.

Cole later left an account of the travails of the journey, recording that:

> We got along slowly following tracks roughly in the right direction, just cross-country wool waggon tracks . . . between station homesteads, with occasional party telephone lines erected along a fence to save posts, which we used as guides to settlements. We sometimes took two days to go a mile, and wore out four tyres, all our blankets, and half our clothes in getting that distance. The Fleur-de-lis of the Prince of Wales on each door soon disappeared under a coat of mud.

After floating the car across the flooded Wooramel River south of Carnarvon, Cole and his companion found themselves in an increasingly desperate situation. Delayed by repeated boggings which eventually saw them run out of food, they went to sleep in the car one night 'thinking we were set for an early start in the morning':

> It rained steadily, with heavy thunderstorms all night. At daylight next morning the car was sitting in a lake—about twenty miles wide—which reached up to the foot-boards. Our suitcases, which we had placed outside the car to make room for us to sleep, had opened and floated away. Collars and shirts were floating amongst the mulga scrub, there was no breakfast, and the situation was grim. We had had no food for two days and nothing seemed to be stirring in that vast country.

With no alternative but to walk, the two men went on for another two days before the driver began to fail with an injured foot. Cole pressed on alone, fortunately finding help next day at Glenburgh station, 250 kilometres east of Carnarvon, which led to the recovery of both men and their vehicle. Eventually Derby was reached, where Cole and his companion spent a fortnight waiting for a ship to take them and the car back to Perth. His dismay can only be imagined when: 'As a final stroke I received a telegram informing me that the Government had changed its mind and decided to give the contract for the North-West air mail to a civilian, with a subsidy of £25 000 a year. I had been on a joy-ride, after all.'[97]

In fact, the possibility that the airmail service might end up being operated by a civilian contractor had been considered even before Cole set off from Melbourne. On 3 May, Cabinet had decided that Geraldton to Derby was the preferred route for the first such service established, whether run by the Air Force or by a private company receiving a government subsidy. After a protracted Cabinet meeting on 17 May it was actually announced by the acting prime minister, Sir Joseph Cook, that tenders would be called from private aviation interests, so that it was clear from this moment that Williams had won his point.[98]

Even though attempts to utilise the new Air Force on a task which might conceivably detract from its defence role had been averted, it seems there was never a doubt that the RAAF would still be involved in some capacity. On 7 May, for example, Brinsmead had written to the secretary of the Defence Department, for the Minister's information, that it seemed reasonable to believe that the air route up the west coast of Western Australia to the Northern Territory would be identified as 'one of the strategic lines' required by the RAAF: 'and if they can be induced to give priority to the ground preparation of this line it would be of the greatest value, and go far towards ensuring success to the experimental service'.[99]

Brinsmead evidently hoped that the Air Force's need for aerodromes and hangars along strategically important routes which ensured mobility for defence purposes would provide the facilities also essential to any civilian firm which might be contracted to run the airmail service. Pearce referred the suggestion to a meeting of the Air Council on 10 May, and it was recommended that the Air Force should prepare such landing grounds along the Geraldton–Derby route 'as are considered necessary for Defence purposes', provided that the cost of these were not more than £5000; any others required by the airmail service should be the responsibility of the Civil Aviation Branch (CAB). Pearce approved this recommendation on 12 May.[100]

The decision thus taken ensured that the RAAF remained obliged to make Western Australia an early focus of attention. In June 1921 press reports stated that Flight Lieutenants Frank Lukis and Frank McNamara were about to embark on a survey of the route between Melbourne and Perth.[101] While in the west these officers also surveyed a stretch of the north-west air route between Geraldton and Hamelin Pool.[102] In September Williams himself visited Perth, primarily it seems to interview applicants from the State seeking appointment to the Air Force but also to conduct discussions with the local Lands Department about acquiring land suitable as a site for an air station, and for landing grounds along the Geraldton–Derby route.[103]

Lukis was still engaged on the survey of the route between Port Augusta (SA) and Perth in February 1922, and the next month a reconnaissance between Adelaide and Melbourne was reported on by Squadron Leader I.N.C. Clarke and Flight Lieutenant Ernest Mustard.[104] In June of that year Flight Lieutenant Allan Hepburn was sent to Perth in DH9 A6–14 to inspect landing grounds between Geraldton and Derby. He was carrying out this work on 15 August when, on a flight from Broome, he crashed while landing at a local station. The aircraft was swung around on the ground after striking a concealed fence post, breaking the undercarriage, a lower wing, longerons, engine bearers, radiator and propeller. Hepburn returned to Broome by car to make arrangements for the aircraft to be shipped south. He finally arrived back in Melbourne in October.[105] When in 1923, it was proposed extending the airmail service beyond Derby to Wyndham, it fell to Flight Lieutenant Harry Cobby to examine this route for suitable landing grounds. Leaving Melbourne by train at the end of August, he performed the survey on the ground, by car, and departed Perth to return to the east on 20 October.[106]

At the same time as these efforts were underway in the west, similar work was being carried out in other directions. In November 1923 Flying Officer George Kindred and two airmen, AC1s Kelly and Newson, traversed the route between Cloncurry (Qld) and Darwin preparing and marking out landing

Flying Officer G.H. Kindred led a RAAF ground party on a route survey between Cloncurry (Qld) and Darwin in 1923. Here the party prepares to leave Camooweal (Qld) in their Model T Ford lorries, which carry stores, 218 litres of petrol, and five bags of mail for outback stations passed along the way. Route surveys were a major part of the early activities of the new RAAF. *(RAAF Historical)*

grounds.[107] This northern survey, covering 2700 kilometres, was accomplished in two Model T Ford lorries (AF401 and AF402) specially purchased by the RAAF for the purpose.[108]

Air Force Fairey IIID seaplanes were also active on tasks of this nature. Wing Commander Goble was piloted by Flight Lieutenant Ivor McIntyre while carrying out a reconnaissance of landing places along the east coast between Melbourne and Sydney in November 1923.[109] In February 1924 these same officers, accompanied by LAC George Gottschalk (fitter aero), made a similar survey between Melbourne and Hobart in Fairey A10–2 (see chapter 7). Flight Lieutenant Allan Hepburn had been sent to Hobart in advance of this flight, to make arrangements for the reception of the aircraft and crew, and while in Tasmania he also made a survey of landing places in the State.[110] In October 1925 two seaplanes, piloted by McIntyre and Flight Lieutenant Arthur Hempel, carried out a survey of Bass Strait to examine suitable landing grounds in the event that a proposed commercial air service was established between the mainland and Tasmania.[111]

The country between Sydney and Brisbane was another area of special interest to the Air Force, thereby completing a network of strategic air routes connecting all the mainland capitals. Every opportunity was taken to identify suitable landing places along it. In his capacity as Director of Works and Buildings (DWB) Hepburn was particularly busy, and in March 1925, after he was sent to Sydney to look for a building suitable for use as a CAF drill hall, he followed up this visit by proceeding to Cootamundra (NSW) and then along the route prepared by Civil Aviation Branch to Brisbane, inspecting landing grounds as he went.[112] In April the next year Hepburn reported on the route from Narromine (NSW) to Brisbane which he had inspected.[113]

Shortly after this other efforts were mounted from Richmond, which was now the base of the RAAF's No.3 Squadron. On orders from Headquarters a DH9 crewed by Flying Officers Arthur Murphy (pilot) and Leslie Sutherland (observer) began an aerial reconnaissance at the end of June 1926 to help identify the most suitable and direct route between Melbourne and Toowoomba, outside Brisbane. Murphy's aircraft was damaged, however, while attempting to take off in muddy conditions at Blayney (NSW) on 30 June and another DH9 (A6–15), crewed by Sutherland (pilot) and Flying Officer Val Augenson, had to be sent to complete the reconnaissance in August.[114]

Although an important point had been established about what comprised the proper sphere of activity for the newly inaugurated Air Force, the first years of its existence equally established the clear fact that the RAAF could not expect to be entirely divorced from or indifferent to government objectives relating to civil aviation. As the involvement of defence personnel and aircraft on aerial surveys in Western Australia showed, for better or worse the linkage was undeniable. The problems arising from this situation and the nature of the relationship between the Air Force administration and the Civil Aviation Branch of the Defence Department are examined further in chapter 10.

3

The fight for separate identity

As described in previous chapters, the birth of the RAAF was accompanied by bitter disputation between Australia's two established defence services. Competition for funding was behind the initial government decision to create a single new air corps rather than separate naval and military air services, but subsequent moves by the Army high command to dominate this force were matched by a determination by Navy that sight should not be lost of its interests in the aviation field. Sadly the competition of 1919–20 for a controlling hand over the newcomer to the local version of Douhet's 'great military family' did not simply cease with the RAAF's inauguration, but became an ongoing family row (even after Navy was re-incorporated within the Defence Department at the end of 1921). As this chapter shows, during virtually the whole of the inter-war period there were recurring attempts to deny the infant Air Force the status of separate and equal membership of the defence family. Periodically, attempts were even made to dismember and swallow it up.

Legacy of 1914–18

The primary cause for rejection of the idea that the Air Force should be a separate and independent third service is to be found in the British experience of air power during the First World War. Although formation of the Royal Air Force (RAF) in 1918 ought to have served as a clear and incontrovertible argument in favour of the course adopted by Australia, this was simply not the case. In Britain itself there was bitter and acrimonious debate about the RAF's continued existence beyond the armistice.

The fact was that formation of the RAF owed everything to political imperatives and very little at all to concepts for the conduct of military operations. Perceptions that Britain's defences against German air attack were ineffective, and the pressures generated on the government, had by 1917 made amalgamation of the two air forces operated by the British Army and Royal Navy a necessary political step. The home defences had faced similar pressure before, after German dirigible-type airships began a strategic campaign against the

British homeland in January 1915. These attacks, more important for their effect on public morale than actual damage done to war capacity, declined throughout 1916 as the fighter defence steadily mastered the threat they posed. In mid-1917, however, the Germans resumed their offensive using Gotha bombers, each capable of delivering 1100 pounds of bombs.

Two brief air raids on London, carried out in broad daylight with maddening impunity, proved far more devastating than all the previous Zeppelin attacks, and a huge public outcry followed. Squadrons were recalled from France to bolster the London defences and reprisal raids against the German city of Mannheim were ordered. A review of the overall direction of the air war was also undertaken on behalf of the British government by the South African General Jan Smuts, a member of the War Cabinet. Smuts' two reports, most notable now for their prediction that 'the day may not be far off when aerial operations with their devastation of enemy lands and destruction of industrial and populous centres on a vast scale may become the principal operations of war, to which the older forms of military and naval operations may become secondary and subordinate', led to the creation of a separate Air Ministry as an equal of the Admiralty and the War Office, and the amalgamation of the RFC and the RNAS into a single air service. Smuts' recommendations were approved by Cabinet in August and legislation was passed in November, but for purely administrative reasons the RAF did not become an official entity until 1 April 1918.[1]

These were events quite unrelated to the air war being experienced by British forces outside England, principally on the Western front but also in the Middle East and on the high seas. Here the employment of aircraft was directly related to supporting the traditional tactical operations being carried on by naval and military forces. While the combat potential of aeroplanes had already been pioneered before August 1914,[2] the war had begun with the RFC and the RNAS which split from it in June 1914 (the Army simply dropped the appended 'Military Wing' from the former title) designed for little more than tactical reconnaissance. In the RFC's case this was hardly surprising, given the close involvement in its development of Brigadier-General David Henderson, the British Army's leading exponent on field intelligence.[3]

All this had changed with the actual experience of air war in France. The appearance of armed German aircraft quickly created the need for something more than unarmed aerial scouts. New roles—long-range surveillance, artillery observation, message dropping, aerial photography, tactical bombing and strafing among them—also emerged. Along with these arose the requirement to maintain control of enemy airspace while such activities were carried out, and to prevent the enemy performing the same over one's own territory, so that the scene was set for the free-wheeling clashes by masses of fighter aircraft popularly known as 'dogfights'. The purpose of all these operations remained, however, purely tactical in nature. Even the doctrine advanced in 1916 by Major-General Hugh Trenchard, commander of the RFC in France, for interdiction well behind enemy lines, was seen as the best means of advancing the ground war even though he considered such tactics as the best means of diverting the Germans from bombing London.[4]

While the possibility of operations of a strategic nature were certainly considered, technological limitations on the range and carrying capacity of aircraft restricted the development of this aspect of air power for much of the war. Though the Germans had plans at the outset of hostilities to attack British industrial targets, which failed to materialise,[5] and the RNAS actually carried out bombing attacks on airship bases at Dusseldorf as early as September 1914,[6] the offensive application of air power remained secondary. The RNAS assigned one of its three wings to making long-distance raids on German munitions centres in early 1916, but this was disbanded in May 1917 to support the hard-pressed RFC which began to suffer heavy tactical losses culminating in what was termed 'Bloody April'.

Even the long-range bombing campaign against German centres ordered in response to Gotha attacks on London provided a less-than-conclusive proof of the value of strategic operations. Carried out by the Independent Air Force (IAF) formed under Trenchard at Ochey, France, in June 1918, this campaign produced only meagre results and heavy losses of aircrew before the armistice brought it to a close—despite inflated claims to the contrary.[7] As Trenchard disparagingly observed: 'Thus the Independent Force comes to an end. A more gigantic waste of effort and personnel there has never been in any war.'[8] Ironically the machines which might have validated the concept of such operations were only just becoming available from British factories at the time of the German capitulation.

The evidence of the war was therefore actually against those arguing that strategic air power had come of age. The value of independent operations capable of significantly hastening the outcome of conflict on their own, while not actually discredited, had not been proven either. For those looking for 'lessons', it seemed true that there was no operational justification for separating air assets from Army or naval control. Although aircraft had established themselves as valuable instruments of war, their use was primarily as adjuncts to existing forms of sea and land operations—in much the same way as other major technical innovations, such as submarines and tanks.

It was not to be wondered at that the existing services disapproved of developments which removed from their control the air elements they had come to regard as an integral part of their own organisations. The amalgamation of the RFC and RNAS into the RAF set the scene for conflict in the early life of the new British air service in the immediate post-war period. Though legislative identity had been readily granted in 1918, largely to counter wartime difficulties of divided naval-military control, the fact that this was not a step which had grown naturally out of some compelling conceptual reason for a separate air entity ensured that there were throughout the 1920s continual challenges and questioning of the RAF's right to existence as an independent service.

An imported quarrel

It was entirely to be expected that this conflict in the British sphere would have a flow-on effect or at least be reflected in Australia. Indeed, such an occurrence was virtually guaranteed by the imperial orientation of the high command of Australia's Army and Navy. Given the operational experience of the Army's top generals, gained mostly with AIF formations serving as part of the vast British military organisation fielded during the war, it was quite inevitable that their thinking on this issue would be the same as their British colleagues. It thus comes as no surprise to note the public and private utterances of such figures opposing the RAAF's creation and continued existence. In November 1923, for example, Lieutenant-General Sir John Monash (a member of the Council of Defence despite his return to civilian life after the war) made a scathing denunciation of the government for starving the defence forces of funds, in the course of which he took the opportunity to decry the Air Force as a sham.[9]

More important was the attitude of Lieutenant-General Sir Harry Chauvel, who remained in uniform as the Army's Inspector-General. He gave his opinion in a private letter to Brigadier-General Thomas Blamey in mid-1924, stating that:

> I am not at all sure that in a Force of our size it is entirely suitable to have the Air Force a separate service entirely from the Army, and it is worth serious consideration whether economy and efficiency might not be served if it were brought under the control of the Army, for administration at all events.[10]

Such questioning of the independent status of the RAAF was equally evident in the leadership of the RAN. As the naval service had RN officers on loan in its top posts, this situation must be regarded as even less a surprise because inevitably these officers saw it as incumbent on themselves to continue within a Dominion sphere the battle of principle which they knew to be raging in London. The importance of the British connection in this regard is quite dramatically illustrated by events in 1926.

In February 1926 Ramsay McDonald, the Labour leader in the House of Commons, asked Prime Minister Stanley Baldwin whether the government had any intention of reopening the question of the independent status of the RAF and the Air Ministry. In his reply, Baldwin made clear his government's intention 'to pursue the organization of Imperial Defence on the existing basis of three co-equal Services', and declared it to be 'in the interests of the Fighting Services that controversy upon this subject should now cease':

> We are convinced that the way to secure the higher co-ordination in our Defence machinery indispensable to full efficiency and, indeed, to economy, lies not in the abolition of any one of the three established arms of His Majesty's Forces, but in combined action between all three through the machinery of the Committee of Imperial Defence and the agency of the recently instituted Committee of Chiefs of Staff. We are sure that we can rely upon all concerned to devote themselves loyally and wholeheartedly to this end.[11]

In late April of that year, Williams wrote to the secretary of the Defence Department drawing attention to the Baldwin statement. For some years, he wrote, 'certain agencies' had been at work in England trying to upset the decision to establish the Air Force as a separate service. Successive governments had enquired into it, and the British prime minister's statement was the result. He then went on:

> Since the formation of the Royal Australian Air Force similar questions have been raised from time to time regarding the status of the Air Force as a separate Service. This has resulted in much waste of valuable time and has not been to the benefit of any Service. It is recommended that a decision on lines similar to that of the British Government be given in regard to the Air Force in Australia.[12]

Williams had been prompted to raise the matter in this fashion at this particular time because the recent airing of the issue in the British parliament had caused 'certain agencies' in Australia to mount a fresh challenge. On 5 March the Australian CNS, Rear-Admiral Percival Hall-Thompson, RN, had written a memorandum questioning whether the RAAF was 'a separate service co-equal with the Navy and the Army', and also querying the composition of the Air Council and whether this body ought to be retained, along with the method of appointing members to both the Council and the Air Board. The reply prepared for the secretary, presumably by Williams, pointed out that 'any doubt of the intention in this matter' could be removed by referral to the existence of the Air Force Act 1923, and particular gazettal notices and departmental files. It went on:

> One of the main reasons for creating a separate Air Force in England (and this reason applies equally here) was to prevent the development of aeronautical science and skill being controlled by authorities principally concerned in the development of the older services, and who were therefore inclined to consider aeronautical development purely as an auxilliary to those older services. Another strong reason is that of economy. The Air Force is continually referred to by the other services as being costly. England has examined this question of a separate Air Force from every angle and in every case it has been decided to continue it but there have been two Committees on the matter whose object has been to cut down costs and both these committees (The Geddes Economy Committee and the Colwyn Committee) have recommended the continuance of the Air Force as a separate service.[13]

Despite Williams' tactic of invoking the British precedent in answer to the RAAF's opponents, he was not successful in eliciting a similar categorical statement from the government enjoining the 'Fighting Services' to cease fighting among themselves and put this issue behind them. Although seen by the Minister (Major-General Sir Neville Howse, VC), his submission was returned without action. More on this incident can be found in chapter 12, dealing with administrative aspects of the fight for separate identity.

Adding to Williams' eagerness to have this question finally settled in its Australian context in 1926 was a fierce battle being waged with the RAN at this time over the latter's attempts to establish a fleet air arm in the same way that the RN had recently done. This proposal had emerged in October 1924, while Williams was at staff college in England, and on his return to Australia in February 1925 he found that Goble, who was acting CAS, had broadly gone along with the proposal which had received ministerial approval a few days before he returned to duty. But Williams had seen the threat in such a development which Goble, with his RNAS background, had either missed, ignored, or welcomed. Allowing the Navy to develop its own air arm demolished a central plank in the rationale for having a separate air force. He therefore set about returning the naval air arm genie to its bottle.

By November Williams had succeeded in getting the agreement of Navy to put the scheme on hold pending further examination. This arrangement went by the board, however, when Hall-Thompson decided to press ahead and submitted a scheme to the Minister on 4 June 1926. The terms of his memorandum made all too clear what the Naval Board had in mind, declaring: 'In Australia . . . there is no justification for a separate Air Force, with its additional overhead expenses[,] and the correct objective should be the establishment, as soon as practicable, of two separate and distinct Naval and Military Air Arms'. Referring to recent developments in Britain, the submission claimed that the Admiralty had long realised the fallacy of the policy which had led to the establishment of the RAF and had been striving since to obtain a naval air service under its own control. Rather declaring his hand, Hall-Thompson declared that 'The present Fleet Air Arm organisation is but a stage in the renaissance of a separate Naval Air Service'. Clearly, the RAN's scheme was intended to achieve the same result locally.[14]

Williams wasted no time in attacking CNS's proposals. In a submission to the Minister dated 25 June he responded that, given the present state of development of Australia's aerial forces, the suggested establishment of air services by the Navy and the Army 'cannot be taken seriously'. He pointed out that what Hall-Thompson proposed could not be justified on the basis of adopting applicable British precedents. The RAN did not have an aircraft carrier and was not likely to get one while the limitations of the Washington Agreement of 1922 remained in force. It was proposed to add a seaplane tender to the Australian Navy in the near future, but this would not provide a counterpart to the type of operations practised by the RN unless the new cruisers under construction for Australia were to be equipped to carry aircraft. 'This is still doubtful and in any case there would not be more than four carried'. The RAN's air requirement for as far as could be foreseen was not expected to exceed twelve aircraft, an equal number of pilots, and probably 40 or so ratings. 'To set up a separate organisation to deal with the highly technical business of aircraft supply and flying training for a force of this size could not be seriously contemplated'.[15]

These counter-arguments were sufficient to cause the Minister (Howse) to step cautiously. On 2 July he referred the question of establishing a fleet air arm to the Defence Standing Committee for consideration and report, warning that no additional expenditure would be considered, nor duplication of existing Air Force machinery or establishments. Howse also explicitly spelt out that government policy in the matter of defence organisation was 'on the existing basis of three co-equal Services'.[16]

Notwithstanding these ministerial guidelines, Williams could foresee the outcome of this particular exercise. On 11 August he wrote to Sir Hugh Trenchard, the British CAS, remarking that 'the Army appears to think the occasion one suitable for trying to cut the Air Force up', and in this prediction he was correct.[17] When the committee met to debate the issue, the other two service chiefs would not be dissuaded from deciding that they were 'firmly of the opinion and recommended that the principle of the Fleet Air Arm already approved by the Minister in January 1925 should be confirmed'.[18]

Although Williams told Trenchard that he had few fears of the moves against the Air Force succeeding, he nonetheless ensured that the British CAS remained in a position to serve as a possibly decisive ally should advice be sought by the Australian government from that quarter. In August he accordingly sent Trenchard background papers, along with a request that 'If you get an opportunity of stressing the importance of a separate air service from the point of view of economy as well as from the point of view of air defence as such I think it would be an advantage to us'.[19] He had cause to appreciate his own foresight when the Minister decided, just before leaving to attend the Imperial Conference in London, that he would make no decision until he returned and proposed to make inquiries into the fleet air arm issue while in England. On 2 September Williams sent off further briefing material to Trenchard, adding: 'I should wish this communication to you to remain confidential, it being sent purely for your own personal information in case Sir Neville Howse should approach you on the subjects mentioned'.[20]

Howse subsequently did ask the Secretary of State for Air for information on the state of RN–RAF relations and the operation of the fleet air arm. A general note on the subject was duly sent to him and it was suggested that Howse might care to call on Trenchard for verbal amplification. As Trenchard explained to Williams in December:

> Unfortunately your Minister became ill, and I never had an opportunity of seeing him.
>
> Since then, we have had a request from Howse through Goble [then in England as Air Liaison Officer at the Australian High Commission] for the views of the Air Staff on this question. It has not however, been decided whether we can supply these at the present moment, owing to our having made peace with the Admiralty, and we do not want to rake up old bygones.

In the event, therefore, opinion in London was unable to exercise a decisive influence in settling the issue and it was up to Williams to chip away at the fleet air arm issue on his own. Not until 1928 was he eventually able to tell Trenchard that the proposal had been finally buried when the government decided on 18 January not to approve a fleet air arm for the RAN.[21] By this stage he was also able to report the overcoming of some administrative drawbacks which had long affected the RAAF's status vis-a-vis the other two services (see chapter 12).

Episodes from a running battle

An example of the sorts of pressures which would be exerted on the infant service by its older 'brothers' arose even as Australia's Air Force had been preparing to enter the world, separate and whole. This concerned the circumstances surrounding the despatch of the new force's initial trainees on a course at the Army's Central Training Depot (CTD) at Holsworthy. The Air Board had recognised early that a requirement existed to train and induct the large number of personnel approved under the RAAF's establishment, and for this reason planned to establish its own Recruit and Non-technical Training Depot. The proposed location was Portsea, where barracks furniture and instructional equipment had been retained at Fort Franklin since the short-lived School of Military Aeronautics closed in 1919, in the expectation that they would be required precisely for such a purpose.[22]

In line with its intention, and in accordance with the departmental policy that existing defence resources should be used wherever practicable, in February 1921 the Air Board asked—through the Air Council—for Army personnel to staff the proposed depot. The initial requirement was for an officer to take charge of the depot, and here the Board specifically mentioned the name of Garrison Sergeant-Major W.H. Brittain as a man it believed suitable. It was proposed to second Brittain until the Air Force was officially gazetted, and then commission him as a flying officer. In March the Board increased its request to include five drill instructors as well.[23]

Considering these requests, the Military Board responded that it believed the required training could, in the spirit of making do with existing facilities, be carried out at the Army depot at Holsworthy. But, if the Air Board wanted to persist with establishing its own depot, the Adjutant-General (Major-General Victor Sellheim) told his colleagues that he could not support Brittain's nomination, although this had nothing to do with the individual concerned. There were, he said, a number of men in the permanent military forces who, having distinguished themselves during the war, had been relegated to lesser peacetime jobs as quartermasters and who deserved to enjoy the benefits of the higher-paid Air Force post offering. He therefore proposed that the position should be rotated among these—starting with Honorary Major Robert Christie,[24] who had attained the most senior position in the AIF.

Perhaps predictably, the Air Board was unimpressed with this counter-proposal. While it appreciated the offer of using the CTD, the Board said that it felt the best results for the Air Force would be obtained from having its own depot; it held that the ideal for any service was to train its own personnel, and while this might then be impossible in regard to officers 'it is thought that it might be done for Other Ranks'. It was, said the Air Board, important to instil in recruits the spirit of the Air Force in which they were part, to provide instruction in customs peculiar to the force, and to bring them under the influence of their own service as soon as possible. Nor was the Board enthusiastic about Sellheim's scheme; it was clearly irritated by the Military Board's attempt to treat the request for an officer to command the depot as an opening for officers to whom the Army was unable itself to offer more rewarding employment, and responded that it could not afford the time involved in offering the post to a list of persons one by one for refusal or acceptance.

At its meeting on 15 April the Air Council supported the Air Board on this matter, agreeing that it was advisable for the Air Force to have a separate depot. The Military Board pressed its point, however, querying whether 'at a time like the present some sacrifice should not be made in the interest of economy'. It suggested that the question of proceeding with the depot should be referred to a special meeting of both boards with the Minister. Matters did not reach the stage of a joint conference, because at this point Defence Minister Pearce and his assistant Minister, Major-General Sir Granville Ryrie, between them decided that 'this training shall be done at Holdsworthy [sic] for the coming year owing to the necessity for economy'.

In accordance with this decision, establishment of a 'Recruit Depot at Portsea' was deleted from the 1921–22 program and Air Force Estimates and, as already narrated, RAAF personnel underwent training in two courses conducted at the CTD. Despite the unpleasantness unleashed by this incident, some good did come of it. Although Brittain did not come to wear Air Force uniform, other Army officers, including Christie and Quartermaster and Honorary Major Roy Marsden, did make the transfer.[25] Initially seconded to the RAAF, they were subsequently appointed to the Stores and Accounting Branch (later renamed the Equipment Branch).

Further illustrations of the senior services' attitude to the Air Force followed. In conjunction with the 1923 Imperial Conference in London, the British CAS proposed to hold talks with an Australian service delegation about the development of 'your Air Force'. These meetings took place in Trenchard's rooms at the Air Ministry early in November and involved detailed discussions regarding a range of crucial matters (see chapter 4). Williams was actually in England at this time—he was then attending the Army staff college—and duly attended these discussions, but not as the senior Australian representative despite the fact that he was still formally CAS back in Melbourne (Goble held this post purely in an acting capacity). The delegation was headed by Brigadier-General Blamey, then filling the London-based appointment as Australia's representative on the Imperial General Staff, and the other member was the CNS-designate, Rear-Admiral Hall-Thompson.[26]

Such an arrangement was presumably justified on the grounds that Williams, although CAS in Australia, was only a wing commander and it was not appropriate to have an officer of his relatively junior rank speaking as an equal with an officer of Trenchard's seniority. The fact was, however, that not only was Williams denied the status that was reasonably his as head of his service in matters concerning directly and solely the Air Force, but he evidently attended the meetings in no other capacity than observer. Draft minutes sent to Blamey by the Air Ministry recorded Williams' presence but described him merely as 'Wing Commander, Royal Australian Air Force'. In response to Trenchard's request for comment on the draft record, Blamey replied with several suggested changes, among them that Williams' name should be added to his own and Admiral Hall-Thompson's as the Australian representatives. Explaining this change, Blamey said: 'As he was a member of the Air Board in Melbourne it would make clear that [Royal] Australian Air Force interests were represented'. Blamey was, in effect, locking the RAAF into the decisions of the negotiations even while denying Williams the status which was certainly his due.

The question of the status of the Air Force, and of its individual members, proved a source of endless friction. In October 1922 Major Coleman, the secretary of the Air Board, had written to the secretary of the Defence Department to suggest that, following the constitution of the Air Force as a separate service 'distinct from, and on a similar footing to, the Navy and Army', the necessary steps should be taken to ensure that the Air Board and the RAAF were represented at official functions 'in the same way as the Naval and Military Boards and their respective Services'. In due course a reply was received from the Prime Minister's Department stating that arrangements would be made for the inclusion of Air Force representatives in any general functions held by the Commonwealth.[27]

Whether or not there was some delay in making the foreshadowed arrangements to address this matter, the RAAF continued to be treated as the poor relation to the point where it became embarrassingly clear to none other than the Governor-General. On 11 November 1924 Lord Forster wrote a chiding letter to Prime Minister Stanley Bruce, remarking that at a dinner given by the Lord Mayor of Melbourne in the Town Hall the previous evening he had noticed CAS seated in the body of the hall without the precedence accorded to senior naval and military officers present. 'In view of the great and growing importance of the Air Service', it was desirable, His Excellency suggested, for the head of the RAAF to receive definite recognition on such occasions in future. He therefore proposed that the Commonwealth Table of Precedence be amended, and also that the Air Force be added to the traditional toast to the Army and Navy at public gatherings.

It was remarkable that these quite obvious changes had not been made earlier. Nonetheless, the Governor-General's personal interest ensured that the matter was now pursued, and in July 1925 advice was received that the King

The Governor-General, Lord Stonehaven, inspects a guard at No.1 FTS in the late 1920s. Accompanying him on the inspection are Wing Commander W.H. Anderson, Group Captain Williams (CAS), and Flight Lieutenant A.H. Cobby (dodging officer's cane). Stonehaven was himself a former Under Secretary of State for the British Air Ministry and a keen advocate of aviation during his vice-regal term. He escaped unhurt from a crash in a RAAF aircraft in 1926. However, he was not the first governor-general to take a personal interest in the RAAF, as his predecessor (Lord Forster) played a crucial role in having the service's status properly recognised. *(RAAF Museum)*

had approved the amendments which his representative in Australia had initiated.[28] The new arrangements were gazetted in October, at which time the prime minister was reported to have personally expressed his hope that in future the Air Force would be included in any toast to the armed services at public gatherings: 'In this matter the Commonwealth practice would be in accordance with that observed in Great Britain'.[29]

The expectation that these measures might settle such problems was unfortunately not fulfilled. In May 1927, the officer acting as Second Air Member of the Air Board (Wing Commander Rolf Brown) still had cause to complain about his treatment at a veterans' luncheon in Melbourne a few days earlier which had been organised by the commandant of the Army's district base in Victoria. Writing to Major Coleman as secretary of the Air Board, he reported that, although seated at the main table, he had been denied the precedence due to him as CAS's representative at this function.

> As a Member of the Board I take exception to the position in which I was placed ... It was, I consider, a reflection on the status of the Air Board, and I request that you bring this matter to the notice of the Secretary, Defence, in order to obviate a repetition of a similar occurrence in the future.[30]

This incident might be said to show that no opportunity continued to be lost by the two bigger services to 'rub in' the RAAF's lesser status whenever possible. They were certainly less attentive to this matter than could have been expected, given traditional service concern with protocol and punctilio, and in the circumstances members of the Air Force were probably right to be sensitive to and resentful of the treatment they received.

More important than incidents involving matters purely of status were those touching on the issue of operational control of air assets. Williams was very alarmed in 1926 to discover some of what the Army envisaged would happen in the event of an outbreak of hostilities in Australia's area. On 29 April he wrote to the secretary of the Defence Department citing a section of the Army War Book, the document setting out in detail the measures to be taken by the military forces in the event of war, which read:

> Under the arrangements in force the only portion of the Air Force which it is intended to be placed under the Army, consists of certain Army co-operation units ... When mobilization of the air units is complete arrangements will be made for the co-operation of the units concerned during training and at the appropriate time for their transfer to the control of the army.[31]

The basis of Williams' concern was the mistaken notion implicit in the War Book that the RAAF existed merely to mind 'Army' aircraft during peacetime, and that on the outbreak of war these assets would revert to the Army commander-in-chief. As he now pointed out, this approach ignored several fundamental realities. For instance, in the defence of Australia it was entirely conceivable that for a period of operations the whole of the Air Force might be required to operate in support of the Navy—'it is certain that more than merely Naval Co-operation units will be required to assist in what might be termed Naval duties'—and that equally there would be a period when the Air Force might be employed to strike the enemy at a time when the Army was totally uninvolved, as in the period before an enemy expeditionary force actually landed. In such circumstances, he said, it would be essential that air assets not be 'subordinated to less mobile services prematurely'.

Williams went on to reject any suggestion that the RAAF saw itself functioning in war independently of the other services at a time when those services were themselves engaged in operations. 'This is quite a wrong impression—there is no such idea'. But, he said, just as one service alone was insufficient for national defence, 'nor is any one purely auxiliary to another'. He pointed out that:

> One of the main reasons for separating the Air Forces from the Navy and the Army was to enable the powers of aircraft to be fully developed by officers whose principal interest is such development and to prevent it being treated merely as an auxiliary to the older services. For this reason alone it would be unwise to accept a principle which placed the Air Force completely under the Military Commander-in-Chief in such broad terms as it recommended.

It can quite be imagined that the Army leadership resented being lectured on their ignorance of the proper employment of air power for national defence purposes in this way, but whether there was any inclination to change their thinking seems doubtful.

Opportunities to ensure that the RAAF was publicly perceived as an equal among the defence services were rare. One such occasion arose with the impending visit to Australia of the Duke of York for the opening of the Federal Parliament at Canberra in May 1927. As it happened, the duke was the first member of the royal family to have a formal air force association, having transferred from the RN to the RNAS during the war at his own request and then serving on the staff of the RAF college at Cranwell in 1918.[32] Williams was keen to make the most of it.

During the planning stage for the visit, Williams wrote privately to Trenchard in August 1926 candidly declaring his hopes of taking full advantage of the duke's commission as a group captain in the RAF:

> I want to do everything possible to make his visit here an 'Air Force occasion' . . .
>
> On the grounds of the Duke of York being an Air Force officer I have obtained the concurrence of the Navy and Army and Ministerial approval that all personal guards on the Duke's residence, etc., will be supplied by the RAAF and that Bands and Guards of Honour at the opening of Parliament will be under the command of a RAAF officer.
>
> We would be glad if you could do anything which might have an influence in ensuring that HRH wears Air Force uniform on the occasion of the opening of Parliament here and that this uniform be his 'first string' generally.

He also told the British CAS of his desire to have a senior RAF officer accompany the duke as equerry and another more junior one as aide-de-camp, so as to 'provide a little more Air Force uniform', although he admitted that 'We may be a little greedy in this matter'.[33]

Williams was to discover that the other services had no intention of allowing the RAAF to hog the limelight during such an important national occasion. Despite the agreement previously won on the arrangements he had mentioned to Trenchard, he was soon forced to chase his letter with a telegram advising

The Duke of York (*centre*) during the informal visit he made to Point Cook in April 1927, pictured in the uniform of a group captain in the RAF. He was the first member of the British royal family to have a formal Air Force connection, and the Australian CAS (Williams) was keen to make the most of it. In fact, the Duke wore naval uniform for his official engagements, but at least accepted an honorary commission as group captain in the RAAF before departing Australia. The RAAF officers at the Duke's left are (*from right*): Wing Commander A.T. Cole, Squadron Leader R. Christie, Wing Commander R.S. Brown. (*Mrs S. Clerehan*)

that Navy was now disputing that an Air Force officer should command the combined guard of honour for the opening of Parliament House at Canberra. Could the Air Ministry advise if a decision had yet been made regarding the uniform the duke would wear while in Australia, he asked? Williams was tempted to suggest to his British counterpart that the RAF promote the duke to air commodore rank, as a means of ensuring that his air force rank was senior to the rank he would wear in naval uniform, but evidently he thought better of this.[34] One can imagine Williams' disappointment when Trenchard wrote in December 1926 telling him that the duke had decided to wear naval uniform after all: 'He feels that as the Navy are taking him out, it would be a little invidious if he wore Air Force uniform at public functions; also I understand that he finds the Naval full-dress uniform more comfortable than ours for speechifying, as the headdress has not got a chinstrap.'[35]

Williams resigned himself to having lost the round over the duke's uniform, considering the outcome to be 'perhaps inevitable in the circumstances',[36] but he was presumably less philosophical as further developments whittled away his hopes of seeing the RAAF achieve prominence during the visit. In view of the dispute that had erupted between the RAN and the RAAF over command of the guard of honour, the defence organiser for the royal tour, Brigadier-General Charles Brand, took the view that it was not necessary to have an officer in charge of the whole guard and that the contingents from each service would be under separate commanders.[37]

Attempts were now also made to limit the size of the RAAF presence at Canberra in relation to the contingents of the other services. Apart from the opening ceremony, a military review was to be held which would require the assembly of considerable numbers of personnel. During discussions on the forthcoming tour by representatives of the three services on 13 September 1926, it was decided that quotas for the forces to be sent to the national capital would be: Navy 480, Army 1410, Air Force 540. Making up the RAAF numbers were the required guard, a band, and No.1 and No.3 Squadrons (including their CAF component) who were to mount a massed aerial flypast and display as well as take part in the review marchpast. The very next day, however, the Cabinet committee overseeing the royal visit arrangements reduced these quotas. Navy was cut by 17 per cent to 400, but was allowed a detachment of cadets from the naval college which was not included in this figure; Army was cut by 15 per cent to 1200; and Air Force was cut by 63 per cent to 200.

The RAAF attempted to have this limit varied so that members of its band were additional to the 200 personnel allowed, arguing that because of the numbers required to man the aerodrome, mount guard on the vice-regal residence at Yarralumla where the duke and duchess were staying, and meet other miscellaneous requirements, the Air Force would have only about 80 officers and men available to march. Initially these representations were without success, causing the Air Board to complain in October that the RAAF was 'not being adequately represented . . . especially in view of the fact that HRH . . . is an officer of the RAF'.[38] As is explained later (see chapter 13), the threatened imbalance was eventually rectified, but the fact that a problem of this nature should have arisen at all was revealing.

No less indicative of prevailing attitudes was an incident which threatened after the various contingents reached Canberra in preparation for the official ceremonies. For the period of more than two weeks that these contingents were encamped at the national capital, the commander of the composite Army brigade was placed in charge of the entire service presence. On 2 May 1927, however, Squadron Leader Allan Hepburn, commandant of the RAAF camp, advised the Air Board from Canberra that orders had been issued by the local commander that the RAAF quarter guard on Yarralumla was to be inspected by a field officer of each service on successive days. Was the Air Board in agreement with this arrangement, Hepburn asked—presumably in the expectation that it certainly was not. The reply went back that the Board considered that Hepburn should be responsible to the brigade commander for the quarter guard, and asked him to get in touch with Wing Commander Rolf Brown, the acting Second Member of the Air Board, who was then already at Canberra.[39]

Though many of the instances cited here as examples of the rivalry which regularly operated to the RAAF's disadvantage may appear petty in nature and of no great consequence, being possibly the result of oversight as much as bloody-mindedness, the underlying motive in practically all cases seems sadly clear. No matter how petty or niggling the incidents themselves may have been,

Experts point out that Australia has only a few defence aeroplanes ready for use. After all, our national bird, the emu, cannot fly at all.

"NOT A FEATHER TO FLY WITH"

Air-Marshal Sir J. Salmond's report on our air force pays a tribute to our airmen who are gallantly carrying on despite a low standard of training due to the obsolete type of machines, entire absence of reserve equipment, all totally unfitted to undertake war operations.
Shade of Cocky Bennett: "I'll fly, my God, I'll fly."

These two newspaper cartoons from the late 1920s/early 1930s provide an indication of the poor public image with which the RAAF had to contend. Such ridicule inevitably had an impact on morale within the service. (*W.D. Bostock papers*) they were an accurate reflection of attitudes held in the other two services, particularly at senior levels. The effect of this situation on the morale of members of the RAAF should not be underrated. The undercurrent of hostility continually evinced towards the Air Force created a collective state of mind in the service which was as unhealthy as it was unnecessary. And if Williams was pushy in his attempts to improve the RAAF profile, who could blame him?

Proposed amalgamation

The belief held by members of the Air Force that they functioned under a constant state of siege imposed by their naval and military colleagues seemed confirmed by events which began to unfold in 1929. Federal elections in October produced a Labor government headed by James Scullin as Prime Minister, and no sooner had the new ministry taken office than it began making significant changes in the area of defence policy. Such changes were initially the product of ideological predilection, but, with the stock market crash in New York at the same time as the new government was sworn in heralding the onset of a worldwide depression, these were increasingly dictated by economic necessity. The previous Bruce government had already indicated that the Air Force expansion recommended in the Salmond Report (see next chapter) could not proceed within the country's existing finances, and that the Navy continued to be Australia's first line of defence. In the new circumstances, however, the RAAF faced an even more uncertain future.

At first it was Army which found itself under attack, with the government's announcement on 31 October that compulsory military training was suspended. Only a week later did the prime minister and his Defence Minister, Albert Green, get around to arranging with the CGS, Lieutenant-General Chauvel, for a new scheme to replace the previous system so lightly discarded.[40] These plans were submitted to the Council of Defence on 12 November, and it was here that the Army's representatives—Chauvel (promoted to the rank of general the previous day) and the Adjutant-General, Brigadier-General Thomas Dodds, along with Monash (now a general too) and Lieutenant-General Sir Brudenell White—revealed the extent of their resentment at the way the military forces were being cut back.

The Defence Committee (comprising the three service chiefs and a finance member) had been previously charged with examining a proposal for the amalgamation of the RAAF with the other two services, and the report of its findings was also made at this meeting of the Council of Defence. Not surprisingly, the Defence Committee had concluded, with CAS dissenting, that it could 'see no reason for maintaining a separate organisation for the Air Force in Australia'.[41]

When the RAAF's position was considered in the Council of Defence, Chauvel, Monash and White all voiced their previous objections to the RAAF's status as a separate service, repeating arguments regarding the absence of a role for an independent air force and the limited resources available for defence. White urged that the Air Force should be no more than a corps administered by the other two services.[42] Such views had the support of the three naval representatives present. Although clearly outgunned, Air Commodore Williams—the sole RAAF voice in this select gathering—put up a vigorous and skilful defence which succeeded in stalling the push which was developing against his service.[43]

In his autobiography Williams claims that he heard no more of such proposals,[44] but in fact he did. He was simply confused in his chronology when he stated that he had first heard the startling news of the renewed threat to the RAAF's future while listening to the Governor-General's speech at the opening of the new Parliament on 20 November. The latter event plainly followed the Council of Defence meeting which he describes in detail, and the Governor-General's remarks actually reflected his success at that meeting, reporting that:

> The question of maintaining a separate organisation for the Air Force, having been brought under the notice of my Ministers by officers of the Defence Department, the Defence Committee was instructed to investigate and report to the Council of Defence. *No decision was arrived at by the council* [emphasis added], the matter being postponed pending a comprehensive review of the position of the Air Force. My Ministers have, however, asked for a report on the possibilities of co-ordinating the work of the Air Force with that of civil aviation.[45]

There were some listening to the Governor-General's speech who, no doubt like Williams, wondered where the suggestion emanated which had actually brought the matter to the notice of ministers. Questioned on this point in the House on 21 November, the prime minister stated that the suggestion had been made by officers in the Army and Navy with the officers of the Air Force dissenting.[46] This accurately reflected events in the Defence Committee and the Council of Defence, but avoided the question of why the matter was referred to the Defence Committee in the first place. Williams clearly suspected the secretary of the Department, recounting a subsequent incident in which Shepherd expressed intense dislike of him to a mutual acquaintance, in a manner which could be interpreted as irritation at the fact that CAS had frustrated some plan which he had afoot in conjunction with the admirals and generals.[47]

As a result of this public discussion of the idea of dismembering the RAAF and amalgamating parts with the Navy and the Army, there was—predictably—a new slump in Air Force morale. Williams records that: 'Most of our officers were very depressed at the thought of a possible splitting of the Service but I must say that I did not find it worrying me and my advice to those round me was "Don't worry, it hasn't happened yet".'[48] The CAS presumably staked his faith in the fact that for years there had been Labor Party support for the idea that an air force should be the main element of Australia's defence, rather than an expensive Navy or large military forces. Not only was Defence Minister Green an early campaigner for a strong Australian air force,[49] but other leading Labor figures—Matthew Charlton, Frank Forde, Frank Anstey—were of similar mind.

As early as 1914 John Curtin, who became Labor leader on Scullin's resignation in 1935, had campaigned (unsuccessfully) for the reduction of expenditure on military organisation and Australia's Navy by the creation of a

Commonwealth fleet of airships.[50] At the 7th ALP Conference at Perth in 1918, Curtin had again advocated a defence policy which resisted placing too much reliance on Army or Navy for defence of the nation's coastline from invasion, contending that in time of war or international crisis 'aircraft stationed at strategic coastal points could keep a watchful eye on the surrounding seas and report the sighting of any convoy movements to a bomber defence base'.[51] He, too, remained a supporter of the RAAF's continued existence, both now and later.

Williams' confidence that the future of the Air Force was actually quite secure proved well placed. Although Green told Parliament on 13 December 1929 that amalgamation proposals would continue to be considered if these promised to save money while preserving efficiency, he indicated that the government's basic inclination was to retain the Air Force in its existing form.[52] If nothing else these remarks showed that the Army–Navy move against the RAAF on this occasion was likely to fail because it could not show any significant economy resulting: in making its recommendations to the Council of Defence, the Defence Committee had not talked of the RAAF's budget being saved. In fact, Army and Navy were simply hoping to divide the spoils between them.

Quite apart from the somewhat ambivalent remarks by Green, the prime minister had gone on record nine days earlier as saying that government ministers present at the Council of Defence discussions had 'urged delay and further investigation, because we were not satisfied that the change was desirable'.[53] Adding to the doubts which Scullin evidently held was a formal submission against the proposal presented by a deputation from the Australian Flying Corps Association on 2 December 1929, which prompted him to state: 'The present opinion of the government is that separate control should be retained for the Air Force unless a strong case can be made out for the new proposals'.[54]

Finally, in June 1930, it was clear that the threat had passed. Delivering a statement on the retrenchments being implemented in his department, Green said on 12 June that RAAF units and establishments would be maintained. 'Since no difficulty has been experienced in keeping the two Citizen Force units at full strength under the voluntary system, no reduction in air force activities has taken place.'[55] By this time the Army and Navy had begun tearing at each other in their bids to stave off further reductions. At a meeting of the Defence Committee on 18 March, Chauvel now turned to asking whether Australia could afford her own Navy, a question which he repeated in his annual report as Army Inspector-General. The fact that the existing naval forces were swallowing up half the annual defence budget had, he said, compromised the efficiency of local defence by reducing resources available to Army *and Air Force.* Chauvel's biographer queries whether he was serious in propounding this view, or whether he was 'only using an extreme form of argument to impress the Government with the plight of the Army and the RAAF'.[56] The present-day reader, however, might be excused a twinge of cynicism about the shifting nature of allegiances between the three services,

One cartoonist's view of the Scullin government's intention to make savings in expenditure on the Air Force following the Labor Party's election to office in October 1929. Air Force morale slumped as a result of threats to its existence, but by June 1930 it was clear that the RAAF would survive as a separate service. (*G. Packer papers, NLA, MS 780*)

and the willingness of the Army leadership to invoke the RAAF (with or without consultation) as a junior partner in attempts to improve its fortunes at the expense of the RAN.

Williams was certainly not deluded into imagining that the Air Force might now actually have an ally. For his part, he still took the precaution of writing to Trenchard once more, in the expectation that he would probably meet with Prime Minister Scullin when the latter was in London for the 1930 Imperial Conference. Although Trenchard had now retired from the RAF, he was still a formidable authority on air defence issues and Williams sent him briefing notes, 'believing as I do from experience that you are interested in making the best use of air defence where a Force is provided for that purpose and whether it be the responsibility of the Imperial Government or of any other part of the British Empire'.[57]

Although dismemberment and amalgamation had now been rejected, the RAAF still faced the other threat to its identity which had been foreshadowed in references to 'co-ordinating the work of the air force with that of civil aviation'. What this really meant was that the government was attracted to the idea of making a saving on the subsidies paid to airlines by having the RAAF carry the air mails. Like its predecessors in 1924 (see chapter 12), the Scullin ministry looked seriously at such a proposal, but after intensive lobbying by the aviation industry such thoughts were again abandoned in 1930.[58]

Even this was not the end of talk of changing the RAAF's organisation. When the Scullin government lost office in 1932 and was replaced by the United Australia Party ministry with Joseph Lyons as Prime Minister, the subject almost immediately arose once more. On 21 March the new Defence Minister (Sir George Pearce) appointed a committee comprising the secretary of the department and the three service chiefs to investigate the question of amalgamating the military and air services and boards, and provide a comprehensive scheme on how this could be effected. Under arrangements approved by Pearce, sub-committees were formed to examine the specific aspects of personnel, training, stores, finance and equipment.[59]

When the reports of all these various bodies were received, the picture which emerged was discouraging. The problems were well illustrated by the findings of the Finance secretary, who canvassed expected economies in a report dated 13 May. Assuming that the number of RAAF units would remain the same, then the only significant saving which could be achieved was through reduction of permanent establishments and a greater reliance on citizen personnel. If this proved impracticable, the only scope for cutting expenditure remained a proportion of the cost of maintaining a separate RAAF headquarters when this was abolished—about £34 000 in all. On the basis of such information, Pearce decided on 25 October to take 'no action at present, except that the present arrangement as regards Finance Branch of the Air Force [whereby it also provided finance and accounting services for Civil Aviation Branch] is to continue'.

Other support for such an outcome had meanwhile come from London. During a visit to England in April 1932, John Latham (whose portfolios included External Affairs), had told the Secretary of State for Air, Lord Londonderry, about the government's revival of the question of abolishing the independent air organisation. He had explained that, in view of the urgent need to further restrict expenditure, the proposal being examined was not to have two air services, one for Army and one for Navy, but to 'put the whole thing under the Army'.[60] Londonderry counselled in a letter on 3 May that such a move was seen by the Air Ministry as a retrograde step. Recounting the exhaustive consideration given to the same idea in Britain on no fewer than eleven occasions, all of which had resulted in rejection of proposals to dismember the RAF, he urged that depriving the RAAF of independent status could only prove 'a false economy' which could well hurt imperial defence cooperation.[61]

Williams had also been lobbying hard to ensure that Latham received precisely the advice that he did by writing to his RAF counterpart, Air Chief Marshal Sir John Salmond, on 1 April. Salmond responded on 28 April by sending Williams a copy of the briefing note being provided to Londonderry on the subject 'for your own personal use', and remarking: 'I have not yet heard the result on Mr. Latham's mind. There is not the slightest doubt it would be uneconomical and bad strategy to divide up the RAAF, and I think the note should convince any unbiased Minister.'[62]

Anything but reassured by reliance on a lack of bias on Latham's part, Williams cabled the British CAS to ask him to ensure that the Minister received the benefit of 'your interest and advice'. Salmond signalled back on 25 May that he had noted Williams' comments and would 'take steps to deal with them as opportunity offers, which I think will occur shortly'.[63] Although this additional pressure was probably unnecessary in view of the remarks Londonderry had already made to Latham, it is unmistakable that Williams was growing increasingly panicky in this period. On 10 March he had pressed Pearce to make a decision once and for all regarding the fate of his service, to put an end to the years of uncertainty which had dogged the RAAF.[64]

Adding to the threats piling up against the Air Force at this time was a more remote but no less dangerous peril arising from the disarmament conference at Geneva. The question which arose concerned the limits likely to be imposed on the size of national air forces. Williams was evidently fearful that if agreement was reached on the basis of restricting aircraft numbers to those already in being, the RAAF's existing strength of 55 machines would condemn his service to permanent impotence. He felt a level of 149 aircraft would provide sufficient scope to develop the Air Force to a credible level, and in April 1932 he sought Air Ministry assistance in getting Latham to accept this approach. His case was readily taken up by the Air Ministry, which actually suggested raising the figure further to 234, but before this question got any further the British government decided in May—without consulting the Dominions—to present proposals at Geneva calling for complete abolition of all naval and military aviation and the imposition of strict controls over civil flying.[65]

This planned initiative was purposely kept in strict secrecy by the British authorities, the Australian government being warned of the possible impact on 'the morale of the air forces concerned' if the move became widely known. Fortunately, the proposal was recognised as impractical before it was tabled, and in October 1933 the Geneva negotiations were brought to a halt anyhow by Germany's withdrawal from them. Had the proposal gone ahead and been implemented, however, its impact on Australian defence arrangements must have been crucial. The Lyons government had quite readily assented to supporting the British proposal, accepting London's injunction to preserve utmost secrecy by failing to consult even its most senior defence advisers about the possible implications for Australia's security of complete renunciation of air power.

Had it taken its defence experts into its confidence, the government might just possibly have received some non-partisan advice on such a crucial issue. In mid-1926 Chauvel (who had retired as CGS in April 1930) had actually opposed the merging of Australian Army *and air* assets in the British totals when national limits were being debated in disarmament negotiations then underway, in the way that RAN ships had previously been counted at conferences in Washington and London. On that occasion Chauvel had taken the view that:

> It must not be overlooked that the strength, organization and control of the Australian military and air forces are subject to the will of the Australian Government and Parliament, and they are not available automatically for the attainment of any general Imperial object . . . The composition and strength of the Australian forces must be governed by the necessities for [sic] local defence . . . They must therefore be considered entirely separately from the other forces of the Empire.[66]

Six years later Australian politicians were apparently quite willing to trade forces away in this fashion, unlike Canada and South Africa which did not agree to the British proposal. One suspects that the leadership of the RAN and Australian Army, however, though they fought to obtain control of Air Force assets, would have still resisted these assets disappearing entirely. All in all, the whole 1932 episode provided—as an Australian air historian has remarked—an illuminating comment on the government's attitude to the RAAF.[67]

Bitter taste of the Ellington Report

Although the events of 1929–32 represented the last significant threat to the RAAF's continued separate existence, the ambitions of at least the Army that it might ultimately have control of its own air force were not entirely extinguished. In May 1935 the CGS, Colonel John Lavarack, disclosed to a colleague that he still firmly believed that the RAAF would eventually come under Army command.[68] That real tensions continued to exist was revealed in a briefing paper prepared by the British Air Staff in readiness for the 1937 Imperial Conference. Here it was noted that Australia's defence program was out of balance in that:

> owing to the demands of the Navy and Army, the value of air forces both for local and Imperial defence is not fully recognised, the strength of the RAAF is out of proportion to that of the other two Services, and its organisation is not entirely suitable for Australia's needs.

The situation had, it was stated, reached a stage where the Air Force was forced to adopt deliberately misleading titles for new squadrons being formed under the expansion taking place. Presumably this referred to these units' roles being described as 'General Purpose' and 'General Reconnaissance'. A letter was cited from a member of the Air Staff in Melbourne to the RAAF liaison officer in London indicating that these units were 'to be trained in duties other than that which their title indicates':

> The reason for this is that the Army and Navy would not admit of any Air Force role in Australia except as auxiliaries to themselves, and in order to obtain their support for our programme the titles shown were given to the squadrons, and in the Defence Committee discussions the squadrons were regarded as being all for naval and military purposes.[69]

In the light of this, it was hardly surprising that there should be a final skirmish in this protracted and nugatory struggle. This concerns the circumstances which resulted in Williams' removal from the post of CAS in early 1939, after an unbroken tenure in that office of nearly 17 years. The facts of Williams' departure are covered in greater detail in the next chapter and chapter 11, reflecting the complications of this extraordinary episode. The importance of inter-service rivalry as a factor in this sorry saga is arguable, the evidence for such a conclusion being strong but something less than conclusive, because it is based primarily on material in Williams' private papers which is unlikely ever to be properly substantiated.

When Prime Minister Joseph Lyons announced in April 1938 that a senior RAF officer was being obtained to report on Australia's Air Force, this was a decision taken against a background of an unfortunate series of air accidents which had made RAAF safety a hot political issue. Although the Air Board felt considerable disquiet at this sign that the service was about to be placed under critical scrutiny for political purposes, there was no reason to suspect that either of the other services had played a hand in bringing about such an outcome. Indeed, in making his announcement, Lyons had also stated that the services of a senior British general were being sought at the same time to report on the state of the local military forces.[70]

Marshal of the RAF Sir Edward Ellington arrived in Melbourne to begin his tour of inspection of the RAAF on 16 June 1938—the same month in which Lieutenant-General Ernest Squires took up duty as Inspector-General of the Army, a post vacant since Chauvel's retirement. Ellington's report, made public at the end of August, became the subject of bitter contention on account of his conclusions regarding the air accident issue. An announcement by Prime Minister Lyons in January 1939 that Williams was being replaced as CAS was, to all outward appearances, a consequence flowing from Ellington's adverse findings.

Williams, however, had learnt *at the time the Ellington report was publicly released* that the government planned to remove him from the CAS post and send him on exchange to the RAF. Though he must have been fully aware that during his long tenure of the RAAF's top post he could hardly have avoided treading on powerful toes while defending the interests of his service, his removal now was an outcome which he had not anticipated and considered could not be justified on the basis of the Ellington report. He assumed—quite rightly on the evidence—that there had been, all along, a hidden agenda behind the government's decision to bring Ellington to Australia.

On 5 September 1938, Williams notified the Air Liaison Officer at the Australian High Commission in London, Group Captain Frank McNamara, VC, that he had been informed that the idea of the proposed exchange was Ellington's 'verbal recommendation to the Prime Minister'. He asked McNamara to call on Ellington, who had just arrived back in England, and ascertain the truth of this claim. When McNamara cabled back the results of a meeting with Sir Edward on 12 September, Williams knew his suspicions were well founded. Ellington stated that the 'question of change of command was initiated by certain Minister in Australia', and he had done no more than recommend that an exchange be discussed with the Air Ministry. If Williams wanted to know more, Ellington suggested that he speak with Chauvel, with whom the proposal had been discussed—presumably because of his continued membership of the Council of Defence.[71]

Having ascertained this much, Williams indeed wanted to know more. On 16 September he tasked McNamara by signal with finding out which Minister had initiated the proposal. He had, he said, spoken with Chauvel, who maintained that Ellington had already made up his mind before consulting him that Williams had been CAS for too long; they merely discussed how to make the change. On 29 September McNamara cabled that he had succeeded in obtaining another interview with Ellington but the latter had refused to give names, stating that this 'matter was confidential between him and Ministers'. With regard to what Chauvel had said, Ellington repeated that the exchange proposal had been 'initiated by certain Ministers' and not by him. In the course of the interview, McNamara also elicited the information that the Marshal had also discussed the matter with Sir Brudenell White, also still a member of the Council of Defence.[72]

Williams seized on the changed reference to 'Ministers' rather than a single minister, and he also told McNamara in a private letter dated 7 October of the special significance he attached to the involvement of Generals Chauvel and White in the whole affair. The fact that White 'has had a finger in the pie' he regarded as evidence of old animosities having surfaced: it was, he claimed:

> the old business of the Army cropping up again, they having been rather annoyed that the Air Force should have been left alone when the Army had an Inspector-General put over them. They at that time recommended that the Inspector-General should act in that capacity for both Army and the Air Force.[73]

McNamara tentatively added another name to the growing list of figures with military connections who appeared to have had an involvement in the coup against Williams—that of the Governor-General in Australia, Lord Gowrie (himself a former Army brigadier and VC holder). Recalling that when the Governor-General had returned to England on six months' leave earlier in the year, McNamara had called on him and been asked to convey to Ellington that Gowrie wished to see him before his departure for Australia. 'I have the impression (I may be quite wrong, of course,) that that might have been one of the subjects of discussion between Gowrie and Ellington'.[74]

After a 50-year interval, it is impossible to prove or disprove the motives which Williams attributed to Chauvel and White for the part they played in bringing about his downfall. But whether their contribution had been an initiating one or merely reactive to Ellington, and whatever the merits of the argument that Williams had been in charge of the Air Force for too long, it is impossible to shrug off the strong imputation that their advice to Ellington had been influenced by personal views regarding the proper role of an air service within the Australian defence sphere, and their perception of Williams as someone who had to be removed from the scene for the more likely achievement of this end. The only wonder of it all was that such a senior figure of the RAF should have so artlessly lent himself to being their catspaw, and that of politicians eager to extricate themselves from their present difficulties. But, as narrated later in connection with this episode, given the character of the man this aspect should really have come as no surprise.

4

The Imperial connection

When Australia's Air Council met for the first time on 23 December 1920, the first item on its agenda was a proposal from the British government to offer cadetships to Australian youths wishing to gain admission to the RAF College at Cranwell.[1] Under this scheme two cadets could be nominated by the Governor-General of Australia to each entry at the college, for the purpose of gaining a permanent commission in the RAF. On considering the offer, the Air Council recommended its acceptance, with the result that advertisements were placed for applicants for the next course in February 1922.[2] In the priority accorded this matter can be seen a neat symbolism of the importance to the RAAF of links with its British counterpart.

For Australia's small Air Force, operating far distant from the mainstream of technical development in aviation and in a nation lacking an indigenous aircraft industry, the RAF provided benefits of vital sustaining importance. Not only was the larger British service a source of training expertise and doctrine, but British factories were the source of the machines and other equipment essential to keeping an air force going. As this chapter shows, there were drawbacks in the relationship—particularly in the expectation (found principally but by no means wholly in London) that the RAAF's obvious practical dependence on the RAF should extend into vital areas of policy as well. But, in the final analysis, one is forced to ask: what sort of force would the RAAF have been in the inter-war period without its links with the RAF?

In this regard, it is significant that Australia's dependency on Britain in air matters was by no means singular, but was matched by that of the other Dominions. While the bonds extended out from London like spokes of a wheel to Canada, South Africa and New Zealand, as well as Australia, it is remarkable that there were practically no lateral connections between those spokes. Even with a near-neighbour like New Zealand, the cooperation between the RAAF and other Dominion air forces was minimal.

In part, the absence of firm links with New Zealand stemmed from the fact that defence aviation there was even less developed than in Australia. The air arm in that country formed part of the military forces until, in 1936, after the visit by an RAF officer, a decision was taken to create a permanent air force as a service separate from the Army, under the control of an Air Board. On its formation in April 1937 the Royal New Zealand Air Force looked remarkably like the RAAF sixteen years earlier, with a strength of 24 officers and 156 other ranks, just two bases initially, territorial squadrons (the equivalent of CAF) in four cities, and a reserve of obsolete biplanes obtained from the RAF with some newer trainers and bombers on order.[3] In the circumstances, it was hardly surprising that by this time the RAAF continued to focus primarily on the benefits of cooperation with the rapidly expanding British service.

Borrowed traditions

On the formation of the Australian Air Force, there was a clear expectation both locally and in England that the new service would model itself almost exactly on the RAF, right down to copying its trappings and traditions—even though these were hardly ones of longstanding given the British service's own relatively brief existence. Thus, even before the AAF officially came into being, the Air Board had before it on 22 December 1920 a submission from Goble as Director of Personnel and Training that the RAF nomenclature for officer ranks be adopted in the Australian service; this proposal was withdrawn and held over for later consideration, but duly came into effect on 31 March 1921.[4]

Such was the strength of the expectation that the RAAF would act like a clone of the imperial service that there was frequently little scope for even minor variations to give expression to a local sense of identity. On 27 July 1921, the Air Board considered a submission from Williams proposing that the RAAF adopt the same distinguishing flags and lamps (shown on vehicles at night to denote ranks), and an ensign which was the same as that of the RAF but with the Southern Cross superimposed on the RAF rondel.[5] The Air Council decided to refer the question 'to the proper authorities', resulting in the Air Ministry in London being asked whether it had any objection to the proposal. On 27 October 1921 the British CAS (Trenchard) wrote to the Australian Minister for Defence (Pearce) referring to a Canadian suggestion that all the Dominions should take the RAF ensign and use it 'without difference' as their own: 'I feel very strongly that this proposal that the same ensign should be used is in the best interests not only of the Royal Air Force but of the Dominions'. When this letter was considered by the Air Council on 3 March 1922 it was inevitably decided to recommend to the government 'the adoption without "difference" of the ensign of the Royal Air Force for use in the Royal Australian Air Force'.[6] This was duly approved by London in October that year.[7]

On 11 April 1923 a despatch was received from London suggesting that the RAAF should adopt the uniform of the RAF. This proposal was apparently prompted by the fact that arrangements were then being made for the Royal Canadian Air Force (RCAF) to adopt RAF Dress Regulations, 'modified only so far as the lettering on the Royal Air Force badges is concerned': 'The Air Council attach some importance to the wearing of a common uniform by the various Air Forces, and . . . they would be happy to assent to any proposal put forward by your Government for the adoption of the Royal Air Force uniform'.[8]

As it happened, uniform had been one of the few areas where there was a local disposition towards independence. According to Williams' autobiography, at the time that uniform was being considered for the Australian Air Force the British service had not settled upon a final solution to this question for itself. He recounts that the RAF, having started its existence in 1918 with khaki dress, had a few months later issued a pale blue uniform to a small number of personnel on a trial basis. This had proved highly unserviceable, with the result that experiments were continuing into what should replace it: 'It was not possible for us to wait and see what the RAF was going to do before deciding on a uniform for our own Service for it seemed to me to be essential that a new Service should have a distinctive uniform at the time of its formation'.[9]

In fact, the RAF had already settled upon and generally introduced a new blue-grey uniform in September 1919,[10] long before planning for the Australian air service which became the RAAF entered this level of detail. Despite this inconsistency in Williams' account, no evidence has yet emerged on which to doubt the rest of his claims. Williams said that he wanted the RAAF's uniform to be 'blue, dark but readily distinguishable from the Navy'. After having Melbourne and Sydney scoured for material of a suitable shade without success, he chanced to be invited to look over the Commonwealth Woollen Mills at Geelong. While there, he saw a batch of serge material being prepared for naval uniforms:

> This material was made up in the natural wool and was then put through five machines in each of which it was dipped into indigo dye and when it came to the fourth machine (that is when the material had had four dips) I found the colour I had been looking for. We adopted that colour—so it was in fact navy blue, less one dip.[11]

The resulting RAAF uniform introduced early in 1922 (see chapter 2)—consisting of dark blue peaked cap, tunic, and breeches worn with black leather leggings and boots—was extremely smart, and as one historian of militaria has noted: 'This gave the new junior service a distinctive appearance, different from any of the other services, and being a complete departure from the blue-grey of the Royal Air Force'.[12]

Despite this initial display of independence, the RAF remained the model in most other respects. RAAF Mess Dress, the wearing of which was approved by the Air Council in November 1922,[13] followed the pattern of the British service with only colour variations, consisting of an Eton style dark blue jacket with pale blue lapels, matching trousers with a pale blue stripe down the leg, dress shirt and black tie. In this case the colour scheme had to be changed later, when it was reportedly realised that it too closely resembled the dress of aides-de-camp to the Governor of Victoria.[14] And, when it came to choosing a 'full dress' uniform for officers, the RAAF naturally looked to the RAF pattern, consisting of a high-buttoned skirted tunic of Venetian cloth and matching trousers, wellington boots, and quaint plumed busby. Samples of this uniform were not received until 1938–39, with the result that—due to the outbreak of war—few officers ever actually acquired these uniforms.[15]

Williams also claims the credit for having initiated the design of the first cap badges for the RAAF, calling on the services of Lieutenant Oscar Coulson, a former AFC photographer. Coulson's design incorporated wattle sprigs instead of oak leaves as on the RAF pattern. This was, Williams says, quite successful in a brass version for other ranks but producing it in gold embroidery for officers presented considerable problems. For this reason the RAAF initially adopted the RAF badge.[16]

When, in 1929, the RAAF looked at adopting a motto, it reportedly turned to two Latin experts at the universities of Sydney and Melbourne to provide suggestions. A number of ideas were presented as possibilities but failed to find favour. When the Melbourne consultant, Professor George Cowling, suggested adopting the RAF motto *Per Ardua Ad Astra* (Through Labour to the Stars), observing that this was the 'only one . . . worthy of the RAAF' and had been adopted by the air forces of Canada and New Zealand, this course was duly followed.[17] On 9 October 1929 the Air Liaison Officer in London, Squadron Leader Henry Wrigley, advised the Air Ministry that the Air Board wanted to adopt the RAF motto as that of the RAAF also, and inquired whether there was any objection to this by the Air Council. A reply dated 16 November indicated that the Council had no objection, but that a formal approach from the Australian government through the Secretary of State for Dominion Affairs would be necessary to initiate the process of obtaining formal consent of the Council and the approval of the king.[18]

As was soon realised in the Air Ministry, circumstances—namely, the election of a nationalist Labor government in Canberra during October—had made this an inopportune time for such a request to be initiated. One official noted: 'Owing to the political situation in Australia, it is improbable that formal application to use the RAF motto will be made on behalf of the Australian Air Force'. There was, however, a desire that no impression should be given that there was any difficulty raised by the Air Ministry, and for this reason a further letter was addressed to Wrigley on 24 March 1930 advising that:

> . . . as it is understood that in present circumstances formal application . . . is unlikely to be made by the Australian Government, the Air Council are prepared for their part to give informal permission now for the adoption of the motto . . . on the understanding that the position will be regularised by a formal application as soon as conditions permit.[19]

Notwithstanding the Air Ministry's reading of a Labor government's likely attitude in this matter, in fact, a formal approach to London *was* forthcoming on 23 November 1931—signed on the prime minister's behalf by Arthur Blakeley, Minister for Home Affairs—asking that the king's approval for adoption of the RAF motto be sought.[20] A reply was sent to Australia on 1 February 1932 stating that the king had given his approval, but by this time the Scullin Labor government was out of office—having been defeated in December elections by the United Australia Party under J.A. Lyons.

The RAAF's third most senior officer, Air Commodore W.H. Anderson, shown in the full dress uniform adopted directly from the RAF pattern. These uniforms were introduced into the RAAF only in 1938–39 and few officers acquired them before the outbreak of war. (*Mr K.M. Bulow*)

Later, in July 1932, a separate approach was made regarding the RAAF's wish to use RAF transfers on drums and similar items of equipment. This was approved by the Air Ministry in October, subject to royal assent being sought through the Dominions Office.[21] The desire to keep the RAAF in conformity with the British service also extended to unit badges. From late 1929 RAAF squadrons and units had been permitted to adopt badges and mottos, provided that these were submitted to the Air Board for approval, and under this arrangement, units operating aircraft were allowed to mount unit badges on either side of the cowling, about halfway between the engine and the front cockpit.[22]

In March 1936, however, the ALO in London, Squadron Leader Herbert Austin, learnt that the RAF had revised its procedures in the matter of badges and mottos, and enquired about the new arrangements. The reply received explained that, to ensure badges designed in units were not at variance with the principles of heraldry, the Air Ministry had engaged Mr John Heaton-Armstrong, Chester Herald at the College of Arms, to act as Inspector of RAF Badges (IRAFB). The next month Austin spoke to Chester Herald and established that he would be prepared to act in this capacity for the RAAF, too, before seeking advice from Headquarters if this was desired. The reply sent to Austin on 30 April stated that the Air Board had decided to adopt the RAF procedure.

A year later RAAF units were asked to compile brief unit histories for Chester Herald's use, along with photographs of any badges already in existence and brief statements about the origins of these. In July 1937 this documentation was sent off to London, along with the design of a general badge for the RAAF itself. The Air Board pointed out that the badges had been designed without the benefit of authoritative advice on the principles of heraldry, and it was therefore agreeable to new badges being designed, should that be considered necessary, after further consultation. At the same time the Board did not want to see radical changes to badges that had been in use for many years if that was avoidable; it did not necessarily want badges that were exactly similar to RAF badges; and it did want badges that were distinctive to Australia and to the type of work performed by the units depicted.

Heaton-Armstrong worked on the designs throughout the second half of 1937. For the RAAF badge, he encountered an initial difficulty in finding some pictorial basis for preparing artwork of the Wedge-tailed Eagle, which was to be the centrepiece and would distinguish it from the RAF badge. A photographer was sent to the Melbourne zoo to capture the bird's 'likeness'. This evidently did not meet Chester Herald's needs, however, and a second photographic foray was made to the London zoo. As this only managed to provide pictures of 'a poor specimen', matters were not proceeding very smoothly; it was reportedly from a conglomeration of the photographs and suggestions from Mr Archibald Campbell of the Bird Club of Victoria that a design was finally produced by January 1938.[23]

The following June a recommendation was put to the Defence Minister (Thorby) back in Australia that Heaton-Armstrong should be formally appointed in a separate capacity to his post as IRAFB and become Inspector of RAAF Badges (IRAAFB). Thorby approved the proposal on 19 July, with the result that Chester Herald formally assumed that appointment with effect from 1 January 1938.[24] Heaton-Armstrong remained as IRAAFB until his death in 1967.

Exchanging personnel

In November 1923, during the Imperial Conference being held in London, a meeting with Australian defence representatives took place in the offices of the British CAS (Trenchard) to discuss proposals of great significance to the fledgling RAAF. The RAF was proposing a system of reserve training, the attendance of RAAF officers at training courses in England, and an offer of short service commissions in the RAF for a proportion of the pilots trained annually at Point Cook. These were measures from which the RAAF not only stood to benefit enormously, but which gave practical expression to the 'guiding principle' accepted by the conference regarding the desirability of developing the air forces of the Empire along common lines as far as practicable, by adopting uniform systems of organisation, training, arms, equipment and stores—all except for aircraft types.[25]

The package of proposals continued to be negotiated during 1924, until finally—on 10 March 1925—the Australian CAS (Williams) brought them forward for ministerial approval. The key scheme was that for the transfer of a portion of the annual output of Point Cook pilots on short service commissions with the RAF. This entailed the despatch of ten Australian pilots annually who were to serve four years in the RAF before being returned and placed on the RAAF Reserve. The Air Ministry would pay for the cost of their initial training—£1500 a head, credited to a special fund in London from which Australia could pay for other goods and services provided to the RAAF by the Air Ministry.[26] While Williams supported this part of the scheme he was opposed to a separate proposal for Australian officer candidates to be sent to the RAF Cadet College at Cranwell, and after graduating from there to serve with the RAF for four years before returning to Australia. He argued that:

> . . . should it become desirable or necessary to send young officers to England for training it would pay the Commonwealth better to pay the full expenses of such training and retain control over the officers concerned rather than to allow them to spend 4 of the best years of their flying life with the RAF before they come in contact with any Australian units.[27]

With the Defence Minister's approval of the short service scheme, the first cadets from the course completing training at No.1 FTS at the end of October 1926 were selected. In the event, only seven not ten graduates were chosen, and these sailed for England the following month to begin their RAF service. Although the original agreement involved RAAF transferees receiving only five-year short service commissions in the RAF (with their training course counting as the first year of service), the RAF was soon showing interest in retaining them for a longer period—in some cases even permanently. In November 1928 the question arose whether the Commonwealth would permit an officer from the first batch sent, Flying Officer Lee Murray, to transfer to the permanent RAF instead of returning to go onto the RAAF Reserve. The Defence Department replied in April 1929 that there was no objection, subject to a satisfactory financial adjustment. By this time, however, Murray was being transferred to the RAAF Reserve on compassionate grounds and was no longer available to make the change, but the Air Council notified the Australian authorities that it would 'bear this correspondence in mind should another candidate come forward'.[28]

In October 1930 the RAF had another candidate in Flying Officer G.K. Tulloch, and advised Melbourne that others had also applied for specialised long courses with the same end in view. The Commonwealth was asked whether it wished to limit the number of officers to be accepted for permanent transfer, but—with the effects of the Depression being felt in retrenchments in the services—the RAF was told that Australia was happy not to place any restriction on numbers.[29] In 1935 the British Air Ministry also raised the question of extending the duration of the short service commissions for RAAF pilots transferred to the RAF from five to six years. Again, the Air Board had no objection, provided this was on the basis that the year spent at Point Cook counted as the first year of an officer's service.[30]

In later years there was considerable criticism by RAAF officers of the existence of such a scheme. Remarks one, 'It was a big mistake in Australian policy, because it resulted in Australia's best being sent to the RAF'.[31] Another described the arrangement as a Sword of Damocles hanging over the heads of Point Cook cadets, and adds, 'I seem to recall one or two people who were a bit bitter about going'.[32] The financial aspect had particularly unsavoury connotations, resulting in references to Australian pilots being sold on 'the Slave Market'.[33]

In fairness, however, there could no suggestion that Australians were being dragooned into the RAF. In fact there was great eagerness among cadets to be selected to go to England, this being seen as a golden opportunity. In any event, cadets were required to sign an undertaking on selection acknowledging their willingness to make the transfer if required. The plain truth was, therefore, that the overwhelming majority went willingly, with no hint of coercion, although occasionally there was unhappiness with the situation found upon arrival. One of those who made the change recalls his experience of a RAF squadron:

> People who didn't have friends or relatives nearby didn't have much to do at weekends except sit around the mess. I was well and truly fed up, with the weather and with the lack of hospitality of my colleagues. I received absolutely no hospitality at all from any RAF officer and I began to think that I didn't like the bloody thing at all. It was just complete thoughtlessness—not lack of acceptance because I was Australian. From what I could see Australians were generally popular. But at that stage, if I'd been offered the chance to return to the RAAF, I'd have taken it.[34]

Somewhat regrettably, it was also not true that the RAF was skimming off the 'cream' of the RAAF either—at least the British service did not see it in that light. The fact was that the RAF had some doubts about the quality of the product it received from Point Cook courses. Even though it declared in June 1927 that it was keen to receive further batches of Australian pilots under the scheme, the Air Ministry noted from the outset that there were distinct differences in the standards and thoroughness of training, with some parts such as actual flying capabilities being 'extremely satisfactory' but others, especially navigation, needing much more attention. The RAAF responded that it was altering the ground training syllabus, and trebling practice in navigation.[35]

Despite attempts to rectify shortcomings of the type discovered, the fact plainly remained that RAAF pilots sent to England were sometimes less than top-quality. When the seven cadets sent to England in January 1931 were found to include two who had failed to pass the ordinary graduating standards and been required to sit supplementary tests to obtain a pass, there was naturally some adverse comment from the Air Ministry. Williams was no doubt severely embarrassed to receive a letter from Thomas Trumble, the former secretary of the Defence Department and now Defence Liaison Officer at Australia House in London, chiding the RAAF at the situation which had arisen and suggesting that 'the matter wants seriously looking into at your end'.[36] One of the two officers concerned had to be sent on a further flying training course and eventually his service was terminated on grounds of ill health, with the RAAF having to meet the cost of returning him to Australia where there was no vacancy for him either. One other member of this same group was cashiered from the RAF as a result of a general court martial.[37]

Deficiencies of the Point Cook course were evidently viewed as sufficiently serious that in later years it was standard procedure to immediately send newly arrived Australian pilots to undergo further flying training. As one young RAAF graduate who transferred in 1938 recalls:

> We considered that we were well trained and ready for anything in the way of flying, but were disillusioned with our RAAF wings after arrival in the UK when we were questioned as to where we had trained. On explaining that it had been Point Cook we were politely told that such training was considered inadequate, and we were all posted to RAF flying training units to complete the final term of the RAF course. The need for us to undergo this further training certainly indicated shortcomings in the RAAF training program at the time, its training equipment or its training personnel. The flying instructors at Point Cook during 1937 were—by RAF standards—lacking in experience and flying obsolescent aircraft in perfect conditions that contrasted greatly with the adverse weather of Europe.[38]

Whatever the deficiencies of Point Cook-trained pilots may have been in the RAF's eyes, the latter was eager to have as many of them as possible once Britain announced measures to build up her air strength in 1935. In personal correspondence with his RAAF counterpart the British CAS, Air Chief Marshal Sir Edward Ellington, enquired about obtaining increased numbers of Australian pilots as well as other ranks. Williams proposed building on the existing scheme, an approach which the Air Board preferred since this justified improvements in the RAAF's own training establishment and at the same time earned increased credits in the London fund with which to pay for more aircraft. In reply Ellington said the RAF would like to increase its intake of Australian pilots to 20–25 a year, but that it was looking for as many as 50 pilots in 1936.[39] Within three years the RAAF's own expansion program would create a pressing need to retain all the pilots it trained, and the arrangement with Britain would be suspended. Up to July 1938, when the last eight RAAF-trained pilots sailed for England, a total of 149 officers had been passed across to the RAF under the scheme.[40]

Notwithstanding that there were some 'duds' among the pilots sent to the RAF, there were unquestionably many of considerable ability. This fact is clearly attested to by the number who went on to attain the senior ranks of their adopted service—men such as Air Marshal Sir Ronald Lees (Point Cook graduate of 1930), Air Vice-Marshals Eric Bates (1930), Donald Bennett (1931), John Hobler (1932), and Ronald Rae (1931), and Air Commodore Sir Hughie Edwards, VC (1936), later to become Governor of Western Australia 1974–75.

While the short service commission scheme helped build the RAAF Reserve, it did little in the way of injecting skills and experience into Australia's permanent Air Force. In this respect the attendance of officers at the RAF Staff College at Andover was far more important. The RAAF availed itself of the opening provided by the college as early as 1923, when the attendance of

Squadron Leader Adrian Cole was approved.[41] Cole had particular reason to remember the dinner which marked the end of his course, finding that the seating arrangements placed him alongside the British CAS, the formidable 'Boom' Trenchard. The brief exchange comprising the evening's conversation resulted in Cole being invited to spend a month with the RAF in Egypt on his way back to Australia, which in the event proved to be a unique experience taking in the RAF garrison in Iraq also.[42] Some 30 RAAF officers passed through the college before the Second World War. In addition to this course, Wing Commander Williams attended the 1923 course at the British Army's staff college at Camberley,[43] before also attending the last six months of the Andover course.[44]

Andover provided valuable educational opportunities, although sometimes the cultural differences which emerged were an interesting reflection of the already diverging paths of Britain and her former colonies. Air Marshal Sir George Jones, who attended the 1929 course as a squadron leader, records that: 'The social life was new to me, as was the etiquette . . . I learned as fast as I could, but I have no doubt that I unwittingly committed many a social blunder'.[45]

One who was untroubled by matters of this kind was Flight Lieutenant Bill Bostock who, when written to by the commandant of the college about the fact that he did his own gardening and had sent his daughters to the wrong school, reportedly sent the letter back after writing on it 'Noted and ignored'.[46] On the other hand, there were Australians on the course who demonstrated willingness and facility in becoming replica English gentlemen.

Another avenue for higher training in England opened up in the late 1920s with the establishment of the Imperial Defence College (later renamed the Royal College of Defence Studies). Formed for the purpose of providing for the training 'of a body of officers and civilian officials in the broadest aspects of Imperial Strategy', each year-long course was expected to comprise 30 students—including two from Australia. For the first course in 1927, the Australian students were Commander C.J. Pope from the RAN, and Wing Commander Goble of the RAAF.[47] It was 1933 before another RAAF officer attended the IDC. This was Williams, who professed himself somewhat disappointed at aspects of the course.[48] Despite this, in subsequent years another three senior officers (Group Captains Anderson, McNamara and Cole) followed him, along with the civilian secretary of the Air Board (Coleman).

In addition to these college courses, RAAF personnel also attended courses at RAF schools. These embraced a wide range of special subjects, including engineering, air pilotage and navigation, signals, armament, photography, army cooperation, maritime observation, flying and ground instruction, medical, and parachuting. Over 40 officers received training of this kind, usually followed by a further period of attachment to appropriate RAF units. Not just officers received such opportunities moreover; by 1939 there were some half-dozen warrant officers also attached to aircraft factories in England, and also the United States, on special duty.

The value of these courses in keeping the RAAF abreast of developments and maintaining existing skills was incalculable. Thus, when an application to join the RAAF was received in 1922 from Sub-Lieutenant Ellis Wackett, an

The Imperial Defence College provided an avenue for the training of officers in higher issues of national and war strategy. Group Captain A.T. Cole, shown here in full dress uniform arriving with his wife at an unknown formal occasion in London, was a student at the College in 1938—the last full course before the outbreak of the Second World War. *(Mrs S. Clerehan)*

Australian naval officer then studying at the Royal Naval Engineering College at Keyham, the Air Board was understandably keen to keep him in England until he had finished his course. Wackett consequently remained at Keyham until August 1923, then underwent pilot training on Salisbury Plain, and finally attended a one-year post-graduate course at the Imperial College of Science and Technology, London—all before he ever had anything to do with the Air Force in Australia![49]

An attachment to the RAF afforded Australians unique opportunities to see at first hand and keep abreast of developments, particularly in the last few years leading up to the Second World War when enormous technical strides were being achieved. Air Commodore William Garing recalls that, while attending a specialist air navigation course in 1935, he and other course members were given a top secret briefing on a new device soon to become known as radar.[50] For others there was the chance to examine, and frequently fly, the newest aircraft types being produced by British industry.[51] Such exposure to the latest developments in aircraft design was of particular relevance given Australia's own entry into the field of local manufacture, but the RAAF at times appeared strangely indifferent to making the most of such insights. Recalls Air Commodore Patrick ('Paddy') Heffernan, who spent two years in England from mid-1936 on a test pilot course and in the space of a month flew ten brand-new types:

> I was pretty disgusted with the RAAF's reaction generally to all this experience I'd obtained. Having got to Melbourne, I reported into Headquarters and fronted Warrant Officer Relf in charge of postings to ask where I was to go. He said I was off to Point Cook. I asked if I was wanted for anybody to talk to me, and he said no-one had asked to see me. So I went to Point Cook and no-one asked me what I'd been doing in England for two years. It made me wonder what the hell was going on—nobody could have cared less.[52]

Close association with the latest in aviation had tragic results in at least one case. When the airship R101—the pride of the British airship service—

The machine shop at the RAF School of Aeronautical Engineering (later called the RAF Technical College) at Henlow in 1937 or 1938. The officer at far left is Flying Officer E. Hey of the RAAF. Attendance at courses in England was an extremely important means of injecting service skills into Australia's small air force. (*AVM E. Hey*)

crashed near Beauvais in northern France on 5 October 1930 on its maiden flight to India, Squadron Leader William Palstra of the RAAF's liaison staff in London was among the dead.

Another highly significant form of experience which a few RAAF officers received came from the chance to hold commands within the RAF while on exchange. In April 1936 Squadron Leader Joe Hewitt was appointed to command No.104(B) Squadron, a unit of No.1 Bomber Group at Abingdon, Berkshire. Coincidentally, in July of that year Wing Commander Bill Bostock joined the headquarters staff of this same Bomber Group, which was then under Air Commodore Sydney Smith, previously the air commander at Singapore. On 1 September the command of No.2 Bomber Group at Andover, Hampshire, was taken over by Air Commodore S.J. Goble of the RAAF, who had been previously filling the post of Deputy Director of Air Operations at the Air Ministry. In his new command Goble was heading a formation which in size and importance was far beyond anything available to him in the RAAF, as at that time No.2 Group accounted for 15 of the 58 squadrons in RAF Bomber Command.[53]

Promoted to temporary rank of air vice-marshal in his capacity as Air Officer Commanding, it fell to Goble to control his group during the annual Home Defence Exercise in August 1937, the first in which it participated. The Fairey Battles and Vickers Wellesleys of his formation were involved in raids on targets which included Fighter Command Headquarters at Bentley Priory, the London docks at Tilbury and the RAF Depot at Uxbridge. In October he played host to a visiting party of German air force officers, including Milch, Stumpf, and Udet, who came to Mildenhall to view the latest British aircraft and to witness a display of air power intended to impress them.[54]

Interestingly, the benefits of these appointments was by no means solely in Australia's favour. The RAF also gained, as is evident from the fact that it was No.2 Group's Australian commander who decided the RAF's system of annual inspections was not really effective in testing the operational readiness of units. With the approval of the AOC-in-Chief of Bomber Command, Goble developed a plan for unannounced inspections within his group. Recalls an RAF squadron leader (later an air marshal) charged with helping to put this scheme into effect:

> To cut a long story short—we descended on a station without warning armed with a series of envelopes all marked 'For Exercise'. These in effect required the squadron to get cracking as soon as possible to attack a target already arranged on one of our ranges in Scotland. These exercises produced some extraordinary results, especially getting live bombs fitted to aircraft—too many units had become too used to practice bombs. Today these operational tests without notice are common practice. I greatly enjoyed my few weeks with AVM Goble, who personally I found to be a most charming man in every way—a man I shall always remember.[55]

Apart from the knowledge and experience gained by RAAF personnel in England, the Australian service also derived great benefit from RAF officers sent out on exchange or under loan arrangements. Only seventeen officers were involved, and the majority of these (13) held rank below that of squadron

Wing Commander W.D. Bostock is introduced to Lord Swinton, Secretary of State for Air, by the AOC, No.1 Bomber Group, RAF, Air Commodore S.W. Smith. Bostock served on the air staff of No.1 Group at Abingdon, Berkshire, in 1936–37, while on exchange with the RAF. At the same time, the AOC of No.2 Bomber Group was Air Commodore S.J. Goble of the RAAF. (*Mrs N.M. Bostock*)

leader. Only two were wing commanders, and there was one air vice-marshal (H.R. Nicholl) and one air commodore (J.C. Russell). Regardless of their rank, these officers had considerable impact by virtue of the positions they held within the structure of such a small service as the RAAF. Nicholl and Russell both served in the key post of AMP in 1935–37 and 1939–40 respectively, Wing Commander Malcolm Spicer was Director of Supply and Research in 1927–29, Wing Commander Sydney Toomer was Director of Staff Duties at RAAF Headquarters in 1936–38, Squadron Leader Roy ('Peter') Drummond was Director of Operations and Intelligence 1925–29, Squadron Leader Harry Smart was Director of Training 1927–29, Squadron Leader Thomas Thompson was Director of Organisation and Staff Duties 1927–29, and Squadron Leader Grahame Dawson was Director of Technical Services 1928–29.

Others held unit commands, in particular the naval cooperation element which had a number of RAF COs—Squadron Leaders Victor Scriven (1929–30), Cyril Wincott (1936–38), and James Brown (1939–40). Others rendered valuable service at the FTS, especially Squadron Leader George Banting, who was Chief Flying Instructor (1934–36).[56] His skills as a pilot excited widespread admiration. One cadet of the period recalls:

> George Banting was a beautiful pilot. He used to give us demonstrations, flying around the airfield in one of the DH60 Moths on one wheel with the wingtip nearly touching the ground. With his beautiful 'hands' [that is, feel for the aircraft controls] he gave superb displays.[57]

Air Commodore H.R. Nicholl, shown here inspecting a guard of air cadets at No.1 FTS in 1936, was the most senior of seventeen RAF officers sent to Australia on loan or exchange and was promoted to Air Vice-Marshal while still filling the appointment of Air Member for Personnel at RAAF Headquarters. Accompanying him on the inspection is Squadron Leader F.M. Bladin, OC of Cadet Squadron. (*Sir Richard Kingsland*)

Interestingly, three of the RAF exchange officers Drummond, Squadron Leader Eric Bates and Flight Lieutenant Wallace Kyle—were actually West Australians. Drummond had enlisted in the AIF straight from school, qualified as a pilot in 1916 and served briefly with the AFC before being transferred to an RFC squadron.[58] Kyle went into the RAF after gaining admission to the RAF College, Cranwell, while Bates had received his pilot training with the RAAF before accepting transfer to the RAF in 1930. His return to the staff of the FTS in February 1939 was a unique occurrence—the only occasion on which a Point Cook 'old boy' came back to the RAAF on pre-war exchange.

The British officers sent to the RAAF were, on the whole, an impressive group. A couple were clearly 'duds', or officers of whom it could only be said they had already reached the peak of their career progression. Among these were Russell (see chapter 15) and Scriven (see chapter 7). However, many of the others were indisputably men of great ability who reached the top ranks of their service. No fewer than seven (Banting, Bates, Dawson, Drummond, Kyle, Smart, and Toomer) attained the rank of air vice-marshal or higher, quite apart from Nicholl who held that rank while in Australia. Kyle became an air chief marshal, Drummond an air marshal (he was, in fact, invited to become CAS in Australia in 1944 but was not released by the RAF to take up the post);[59] both were knighted, along with Nicholl.

The Singapore strategy

At the end of the First World War the British government adopted a plan to anchor the defence of the far eastern reaches of her empire on a major naval base at Singapore. It was to be an empty base, with few if any ships home-ported there but garrisoned and equipped for the day when a great Royal Navy fleet might need to be despatched from elsewhere to protect imperial interests in the Pacific. The imperatives of what became known as the Singapore strategy were to dominate Australian as well as British defence policy for almost two decades. The issues involved are beyond the scope of this work to delve into in depth, yet it is important to recognise what the strategy meant for the RAAF in the period.

As a presumed beneficiary of the base it was inevitable that Australia would be drawn into the scheme, especially when, at the 1921 Imperial Conference, Dominions were invited to contribute towards the cost of construction.[60] This Australia declined to do, though its government repeatedly acknowledged throughout the 1920s and 1930s the central role which Singapore would, once—and if ever—completed, play in underwriting Australia's own security.[61] Accordingly, the central plank of Australian defence policy during this period became to provide a local naval force of sufficient strength to 'hold out' in the face of invasion until the promised British fleet could reach Singapore, and to cooperate with a superior British force once this had arrived against what was expected to be a common foe. Military and air forces were only secondary in this scheme of things, being necessary purely to deal with lesser local defence contingencies such as raids.

The emphasis put on the Singapore strategy and naval defence by successive governments in Australia was strongly and at times bitterly contested by Army authorities. Quite apart from doubts about whether the protracted construction of facilities at Singapore would ever be completed, khaki critics questioned whether an empty base might also hold only empty promises. Could Britain be relied upon to despatch a fleet of sufficient size to deal with an adversary like Japan, or a fleet at all if—as was most likely—Japan chose to strike when Britain was distracted by a European war; what relevance did Singapore have when it was so far from the direct line of communications between Japan and Australia; and would the garrison at the base itself be able to hold out until relieved?

The point of such questions, posed most volubly by Generals Chauvel, Bruche and Lavarack, and by such less senior but significant thinkers as Lieutenant-Colonel H.D. Wynter, was to challenge the by-now traditional wisdom of governments that protection against an invasion of Australia was solely a matter of naval strategy. They argued instead that invasion was a distinct possibility, notwithstanding the best intentions and vague promises of Britain, and it behove Australia to be self-reliant and make its own adequate provision against such circumstances. The steps required, in Army's view, were the bolstering of *both* military and air forces 'to a proper degree of power'. What made this proposition an even more contentious issue was the argument that the previous priority accorded to the RAN was quite erroneous, that the force which had soaked up such a high proportion of limited annual defence votes was actually next to useless, being too small and wrongly structured to make the slightest difference in any likely future conflict.[62]

Although the Army deigned to speak for the RAAF in this debate, the attitude of the Air Staff—in effect Williams—was actually quite 'detached', as one account terms it. When CNS complained at a renewed Army push in the wake of the 1934 visit to Australia by Sir Maurice Hankey, secretary of the Committee of Imperial Defence in London, he could point to only partial support being lent by CAS to this campaign.[63] Whether Williams was deliberately distancing the RAAF from the sort of acrimonious inter-service feud on the receiving end of which the RAAF had repeatedly found itself since its establishment, or whether his own thinking on the issues was confused or not fully formulated, are questions which are impossible to answer. As one historian of Australian air and sea power in the inter-war period notes: 'It is difficult to know precisely what Williams thought: there is a certain timidity in his appreciations'.[64]

Under Williams' tutelage, the Air Staff agreed with Navy that raids rather than full invasion were the primary contingency that Australia needed to prepare for, though holding that aerial rather than naval means was the best form of defence in such circumstances, but sided with Army in pressing the prime necessity of measures for local defence. At the same time as stating the view that Australia's first line of defence lay at Singapore, he resisted British attempts to encourage the structuring of the RAAF so as to be able to 'safeguard Australian interests in any part of the world'; such squadrons as the Air Force had were, he considered, essentially to meet local defence requirements, a view which he believed simply reflected government policy.[65]

The fact must be faced that Williams was perhaps only interested in consolidating his own personal position and that of his service during this dispute. By 1935 he already had government agreement to a program of Air Force expansion predicated on an improved capability to deal with raids, so that—several authors have suggested—it seems probable that he judged there was no good purpose to adding his voice to Army's regarding the need to make increased provision against invasion.[66] As one notes: 'He no doubt considered that if the general staff got their way they would secure the lion's share of the defence vote for the army, and the RAAF could be worse off than under the Hankey proposals'. Accordingly, when invited to attend the official opening of the graving dock at Singapore, finally completed in 1938, this occasion meant little more to Williams than an opportunity for some personal grandstanding by making another epic flight (see chapter 13).

Realisation that he had possibly misread the debate over Singapore came only belatedly to Williams in the second half of 1938, following the inspection and report on the RAAF carried out by Sir Edward Ellington of the RAF. When Ellington criticised the Air Staff's failure to make plans for the reinforcement of Singapore's air defences, CAS found himself admonished by the government on this point. He could only tartly enquire when the government's policy had changed, and was referred to a 1937 public statement by Prime Minister Lyons which only by a stretch of imagination could be construed as signifying a shift towards commitment of forces to Singapore.[67] Although not directly a part of the story of the pre-war RAAF, the Singapore strategy was to have its logical consequence in the despatch of a RAAF group to Malaya in 1940.

The Salmond Report

In August 1925 R.G. Casey, the Commonwealth liaison officer at the Foreign Office in London, reported to Prime Minister S.M. Bruce a conversation he had with Sir Hugh Trenchard, the British CAS:

> Trenchard . . . seemed a little grieved at the fact that none of the Dominions had ever yet invited a Senior Air Force Officer to visit them and discuss their Air Force or Air Defence generally in the same way as Kitchener, Jellicoe and others had done in the past from the Army and Navy point of view.[68]

Although Trenchard continued to press this view, even writing to the Governor-General of Australia, Lord Stonehaven, to propose a visit by recently retired Air Vice-Marshal Sir Vyell Vyvyan, the government showed no interest in taking up the suggestion.

On 31 December 1927, however, Williams wrote privately to Trenchard to advise him that he had now proposed such a visit and believed that a request was being forwarded by the prime minister to the British Secretary of State for the Dominions:

> You will remember previous discussions and correspondence on this subject. I have felt for some time too that such a visit might do a very great deal towards helping us get our Air Defences more in proportion to our Sea and Land defences. At the rate we are going we will never get the necessary proportion of Air Force units and I find youth and junior rank against me when I try to impress these facts.
>
> Quite apart from this, however, it would be very valuable to me as the one largely responsible for what has been done, to have the opinion of a more senior and experienced officer on our organisation, training, etc. as a whole . . .
>
> When making the recommendation that this should be done, I had in mind recent reductions in our Estimates, and thinking that the Government might hesitate on the question of expense just at the moment, I suggested that if the services of the Air Officer Commanding in India could be obtained the expense of such a visit would not be great.

Admitting that the suggestion of the AOC India, Air Vice-Marshal Sir Geoffrey Salmond, was influenced by personal knowledge—practically the whole of Williams' war service having been under his command—he went on to say that 'if the suggestion does not fit in with your ideas and plans, I do not think you need hesitate in nominating some other officer'. Williams urged, however, that whoever Trenchard nominated should be on the RAF's Active List rather than on half-pay or retired, and the more senior in rank the better. As a footnote he added: 'Throughout the foregoing I have assumed that you would not be available to come out here—of course if you possibly could come it would be the best thing that could happen to us and the Government would, I am sure, be more than pleased'.[69]

As Williams' letter to Trenchard strongly indicates, the reason behind his action in proposing a visit at this time was his concern that the RAAF was fighting a losing battle with Army and Navy in attempting to obtain an adequate provision for air defence requirements. One historian has pointed to the 'Memorandum regarding the Air Defence of Australia' dated 21 April 1925 which Williams had produced in an attempt to rectify the deficiencies of the five-year defence program to 1929: 'With considerable vigour he argued that it was his "duty" to draw the government's attention to "the inadequacy of the provision for and the lack of proportion in the size of the Air Force contemplated as compared to the other services" '.

The thrust of Williams' arguments in 1925 had been to gain acceptance for a nine-year program which would increase RAAF strength to some eighteen squadrons and twelve flights, and see an Air Force presence in all States.[70] Having failed in his bid to carry the argument by simple logic, he was turning to the tactic often favoured by Australian governments of seeking recourse in reputable British opinion. As he later recorded, 'I realised . . . that no senior RAF officer could possibly refrain from recommending a substantial increase in the Service then existing if our defence needs were to be fulfilled'.[71]

Despite Williams' hopes—as he admitted later—of obtaining a critic that he knew he 'could talk to',[72] it transpired that Sir Geoffrey Salmond was not available. As Casey reported to Prime Minister Bruce from London on 19 January 1928:

> [Trenchard] is delighted that you have asked for a Senior Air Officer to be sent to you on a visit. For domestic reasons it is apparently very difficult for him to spare Sir Geoffrey Salmond, Sir Edward Ellington [AOC Iraq] or Vice-Marshal Tom Webb Bowen [AOC Middle East], but he hopes that his Secretary of State will let Sir John Salmond go. He is presently commanding the Air Defence of Great Britain, is a great name and a great figure in the Air Force . . . [73]

For a time it seemed that this choice might fall through, but by March it was certain that Sir John Salmond would be the one making the trip.[74]

Salmond proceeded to Australia by sea and arrived at Fremantle (WA) on 26 June 1928, where he found Williams on hand to meet him. He remained for almost three months, leaving for New Zealand on 20 September—the date on which he formally submitted his two-part report. The terms of reference received from the Australian government asked that he examine the equipment, training, organisation, administration, and development policy of the RAAF, and advise on the employment of the Air Force specifically in defence of the Commonwealth. This last aspect was the subject of the second part of his report and remained secret, but the rest was all made available publicly.

The picture which Salmond presented of the RAAF was far from flattering or reassuring. Judging that the force was totally unfit for war, he went on to produce a catalogue of deficiencies: low standard of training, worn-out equipment, and poor conditions of service which had caused the exodus of 15 per cent of officers and men in the previous two years. The development program he advocated was a modest and realistic scheme to form three permanent and one CAF squadrons, along with two coastal reconnaissance flights and two flights of fighters, another stores depot, a training wing at the Royal Military

College which was then training officers for the RAAF as well as Army, and a Recruit Training Section. These additions to the Air Force structure were to be spread over nine years.[75]

In the second part of his report, in which he attempted to argue how the increased force he recommended should be employed, Salmond outlined a scheme requiring some additional units which—if the government chose to implement it—would create what he considered the 'ideal' air defence for Australia. Taking heed of Trenchard's injunction before he left England, however, he carefully avoided preaching preconceived ideas such as the Singapore strategy:

> My terms of reference confine me to advice on 'the employment of the Royal Australian Air Force in the defence of the Commonwealth,' and I will venture to exceed these limits only by suggesting that where it is a question of re-equipping the fixed coast defences, the interests of economy demand that serious consideration should be given to modifying any previously designed programme which has been drawn up without a full realization of the increasingly important part which Air Forces are destined to take in the general defence of a country so admirably suited to their functions as Australia.[76]

In June–September 1928 the RAAF came under the critical scrutiny of Air Marshal Sir John Salmond of the RAF, shown here inspecting a guard during a visit to Point Cook. The officers behind him are (*at left*) Wing Commander A.T. Cole, the base CO, and Squadron Leader W.D. Bostock, OC of Flying Squadron. Salmond's report remained an important factor in Air Force planning for the next decade. (*Mr R. Austin*)

The thinking embodied in the Singapore doctrine was, however, evident in his assessment of the scale of attack which Australia might be called upon to meet. There was, he said, little likelihood of large-scale invasion, which left only raids by cruisers and submarines either against trade routes or particular land targets, or air attacks from small carriers or seaplane tenders. The implicit assumption in this was that anything more serious would be dealt with by some unspecified imperial force.

The public half of the report drew a fairly predictable response from the Australian press. The Melbourne *Age* on 9 October declared Salmond's comments to be 'a disturbing exposure and a grave and timely warning': 'This is not a reflection on the officers and men who comprise the organisation, but it puts on the defensive those responsible for policy and administration . . . The public at least expect that what little was possible was being done well.'

As one of those responsible, Williams was very definitely on the defensive. On 11 October he wrote a minute to the secretary of the Defence Department drawing attention to the press criticism, remarking: 'It is natural that the press will only publish extracts from a report such as this. In this case, I think it is true to say that extracts made by the press in most cases give a quite wrong impression as to the nature and tone of the report.' Williams was plainly relieved to find, and quick to make use of, the caveat which Salmond had thoughtfully written into the report's covering letter to the prime minister, stating:

> In the course of my Report, I have made criticisms which suggest that in some respects improvements could be made in the administration, organization, and equipment of the Force, but I wish it to be understood that these defects are largely due to the immense difficulties inseparable from the task of building up an Air Force in its initial stages, and without any properly established organization or bases.
>
> As it is to-day, the Force is established on a firm basis, and has been developed along sound lines, and I consider credit is due to those responsible for the great work achieved in bringing the Royal Australian Air Force to its present stage.[77]

On his return to England in late November, Salmond was greeted rather like a conquering hero. Even Trenchard, his legendary inarticulateness deserting him, wrote a fulsome letter of welcome when Salmond's ship berthed at Liverpool:

> I am more than pleased with . . . our 'Ambassador' of the Royal Air Force. He has done this work as we all expected—splendidly, and it ought to give you a lot of useful knowledge too. On behalf of the Royal Air Force I must say I do not think it could have been done better and I congratulate you most heartily. You have really obtained great results.[78]

Back in Australia, however, the response to Salmond's findings and recommendations was rather less warm. In January 1929 Salmond (now promoted air chief marshal and within a year of succeeding Trenchard as CAS) received from Squadron Leader Roy Drummond, the RAF loan officer who acted as his staff officer during the visit, 'specimens of the Naval, Military and Air Staff criticisms of your Australian report'. These had, Drummond said, all been referred to himself as 'acting counsel for the defence': 'The gist of the criticisms is that the Navy say you have recommended too much and the Army say too little and the difficulty has been to compose a combined report which the Minister wants signed by the three Chiefs of Staff'.[79]

Although Salmond did not hear directly about it, the man who started the whole business also had few praises. In letters to Trumble at Australia House, Williams sulked that Salmond's report had been disappointing:

> Sir John Salmond's was an extraordinary report in many ways and I do not think Australia got real value for its money. There is no doubt, to my mind, that we got the wrong man . . . My experience with Sir John Salmond makes it clear to my mind that we should not leave any developmental programme to anybody but Australian officers who have Australia's interests at heart . . . there was much he could have said which would have given a truer picture than he gave. He could, in my opinion, have helped a good deal more than he did. Of course, much of what he said and recommended is what we have been saying and recommending for some time without result—on the other hand, some statements are absurd and misleading.[80]

Despite his unhappiness with aspects of the report, Williams was to find the document had its uses—even though the onset of the Depression barely a year later caused the Australian government to shelve Salmond's proposals for a moderate expansion of the existing force, and to actually reduce the size of the RAAF. Not until 1934 was a rebuilding of the defence forces commenced, using the Salmond Report as the basis for Air Force policy (see chapter 15). In the meantime, Williams found it extremely useful to be able to base his bids for resources on Salmond's recommendations. In 1933 he was alarmed to find that the failure of the British Air Staff to continue to refer to the report had induced a belief in the Australian government that the report was no longer considered relevant to the country's needs. He therefore wrote to London to tell the Air Staff that 'his position would be very much strengthened if he could continue to base the demands for money on this report . . . [and asking] to have the report by Sir John Salmond brought back as the recognised basis of the expansion of the Australian Air Force'.[81]

Williams' dissatisfaction at the ultimate outcome of the visit must have been greatly compounded had he known about Salmond's findings regarding himself. Writing to Trenchard from the Melbourne Club on 15 August, Salmond had reported his three or four interviews with the Australian Prime Minister, Stanley Bruce. As a result of these talks, he said, Bruce 'now wants a RAF officer to command here for a year or two. At the same time he is most anxious that Williams should get experience in a command at home':

> Williams is due for the Imperial Defence College next January and to my mind it will be a sheer waste of time to send him there. He is very short of practical experience in command of anything more than a small unit, he is rather theoretical and does not realise where or how to get a move on, and does not like making a decision. He has plenty of brains & is a practical flier but is at sea at the present time. I strongly recommend an exchange if it can be managed. I have recommended a lot of changes in the RAAF and it wants a good man to put them through and to ginger up the show generally. The CAS here should have as much

> command as office work and it is in the former & in the working of his staff that Williams is short of experience.[82]

Hints of Salmond's thinking regarding Williams actually reached the Australian press after the visitor's report was released. One Melbourne newspaper mentioned that changes to the membership of the Air Board were expected to follow, with Williams to go to England to 'take up a temporarily responsible position with the Royal Air Force. The Ministry has not decided whether it will promote a member of the board to act in his place or will borrow an officer from England.'[83] Back in London, Trenchard instructed his AMP to begin looking into the question of an exchange at the rank of air commodore 'or thereabouts', at the same time sending Salmond's letter to the Secretary of State for Air, Sir Samuel Hoare, for information. When returning the letter a few days later, the Minister asked CAS: 'What do you propose to do about Williams?'[84] This was a question not immediately resolved once political and financial circumstances combined to rule out the possibility of sending Williams to England at this time, but it lay—unanswered, like a fizzing bomb waiting to go off—for fully another ten years, until a successor to Salmond also passed his verdict on the Australian CAS.

A convenient destination

The distance of Australia from Britain made it a suitable destination for epic flights aimed principally at 'showing the flag' for British air power. In a briefing paper prepared on 6 August 1926 for the Australian delegation to the Imperial Conference later that year, the Air Board in Melbourne commented on recent RAF flights from Cairo to Cape Town, and from Cairo to West Africa, and said that it 'would welcome the extension of such flights towards Australia'.[85]

The first such flight took place in 1927–28, when four RAF metal-hulled Supermarine Southampton flying boats forming a 'Far East Flight' left Felixstowe on 14 October 1927 under the command of Group Captain H.M. Cave-Browne-Cave. Their 43 000-kilometre trip, taking them through the Mediterranean and India before coming on to Australia and back to Singapore, then on to Hong Kong before returning once more to Singapore, has been described as 'the finest example of organized formation flying ever seen at that time'.[86] The big machines did not reach Australia until 1 June 1928, arriving first at Broome (WA) and proceeding in an anti-clockwise direction around the coastline to call at mainland State capitals, before departing again from Darwin on 1 September.

Although primarily an RAF show, the event held considerable significance to the Australian service. After the arrival at Broome, Cave-Browne-Cave was greeted with a telegram from the Minister for Defence (Glasgow) expressing the hope that 'the continuance of your flight round our coast will be an indication of the closest and most amicable co-operation between the fighting forces of the Empire'.[87] Adding substance to this pious sentiment was the presence of the RAAF's Widgeon amphibian, which had been flown from Sydney via Darwin by its designer, Wing Commander Wackett, to meet and escort the British machines (see chapter 9). As the visiting flight neared

Adelaide, the RAAF prepared to provide an additional escort in the form of one of its own wooden-hulled Southamptons. Two of these aircraft had been purchased by Australia, delivery being taken in January 1928, even at which time the idea had clearly been that these should be seen to 'cooperate' with the RAF planes in June if at all possible.[88]

Unfortunately, the impact of this gesture was totally lost when the RAAF Southampton S1158 (later numbered A11–1) was overturned by a freak wind while moored in the Torrens River, near the quarantine station at Osborne, on 22 June. The radio mechanic, Corporal Douglas Endean, was on board at the time, and he was rescued unhurt by the harbour launch. But the disappointment of Wing Commander Goble and Flying Officer F.A. Briggs, who had flown the machine from Point Cook, can only be imagined. They, accompanied by Squadron Leader De La Rue, were just leaving the shore, to prepare for take-off in order to meet up with the British aircraft 160 kilometres down the coast from Adelaide, when the wind struck. Instead of a triumphal meeting of fraternal services, there was for the RAAF men only the business of salvaging their own Southampton for return to Melbourne. Meanwhile, the Far East Flight flew on to Point Cook, where it landed on 29 June and remained until 30 July. Eight thousand people flocked to the airbase in one weekend alone to inspect the British boats.[89]

The RAAF unfortunately suffered a second indignity when the Far East Flight moved on to Sydney. For its arrival on 1 August No.3 Squadron provided an escort of one DH9, two DH9As and a Moth, but one of the DH9As (A1–25)—flown by a CAF officer, Flying Officer Sturt Griffith, and carrying the official cinema photographer filming the event—ditched into the harbour after the machine's engine failed. There were no injuries and the plane was later salvaged.[90] After ten days in Sydney the RAF Southamptons went on to Brisbane, and from there proceeded up the Queensland coast. At Bowen they called on the RAAF's 101 Flight, then engaged on the Barrier Reef survey (see chapter 14); 'The eight days spent there by the crews of the Southamptons were regarded by them as amongst the most enjoyable of the whole cruise'.[91]

After the Far East Flight was disbanded and reformed as No.205 Squadron, RAF, in January 1929, this unit made another 'show the flag' visit from Singapore in 1932. On this occasion the Southamptons, commanded by Wing Commander Wright, arrived at Darwin on 22 March for a stay lasting just over a week. On hand to welcome them this time was a contingent of landplanes sent from Melbourne: three Wapitis from No.1 Squadron (A5–37: Squadron Leader Allan Hepburn (CO), Sergeant C.H. Scott; A5–23: Flight Lieutenant F.J.B. Wight, LAC R.T. Doran; A5–34: Flying Officer H.W. Berry, Flight Sergeant T.H. Stewart) and three Bulldogs from No.1 FTS (A12–2: Squadron Leader John Summers (OC Fighter Squadron); A12–1: Flight Lieutenant R.H. Simms; A12–6: Sergeant L.M. Diprose). These machines left Melbourne on 14 March and reached Darwin after what proved to be an eventful flight north.

One of the Wapitis had problems with its oil pump immediately after take-off from Charleville (Qld) on 16 March, and suffered minor damage on making a forced landing after the engine seized. A further misadventure

When the RAAF sent one of its two Southampton flying boats to Adelaide to meet and cooperate with the RAF's visiting Far East Flight in June 1928, this plan was upset by a freak wind which overturned the machine in the River Torrens. Instead of a demonstration of RAF–RAAF cooperation, the RAAF boat ended up a pile of wreckage, shown here after it had been raised by a 60-ton crane, conveyed up-river and lowered onto the Shell Company's wharf. (*Cdre J.D. Goble*)

occurred at Brunette Downs (NT), essentially as a result of the bulky parachutes which crew members had to wear. As Len Diprose, the pilot concerned, later recalled:

> Nine times out of ten you were out of the cockpit when you started the engine. To get in after swinging the propeller, the parachute often knocked the throttle open before you were properly into your seat. This happened to Bertie Simms starting off at either Cloncurry or Charleville, but luckily he had a clear run ahead and managed to switch off.[92]

Diprose was not so lucky at Brunette Downs (NT), as before he could switch off his runaway Bulldog he collided with a DH50 of Australian Aerial Services, which had just arrived from Camooweal (Qld) with mail bound for Daly Waters (NT). The collision only put a dent in the outer tip of the Bulldog's mainplane, but the struts of the DH50's starboard wings were knocked out and the machine disabled. The RAAF contingent was obliged to fly the commercial airliner's pilot, Alf Ashley, his mechanic, and the mail, on to Daly Waters in Flight Lieutenant Wight's Wapiti.[93]

Landing at Birdum (NT) on 18 March provided still more moments of drama, with Summers crashing when he touched down in A12–2 too close to the right edge of the runway and one wheel sank in soft ground. The machine swung around violently, damaging its undercarriage and just about everything else. Diprose, who was following him in to land, recalls:

> It was the end of the wet season, and Birdum aerodrome was just a strip cut in the tall grass. The strip was solid enough but Summers got just off it and spun into the tall grass and turned over on his back. I landed and managed to lift up the tail of the aircraft to let him get out. He was trapped underneath, though he wasn't hurt.[94]

Observing this mishap, Hepburn attempted to warn off the other machines of his Wapiti flight by radio but without success. As he later noted in his report, however, 'the sight of the Bulldog in a quiescent attitude was sufficient to send them on to Darwin direct'.[95]

The members of the Bulldog flight were obliged to stay overnight at Birdum, in what Summers disgustedly described as the 'one small building masquerading under the name of an hotel . . . a tin shed roughly sub-divided into a Bar, 4 bedrooms (2 occupied) and a verandah dining room'.[96] Recovery of the crashed machine posed a considerable problem, as there was only a weekly train service and the train for that week had left Birdum earlier on the day of the RAAF's arrival. Summers consequently decided to take A12–6 and press on, leaving Diprose to dismantle the wrecked machine and arrange for its shipment by rail to Darwin, there to await the next freight ship (in six weeks' time) and accompany the aircraft back south.[97]

Celebrations marking the centenary of Melbourne in 1934 were the occasion for the next RAF flight, this time involving three Rangoon flying boats of No.203 Squadron based at Basra in Iraq. These big three-engined machines, under the command of Group Captain R.E. Saul, arrived at Darwin in late September and proceeded down the east coast, stopping at Brisbane on 2 October and reaching Melbourne six days later.[98] Similar celebrations for

Three Wapitis from No.1 Squadron and two Bulldogs from No.1 FTS pictured at Darwin on 22 March 1932. The RAAF machines, with a third Bulldog which crashed at Birdum (NT), had been sent from Melbourne to meet up with flying boats of No.205 Squadron, RAF, on a 'show the flag' flight from Singapore. (*Crome collection, NLA*)

Sydney's sesquicentenary in 1938 also provided the inspiration for the next flights. In August 1937 press reports referred to proposals for a non-stop flight occurring in conjunction with the anniversary, although these reports were initially denied by the Air Ministry.[99]

In fact, formal approaches were made to London later that month by the Commonwealth, on behalf of the New South Wales government. Eventually, in December, the Australian government was advised that an attempt to break the world record for long distance flight would be made, probably in the following March, using modified RAF Vickers Wellesley bombers. These planes had been formed into a special unit called the Long Range Development Unit (LRDU), which was based at the RAF station at Upper Heyford under Wing Commander O.R. Gayford.[100] Planning for the attempt on the record continued, and in February it was decided that the flight would have Ismailia, Egypt, as its starting point and was expected to take place in April. In March, however, the Air Ministry advised that the attempt had been deferred until October.

In the meantime other arrangements were put in train to have an RAF presence in Sydney for the actual anniversary celebrations in January 1938. On 2 December 1937 five Saro London II flying boats of No.204 Squadron left their base at Mount Batten, in Plymouth Harbour, under the command of Wing Commander K.B. Lloyd. On hand to farewell the formation was Air Vice-Marshal Cave-Browne-Cave, the leader of the first squadron flight to Australia ten years earlier. Accompanying Lloyd as a passenger was Air Commodore Goble, who was returning to Australia on completion of his period of exchange with the RAF. After 20 refuelling stops, the formation arrived on schedule in Sydney on 25 January.[101]

There were, however, now just four Londons, as one had force-landed in the Bay of Bengal and been towed to a port in Burma. A replacement plane was flown out from Britain, and it was reportedly this aircraft which also suffered a mishap when the squadron reached Melbourne on 7 February. Met and escorted by five RAAF Ansons of No.21 Squadron led by Squadron Leader Charles Eaton, the Londons were in a formation flypast over central Melbourne when the starboard propeller of one of the flying boats broke into pieces and fell to earth. One section of blade hit Flinders Street railway station, crashing through an iron roof and embedding itself 90 cm into the platform within 2½ metres of commuters. Another blade hit the roof of a building in Collins Street, while a third piece of debris was heard to fall into the Yarra River. The damaged London put down safely in the bay where it was discovered that the aircraft was sporting a large hole in the top wing.[102] With an obvious fault in their wooden propellors, the Londons remained in Melbourne until a shipment of replacement propellers were received. On 4 April the squadron set off on the return flight, arriving back at Mount Batten on 29 May after a cruise totalling 48 000 kilometres.[103]

In September the LRDU's planned flight was put back from October until November. To support the RAF machines a RAAF support unit was assembled comprising four Ansons and crews from No.21 Squadron, to be under the command of a liaison officer.[104] Originally it was intended that this officer should be Wing Commander George Jones, but following the further deferral of the date of the flight it was found that Jones was no longer available and

Flight Lieutenant Patrick ('Paddy') Heffernan, Chief Flying Instructor at No.1 FTS, was appointed instead. One of the four Ansons was for Heffernan's use and the other three (piloted by Flight Lieutenant Peter Lavarack and Flying Officers Robert Hitchcock and William Keenan) were to transport the RAF advance party responsible for maintaining the Wellesleys on an Australian tour following their arrival.[105]

On 29 October Heffernan's four Ansons left for Darwin, checking along the way that pre-positioned stores of fuel at Cloncurry, Daly Waters and Katherine were, in fact, available in case the Wellesleys continued on beyond Darwin in their record-breaking attempt. On 4 November, advice was received that three Wellesleys under Squadron Leader R. Kellett would set off from Ismailia the next day. In the event only two aircraft completed the attempt on the record, as one was forced to land on Timor after running short of fuel. The arrival of the remaining planes at Darwin on 7 November created a world distance record (11 526 kilometres) which remained unbroken for eight years.[106]

After the RAF aircrews spent several days recovering from the flight, during which time the three Wellesleys were serviced by their ground crews, a month-long tour to all the State capitals began which was to take a toll on their numbers. On 28 November one of the bombers was badly damaged in a forced landing at Richmond during their two-week stay at Sydney, and had to be dismantled for shipment back to England. On the final leg of the tour (a non-stop flight from Perth to Darwin on 16 December) another of the bombers made a forced landing in the Leopold Ranges some 200 kilometres north-east of Derby (WA). The distress signal from this machine was picked up by Heffernan, whose Anson was then refuelling at Derby, and he proceeded to find it and arranged for the rescue of the crew. The wrecked Wellesley was left where it was and not finally salvaged until late 1939. The sole remaining aircraft returned to Pearce airbase outside Perth, where it was dismantled and shipped back to Egypt.[107]

Despite this rather dismal end to the enterprise, the flight had been a tremendous success. The LRDU received the 1938 Brittania Trophy for the most meritorious aviation achievement by British airmen and six of the nine crew of the Wellesleys were decorated; Heffernan was also awarded the Air Force Cross for his work as liaison officer.[108] One RAF history remarks: 'This was the last of the many great pioneering flights by the RAF, however, for by that time rearmament was in full spate again to prepare for the war that everyone in the RAF realised was on the way. There was no longer time for epic journeys'.[109]

Not all the epic flights coming from England, and which involved Australia's air force, were RAF undertakings. Alan Cobham made an historic passage to Australia and back to England in June–October 1926 which earnt him a knighthood. He recorded that when he arrived at Melbourne on 15 August, his DH50J plane was rushed by welcoming spectators: 'police and Air Force literally had to extract me from the cock-pit and carry me through that astounding throng towards the hangar'. Subsequently his aircraft was moved out to Point Cook where Air Force facilities were made available for its overhaul. It was from the RAAF base on 29 August that Cobham set off for Adelaide on the return journey.[110]

Vickers Wellesley bombers of the RAF's Long Range Development Unit near Windsor (NSW) on 29 November 1938, during their round-Australia tour after breaking the world record for long distance flight. The number of Wellesleys had already been reduced from three to two by a crash near Windsor the day before this picture was taken, and another forced landed in Western Australia on 16 December. The sole remaining machine was shipped rather than flown back to Egypt. Such 'show the flag' flights were not always the display of air power intended. (*Mrs Y.C. Bonwick*)

Ten years later the RAAF also played host to Jean Batten as she made her solo flight from England to New Zealand. Reaching Darwin on 11 October 1936, she arrived at Sydney's Mascot airport three days later. After two days she received a weather report which decided her to make the Tasman crossing the following morning:

> Permission had been granted for me to use the large Air Force aerodrome at Richmond, where the Gull would have sufficient room for the full-load take-off. Everyone was most helpful, and before I took off for Richmond I received word that arrangements had been made for me to be accommodated at the officers' mess overnight.[111]

The evening Jean Batten spent with members of the Richmond mess were recalled by one officer who met her: 'She was extremely attractive and personable. We had a lot of interest in finding out what made her tick and why she wanted to take on these hazardous flights.'[112] Batten's memories of her stay were somewhat different, recounting that the heavy-footed pacing of the sentry posted outside her door by the station commander, Group Captain Cole—presumably to keep the large press corps at bay—caused her to get little sleep that night. She also recorded that, shortly before taking off a few minutes after 4.30 am on 16 October, an officer handed her a life-jacket, acting on her mention over dinner the previous night that before leaving England she had thought to take one on the Tasman leg of her flight.[113]

Alan Cobham's DH50J aircraft pictured at Point Cook in August 1926 during his flight from England to Australia and back again. The RAAF regularly became associated with pioneering ventures of this kind, providing hangar accommodation and servicing facilities even where the aviator had no Air Force connection. (*RAAF Museum*)

Empire Air Mail Scheme

During 1934 there had been a significant development in the field of civil aviation with the formation of QANTAS Empire Airways (QEA) and this company's successful tendering for a five-year contract to operate an airmail service between Brisbane and Singapore, linking up with Imperial Airways Limited (IAL) flights from England. QEA's weekly service, carried on with DH86 landplanes, was a step on Australia's part which—CAS noted soon afterwards—was applauded by the British government 'as adding to Empire activities and developing valuable communications'.[114]

In October, however, the Australian government received from London the cabled outline of radical new proposals for an Empire Air Mail Scheme. Full details of the proposal were not forthcoming until February 1935, when it was learnt that the new scheme was to supersede the existing service so recently established. What was more, it was to be run solely by IAL under a 15-year contract. Under the new arrangements all first-class mail, and passengers, would be carried between London and Sydney on Short Empire flying boats. These big, four-engine machines were a new and untried type, yet so confident were the British authorities of Australian acceptance of the scheme that the flying boats were already on order.

In fact, as was soon discovered, there were serious objections to the scheme and the Australian government did not automatically give consent. Although the issues involved were not primarily ones of a defence nature, there was nonetheless an important aspect to be considered. When consulted by the Minister for Defence (Parkhill), CAS did not hesitate to point it out. As he expressed matters in one of his later blasts against the British proposal:

> The air route from Darwin is important to us from a defence point of view. It is important that Australian flying personnel should know it and, in my opinion, from a defence point of view it is sound to have that area covered in peace by an Australian company whose aircraft and personnel would become available to Australia in war.

He could see no benefit in the IAL proposals requiring QEA to give up control over its existing route and which confined Australian aviation generally to within its own borders, yet involved Australia in meeting the costs of facilities for the flying boats. In fact, from the time of first hearing of the flying boat service he had been 'unable to think of any sound reason for the project'.[115]

To examine the merits of the British proposal, an inter-departmental committee was established in November 1935 comprising the secretary of the Defence Department (Shepherd), Williams, the Controller of Civil Aviation (Johnston), and the PMG Department's Director of Postal Services, H.P. Brown. The Defence representatives on this body reported to their Minister (Parkhill) on 9 December that Australia should retain control of the Singapore–Sydney section of the air route and recommended that the Commonwealth 'advise the United Kingdom Government that it finds itself unable to participate in the Empire Air Mail Scheme as put forward'. The committee also noted, however, the unexpected inability of the existing service to cope with the volume of mail, and accordingly presented alternative proposals to accelerate and improve it.[116] In February 1936 the British proposals also received consideration in the Defence Committee, and here Williams reported to Shepherd that 'the Naval view (put forward by RN officers of course) was in favour of the Imperial Airways scheme'.[117] Despite this element of support, with the weight of advice clearly against the new air mail scheme the Australian government decided not to acquiesce in it.

While initially rejecting the scheme the government was still willing to investigate the suitability of flying boats on the section of the route from Singapore, and in March 1936 it was accordingly decided to use a RAAF Southampton to conduct a survey of proposed landing places and mooring sites along the Australian coast between Sydney and Darwin. On 26 June 1935 Southampton A11-1 left Point Cook piloted by Squadron Leader A.E. Hempel and Flying Officer L.E. Burt; the other crew members were Corporal Percy Neale (rigger), LAC John O'Donnell (fitter) and LAC Alexander Clark (radio operator). The aircraft and crew were away for more than seven months, due partly to a diversion late the next month which took them to New Guinea for some weeks, and ultimately entailed them flying around the Australian continent (see chapter 10).

Hempel's report, submitted from Perth on 21 January 1936, recommended that the airmail service should follow the route Sydney – Brisbane – Gladstone – Bowen – Karumba – Roper River – Darwin.[118] Even though the survey did identify a route, Hempel actually condemned the whole flying boat concept.

The problem, as he explained in a letter to the managing director of QEA, Hudson Fysh, on 6 April 1936, lay in the Imperial Airways specification for landing places:

> The requirements are such that no place in Australia conforms to the specification ... I believe that the flying boat under construction for the Empire routes requires such a run of smooth, deep water into wind to take off, and so flat an approach to alight, that the rivers and most other places are out of the question.[119]

Evidently not accepting that Hempel nor any other RAAF officer might know what he was talking about, Imperial Airways despatched their own expert to conduct a further survey. The man they sent was Major H.G. Brackley, air superintendent of the company, and it was arranged that he would use a Short Singapore III flying boat on loan from No.205 Squadron, RAF.[120] Arriving in Darwin on 15 May 1936, Brackley was joined by Hempel and A.R. McComb of the Civil Aviation Board. These two officers accompanied him down to Sydney, and also as he recovered the route all the way back to Singapore. At the latter place the most extraordinary scenes now took place.

As a result of his survey Brackley was satisfied he had found a viable route but Hempel was still flatly opposed, considering that Brackley had 'absolutely disregarded the specification' laid down for it. Even though modifications to the specification had been received on 30 May, the route still did not comply 'and it is not clear how in most places this non-conformity can be overcome'. As Hempel went on to record, on arrival at Singapore the AOC there, Air Commodore Sydney Smith, convened a conference which appeared to have the sole purpose of pressuring him to change his view of matters:

> The Air Commodore endeavoured, in strong terms, to convince me that my adverse report was wrong ... and that it was in the best interests of the Empire and myself to report that, in the light of my experience in the Singapore boat, the route was suitable ... [121]

It appears that Hempel was in no way exaggerating what transpired, as McComb (who was also present) told the Controller of Civil Aviation (Johnston) in a personal letter dated 19 June: 'Hempel got rather a bad reception and I thought some of the Chairman's remarks were distinctly of a brow-beating nature ... Hempel is furious about the matter and when one thinks about it, one wonders whether the AOC of the station was justified in butting in at all.'

Williams also made sure the Minister got to hear about this 'conference', reporting that:

> Air Commodore Sydney Smith's attitude was such that some of his staff apologised to Hempel after the conference and during his stay at Singapore Air Commodore Sydney Smith ... took every opportunity including such occasions as when at dinner with the Senior Naval Officer there, to tackle Hempel on this question. This went on to such an extent and Hempel was made so uncomfortable that he left Singapore a week earlier than he otherwise would have done ...

When the prime minister received several approaches from parliamentarians to invite Air Commodore Smith to make an official visit to Australia, CAS stated the unequivocal view that he could see no justification for doing so.[122]

Williams also came in for some attention, though in very different vein to that received by Hempel. As it happened, QEA shared none of Williams' objections to the British proposals and were effectively allied with Imperial Airways against the Australian government on this issue. Given the relationship between Williams and Fysh, it was not surprising that the latter wrote on 11 June in an attempt to bring CAS around. Williams replied on 17 June, pointing out that Fysh seemed to have got 'quite the wrong impression' about the basis of his objections. He had no qualms about QEA's contacts with Imperial Airways; on the contrary he considered their present relationship ideal. All he objected to (apart from the use of flying boats) was the proposal to give total control to London, which would make QEA nothing more than the local representative of IAL:

> The proposed flying boat route on the Australian coast would serve nobody and would achieve nothing. A landplane route is already operating and must be developed for Defence purposes anyway and can be developed to meet the needs of the overseas air mail ... Why should it be changed now, and the Commonwealth Government put to the expense of providing the ground organisation for a flying boat service? For what?[123]

Agreement on the Empire Air Mail Scheme was announced in February 1937 after some significant alterations had been made to the original terms, including that QEA would operate the flying boats over the Sydney–Singapore section of the route. Williams' objections had been partly met, though the RAAF still had no enthusiasm for the scheme as a whole. As an Air Staff note apparently from 1937 remarked:

> The Empire Air Mail scheme as now operated from Singapore to Brisbane with landplanes is of the greatest value for air defence purposes. It is indeed a vital part of the air defence scheme for Australia ... With the change over to flying boats between Singapore and Sydney, the value of the mail route to Australian defence has now become seriously limited, and in fact in many respects has become a liability.

Adducing one of London's own arguments, the note pointed out that 'Air re-inforcement of Singapore from Australia, and vice versa, is likely to be by means of landplane squadrons using, if the Netherlands Indies is friendly, aerodromes in that country'. More than this, however, multi-engine landplanes used on the mail routes might be useful for carrying supplies and troops, whereas flying boats were not much use for such purposes, and the workshop facilities formerly built up along the route to Darwin must gradually decline too. 'It is obvious, therefore, that the change over to the use of flying boats between Singapore and Sydney is against the best interests of this country as regards local air defence, and also in the wider sphere of Imperial Defence'.[124]

Predictably, the British Air Staff did not agree with this assessment. Asked by Australia for advice on the air defence aspect of the Empire Air Mail Scheme, the RAF response focused only on the notion that development of the route would open up large portions of the northern coastline while the establishment of facilities would have definite strategic value for defence. Flying boats, it was argued, were actually very suited to the conditions of Australia's north, and for the most part the route offered few attractive targets for sporadic raids such as Australia had reason to fear from Japan in case of war: 'In general, therefore, it can be said that the air mail route will be neither particularly vulnerable nor attractive to enemy sea-borne raids'.[125] This was another example where advice tendered at a distance of 20 000 kilometres rarely took much account of local circumstances.

The Ellington Report

A decade after the Salmond visit, the Australian government decided to invite another senior RAF officer to report on the state of the Air Force. As narrated in the previous chapter, this was an action taken in the context of a controversy over the RAAF's high accident rate (see also chapter 11). Consequently, this second visit took place in a very different atmosphere to the first, being regarded by the Air Board with concern and apprehension—especially in view of a lack of consultation with either the Board as a whole or privately with the CAS, Williams, who later made much of the fact that the first he heard of the plan was from press reports.[126]

The man called on to inquire into the efficiency and management of the RAAF was an Inspector-General of the Royal Air Force, Marshal of the RAF Sir Edward Ellington. Based on his performance as Britain's CAS in the years 1933–37, however, his suitability for this task was questionable. Historians have largely come to regard him as a second-rater quite out of his depth during the crucial period of expansion of Britain's air service before the Second World War, a commander who repeatedly showed himself as out of touch and even inept. In the opinion of a colleague on the Air Council, Ellington was 'the worst [CAS] we ever had', a misanthrope who pretended to despise politicians but was actually frightened of them, and other contemporaries have left a picture of him as being a cheerless, ungracious personality.[127] Lord Trenchard, a former CAS who is regarded as 'The Father of the RAF', recognised that Ellington had a great brain, but believed its value was diminished by having an appallingly shy owner more concerned with obtaining advancement. Even a modern writer who suggests that Ellington did not deserve the 'bad press' he has received, acknowledges that more favourable opinions that he had an acute mind and a remarkable memory, was quick to embrace new ideas, and in no way lacked understanding, must be weighed alongside his retiring and self-conscious nature.[128]

From all this it must be concluded that Ellington did not lack the professional ability to produce a fair and accurate review of the needs and problems of the RAAF. But serious doubts surround his capacity to operate within, or to appreciate the pitfalls of carrying out this task in, a delicate and politically charged situation such as awaited him in Australia. Among members of the service due to receive his attentions, there was already a belief that Ellington's visit had nothing to do with an open-minded review of the whole RAAF but instead had as its purpose the finding of a scapegoat for a problem that would not go away politically. Air Marshal Sir George Jones, himself a later Australian CAS but then at RAAF Headquarters as Director of Recruiting, is among those who believed that Ellington was 'invited to Australia in order to inspect Williams rather than the air force and to recommend his removal from the post of Chief of the Air Staff if necessary'.[129]

Ellington's visit went badly from the beginning. Newspapers reported sweeping pronouncements he made on his arrival about the primacy of imperial defence arrangements and the secondary importance of provision for local defence.[130] To those vitally concerned with the formulation and implementation of Australian air policy, these statements were evidence that Ellington was undertaking his task with preconceived ideas. This was an impression which would appear fully justified in the light of criticism in his report regarding RAAF abstention from planning principles he had espoused as CAS in May 1936—involving the commitment of Australian air squadrons to the prospective reinforcement of Singapore, and the decision to embark on production in Australia of the NA-33 variant of a US-designed advanced trainer, known locally as the Wirraway (see chapter 9).[131]

A far more adverse impression was created by Ellington's personal style during his visit. Although Williams accompanied the Marshal on his tours of RAAF establishments, he found that the only input required of him was answers to particular questions. There was very little general discussion of issues such as the RAAF's problems and no interest at all in ascertaining the views of local air officers regarding the matters submitted for his consideration. Ellington scorned even to watch a demonstration of the Wirraway's performance arranged for him during his visit to Laverton.[132] Whether or not Ellington's personal style was inspired by a desire to demonstrate the objectivity of his investigation, fairly predictably Williams resented being relegated to the status of mere 'colonial'. Such treatment was certainly dismissive of his unparalleled knowledge of the RAAF, a service in which he had played a pivotal role since its foundation 17 years earlier. By 1938 he was a widely respected figure, seen by many as Australia's Trenchard; probably even Williams saw himself as fully entitled to be regarded as 'Father of the RAAF'.

But while Williams might have been irritated by Ellington's attitude to himself and his officers, he was to be outraged at the findings in the report submitted by Ellington regarding the cause of RAAF air accidents. While the Air Board would subsequently take strong issue with Ellington's conclusions and the methodology used to arrive at them (see chapter 11), no less important at this juncture was the strange manner in which the report was made public. Having been shut out of Ellington's thinking during the actual visit, Williams maintains that he was not permitted to see the report until shortly before its release. He had, however, been told at one stage by the Defence Minister (Thorby) that although Thorby was not at liberty to discuss the report there was nothing in it that the CAS need worry about. He was eventually handed a copy of the document—on 29 August, just two days before it was made public—but even then with instructions not to share its contents with other members of the Air Board until it was published.[133]

Williams had been particularly shocked by the critical press coverage following its release. Considering that there was nothing in the published part of the report to support comments of this kind, he was convinced the report was in two parts and the press had been given access to a confidential section in writing up their articles. Added to this was the humiliation of learning first from newspaper reports about his proposed secondment to the RAF.[134] Williams believed that the government had gone out of its way to deceive him, and privately told the ALO in London, Group Captain Frank McNamara, that he considered himself to have been:

> discredited from one end of the Commonwealth to the other and before the whole of my Service ... The Ellington Report is a damnable thing which ... contained statements which really are vicious ... If the Ellington Report were really true a Court-Martial would be much more suitable than an exchange.[135]

Writing years after the event, Williams contended that his removal from the CAS post was engineered from the outset by members of Cabinet, one of whom he identifies as the Treasurer, Richard Casey (later Lord Casey and Governor-General of Australia 1965–69). Williams claimed to have been told later about an approach by Casey to the two RAF wing commanders who accompanied Ellington to Australia, requesting that they suggest to their chief that he should include in his report a recommendation for Williams' exchange; the two officers had responded that Casey ought to make any such suggestion to Ellington himself. The reason which Williams advanced for this action being taken against him (and by a Minister not having the portfolio responsibility for defence matters) was disputes in the Council of Defence in February 1938 over the pace of expansion of the armed forces. The Treasurer's claim that the government had been making sufficiently large budgetary allocations to finance rearmament had been regarded by Williams as a cynical attempt to absolve the government from responsibility for its own policy by shifting blame for perceived shortfalls onto the service chiefs.[136]

By October, McNamara was able to pass on to Williams the news that London had finally submitted the name of Air Marshal Sir William Mitchell, an Australian-born officer then serving as AMP in the Air Ministry.[137] At this stage the Australian government's interest in pursuing the idea of an exchange had cooled somewhat, following a vigorous campaign in defence of Williams—a campaign which Williams himself had no small part in orchestrating. Material in his papers makes clear that Williams was fully consulted by the 1000-strong AFC Association when that body rallied to his support, even though its efforts seem a little misguided by including an attempt to lobby 'our Ministerial friend Dick Casey'.[138]

A special committee within the AFC Association was set up, comprising Alan Murray Jones, G.N. Mills and A.W.L. Ellis, to 'investigate' the Ellington report section by section and to have its findings published in the press. This effort produced some quite sensational headlines suggesting that the government was seeking to have an 'Innocent Aussie' take the blame for its own errors, and that Williams' difficulties represented a clash between pro-English and pro-Australian cliques involving a plot by the former to have the last of the three Australian services placed under British control.[139] At the same time Williams passed around copies of his critique of the Ellington report to well-placed personal contacts, such as the Governor of Victoria, Lord Huntingfield, who had taken an active interest in the RAAF and become the Honorary Air Commodore of No.21 (City of Melbourne) Squadron.[140]

The effects of Williams' efforts were plainly visible by late October, when press reports indicated that the government had decided to refer the Air Board's rejoinder to Ellington for a further report and might modify the exchange proposal.[141] Williams told McNamara on 26 October that the Minister for Defence had told reporters verbally that he (Williams) would now not be sent to England. The attempt to engage Ellington in some sort of dialogue over his report did not get far, however, as the Marshal had evidently had quite enough of the whole business. As Hudson Fysh, managing director of Qantas Empire Airways, wrote to Williams from London on 13 January 1939, Ellington was telling contacts there that he was 'a surprised man indeed at the effect in Australia of his report—an effect which he did not expect or desire . . . he might have made the same report on many an RAF Station'. Having physically collapsed in mid-October, he was not interested in reacting to the Air Board's comments; McNamara reported that he was sticking by his findings.[142]

Further reports from McNamara nonetheless seemed to confirm that the exchange proposal had been shelved. By early December it was definite that Mitchell's nomination had been dropped—he would, in fact, take up an upgraded RAF command in the Middle East, succeeding Air Vice-Marshal H.R. Nicholl. Writing to Nicholl himself, Williams attributed the government's decision not to accept Mitchell to a lack of pluck and a fear of criticism. It seems, however, that Cabinet's resolve to see through Williams' removal was once again re-ignited by a speech in Parliament by a government backbencher, James Fairbairn.

Using the 1938–39 Budget debate, Fairbairn—popularly recognised as an authority on aviation—delivered some stinging criticism of his party's handling of the Ellington affair.[143] One writer has suggested that this occurrence only served to open up 'the whole question of the Ellington report as it reflected on Williams as Chief of the Air Staff'.[144] It is very likely that Williams over-stepped himself in this instance, as it is beyond doubt that collusion took place between CAS and Fairbairn over the speech, and ministers must have known it. Williams later stated that Fairbairn came to see him before the Budget debate, looking for accident statistics with which to refute Ellington, and admits that he showed him the classified figures even though explaining that these could not be cited publicly.[145] Williams' briefing had plainly extended beyond this aspect alone, however, as Fairbairn's references to journalists having been provided information additional to the Ellington report on which to base attacks on CAS directly reflected Williams' own beliefs on this matter.

On 16 January 1939 the prime minister issued a press statement announcing that Williams was being sent on an exchange with the RAF for two years and would be replaced by Goble as acting CAS. According to this statement, the government considered that notwithstanding the great credit due to Williams for his past work in building up the RAAF, and the difficulties acknowledged to have arisen from recent expansion, when it came to the matter of air accidents 'the Air Board cannot be absolved from blame for these conditions and . . . the main responsibility rests on the Chief of the Air Staff'.[146]

One historian has concluded that the government was quite justified in moving Williams from a position he had so long occupied—seventeen years, he says, is 'too long for any one man to hold such an office'—but that such a change did not need a report like Ellington's as a pretext. The government, it is reasonably argued, should have realised that a change at the top was overdue and acted quite independently of Ellington; instead, Ellington's prestige as an imperial representative was invoked as evidence that Williams had to go.[147]

With the prime ministerial statement of January 1939, Williams' future was finally settled. Despite the support of friends and sympathisers, Williams had to reconcile himself to having lost the battle. When Sir John McLaren (a former secretary of the Prime Minister's Department) wrote to commiserate with him in his troubles, Williams responded on 3 February with a lengthy recounting of his version of events which ended with the tart lament:

> As a Commonwealth servant of long experience you will appreciate that we must be loyal to the Government, but that in return we have a right to expect that Government or our own Minister will stand between us and the public against any unfair or improper criticism. We have a right to expect that the Government will stand up to its own decisions if they are the subject of criticism . . . Instead of this, we find the Government indicating its policy or its views on certain matters to the Air Board for the first time per medium of the public press and going out of its way to censure the Board publicly and to put the blame (if there is any) not on the Board but on an individual . . . In all the circumstances I shall be glad of a change.[148]

On 28 February Williams sailed for England on board the P&O liner *Orcades*.

5

Base infrastructure

When the RAAF came into being in 1921, there was a total of just 24 aerodromes and emergency landing grounds throughout Australia—nine aerodromes and five landing grounds owned or under the jurisdiction of the Commonwealth, and another 10 controlled by other authorities.[1] Most of the Commonwealth grounds had been constructed for the use of competitors in the 1919 air race from Britain, but included in the total was the sole military airfield at Point Cook.

The number of aerodromes and landing grounds built by the Civil Aviation Branch underwent a dramatic increase over the next three years to total 120 by the end of 1924,[2] but the RAAF continued to operate just one station during this period. In 1925–26 expansion finally occurred with the opening of new bases at Richmond (NSW) and Laverton (Vic). For more than a decade Point Cook, Laverton and Richmond were the RAAF's only permanent airfields and remained the backbone of Air Force infrastructure, although an experimental section operated during 1924–30 in Sydney (see chapter 9), and a temporary base was established at Bowen, North Queensland, for the detachment engaged on the Barrier Reef Survey 1926–28 (see chapter 14).

In the late 1930s the situation underwent dramatic change as part of schemes to expand the size of the RAAF (see chapter 15). In the two years before the outbreak of the Second World War, new air stations were opened at Bullsbrook (renamed Pearce) outside Perth (WA), and Darwin (NT). In Queensland, a temporary station also opened at Brisbane's Archerfield airport pending construction of a new permanent airbase at Amberley, near Ipswich. Construction also began on a new maritime air base at Rathmines on Lake Macquarie (NSW), and planning was well in hand for yet another station at the national capital, Canberra.

Point Cook

Long regarded as the 'cradle' of Australia's air force, Point Cook is also one of the world's oldest military airfields. Acquired for defence purposes in 1913 (just a decade after the Wright brothers made their pioneering flight in the United States) the area saw the first planes take to the air early in 1914 and has been in continuous use ever since—unlike many of the original aerodromes elsewhere in the world which have been closed down or built over.

At the time the RAAF was formed and took over the CFS from the Army, Point Cook was already a well-developed facility occupying a site of 720 acres (291 hectares). A row of permanent hangars, workshops, stores, offices and a seaplane jetty 177 metres long stood along the foreshore of Port Phillip Bay. Three-quarters of a kilometre to the north-west a single row of officers' houses built in 1915–16 faced towards the airfield, sitting behind picket fences and wide front verandahs as in a country town. An ambitious masterplan drawn up in 1917–18 actually envisaged developing the residential part of the base into an attractive 'garden city' suburb, a planning approach not adopted on British air stations until the 1930s.[3] This plan, prepared by the Director-General of Works, made provision for an eventual population of 800 people. Over the next two years sewerage and electricity supply works were completed in accordance with this scheme, and construction carried out on the new officers mess, NCOs married quarters, and some of the accommodation buildings required.[4]

Although suspended at the end of the war while the government decided its future aviation policy, the masterplan remained the blueprint for subsequent development of the RAAF's 'No.1 Station'. In December 1921 contracts were let for the construction of further buildings under the program—a hospital, a motor garage, an Aeronautics School and additions to the workshops. In that same month proposals for airmen's messing and accommodation buildings (replacing wartime iron huts which were little better than long dormitories) were referred to the Parliamentary Standing Committee on Public Works.[5] With the blessing of the committee, the first stage of construction got underway in December 1922 and continued for the next three years. Among the permanent buildings dating from this period were recreation rooms, a gymnasium, and community facilities such as a children's school, canteen, and a meeting hall cum theatre. Both the gymnasium and meeting hall were equipped with projection rooms for picture shows. As one study of the base's history remarks: 'By 1924, Point Cook was apparently a flourishing community with a village atmosphere'.[6]

Adding to the appearance of the area was the planting of trees which began in 1921. The original purpose was to provide some protection from the high winds that swept the flat, open country in which the station was sited, but these windbreaks also helped soften the stark lines of houses and other buildings designed in what is described as 'Early Commonwealth Vernacular' style. The first plantings took place after the Air Board approved in August 1921 a proposal by Williams for rows of trees and a hedge along the road running in front of the married quarters (now Cole Street), with further trees on the western and northern sides of the area eventually planned to be covered by future buildings.[7]

A second bout of plantings in 1923 owed something to the fact that Flying Officer Ulex Ewart, then on secondment from the Army, was the son of the professor of botany and plant physiology at the University of Melbourne,

This view of Point Cook clearly shows the small size of the base community in the early 1920s. Wartime huts in the foreground was the airmen's accommodation; the officers' accommodation is in the grove of the trees at the centre left of the picture; the number of married quarters is plainly not very numerous. The last of the row of Bessoneau hangars lining the road down to the jetty area was dismantled in 1928. (*Mr J.C. Miles*)

Professor Alfred Ewart. As the earlier tree planting attempt had produced only a few straggly specimens of sugar gums, CAS decided to seek advice on what would produce the desired effect with quickest results. The professor came out to Point Cook, and after a visit which included a flight with Flying Officer Ewart at the controls, took away soil samples to be analysed at the university. The professor's son recalls: 'When the analysis of the soil was complete, father recommended to Dickie that cypress pine would best serve the purpose. These were soon planted, grew steadily in a favourable environment, and may be seen today in their full maturity, mostly in the form of hedges.'[8] As noted by the report on the base's history cited earlier, the comprehensive landscape planting carried out in 1923 'helped establish the special character of Point Cook which has remained almost intact to the present day'.[9]

The erection in 1927–29 of a CO's residence, further married quarters, accommodation blocks for single cadets and airmen, and an administration building fronting a formal parade ground, completed a decade of construction at the station. The jetty was also lengthened by 61 metres, and strengthened, and a new seaplane hangar added to house the two Southampton flying boats on order. When these big machines—the largest in Australia at the time—were delivered, it was reportedly discovered that the hangar doors were too narrow to admit them and required rebuilding.[10]

With the completion of this first construction phase, Point Cook had acquired a feel of permanence. Also settled in this period was an old uncertainty regarding the name of the place. Point Cook had been originally named after John Cooke, the mate in HMS *Rattlesnake* which charted part of Port Phillip Bay in 1836, and the spelling subsequently appeared as either 'Cook' or

'Cooke'—although the area was just as frequently called Laverton (after the nearest railway station) throughout the 1920s. Around 1927, an arbitrary decision (sometimes attributed to a 'clerical error by typists') was taken to standardise the spelling without the 'e'.[11] The distinctive look of the station was also of significance to the wider RAAF, as a recent conservation study of the base's buildings notes that 'Both the basic form of the residential precincts, and the standard building types developed for Point Cook, were later employed at other RAAF bases'.[12]

Despite the scale of the building program carried out during the 1920s, for much of this time there was still a shortage of quarters for married personnel. Many were accordingly obliged to find accommodation off the base, in nearby towns or even further afield in the suburbs of Melbourne. Ulex Ewart records the travails of travelling to and from Werribee in 1923. Trying first a motorbike, he discovered that traffic slowed to a crawl along the Point Cook road in winter, with base instructions limiting car speeds to 30 kilometres an hour and motorbikes to just 15 due to the deterioration of the surface.[13] In any event, Ewart recalls, 'The road's clay surface became not only slippery but clung tenaciously to the tyres, clogging the mudguards and preventing the wheels turning'. He then joined the ranks of married men who took pushbikes on a strenuous ride across paddocks, negotiating innumerable fences and at one point a shallow creek:

> Flying Officer Bert Austin and another officer had managed either to buy or hire this horse and trap, sometimes called a 'Jinka'. We, struggling along on our bikes, used to envy them as they drove sedately past. Eventually Flying Officer Hempel bought himself a second-hand motor car, and I managed to secure a seat in it

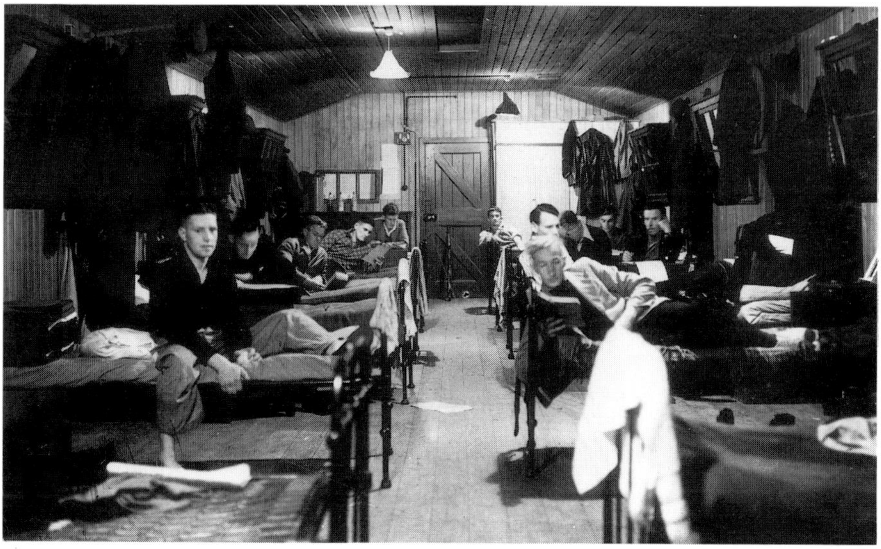

A view of the airmen's quarters in the old wartime huts at Point Cook before the new accommodation blocks were constructed. (*RAAF Museum*)

The line of married quarters—dubbed 'Harmony Row'—along what is now Cole Street, Point Cook. The tree planting program begun (not very successfully) in 1921 is not in evidence in this picture, which graphically illustrates the need for such wind-breaks. (*RAAF Museum*)

> along with two other officers. Naturally we each contributed to his expenses but the luxury of travel in it was well worth it.[14]

Personnel driving from homes around Melbourne had other difficulties. One officer recalls that Flight Lieutenant Bostock, to whom he was related by marriage, 'used to leave home at 4.45 am to get down to Point Cook by 8 am, and got home so late in the evening that his children seldom saw him in daylight'.[15]

The presence of families on the base itself played a large part in shaping the pattern of life on the station. A great deal of formality was observed, such as the practice of 'calling' carried on among officers—not just at Point Cook but throughout the RAAF—requiring new arrivals to observe the protocol of leaving cards at the home of their CO and other senior unit officers. Junior officers were actively discouraged from marrying by the withholding of marriage allowance until the age of 25, but even then such a step was truly a momentous career move. As one Air Force wife of the period recalls:

> Needless to say, the lady was vetted—although she didn't know it—usually by the CO and his wife . . . The whole thing that made it bearable, liveable, was the idea that one was joining a big family. One had an extended family which was a marvellous thing when you went to a new station.[16]

For those women who did 'join' the RAAF as wives, there was also an unwritten code to be observed. The daughter of Air Vice-Marshal Bladin recalls: 'Back-stabbing was especially frowned upon, as the community on the base was too small and isolated for this to be tolerated'. The purpose of such rules was plainly to ensure that life on 'Harmony Row'—the name given to the row of officers' married quarters remained truly harmonious.[17]

Definitely in contravention of the rules was one famous occasion during the early 1930s when a certain squadron leader came home early from a dining-in night at Laverton to find his wife in bed with another officer. The latter decamped through the bedroom window (without opening it) and made off, naked and bleeding, in the direction of the Officers Mess. When the outraged

The isolation of the Point Cook community meant that it was largely dependent on its own resources for entertainment. Mess functions, such as this one in the Officers Mess about 1936, were important in the life of the place. (*Mrs J. Gray*)

husband began searching for his service revolver, the wife also hastily left and dashed along Harmony Row looking for refuge. The house she chose happened to be the residence of Squadron Leader Fred Scherger, who was in the front room studying for the entrance examination to the RAF Staff College. Confronted first with the scantily clad wife pleading for rescue, then the enraged husband brandishing a pistol and declaring his intention to 'shoot the bitch', Scherger did the only thing possible in the circumstances. Reaching for a poker, he laid the man out on the floor with a single blow. While the husband was taken, still unconscious, to a guardroom cell to come to his senses, arrangements were made to get the disgraced wife off the station as quickly as possible in case he should again come after her once released. The other officer involved was permitted to quietly resign his commission forthwith.[18]

Inevitably a community in this position was reliant on its own devices for entertainment. Concerts and picture shows were popular amusements, along with the normal round of functions at the various messes.[19] Formal occasions included the annual Naval, Military and Air Ball held under the auspices of the Defence Department Social Committee and also the RAAF Ball, both held for many years in the St.Kilda Town Hall.[20] Another function was organised by the Air Force Officers' Wives and Mothers Association, of which one wife of the late 1930s recalls: 'The irreverent officers used to call it the 'Dirty Babies Ball', and you went or else!'.[21]

Also important as a form of entertainment were sporting activities. Recalls the wife of the base adjutant in 1927, Flying Officer Eric Woodward: 'We used to play golf on the aerodrome at the weekends, there was no weekend flying'.[22]

For actual members of the service such activities were more than simply recreational, as athleticism was very much encouraged for all ranks. From the first days at Point Cook, football and boxing matches were organised among station personnel.[23] This emphasis on sporting prowess acquired a keen edge after annual inter-service competitions were instituted in 1925 between personnel at the Flinders Naval Depot, the Royal Australian Garrison Artillery (Queenscliff) and the RAAF station at Point Cook. After two years of coming second, the Air Force succeeded in winning one of these triangular sports meetings in 1929.

The desire to do well in these contests lends credence to the claim by AC1 Bert Gardiner, a keen athlete who enlisted in 1925, that bandsmen and athletes were actually given preference in Air Force recruiting, with the result that a number of leading Australian sportsmen were to be found in blue uniform.[24] The opening of the base at Laverton later enabled the holding of inter-station and inter-unit athletic competitions.

The presence of wives and families on base could sometimes give rise to difficult pressures when it came to dealing with the accidents inevitably associated with RAAF flying. An example of this can be seen in the case of Eric Woodward, a seconded Army officer who elected to return to his original service as an indirect consequence of the air disaster over Melbourne in April 1927 (see chapters 11 and 13). He had already survived one crash when a fellow pilot, Alan Charlesworth, overturned an aircraft on landing and came to rest on top of a fence, leaving both occupants hanging upside down but fortunately without major injuries. But when he took part in the aerial display mounted for the arrival in Melbourne of the Duke and Duchess of York in

V.E. Hancock, a later CAS, competes in the high jump during an inter-station and inter-unit sports meeting during the early 1930s. (*Mr C.H.G. Gardiner*)

E.W. Woodward was the most notable of the seconded army officers who elected not to remain with the RAAF. He returned to military uniform and rose to become a lieutenant-general and Governor of New South Wales 1957–65. (*Sir Edward Woodward*)

April 1927, he was a husband of less than three months. His bride later recalled the spate of accidents which hit the RAAF in that year, especially the mid-air collision of two planes:

> We heard that there there had been a crash but it was some hours before we knew who was in them—not until our husbands had landed and could ring us up. I'm probably wrong in saying hours, I suppose it was about an hour. We were living in a flat in East Melbourne then; we moved to the station [Point Cook] a few weeks later. 'King' Cole [the CO] tried very hard at the end of the year to persuade Eric not to go back to the Army . . . but I had had enough of people being slain all around me.[25]

Although the future Lady Woodward denied responsibility for her husband's decision to resume his military career in preference to the Air Force, the belief that this was so remained widespread. According to Sir Frederick Scherger: '[Woodward] was a good pilot, a keen pilot, but he did that fatal thing—he got married. His wife didn't like flying, so he went back to the Army and finished up as Governor of New South Wales which I suppose was not so bad anyway.'[26]

Because of its role in the training of new pilots, Point Cook was scene of many fatalities (see chapter 11). By and large these had little direct impact on families on the base, involving in the main young unmarried men with few local personal ties. One tragedy felt more keenly than most concerned Henri Heememe, a Belgian waif brought back from the war with No.4 Squadron, AFC. Having attached himself to the Australians at Christmas 1918, Henri—or 'Digger' as he became known—was adopted as the unit mascot and was smuggled into England, and subsequently Australia, in a kitbag when the squadron returned from the war. He was initially raised in the home of a squadron mechanic, T.W. Tovell, at Jandowae, near Kingaroy (Qld).[27] In 1926 he was sent down to Point Cook, reportedly because he had set his heart on pursuing a career in the RAAF. According to Tovell, it was arranged that he would stay in the home of Hector Wilson, a former member of 4 Squadron, and his wife: 'Digger . . . started off working for the RAAF as a civvy. That was specially arranged. The idea was that he'd do this work until he could go in as a trainee. In the meantime, the No.4 chaps would be keeping an eye on him'.[28]

Charles Gates, then an NCO who was a father figure for many young airmen who came under his charge in the Air Force, takes up the story. He recalls that Henri had become increasingly unmanageable as he grew up, and it was primarily for this reason that the RAAF was asked to take responsibility for him:

> Henri was living with the station armament officer, Hec Wilson, and his wife. I don't know if he had troubles with them or not but he came down to us one day—apparently somebody had sent him. We were the only couple in married quarters with no children and we took him. I tried to get him to do a bit of study—obviously he had no schooling—but he was quite hard to manage. He got a pea-rifle and one time he went out duck-shooting, so he thought—came back with a chaff bag full of shags. He was quite a trial. My wife also found him a bit tiring too, so I went to the station commander and told him we couldn't cope. He was then nearly eighteen and they put him in the airmens' quarters. He bought a motorbike. I remember 24 May 1928. The Orderly Sergeant knocked on my bedroom window about 4 o'clock in the morning—Henri had been killed in Nicholson Street, past Parliament House. A taxi came out of a lane and Henri, who had been speeding on his bike, went clean through it.[29]

Henri was buried in Fawkner Cemetery with semi-military honours, and a subscription raised by Major Coleman (secretary of the Air Board) for a monument over his grave.

After the development which had gone on unabated at Point Cook for most of the 1920s, there was an hiatus in building during the Depression years of the early 1930s. By 1934, however, work slowly got underway again, culminating in the rapid completion of the original masterplan in the years before the Second World War. In a sense Point Cook was spared many of the expansionary pressures evident at other RAAF stations in the immediate pre-war period. This was because the principal unit in residence at the base remained only No.1 FTS. Although 'Fighter Squadron' and 'Seaplane Squadron'—both really no more than flights—were located alongside, these formed part of No.1 FTS until 1935, when the surviving Bristol Bulldog aircraft of Fighter Squadron were sent to join No.1 Squadron at Laverton. In July 1939 Seaplane Squadron became the basis for the formation of No.10 (General Reconnaissance) Squadron. There was, though, none of the dramatic proliferation of units which occurred at Laverton and Richmond as part of the RAAF's development program.

Despite this, the physical appearance of Point Cook did undergo significant change during the late 1930s, with the construction of a headquarters building for Seaplane Squadron, and new Air Navigation and Wireless Schools. An Armament School was also designed. Main entrance gates were completed in 1937, and in the same year construction was commenced on a new officers mess which was a major departure from the building style used previously for the base. As remarked in the recent conservation study:

> Not only was the new building of red brick and terracotta tile; it has pretensions to formality, to style, even (within stringent economic limits) to grandeur! Fortunately it was located at a sufficient distance from the existing buildings to ensure that it did not detract markedly from the earlier character of the base. It was left to enjoy its pretensions in solitude ... [30]

In fact, the style of the new building did no more than reflect the sense of certainty evident throughout the RAAF at this time. As one study points out, a much larger and grander officers mess building was begun at Laverton the following year,[31] and an equally impressive mess was opened at Richmond in 1938 too. The new sense of identity found expression in other forms as well. Paying tribute to the origins of the service (and Point Cook) was a monument to Australians who died while serving in the AFC and other air services of the First World War, which was unveiled alongside the main parade ground by the Governor-General, Lord Gowrie, in November 1938. The monument also symbolised the RAAF's links with the AFC Association which, with Williams as president, had emerged as an important political lobby group on air matters since its birth in April 1922.[32]

Laverton

In September 1921 an area of 160 acres (65 hectares) was acquired adjoining the railway station at Laverton, 8 kilometres from Point Cook on the Melbourne–Geelong highway.[33] The Air Board's intention behind this purchase was to permanently house No.1 AD here, making it the 'warehouse of the Air Force'. A building program extending over four years and expected to cost nearly £300 000 was prepared. This provided for the construction of a complex of substantial brick hangars, workshops and stores buildings, along with timber accommodation blocks and married quarters to house some 300 personnel.

In 1923 the plan came under the scrutiny of the Parliamentary Standing Committee on Public Works. Tours of Point Cook and Spotswood to see existing arrangements for storing the imperial gift equipment left the committee's members in no doubt of the need for such a facility. At Point Cook, aircraft and equipment jammed seven permanent and ten temporary hangars, while at Spotswood similarly valuable items had the scant protection of

Inside the Officers Mess at Point Cook in the early 1930s. The large propeller mounted on the wall is from the Smith brothers' Vickers Vimy flight of 1919, and is still in the same building which is now the Sergeants Mess. (*AVM E. Hey*)

timber-and-iron wheat sheds originally erected by the Wool Commission and now leased from the Railway Department. The program under consideration was not only intended to ensure that stores were housed under proper conditions, but would include a workshop for overhauling and repairing aircraft, aero engines and transport, as well as wireless, photographic, gunnery and other materiel.

Members of the committee were clearly convinced that, in the interests of economy and efficiency, construction of the depot 'should be made to push on . . . as expeditiously as possible', but they were understandably curious to know why the Air Board required to build a second major base so close to the existing station at Point Cook. The explanation given was essentially one of rail access, a major consideration in view of the outsized freight entailed in moving items such as aircraft. Estimates of the cost of connecting Point Cook to the main line pointed to a likely expense of around £45 000, including almost £5000 for a bridge across Skeleton Creek. Giving evidence to the committee, RAAF witnesses argued that—apart from savings to be gained by building at Laverton—there were advantages in separating the stores function of the depot unit from the unrelated activities of the flying school which was the principal unit at Point Cook. The acting CAS (Goble) stated that the

Construction of a new hangar in progress at Laverton in August, 1935. (*RAAF Museum*)

RAAF also intended to relocate the service squadrons, planned to be raised and stationed in Victoria, at Laverton rather than Point Cook.[34]

Before the committee approved the Air Force bid, the inquiry passed through an uncertain stage when it was learnt that the Victorian government had reserved areas in proximity to Laverton for the conduct of 'noxious trades', such as abattoirs. In case these activities might prove inimical to the health of personnel at the depot, the committee sought assurances from the State authorities that such trades would not be permitted within a 5 kilometre radius, but failed to receive such a guarantee. Consideration therefore turned to another site at Hopper's Crossing, some 5 kilometres nearer Werribee. This alternative offered equally good access and was no less suitable for flying, but involved greater expense. Accordingly when advice was subsequently received that the area around the original site was unlikely to be affected by any developments actually in view, the committee's report dated 27 May 1924 duly recommended that the Laverton site be utilised.

By November 1924 the Melbourne press reported that the first steps in the construction of the Laverton base had been undertaken with the building of a special railway siding adjacent to the site, which was required 'for the purpose of receiving material required in the construction of the buildings of the depot'.[35] The following January it was reported that work on the 'great aircraft depot' had commenced,[36] and a year later parts of the base were available for occupation.

The move of No.1 AD to its new home took place on 1 March 1926, at which point the unit came under the command of Squadron Leader Percy McBain. The depot's first permanent residents were initially just four families occupying married quarters on the base. The remainder of the unit's eight officers and 77 other ranks continued to live in their existing accommodation at Point Cook. These commuted daily to Laverton in transport provided by the RAAF, taking with them cut lunches provided from the various messes at No.1 FTS.[37] One of those who remembers the opening of the depot is AC1 William Gilmour:

> After a period, early in 1926 we settled in to the new Depot. Barrack blocks 40 and 41 were completed, but a temporary kitchen—a framework covered with hessian—was in use, and no hot water was available. We hadn't a cook so the boot repairer, AC1 Jim O'Brien, undertook the cooking.[38]

In addition to the personnel involved in the transfer were huge quantities of stores. These had to be shifted not just from Point Cook but also from Spotswood. Arrangements were made for the storage sheds at the latter place to be cleared within a week using both rail and road means. Some 60 truck-loads of stores were brought by road alone.[39]

Laverton underwent its first major expansion in 1927, in preparation for the transfer of No.1 Squadron from Point Cook on 1 January 1928. With the formation of No.21 Squadron on 20 April 1936, Laverton acquired the status of a station with a separate headquarters commanded by Group Captain Frank McNamara. The number of units located at the base further increased on 3 May 1937, when Nos. 2 and 23 Squadrons were brought into being. No.2 was a general reconnaissance unit, initially comrising just three officers—the CO, Squadron Leader John Summers; adjutant, Flight Lieutenant Peter Lavarack (son of the CGS, Major-General John Lavarack); and equipment officer, Pilot Officer Arthur Hocking—38 airmen who were mostly recent recruits, and just two Hawker Demon aircraft. No.23 (City of Perth) Squadron was a general purpose unit raised to provide an air presence in Western Australia (see a later section of this chapter), and was gone from Laverton the following March. Yet another new unit, No.12 (General Purpose) Squadron, was raised at Laverton on 6 February 1939 preparatory to moving north to occupy a new air station at Darwin. In a very real sense therefore, Laverton was—even more than Richmond—a key centre in the program of expansion undertaken by the RAAF from the mid-1930s.

The reason for Laverton's use in this way was not mere coincidence. The fact was that the influx of new recruits, brought about by expansion, resulted in March 1935 in the discontinuation of the system of posting newly enlisted airmen direct to units for training. A Recruit Training Section was now formed as part of No.1 AD, absorbing the School of Armament and Signals which was then also in the process of formation. On 1 September 1936 this section became Recruit Training Squadron.[40] The presence of this training depot at Laverton, in addition to the base's role in receiving and testing aircraft delivered to the Air Force, made this place the logical point at which to initiate new flying units.

The concentration at Laverton of stores, workshops, and other small units—along with squadrons already mentioned—also made this station the biggest RAAF establishment of the inter-war period. The number of personnel working at the base naturally meant that a sizeable Air Force community was centred on the base and its surrounding area, although many married members lived in Melbourne and commuted daily; as one officer later wrote, these

Parts of a Wapiti aircraft are repainted in a workshop at Laverton. (*RAAF Museum*)

'arrived by train in time to scurry on to morning parade at eight o'clock and, like factory hands, ran to catch the evening train home'.[41]

The strong RAAF presence, both at Laverton and nearby Point Cook, not only found reflection in many aspects of the life of the local district, such as membership of local sporting clubs like the Western Golf Club at Altona, but was significant enough to be specially catered to. Air Commodore Rex Taylor, an airman during this period, remembers the enterprise shown by one businessman, Anders Moritz Hanson, who specialised in mass housing construction:

> He acquired a large amount of land in the West Footscray district, and built houses on it which were very well designed and in close proximity to Laverton. These were made available to serving RAAF personnel on a special offer: £776 bought a five-room house fitted with all modern conveniences of the day—gas bath-heater, stove. Hanson carried the finance, which was tailored to the wages that we received. The area became known as 'Five Squadron'.[42]

Richmond

In July 1921 the Air Board settled on Richmond—about 65 kilometres west of Sydney, between the towns of Richmond and Windsor and close to the Clarendon railway station—as the site at which to locate the first units scheduled to be stationed in New South Wales.[43] Action to give effect to this decision, however, lagged some way behind the gaining of ministerial approval. In recommending the site, Williams noted that an airfield closer to Sydney would be preferable but did 'not appear to be obtainable economically'. In its

Laverton air base pictured in June 1939. The view is from the south, from above the Melbourne–Geelong rail-line (visible in foreground) and highway. (*RAAF Museum*)

favour Richmond already offered a large iron hangar containing stores, workshop and machinery, a caretaker's cottage, and wooden huts for living accommodation—all remnants of the aerodrome's prior use as an aviation school run by the New South Wales Department of Education during the First World War.

By 1923 the Commonwealth had action in hand to acquire the Richmond buildings and 175 acres (71 hectares) of land for a price agreed at £9318, despite strong opposition from local residents who wanted to see the area revert to common ground.[44] Although the transaction was not formally completed for another three years,[45] planning was begun for the RAAF to take up occupancy as soon as there was a unit formed for the purpose. Because of economies this was not straight away, and when aircraft were required at Sydney for the welcome in honour of the visiting British squadron in April 1924 these had to be flown up from Point Cook (see chapter 2) and temporarily stationed at the aerodrome. Later that same year the aerodrome was made available by the Defence Department for the 1924 Light Aeroplane competition and Aerial Derby (see chapter 9).

Beginning in October 1924 the Parliamentary Standing Committee on Public Works examined proposals for Richmond to be established as the RAAF's 'No.2 Station'. A five-year program called for the construction of brick stores, hangars and guard-room, and iron-roofed timber buildings for use as headquarters, cottages, barracks, and recreation and mess accommodation, at an estimated cost of £177 400. In a report dated 30 March 1925, the committee declared itself as 'impressed with the necessity of making early provision for the adequate air defence of the Commonwealth' and stated the unanimous opinion 'that funds should be provided at such a rate as to make it possible to complete this station within three years of date of approval'. It even recommended upgrading the material for building the barrack buildings requested from timber to brick![46]

Richmond began its life as an active RAAF base in mid-1925 after aircraft and personnel arrived to occupy it. According to unit records, No.3 (Composite) Squadron was formed at Point Cook on 1 July and gradually made the move north by road and rail (except for the aircraft which were flown in) over the following three weeks.[47] In fact, the transfer of planes occurred before the squadron even had an official identity. Ten aircraft (four SE5As, three DH9s and three Avro 504s) began the trip from Point Cook on Monday, 29 June, following the path of a lone Avro which left two days before to scout ahead and ensure that landing places along the route were clearly marked. Crewed by Flying Officer William Duncan and AC1 Leslie Milgate, this machine developed engine trouble while circling to land at Benalla (Vic), and crashed. Duncan and Milgate escaped injury but their machine was damaged and had to be dismantled for return to Point Cook.[48]

The main body safely reached Cootamundra (NSW) and spent the night of 29–30 June there. Preparing to take off the next day, however, only eight planes were able to proceed—the remaining two delayed by engine or propellor defects. Disaster struck at Gunning, with the number of surviving machines slashed in half through mishaps suffered on a muddy landing ground: one machine overturned and three sank in the bog. Consequently, reported the Sydney *Daily Telegraph* on 1 July: 'Shortly after 2 pm yesterday the impatient watchers at Richmond aerodrome saw the first aerial visitor. Flying through the arch of a rainbow, it swung clear of a heavy cloud bank, and headed for the flying field.'

A signal fire was lit to indicate the strength and direction of the wind, and soon afterwards a DH9 piloted by the officer designated to command the base,

Aircraft of No.3 Squadron pictured at their new base at Richmond in 1925. The hangar, which formerly housed the wartime aviation school run by the New South Wales government, was the principal structure on the site. The machines parked in front are DH9s. (*RAAF Museum*)

The RAAF's long-time chief, R. Williams, shown here inspecting a parade at Point Cook post-1935 (when he was promoted Air Vice-Marshal), was noted for his detailed annual inspections of units. (*Mrs Y.C. Bonwick*)

Flight Lieutenant Frank Lukis, landed and taxied up to the hangar. About 30 minutes later two SE5As, piloted by Flight Lieutenants Garnet Malley and Les Holden (both CAF pilots in civilian gear), arrived together and touched down. This left one machine unaccounted for, it having evidently been forced down somewhere along the way.

Thus, just three out of 11 aircraft had reached destination safely to signal the commencement of RAAF operations from Richmond. According to the *Telegraph* account, two fully laden tenders travelling from Point Cook by road had also suffered in the current wet conditions, having been held up for two days by floods. Nonetheless, it was claimed that there would soon be 58 airmen permanently stationed at Richmond.

Getting the new base organised and habitable was a first priority. According to one story, the unit had not long been in occupation of the aerodrome when the CO (Lukis) received notification of an impending visit by CAS. Williams was already well known by this stage for carrying out inspections with an eye to detail which in time would become almost legendary; indeed, his visits were more like those conducted by an Army battalion commander than a chief of staff. These extended to individual room inspections of airmen's quarters, right down to the contents of drawers, with an award of a half-day off for the man with the tidiest room.[49] Group Captain Wilfred Compagnoni, who was at Richmond in the mid-1930s recalls:

> On all his inspections he paid enormous attention to everything being painted

> correctly, ensuring that there was nothing untidy about the area, and the lawns were in good order. When he inspected the barracks the menswear had to be properly laid out on beds for inspection. He would look at the hangars to see that everything was clean and neat, and he looked very hard at the aircraft, but I don't remember him ever getting in or asking anyone to fly one.[50]

In Lukis' situation, however, he realised he had made 'a serious tactical omission' by not devoting time and attention to an area beautification program. This deficiency was remedied by the purchase of a substantial quantity of potted plants and shrubs, which were then embedded in the station grounds early on the morning of the inspection. The device had the desired effect—CAS 'was happily surprised, he declared, at the rate of growth and that so much had been done so quickly'.[51] Richmond evidently prided itself on the degree of initiative shown for CAS inspections, as it was here, too, that a later problem with two dying trees was allegedly fixed with some cans of green paint.[52]

No.3 Squadron remained in sole occupation of the aerodrome for a year before No.101 (Fleet Cooperation) Flight arrived from Point Cook in August 1926 to make Richmond its home also. The construction program for the base was underway by this stage, with barracks and mess buildings being completed by mid-1927. By the completion of this work, Richmond had lost its raw look. One newspaper reporter remarked of the place in 1935:

> It's a surprising place, Richmond, after the grey-walled Victoria Barracks, and the sprawling exclusiveness of Garden Island. Of course, you expect the huge gaunt hangars, and the mile-square aerodrome. But the quarters and the offices are brick and tile; there is nothing barracky or institutional about them. And there are flower beds and hundreds of trees and a big ornamental lake, and a children's playground and big stretches of sward. It is all modern, and yet intimately homely. And with the waving cornfields at the back, and the protecting sweep of the Kurrajongs, on the surface, at least, the setting is peaceful and bucolic.[53]

Occasionally the events of the wider world pressed in on this outpost of apparent bliss. In the early 1930s, for example, Sydney was witness to the rise of the neo-Fascist 'New Guard' organisation opposed to the State Labor government led by Premier Jack Lang, which was itself in conflict with the Commonwealth over its policy towards payment of overseas debt. According to some accounts there was a fear for several weeks that the New Guard contemplated storming the Richmond base to obtain arms.[54] However, Ulex Ewart—then serving with 3 Squadron—recalls the briefing which officers received from the CO, Wing Commander Bill Bostock. This suggested the greater worry was that, in the event of the confrontation between the State and Commonwealth worsening, the local police might stand idly by while federal instrumentalities became targets for unruly mobs:

> Bostock told us that to meet such a contingency the services were to be ready to adopt a police role. The RAAF at that time had a greater number of regular servicemen available than either the Navy or Army, and so we had to shoulder the major part of the task set. Bostock ordered all leave stopped and put the station in a state of readiness. Extra guards were mounted for the security of the station itself, and a mobile column was organised in readiness to make a dash to Sydney

> to guard specific buildings should such an order be given. We had only $1\frac{1}{2}$ and 3-ton lorries to give us motorised mobility on the ground. Their canvas hoods were stripped off and replaced by a steel mesh to protect crews from sticks, stones, bottles, etc. and on each vehicle a machine-gun was mounted with live ammunition. All this was an exciting change from our humdrum routine.[55]

As squadron CO from 1931 until March 1933, when he assumed the additional authority of station commander until 1936, Bostock exercised a considerable influence over the character of Richmond. Under his stewardship the station held the Stonehaven Cup—a gift from the former Governor-General and awarded annually for the most efficient Air Force unit—for several consecutive years.[56] From his first year in command, it was also the practice to have a Christmas party for the children of personnel at the base, the one in 1934 being attended by 600 people.

Another initiative for which Bostock could take credit was the formation of a station brass band. According to the unit diary preparations towards this end began in April 1932, but it was not until the following year that this took shape. On 20 February 1933 Bostock advised Headquarters that a band had been in existence for two months and was used for the ceremonial parade during the inspection by the State governor on the 16th of that month. Its performance on that occasion was, he considered, 'such that it should be given official recognition'.[57] A request for the issue of band uniforms was approved, and at the annual AFC reunion in Sydney the following April the AFC Association presented a mace for the drum-major's use, the actual presentation being made by Sir Charles Kingsford-Smith. In his speech accepting this gift, Bostock outlined the difficulties experienced in assembling the 22 musicians from the bare 110 personnel available at the base: 'In the end it boiled down to detailing men to the various instruments you take the trombone, and I'll take the trumpet sort of thing', he explained.[58]

Despite having to make do with a narrow pool of musical talent, the 'Aerodrome Band' (as it was apparently known initially) was a conspicuous success. Under its bandmaster, LAC Ernie Keller, the band gave popular Sunday afternoon recitals in Richmond Park before making a social debut at the 7th Annual RAAF Ball at Farmers' Blaxland Galleries in September in front of 1000 guests including the Governor, Air Vice-Marshal Sir Philip Game.[59] There were nonetheless detractors, especially after its appearance at Remembrance Day observances at the Sydney Town Hall in November.

On 30 November 1933 Warrant Officer Hugh Niven, bandmaster of the station band at Point Cook, wrote to AMP objecting to the fact that the Richmond band had been described as 'the Royal Australian Airforce Band' for the Sydney observances. Its performance on this occasion had been, in his opinion, so poor that he felt the need to uphold the reputation of the No.1 FTS Band in case the two should be confused. This complaint resulted in Bostock being told by Headquarters on 7 December that in future 'the Band at your unit should always be designated as the "Royal Australian Air Force Band, No.3 Squadron, Richmond"'. Bostock quickly pointed out the inaccuracy of such a title, since the band already included members of another unit on the base not just No.3 Squadron. A more appropriate name would be 'RAAF Band, Richmond', he said, and this suggestion was accepted by Headquarters later that month.[60]

The suspicion that inevitably attaches to motives in this episode are at least partly allayed by the more generous observations made by Niven about the Richmond band the following year. After a request from Bostock that he visit the station to provide the local bandmaster, LAC Keller, with some instruction, Niven visited Richmond in January 1934. His subsequent report to Headquarters remarked that the band's playing was still not of a high standard, due to the fact that most bandsmen 'knew nothing of music or playing a brass twelve months ago', but the progress made thoroughly justified its existence. Everything apart from the musical standard—the snap and precision of movements, the mace work of the drum-major, and turn-out—was 'quite up to our best standard', and he felt sure time and training would improve present shortcomings. This endorsement at least helped ensure that the band continued to be a feature of the local scene. Recitals in Richmond Park were not only seen as providing the band with valuable public practice but were judged to help 'promote and enhance the good will between the Service and the local community' and provide 'a means of bringing the Air Force before the public in a dignified and suitable way'.[61]

The RAAF frankly needed a boost in its public relations with the local community at this time, as its presence at Richmond was not universally seen as a boon by residents of the surrounding area. In 1934 there were so many complaints across the Hawkesbury district about planes frightening poultry and interrupting church services that in October base authorities announced a ban on low-flying by station personnel.[62] This measure failed to entirely restore peace and harmony, however, as within a year residents had to endure a fresh assault upon their eardrums.

Late in May 1935 the first three new Hawker Demon aircraft arrived at Richmond, and further deliveries followed at short intervals. Air Vice-Marshal Ernest Hey, then a flying officer at 3 Squadron, described the changeover to the new machines which took place at this time:

> The drill was to fly a Wapiti down to Laverton and pick up a Demon from No.1 AD. [Flight Lieutenant Wilfred] Compagnoni and I were sent down together on 8 July 1935 and flew back two days later. I recall that we had [Pilot Officer] Lyle Holswich in B Flight at this time, and the three of us used to practise a program involving circling Richmond base and diving into the centre of the airfield in succession. It must have driven the locals mad.[63]

Indeed it did. Within weeks of the appearance of the first batch of the new type, renewed cries of protest were raised. On 2 July the Sydney *Sun* reported that Richmond had not been the same since their arrival:

> In dive-bombing, frequently practised at the aerodrome, the Demons hurtle down the air at up to 340 miles an hour. When they are pulled out of the power-dive they emit a noise which is something between a horrific roar, an unearthly growl, and a blood-curdling scream. And they come out right over Richmond. The nerves of the townsfolk are jumping up and down in protest. The Air Force is conscious of a cold stare of unpopularity from the direction of the town, but what can the

> Air Force do? It is small comfort to the townspeople for the Air Force to point out that the noise and vibration of the Demons has caused plaster to fall off the ceiling at squadron headquarters and ornaments to totter off potracks.

The squadron's conversion to Demons was only one of several major changes taking place at the station during 1935. New buildings costing £100 000 were also under construction, in preparation for a planned trebling of the base's existing strength of around 140 personnel. Included in the program were two twin hangars to house 36 aircraft, a large store building, a small store, a transport building, and five single-men's quarters to accommodate an additional 200 men.[64] Later that year, it was announced that 30 airmen were being transferred to Richmond during the last week of October from Laverton, where they had been undergoing training, and were to form the nucleus of a new unit to be created early the following year.[65]

On 20 April 1936 this new unit, No.22 (City of Sydney) (General Purpose) Squadron, was raised at Richmond with Squadron Leader Douglas Wilson as CO, and at the same time No.2 AD was formed under Wing Commander Robert Christie. No.101 Flight also achieved squadron status as No.5 (Fleet Co-operation) Squadron; it was renumbered 9 Squadron on 1 January 1939. A station headquarters now came into existence, with Group Captain Adrian Cole as commander. A year later, on 3 May 1937, Wilson was transferred from 22 Squadron to the command of yet another unit, No.4 (General Reconnaissance) Squadron, which was raised the next day; this unit was renamed No.6 Squadron on 1 January 1939. Within the space of two years Richmond had thus grown from being the home of just one squadron and one flight, to three squadrons and an aircraft depot. From 6 February 1939 the base also served as temporary home to a new cadre CAF unit, No.23 (General Purpose) Squadron, prior to its move to Brisbane.

The rapid proliferation of units using the station naturally placed considerable strain on available facilities, so that construction was undertaken in 1937 of new officers and sergeants messes. The Sergeants Mess, officially opened on 15 December 1937, was an immense improvement on the previous building, as was the new Officers Mess. The latter building was opened on 28 August 1938 with a reception for 400 to 500 guests and was claimed to be 'quite the largest and most imposing military Mess in Australia'.[66] This was probably true, as it was singled out during the visit of Sir Edward Ellington later that year for criticism as being 'lavish'.[67] A very comfortable standard of accommodation was provided for 72 officers, with large games rooms and lounges, and polished parquet floors throughout. With these facilities available, the new station commander, Group Captain Cole, gave Richmond a heightened profile within influential Sydney circles. Compagnoni recalls:

> 'King' Cole was a gregarious man, with what you would call social flair. Every month he used to have a dining-in night to which we could invite guests. He always had some of the Who's Who of Sydney down to Richmond and looked after them pretty well.[68]

Under Cole, the Air Force also joined in an annual inter-service sports competition with Sydney-based Army and Navy units beginning in April 1937.

A view of Richmond air base about 1937. The large building in the foreground is the Officers Mess then under construction. Opened in August 1938, the Mess was criticised as 'lavish' in Sir Edward Ellington's report on the RAAF. (*RAAF Museum*)

Pearce

In his 1928 report on the RAAF, Sir John Salmond had recommended locating a CAF bombing squadron at Perth (WA). This unit would, he expected, cooperate in the training of Army units in the State, and also provide added security against sudden sea-borne attack.[69] Like much else in the Salmond report, nothing immediately came from this proposal and it was not until the Air Force expansion program got underway in 1934 that it was finally taken up.

During 1934, the AMP and the Director of Air Force Works and Buildings inspected the country within 55 kilometres of Perth seeking a suitable site for a RAAF station. Finding that most of the soil within this radius was sandy and therefore unsuitable, the choice presented to the Air Board was between an area at Hearnes Hill planted with prize vines and valued at £80 per acre and a site of 640 acres (260 hectares) at Bullsbrook, 44 kilometres north-east of Perth and adjoining the Geraldton road, costing about £4 an acre. The Board's recommendation for the purchase of Bullsbrook was approved by the Minister (Parkhill) on 6 December.[70]

Work on the new airbase was originally scheduled to start during the 1934–35 financial year, but was postponed on the Minister's instructions.[71] Accordingly, construction did not commence until 1936, the aim being to have the permanent station ready for occupation in June the following year. On his return from England in September 1936, Wing Commander Ray Brownell was advised by the CAS (Williams) that he had been selected to command the station once it was opened, and in the meantime he became OC Flying Training Squadron at No.1 FTS.[72]

On 3 May 1937, No.23 (City of Perth) (General Purpose) Squadron officially came into being at Laverton with an allotment of six Demon aircraft and a personnel strength of four officers and 68 other ranks. This unit was formed as a cadre squadron, meaning that it was ultimately intended to include several CAF flights along with one PAF flight, built around a core of PAF administrative and instructional personnel. Although Brownell was appointed CO, he was attached to RAAF Headquarters for some three weeks and Flight Lieutenant Jerrold Fleming exercised temporary command of the unit. The other officers were the equipment officer, Flying Officer Ian Yeaman, and the adjutant, Pilot Officer Colin Hannah—in later years to become CAS (1970–72) and Governor of Queensland (1972–77).[73]

While the unit steadily expanded and grew in succeeding months—by the end of July it had eight officers and 104 airmen, along with an Anson (A4–4) added to its aircraft strength—in other respects progress was far from satisfactory. Accompanied by his adjutant, and Wing Commander Sydney Toomer (an RAF officer filling the appointment of Director of Staff Duties at Headquarters), Brownell flew to Perth on 23 July in A4–4 to inspect the progress of the work at Bullsbrook.[74] He was concerned to discover that delays in construction of the station showed every sign of continuing unless something was done to urgently redress the situation. As a result of his visit, the Air Board agreed that Brownell should move to Western Australia 'and stay right on the spot . . . to keep in constant touch with the Director of Works with a view to expediting the lagging works programme'. A 20-strong party under the unit quartermaster, Flying Officer George Kelly, was also sent across to prepare for later transfers of personnel. Accordingly Brownell took up residence at the still-unfinished base, where he was joined by his wife and two daughters:

> [A]s no quarters were available on the station, we 'camped' in the recently completed Sergeants Mess. There was no water laid on, but among my key personnel there was an electrician who . . . supervised the installation of a power unit so we at least had electric light and power.[75]

With Brownell already at Perth, administration of the unit at Laverton fell to his second-in-command, Wallace Kyle, an RAF flight lieutenant on two-year's exchange duty with the RAAF. Kyle was actually an Australian, having been born at Kalgoorlie and educated at Guildford (WA); at the end of a distinguished flying career, he returned to his home state as Air Chief Marshal Sir Wallace Kyle to serve as Governor (1975–80). During his time with No.23 Squadron, Kyle proved an effective and popular officer who exercised a beneficial influence over the whole unit.[76]

From late 1937 the slow process of ferrying aircraft, equipment and personnel across the Australian continent was begun. The usual procedure followed was for two aircraft to fly from Laverton, with the second plane bringing back the crew of the one left at Bullsbrook. By this stage the new base was already known as 'Pearce Aerodrome' in honour of Sir George Pearce, a senator from Western Australia and Australia's longest-serving Minister for Defence. (The first reference to Bullsbrook as 'Pearce Aerodrome' occurs in the 23 Squadron unit diary on 11 October 1937.)

From October 1937 an Anson was transferred monthly to the west. In an age when air travel is regarded as commonplace, the contemporary significance of the undertaking is easily overlooked. Those involved in the flights, however, saw them as major events. For 22-year-old Pilot Officer Arthur Garrisson, for instance, who had joined the unit in December 1937 on completion of his flying training at Point Cook, his participation in the February ferry flight to deliver A4–24 rated as 'one of the most interesting experiences of my life. Right across Australia and back—nearly 4000 miles—by air'.[77]

In February 1938 planning envisaged that No.23 Squadron would be established at Pearce on or about 9 March, though there were those with serious reservations that this timetable was achievable. On 25 February Williams wrote to the secretary of the Defence Department advising that on a tour of the new base five days earlier he had been 'very disappointed to observe the backward state of its development'. Although work on the station had been in hand for almost two years, he was disturbed to find 'there was not on the above date a single building which was actually ready for handing over to the Service and ready in all respects for occupation'. He went on to provide a catalogue of the base's deficiencies, concluding that—all in all—the station was not in a fit state to receive personnel and he felt he should discuss with the Minister delaying the final move from Laverton.[78]

In the event, the transfer of the remainder of the squadron went ahead as scheduled. On 6 March, 81 other ranks under Hannah and Yeaman left Laverton by train. Two days later, the squadron's six Demons under Kyle, along with the remaining Anson piloted by Flight Lieutenant Charles Pearce, set off by air—the largest group of planes to transit Australia together since a squadron of nine Wapitis made the crossing in 1929. At Tammin, 140 kilometres east of Perth, the air party landed to take on more fuel and to await the arrival in Perth of the ground party on the Transcontinental train.

The RAAF base Pearce at Bullsbrook (WA), still only half-complete, pictured in July 1938. Though plagued with major construction delays, the station was occupied by No.23 Squadron the previous March. (*Mrs J. Gray*)

The arrival of the remaining aircraft of No.23 Squadron at Perth on 10 March 1938 was timed to coincide with the arrival of the main body of the unit on board the transcontinental train. After disembarking, the ground party paraded through city streets, while the air party flew in formation overhead. The first permanent RAAF presence in the west had arrived. *(ACdre A.D.J. Garrisson)*

Taking off again, the planes arrived over the city, coinciding perfectly with a march through the streets by the rest of the squadron. Arthur Garrisson, flying Demon A1–54 that day, wrote to his mother soon afterwards:

> We were due over Perth at 2 pm and at exactly 2 pm we passed dead over the centre of the City in perfect formation. After 2000 miles across Australia to arrive dead on time! We flew around the suburbs and the city for about half an hour, and then dived in formation in salute over Pearce, broke formation and landed, after one of the most successful flights ever made by the RAAF . . . I feel certain that our flight has confounded the critics of Demon aircraft . . . We never even had so much as a puncture or a scratch of paint on the whole trip.[79]

Recalling the event a half-century later, Garrisson adds:

> At that particular time Demons were having forced landings and accidents all over Australia. It was a new type of aircraft and we had a lot to learn. At the Officers Mess at Laverton the betting was very strong that we would not get all six aircraft across to Perth. We made enough on those side bets to open the bar at Pearce for several weeks.[80]

Amidst the novelty of their arrival at Pearce after the protracted period of preparation, members of the squadron soon found that the new base was indeed still far from ready to receive them. This was actually a situation which persisted well into 1939, as evidenced by a three-page report submitted to the Air Board in May of that year by G.S. Cook, the local works director, which showed work still outstanding on numerous buildings, including hangars, airmen's quarters and the Officers Mess and quarters.[81] In March 1938 the

Officers Mess had still to be built, so that officers were housed in the Sergeants Mess, while the sergeants occupied a partitioned area in the huge building which was the airmen's mess hall. These temporary arrangements were, nonetheless, quite satisfactory and attention initially focused on getting the unit—now totalling 12 officers and 135 airmen—settled into its new home amid all the construction still going on.

Not unnaturally, intense public interest in the squadron's arrival served as something of a distraction from this purpose. Garrisson noted on 13 March that the station was 'beseiged with visitors all day' but, as the base was not yet open for inspection, none (with only a few exceptions) were being allowed in. Public curiosity remained high, however, and Brownell recorded that in response to a request from the Mayor of Perth, Charles Harper, he led a squadron formation over the city 'in order to let the citizens of Perth know they now had an Air Force to protect them'.[82] No doubt fuelled by such a display, when Empire Air Day was held at Pearce on 28 May the station found itself awash with visitors—7000 according to the unit diary but 25 000 according to Brownell.[83]

By mid-1938 the squadron was beginning to undertake a normal program of operations. Training flights were made to points and towns around the State, and exercises were conducted in cooperation with local Army units; in September two Ansons and three Demons took part in an air display at Kalgoorlie. Preparations were also made for training the first batch of citizen force personnel. In readiness for this latter role, delivery had been taken of five Avro Cadet trainer aircraft at the end of March. These arrived in disassembled state and were only progressively brought into operation over succeeding months, the first machine being taken on charge late in April.[84]

Life at Pearce offered distinct advantages over that of eastern air stations. For Brownell as CO it meant a degree of independence from the Air Board which was previously unheard of; as he wrote, 'I was as often as not able to carry on uninterrupted in my "separate Air Force" '.[85] Other unit members also felt that distance gave them a sense of separateness rather than isolation. As Garrisson puts it, 'We were part of the Western Australian Air Force, there was no doubt about that'.[86] The squadron found it was adopted by the local community, with officers and other ranks regularly invited into homes and to join in sporting and social activities.

On 1 January 1939 No.23 Squadron was re-numbered No.25 (General Purpose) Squadron, and the following month Pearce reached station status when a station headquarters was formed, along with a second squadron to share the base facilities. The new unit was No.14 (General Reconnaissance) Squadron, which was created by taking personnel and the six Anson aircraft from 25 Squadron; two more Ansons arrived from Laverton on 2 March. Under these changed arrangements Brownell became station commander, while Flight Lieutenant Pearce was posted to command 14 Squadron and Flying Officer Brian Walker assumed temporary command of No.25.[87]

Following the division, 25 Squadron was left with just three officers and 56 other ranks, and had to commence the laborious business of rebuilding itself from new Ansons ferried across from Laverton. It was with a strength of six officers, six cadets and 65 airmen, manning four Ansons, five Demons and three Avro Trainers that No.25 Squadron went to war.[88] At this time No.14 Squadron had eight officers and 83 other ranks, with eleven Ansons.[89]

Darwin

Planning for the first permanent RAAF presence in Northern Australia began in 1937, and by late December of that year had reached the stage where design drawings for an air station to be established at Darwin were sent from the Defence Department to the Department of the Interior.[90] In April 1938 the CAS (Williams) advised Wing Commander George Jones, then Director of Personnel Services at RAAF Headquarters, and Squadron Leader Charles Eaton, then on the station headquarters staff at Laverton, that they were to be appointed commanders respectively of the new northern station and the squadron which it was anticipated would come into being at Darwin as early as September. These two officers worked together over succeeding weeks in developing plans for the station, and in mid-May flew to Darwin in Anson A4–1 to carry out an inspection on the ground.[91]

Despite the timeframe originally envisaged, matters progressed slowly. It was not until 21 February 1939 that the Air Board had before it estimates from the Works Branch of the Department of the Interior for the construction of major buildings of the new base. The Board was horrified to discover that the cost of two hangars was £95 000, but the Minister (Street) approved the project on 8 March 1939, directing that the Board's concern regarding the cost should be conveyed to the Interior Department.[92]

By this time Jones was posted to RAAF Headquarters, but Eaton was still appointed to command No.12 (General Purpose) Squadron when it was formed at Laverton on 6 February 1939. Eaton (promoted to wing commander on 1 March) and his unit equipment officer, Flying Officer Arthur Hocking, were ordered to form the squadron immediately and prepare to proceed to Darwin at short notice. The new unit was rapidly built up with equipment and personnel, and a week after its formation had a total of fourteen officers and 120 airmen, along with four Anson and four Demon aircraft.[93]

Preparations for the move north continued throughout the following weeks. On 8 March 250 beds were shipped on the Burns Philp steamer *Montoro*, followed a month later by 100 tons of barracks equipment on the *Marella*. Three trucks were also despatched, and other advance stores were packed in readiness for shipment. Delay continued to be experienced, however, principally on account of a shortage of aircraft. Four more Ansons were awaited to complete the squadron establishment but it was not until the end of May that a further three machines were allotted from No.1 AD.[94]

On 1 July the move to Darwin finally began with the despatch of an advance party of 30 NCOs and airmen.[95] This group, under Flying Officer Hocking, left Melbourne on board the *Marella*, and was farewelled by five Ansons and three Demons of the squadron flying in formation over the ship. In Sydney the party transferred to the *Montoro* for the leg to Darwin, and was joined before departure by Flying Officer William Froggatt, the unit barracks officer, who travelled up from Melbourne by train. Arriving at their destination on 24 July, the 32 members of this first element prepared the way for the drafts of personnel which were to follow.

Until the buildings of the permanent station were completed by the Interior Department and available for occupation, the squadron was to be accommodated in a temporary camp of field huts and Bellman hangars at Darwin's civil aerodrome at Parap.[96] This camp, too, had to be specially constructed, and it was this work which was the advance party's primary task. This arrangement had allegedly been arrived at with Interior both to expedite the work and to provide RAAF personnel with experience in handling and assembling reserve equipment used by their service.[97] At a ceremony on 29 July the personnel of the advance party paraded for the Administrator of the Northern Territory, C.L.A. Abbott, who formally welcomed them and turned the first sod at the site of the hutted camp.[98]

Meanwhile another party of four NCOs (Sergeants H. Hope and W.C. Hamilton, Corporals J.L. Truscott and J.C. Kane) had been sent by rail to Brisbane on 7 July for instruction in erecting Bellman hangars. Remaining until 3 August, when they flew to Darwin on board a QANTAS flying boat,[99] this group was accommodated in tin huts at an acetate of lime factory at Cannon Hill—which proved to be excellent preparation for what awaited them at Darwin. Until the advance party had finished erecting living quarters, its members were housed at old meatworks built during the First World War by the Vestey pastoral company to process and export beef; the enterprise was unsuccessful and these works had been closed for many years.

As Darwin was then a town with a European population of less than 2500, the range of options in temporary accommodation for a large influx of servicemen were limited. In fact, the advance party did rather well in having the substantial buildings of the meatworks to begin with. An Army artillery unit had been sent to Darwin on HMAS *Albatross* to mount 6-inch guns in defensive emplacements and provide the town with a permanent garrison in September 1932, at which time Flight Lieutenant Joe Hewitt of the ship's RAAF detachment was then:

> intrigued with the lavish layout and potential uses of Vesteys brick buildings. Darwin township, with its decrepit ill-kept corrugated iron buildings haphazardly fronting untidy dusty roads, could not compare with the prospects Vesteys vacated meatworks offered—with a clean up . . . [100]

Despite this, life in 1939 had a distinctly raw, pioneering feel about it for the members of the advance party. The quarters in the meatwork buildings, which were shared with personnel of the Army's Darwin Mobile Force, was very basic. The Air Force possibly benefited from some adverse publicity given by a Sydney newspaper to the 'shocking' conditions borne by the Army personnel before the advance party's arrival,[101] but even so things remained primitive. Airmen slept on the first floor, on beds placed wherever they could be fitted in among the disused machinery and avoiding large holes left in the concrete floor where chutes or machinery had been removed.

These less than salubrious quarters were dimly lit with electricity from a diesel generator operated by the RAAF party's own electricians. The generator

When No.12 Squadron moved to Darwin in mid-1939 to establish a permanent RAAF presence there, Vesteys Meatworks (pictured here in 1940) served as accommodation pending erection of a temporary camp at the civil aerodrome—itself a temporary arrangement until construction of buildings at the permanent air station was completed. (*Mr J.L. Truscott*)

was turned off at 9 pm, making it dangerous to move around after lights out. At 10.30 pm on 9 September AC1 R.J. Hudson fell through one of the holes in the floor, sustaining injuries which led to his death on the 12th. He was buried the next day, in circumstances which graphically illustrated the miserable conditions endured by the men. Recalls George Grendon:

> The cemetery was down below the Botanical Gardens, in from Mindil Beach, and it was raining. When we went down to bury him the grave was filled with water. So we finished the ceremony by going up to the borders of the Botanical Gardens and carrying down rocks with which to sink the coffin.[102]

Another member of the unit, Jim Truscott, still vividly remembers the primitive conditions under which the RAAF personnel initially slept, washed, ate and worked. He recalls that there were no entertainment facilities, and only Fanny Bay beach or camping at nearby lagoons for recreation. Mail (except airmail) was generally a month in reaching Darwin, so that servicemen generally felt marooned:

> Despite all this, there was no serious outbreak of disease, and morale was good. We were transported daily from Vestey's to the civil drome in a light tender to work on erecting buildings. There were no formal parades, we just boarded the tender in any old clothes as work togs . . . Starting early in the morning, we worked until lunch, then had a siesta before returning to put in some more hours later. There was no undue exercise, as we were advised to conserve energy for our own health and to ensure the work—our No.1 priority—was completed. Dark glasses were issued to combat glare while working on the buildings, which were all

> made of angle and galvanised iron. The huts had to be erected by hand, and the girders and rafters of the hangars were assembled and hoisted by block and tackle. There was no crane or hoist equipment to help with this work—we burnt out the clutches of the three trucks raising the frames—but eventually a hand operated winch was procured. The roof iron, however, was still pulled by rope.[103]

On 26 August orders were received from the Air Board for the squadron headquarters and the existing aircraft of A and B Flights (a total of seven Ansons) to make the transfer to Darwin the next day. The personnel duly left on board a civil airliner from Essendon, but the departure of the Ansons was delayed by a day due to adverse weather and these did not reach their destination until 29 August. The planes were housed in the civil hangar and a Bellman hangar already put up by the advance party. The arrival of the flying party and planes meant that the first air patrol out of Darwin was able to be undertaken on 31 August. Back at Laverton, on 1 September C Flight received five Wirraway aircraft to replace its Demons. The next day—with war breaking out in Europe—these planes and another Anson set out on the trip north, and the same day 77 officers and airmen embarked at Melbourne to make the slower sea passage round the coast to the RAAF's northern-most station.[104]

Archerfield

In July 1938 moves were made to acquire land in southern Queensland for the establishment of a second air station in Northern Australia, when ministerial approval was obtained for the purchase of 1100 acres (445 hectares) at Boondall, some 14 kilometres north of central Brisbane, close to the shores of Moreton Bay. This proposal was abandoned, however, after information was received that the area might be prone to flooding under certain conditions. Consequently, further investigations were made, but these produced only one other site close to Brisbane, at Zillmere (adjoining Boondall). This area was also discounted on closer examination.

Following these initial false starts, it was decided to go further afield. An 800-acre (324 hectare) property called 'Amberley', located 42 kilometres from Brisbane and just 4.8 kilometres west of Ipswich, was identified as suitable. A recommendation from the CAS (Williams) that this site was ideal was accepted by the Air Board, which submitted a proposal to purchase it to the Minister for Defence (Street) on 11 November. A month later, on 12 December, the site was acquired and plans began being drawn for construction of an air station.[105]

In preparation for this additional airbase, another squadron was formed at Richmond on 6 February 1939. This was No.23 (General Purpose) Squadron, which took over the previous numerical designator of the Perth-based No.25 Squadron. Formed with four officers, 100 other ranks, and just two Anson aircraft, the new unit was temporarily commanded by Flight Lieutenant Thomas Curnow.[106]

By late June the Minister announced that Wing Commander John Waters had been appointed CO. He also stated that the unit would soon be transferred to Brisbane, initially based at the civil aerodrome at Archerfield pending construction of the permanent station at Amberley. Until permanent accommodation was ready for occupation, the squadron would be housed in field huts and Bellman hangars, in the same way as was being done at Darwin. As in Western Australia, the RAAF presence in Queensland would be a mixed unit of PAF and CAF personnel, with its permanent cadre forming just under half its strength.[107]

At the time Street made his announcement, preparations for the squadron's move north were already well advanced. On 5 May Flight Lieutenant Curnow had flown to Brisbane accompanied by the unit equipment officer, Flying Officer Colin Cleary, on a four-day trip to make arrangements for housing a stores and billetting party soon to follow. This party, comprising Flying Officer Phillip Howson, Sergeant P.A. Ford, LAC V.E. Down, and AC1 R.S. Alexander, set off from Richmond by road on 20 May, and arrived three days later to begin preparing accommodation for further unit arrivals in the acetate of lime factory at Cannon Hill, some 12 kilometres from Archerfield.[108]

On 16 June, the next element made the journey north by road, arriving at Cannon Hill on the 20th. Comprising the unit barracks officer, Flight Lieutenant Arthur Styles, and 22 other ranks, this party was to begin erecting a Bellman hangar. This work progressed over the next few weeks, anxiously reviewed and supervised by groups of officers who flew up from Richmond. On 28 July Wing Commander Waters finally arrived to take over command.[109]

In the last days of August, with the outbreak of war now imminent, the squadron moved to Archerfield seemingly—as one airman recalls—'overnight'.[110] On 28 August the bulk of the unit's aircraft departed Richmond to begin operating from the new temporary station. The squadron now had seven aircraft—four Ansons and the first three DH82 Tiger Moths in RAAF service (A17–1, A17–2 and A17–3)—and of these the Ansons and A17–2 took off that day to make the trip north. In following days, further drafts of personnel travelled by road and rail to complete the move.[111] While a regular RAAF presence had thus been established in Queensland on the eve of the Second World War, the permanent base at Amberley did not become operational for almost a year, with the arrival of station headquarters and No.24 Squadron on 17 June 1940.

Rathmines

Development of a new seaplane station to supersede the facility at Point Cook was also undertaken under the RAAF's expansion program. In his 1928 report, Sir John Salmond had first proposed establishing a Coastal Reconnaissance Flight at Lake Macquarie, south of Newcastle in New South Wales,[112] but it was not until early in 1933, with Williams temporarily out of the country and Goble as CAS, that the question arose of identifying an appropriate site. After CAS made an inspection of Richmond station, a flight of three Seagulls under Squadron Leader Joe Hewitt was despatched to make a reconnaissance. This examination located a promising site on the southern shores, but this was soon realised to be too restricted for a substantial base.[113]

Early in 1936, Flight Lieutenant Leon Lachal received instructions from RAAF Headquarters to investigate and furnish recommendations on the best site for a flying boat base in the area. Travelling by car from Point Cook to Toronto, on the western shore of Lake Macquarie, Lachal carried out a comprehensive ground and water survey accompanied by a local resident. After inspecting various bays and inlets, Lachal's first preference was on the eastern shore, but his second choice was a site on a property called 'Rathmines' on the western side of the lake.[114]

After discussion with the Director of Air Force Works and Buildings, Wing Commander Allan Hepburn, Rathmines was judged by Lachal as the most likely site. The question was still far from settled, however, and alternative areas continued to be examined. On 21 December 1937 Flight Lieutenant John Lerew, travelling as a passenger in a No.9 Squadron Seagull V (A2–14) piloted by Flying Officer Hugh Birch, visited Myall River, Tanilba and Lake Macquarie for what was termed 'general inspection of certain proposed Air Force sites'.[115] Hewitt, by now a wing commander, undertook a similar tour in Seagull A2–14 in mid-1938. Initially visiting the Port Stephens area on 2 July, he identified Salamander Bay as meeting some of the requirements for the project, but, moving on to inspect Lake Macquarie four days later, he concluded that the latter place was far more suitable. To confirm this judgement, a photographic reconnaissance of Rathmines and Kilaben Bay was undertaken on 4 October by Flight Lieutenant James Alexander in Seagull A2–20.[116]

By early 1939 Rathmines had been firmly decided upon as the intended site.[117] Finally, in May of that year, the Defence Minister Street formally announced plans to develop Rathmines as a base to house No.9 Squadron and a new No.10 (General Reconnaissance) Squadron about to be raised to operate Short Sunderland flying boats purchased from Britain.[118] Hewitt had already been notified that he was to be the first CO of No.10 when this unit came into existence on 1 July, and was to proceed to England with squadron aircrew later in the year to take delivery of the new aircraft and ferry them back to Australia. In May, however, he had the misfortune of breaking his neck in an accident while riding home from Richmond on his motor cycle, and was replaced. Nonetheless, upon his recovery, Hewitt was appointed in August to command the Rathmines station and assumed responsibility for overseeing progress at the site. In his words: 'Our chief anxiety at Rathmines was to be ready to receive the flying boats and crews of No.10 Squadron on arrival from the United Kingdom'.[119] In the event, delivery of the Sunderlands was suspended by the outbreak of war in September, both the aircraft and their RAAF crews remaining in England with the RAF (see chapters 6 and 15).

To get the new station open, a four-man party: Warrant Officer W. Bartlett (in charge), Sergeant S. Haggarty (fitter), Corporal J. Fuller (clerk), and Sergeant K. Lord (driver) was sent from Richmond by road on 14 August, with Hewitt following in a Seagull aircraft. As the site was still in the hands of its private owners—it was not acquired until mid-September—this party rented a cottage just outside the base area, one of the many used by the lake's week-end population, while they made arrangements to rent further cottages as quarters for an advance party consisting of a barracks officer and 28 airmen which was to follow.[120]

Station headquarters was established in a cottage on a rise facing the lake while station stores were housed in two rented halls—one of them the local dance hall. With the outbreak of war, plans for the permanent station were cancelled and in place of the brick buildings originally proposed, timber-framed tin huts were substituted. The work of constructing what was inevitably dubbed 'Tin City' was got underway. Tin huts were to constitute not just the base's living quarters, but were also used to accommodate the orderly room, messes, stores and a cook house. Weekenders, however, continued in use for the photographic section, guard room, oil and grease storage for the transport section, and (until August 1940) the Officers' Mess.[121]

No.9 Squadron occupied the base in December, at which time sufficient tin huts had been completed to house the majority of the squadron personnel, although a few lived in tents until the rest of 'Tin City' was finished.[122] On 2 March 1940, the Seaplane Training Flight arrived from Point Cook to complete the original pre-war plan, and in subsequent years Rathmines became the main centre of RAAF maritime air operations until closed down in 1952.

Canberra

One place inevitably considered as a site for a RAAF base was the slowly evolving national capital. From before the First World War, when for a time it had been planned to make that place the focus for the newly raised aviation corps, including the Central Flying School, the idea existed that Canberra held special defence significance. In the strategic and financial circumstances of the mid-1920s there was no urgency attached to such a proposal; the capital was, after all, far inland and secure from anything short of a well-supported military

To establish the new seaplane station at Rathmines (NSW) in 1939 a party was sent by road from Melbourne with first equipment for the station. This party is shown here during a stop at Gundagai (NSW). (*RAAF Museum*)

expedition—even aircraft from carriers then entering naval forces of the major powers posed a minimal threat for a place which, at that time, hosted no government institutions of the slightest importance.

Nonetheless, by virtue of its intended national status Canberra saw an intermittent but increasingly frequent RAAF presence, which accounted for RAAF fatalities there in 1926 and 1927 (see chapter 11). From 1924 an airfield, little more than an emergency landing ground, was used on the town's northern outskirts (now covered by the suburb of Dickson), but at the end of 1926 an alternative site at the western end of the present Fairbairn Airport—the same as that considered for the CFS in 1913—was chosen by the Civil Aviation Branch of the Defence Department.[123]

In 1928, by which time Federal Parliament had finally been transferred from its temporary seat of Melbourne, Air Marshal Sir John Salmond included in his report on the RAAF a recommendation for an Army cooperation squadron to be permanently based at Canberra.[124] The Depression prevented this suggestion being immediately implemented, but it was not forgotten. After calls were made in 1933 for the town's commercial airport to be developed, discussions took place between the RAAF and civil authorities to agree on the location of their respective establishments on the aerodrome, and plans were drawn up for construction of a large hangar which was completed during 1936.[125]

In 1937 the Defence Department promised to establish an air station at the national capital within three years,[126] but there was little urgency evident in giving effect to this pledge. One planning document at RAAF Headquarters in March 1939 showed Canberra as the intended location of a station for three Army cooperation units, but the first and only one of these actually in existence—No.3 Squadron—was still listed as 'temporarily located at Richmond'.[127] In February 1939 the transfer to Canberra of the RAAF's Communications and Survey Flight at Laverton was also approved by CAS (Goble) but no firm timetable was associated with this move.[128] When war finally began in September, a squadron was, in fact, transferred to Canberra barely a week after the declaration. This was No.8 (Reconnaissance) Squadron, raised at Richmond using four Douglas DC-3 airliners commandeered from Australian National Airways, and on arrival in the capital it was quartered solely in a tented camp.[129] A station was not established at Canberra until April 1940, in which year construction of permanent buildings was also begun.[130]

Landing grounds

Although manning only three bases from the mid-1920s, the RAAF was by no means confined to just these. In fact, the RAAF maintained a network of landing grounds for its own needs additional to the facilities constructed by the CAB (which were also available for defence purposes). These grounds were acquired through purchase or lease, although some were simply areas of government land which were conveniently located and others were used on the tenuous basis of obtaining a landowner's permission. In November 1925 there were nineteen such grounds, located at: Corio, Seymour, Benalla, Wangaratta, Ararat and Nhill (Vic); Wagga Wagga and Goulburn (NSW); Cooks Plains and Cook (SA); Naretha and Boulder (WA); and Brunette Downs, Newcastle Waters, Katherine and Darwin (NT).[131]

The problems of using such grounds under permissive arrangements with landowners was, however, graphically illustrated in the case of Wangaratta in 1929. After the owner of the ground, which was conveniently located near the town, withdrew his permission for aircraft to land there the RAAF was obliged to look for another site. This proved no easy matter, and in the meantime, since Air Force continued to use the original ground, the owner erected posts as a hazard to prevent them doing so.

Faced with this situation the RAAF began using another site considered to be the best available, though not entirely suitable for all types of machine. A better ground had been identified near Bowser in the meantime, but although the owner of this land agreed on terms three attempts by the Defence Department to get a formal document signed had been unsuccessful.[132] Air Force machines were accordingly forced to continue using the less suitable alternative ground, and it was here that Moth A7–19 of No.1 Squadron failed to clear trees on take-off on 7 March 1929 and crashed and burned. The passenger, Flying Officer Herbert Warland, was trapped in the wreckage and perished, while the pilot, Flight Lieutenant Thomas Swinbourne (the secretary of the Air Accidents Investigation Committee at the time) suffered disfiguring burns when he went to Warland's aid.[133]

Early in 1931 the RAAF passed all its landing grounds across to the CAB, leaving just the three main bases with the Air Board.[134] In May of that year the Controller of Civil Aviation (Brinsmead) wrote to the Board querying whether the Air Force wanted Katherine 'put in order at the present time', in view of the fact that a civil ground was being prepared at Pine Creek. This place was more central between Darwin and Birdum, he said, and its completion would make unnecessary 'the rather costly maintenance of your ground at Katherine'. Brinsmead also raised the question of three new grounds the RAAF had indicated it wanted established in the following year to meet its requirements. Inspection of grounds at Trida and Bega had been arranged, and he would get back to the Board about these, but a landing ground proposed by municipal authorities at Sale might, he thought, obviate the Air Force's stated requirement for a ground at nearby Bairnsdale. The only problem that he foresaw was the question of money:

> Every effort will be made by this Branch to provide funds towards the establishment of essential grounds along approved aerial highways but for the acquisition and preparation of grounds that do not come within this category it will be necessary for requisite funds to be provided on the RAAF votes.

Once the grounds required primarily for Air Force use had been prepared, however, he assured the Board that the CAB could provide for their upkeep. The Air Force agreed to Katherine being abandoned on the completion of Pine Creek, and on substituting Sale for Bairnsdale.[135] Fifty-five years later the RAAF would set about re-establishing an Air Force facility at Katherine in the form of the Tindal airbase, while the move to Sale at this time presaged the development of a wartime airbase here, too, which was to become a permanent presence after the Second World War.

Bombing and gunnery ranges

In 1935, RAAF Headquarters began to focus on problems being experienced over existing range facilities available to the Air Force at Point Cook and Richmond, and at Liverpool (NSW) where practice and live bombing was also carried out. On 13 May Squadron Leader John McCauley, on the staff of the Director of Training, wrote to the Director of Works and Buildings, Wing Commander Allan Hepburn, suggesting that a central practice range was needed where units would annually go into camp away from their home stations to do advanced weapon and tactical training. The two areas he thought most likely to satisfy geographical and economical considerations were Wakefield (SA) and Deniliquin (NSW). While the latter locality was closer, he pointed out that a large area at Wakefield was already owned by the Defence Department and the longer trip to get there would offer benefits in the form of cross-country and air pilotage training. He therefore recommended an inspection of Wakefield be undertaken, with a view to establishing a practice camp before December that year to ensure its availability 'in time for the new units'.[136]

Hepburn clearly shared McCauley's concern that the problem needed to be addressed. In a letter on 7 June he put the proposition to CAS that 'one cannot be happy' about the existing ranges. The four at Point Cook (air gunnery, towed target, live bombing and practice bombing) had all been modified and were 'an inconvenience if not a danger to Cadets' undergoing training at No.1 FTS; the safety factors on the air gunnery range at Richmond were so small that an accident there was a distinct possibility; and at Liverpool stray bullets from the Army range were reported to be dropping onto the adjacent landing ground used by RAAF planes! He sugggested laying down a policy to remove 'a certain amount of existing uncertainty' and agreed that if the policy favoured a central range for all units there were practical steps that could be taken to establish it.

CAS was receptive to these suggestions, and asked Hepburn and McCauley to look at obtaining a wide and safe area which was unlikely to clash with private interests and where expense could be kept to a minimum. In accordance with this instruction, an attempt was made to enlist the help of civil authorities in the five mainland States, with these being notified that the RAAF was seeking an area of about 13 000 acres (5265 hectares). This process took longer than McCauley had originally anticipated, with a suitable site still not firmly identified by March 1936. By this time the choices for RAAF Headquarters appeared to be between Menindee and Hay (NSW), and Renmark (SA). An aerial survey, carried out between 20 and 26 March in a Gannet aircraft from No.1 AD, piloted by Flight Lieutenant Valston Hancock accompanied by Flight Lieutenant Ivor Lightfoot and a fitter, pointed to Menindie as the area most worth a thorough investigation.

Flight Lieutenant A.R. Tindal, OC of No.1 Armament Training Station at Cressy (Vic), pictured with a Comper Swift aircraft at the station in 1939. (*Sir Wilfred Brookes*)

At this point the process of site selection was complicated by offers from places actually eager to host an Air Force establishment—even a bombing range. In early May the town council of Bourke (NSW) offered facilities for locating the training camp and range there, and this was viewed as a serious possibility. Securing the armament training camp was, however, taken up by others. Among these was Charles Hawker, the Member for Wafefield (SA) in the Federal Parliament, who wrote to Prime Minister Lyons on 9 September 1937, urging that the siting of the facility should be viewed in the context of the government's policy of decentralisation and claiming that his home State was keen to host it. His suggestion was that Renmark be considered, as he understood the RAAF wanted at least two other training centres besides the one sometimes used at Cootamundra (NSW). This was a reference to the camp held at Cootamundra by No.1 Squadron (see chapter 7).

Faced with these conflicting pressures, CAS convened a conference to examine the subject on 4 October 1937. After the conference reviewed the requirements, and the limitations of existing ranges, Williams weighed the desirability of having ranges which were conveniently placed so that armament training was not restricted to a short annual period, and of reducing the associated costs of travel. For these reasons he decided that a temporary armament training camp should be established at Cressy, 66 kilometres west of Geelong, where it could be administered as an outstation of Laverton. Directing that action be taken to lease (with an option to purchase) the areas needed for a landing ground, campsite, and bombing and gunnery targets, CAS instructed the camp to be established under canvas initially, leaving a decision whether it was to become permanent until later. The RAAF's requirement for a moving target for air gunnery purposes would meanwhile be met by buying a standard RAF armoured motor boat, to tow targets in the bombing range off Point Cook.

The facility at Cressy was established in February 1938 when No.1 Squadron sent an advance party to set up tents and hangars, in preparation for a three-week armament practice camp which was claimed to be the first of its kind. From 13 February until 6 March flights from Laverton came for a week in succession, living under canvas while carrying out live gunnery and practice bombing (no high explosive) on temporary ranges at Lake Corangamite. At any one time some 50 officers and airmen were in the camp, which had Flight Lieutenant Thomas Curnow as commandant.[137] Recalls one young officer who attended: 'We mainly practised ground attack—taking off, shooting at targets, and then landing. We just did this all the time, two or three flights a day. The idea was to raise pilots' accuracy in shooting, although it didn't help much shooting at static targets.'[138]

The Cressy camp was made a permanent facility on 24 July 1939 when No.1 Armament Training Station was established under Flight Lieutenant Archibald Tindal, who was later to find a place in history during the Japanese air attack on Darwin in February 1942.

6

Aircraft and equipment

At the time that the armistice ended the First World War in November 1918, Australia was operating a miscellany of British aircraft types at Point Cook. Nearly all were relatively recent acquisitions, having been received since 1916, but in each case they had become outdated by the end of the war and steps were taken to dispose of most of them.[1] Early in 1919 the Defence Department decided to make some of its aircraft holdings available for sale to the public 'in the hope of inaugurating commercial aviation in Australia'.[2] Six planes, along with nine aircraft engines of 35 to 80hp, were offered at auction at Point Cook on 19 March. None of the aircraft attracted any worthwhile or acceptable bids, and in the event a 35hp Anzani engine was the sole item sold. This went for £25 to 13-year-old Wilfred Brookes (son of Herbert Brookes, the managing director of Australian Paper Mills), who planned to instal it in a small two-seater seaplane he was building![3] The six planes were subsequently sold privately to aerial companies starting up in Sydney and Melbourne.[4]

Not all of the CFS's planes were disposed of at this time, as some were carried over to the Australian Air Corps (AAC) and, in at least one case, the RAAF. A BE2E, an FE2B and a Bristol Scout D are known to have been passed across to the AAC in 1920. The BE2E was, in fact, the machine used by Wrigley and Murphy on the first transcontinental flight in 1919. Records indicate that this plane was transferred to the War Memorial in August 1920,[5] but one return from mid-1922 notes a BE2A on the RAAF's strength which is conceivably a mistaken reference to this aircraft (as the two BE2As originally bought for the CFS in 1912 were apparently disposed of in 1918).[6] The FE2B was at Point Cook until November 1920 when it was sold; it appeared on the civil aircraft register in June 1921.[7] Similarly, the Bristol Scout stayed at Point Cook and, indeed, continued to be flown by RAAF pilots at least until January 1923.[8] The Scout's career reportedly ended after it was crashed by Flight Lieutenant 'Kanga' De La Rue.[9]

Aircraft of the Australian Air Corps

In late 1918, with the war in its final stages, the Defence Department placed an order for 20 Avro 504K trainers and 12 Sopwith Pup Scouts to meet Australia's post-war defence requirements. These machines reached Australia during the first half of 1919,[10] and became the primary equipment of the AAC. Shortly after arrival, however, it was discovered that not all the Avros had survived the voyage from England in sound condition. On 27 June, the CGS (General Legge) was advised that an oversight in packing prior to shipment had caused major damage. Damp air had entered the crates causing glues to dissolve and mildew to form, metal parts had sweated and rusted, and timbers had rotted, with the result that six planes were unfit to fly. Legge inspected the problem a month later, ordering new wings to be constructed using materials from the wings of the six worst-affected machines as far as possible.[11]

Though serious, this situation was not ultimately as disastrous as it seemed, as it appears that practically all these damaged aircraft were successfully restored to flying condition. Unfortunately, three Avros were struck off the strength of the AAC in April 1920—one of these as a result of a crash suffered during the First Peace Loan—and another four were written off in October of that year (casualties of the Second Peace Loan), so that of the initial 20 planes of this type received only thirteen were still available to be handed over to the AAF on its formation.[12]

Adding to the problems initially experienced at the time that the Avro and Sopwith planes reached Australia was a controversy caused by press reports that returning members of the AFC were strongly critical of the selection of these types for purchase. According to one Melbourne newspaper:

> Several of those spoken to stated that the Avros had proved useless for practical war work and had been discarded by the War Office within a few weeks after the outbreak of hostilities. The 'Pup' had been superseded in 1916 by the Sopwith 'Camel' which in turn had been improved upon by the makers. The 'Snipe' was the latest type of scouting machine made by Sopwith and its superiority over the earlier machines was undoubted.

Responding to such remarks, the acting Minister for Defence pointed out that the planes the government had bought were wanted for training purposes, and were still used for that role in the RAF.[13]

The acting Minister's reply was certainly sound in the case of the Avro 504, as this type was still the standard primary trainer in British use. Admittedly first produced in 1913 and soon found inadequate for a combat role in the early stages of the war, the lightness of its dual controls made it particularly suitable as a trainer and it had consequently become the backbone of the training effort in the British aerial forces. The choice of the Avro 504 was, therefore, entirely logical for Australia's purposes. The K variant which emerged in mid-1917 had the additional advantage of engine-bearers which permitted the fitting of different power units—either the Monosoupape 100hp, the Le Rhone 110hp, or the Clerget 130hp. These were all rotary engines using pure castor oil for lubrication; as one airman of the 1920s recalls, those working on them were invariably spattered with the oil, the burnt smell of which constantly surrounded the machines.[14]

The Avro 504 was the type principally used in training. The machine shown here is one of the RAAF's two L variants used for seaplane training, pictured on the jetty at Point Cook sometime before August 1921 when the 'A' system of aircraft numbering was adopted. *(RAAF Museum)*

In the case of the Sopwith Pup, the Minister's justification was less persuasive. Like the Avro 504, the Pup had evolved from a design which dated back to before the war (its predecessor, the Sopwith Tabloid, had actually been demonstrated in Melbourne to the Defence Department in early 1914 without attracting any orders), and although highly regarded for its combat record there was little doubt that fighter types produced during the war's later stages had better performance. For all that, the Pup was perfectly adequate for the role of an unarmed fighter trainer intended for it with the AAC. A small single-seater, powered by an 80hp Le Rhone 9-cylinder rotary engine (a remarkable type where the airscrew shaft was attached directly to the crankcase while the crankcase, connecting rods, pistons and cylinders all revolved at high speed around a stationary crankshaft), it was possessed of outstanding flying qualities and was fully aerobatic to within 2500 feet of its ceiling of 17 500 feet. Reputedly devoid of all vices, it was said that 'no pilot flew anything more delightful than the Pup'[15]—some even considered it to be 'the perfect flying machine'.[16]

Imperial gift of 1920

Even as Australia was bringing its purchase of Avros and Sopwiths into service in June 1919, the British government offered to provide any of the Dominions wishing to establish their own air forces with a gift of 100 aircraft from the huge wartime stock which was now surplus to its own requirements.[17] This offer was accepted by Australia, and action put in hand to select and despatch the machines comprising what became known as the 'imperial gift'.

At the time that negotiations regarding the composition of the gift were going on, Lieutenant-Colonel Williams was the Staff Officer for Aviation at AIF Headquarters in London. If Williams is to be believed, the future RAAF was extremely fortunate that he was there at such a crucial juncture, and thus able to intervene when a communication arrived from Melbourne suggesting that Australia wanted the 100 machines offered to comprise some '10 or 12 types, in small numbers, some already out of production':

> If one were forming an Aviation Museum in each of the States this list may have had some merit, but as this was not proposed I was curious to know how it originated. I learned later from Maguire [the aviation adviser to the Naval Board] that it originated with him, with no particular object in view. He was interested principally in seaplanes.[18]

Williams gives as his view that a single type of general purpose machine was best suited to Australian needs, and he therefore sought to obtain Bristol Fighters fitted with Rolls-Royce engines—generally conceded to have been the finest all-round aircraft of the war. Unfortunately, runs his account, he was unsuccessful in this bid. The RAF was fully aware of the merits of this plane and required all of them for its own use, and although Bristol Fighters fitted with different engines were available these were not as good. 'Hence if we were to establish any sort of a balanced force we had to take more than one type'.[19]

Unfortunately for Williams' veracity in this matter, the surviving record suggests that he was less than fair to Maquire in his account. In fact, the choice of aircraft which the Air Ministry was prepared to include in the gift equipment was not initially unspecified in the way that Williams suggests. On 27 June the Australian authorities learnt that the types which were available were limited to Airco DH9 and DH9A, Bristol Fighter with Arab engine, SE5, Sopwith Dolphin and Avro 504; to this list the Sopwith Salamander was later added. It was actually Maguire, in his role as Australian liaison officer at the Air Ministry, who was responsible for advice sent back to Melbourne on 10 July 1919 suggesting that Australia seek 30 SE5A and 35 DH9A to create a squadron each of fighters and bombers, along with 35 Avros for training.[20] The Air Service Committee within the Defence Department, itself mindful that it was logistically desirable to have as few types in service as possible, accepted this advice and decided to cancel previous orders for aircraft resulting from consideration of advice from General Birdwood (see chapter 1). The only aircraft Australia now wanted in addition to the 100 gift planes were four Vickers Vimy long-range night bombers.[21]

On 19 January 1920, Major William Coates (who succeeded Maguire as Australian liaison officer when the latter resigned his RAF commission in October 1919) advised Melbourne that the Air Ministry was now also offering to replace the 41 'presentation aircraft' donated from Australia through public and private subscription during the war with aeroplanes of equal value. The replacements available for this purpose included DH9s, Bristol Fighters with Puma or Arab engines, SE5As and Sopwith Snipes, and Coates said Australia could expect to receive about 36 of whichever type it chose.[22] In response to this second offer, London was notified on 17 February that the newly constituted Air Board—possibly on the advice of Williams who was back in

Melbourne by this time—had decided to seek 24 Bristol Fighters with Puma engines, the balance of 12 to be taken in floatplanes if the Air Ministry had any suitable types available.[23] In the event the Australian high commissioner advised on 11 March that Australia had been allocated 28 Puma Bristols, and this was accepted three weeks later.[24] Subsequently the type accepted for these 28 machines was changed to the DH9, also fitted with the Puma engine.

In effect, the outcome of these negotiations does not appear to have hinged decisively on Williams' input at all, nor was his a sole sage voice in a company of fools. In one related aspect, however, Williams certainly does appear to have made a signal contribution for which—in the absence of any contrary evidence—he is entitled to claim full credit. According to his account, it was he who was responsible for convincing the British authorities that a gift of machines was of little real value on its own. Approaching a friend at the Air Ministry, Wing Commander (later Air Marshal Sir) Leslie Gossage, he argued that:

> I could not see what Australia could do with 128 aircraft unless they were accompanied by spares and other equipment with which to form and operate units . . . I was, of course, asking for equipment more valuable than the aircraft. [Gossage] thought for a moment and then said, 'I'll see what I can do', and in a couple of days he rang to tell me that the Air Ministry would be glad to include in the gift whatever we needed to form units as soon as I was able to advise him of the units proposed.[25]

As a result of Williams' stance, when the 128 gift aircraft began arriving in Australia from the end of March 1921 they were accompanied by 191 aircraft engines; 258 transport vehicles (including Crossley cars and tenders, Leyland lorries, trailers, motor-cycles and sidecars); aircraft, engine and transport spare parts; machine and hand tools; photographic equipment; workshop plant; instruments and testing apparatus; fourteen canvas Bessoneau hangars and other aerodrome equipment sufficient to house three complete squadrons in the field; flying clothing and a range of other materials and items. For about twelve months, shipments of gift equipment arrived in Melbourne every 10 to 14 days, each consisting of 500 to 700 packages. At the end of this period some 19 000 cases had been delivered, many weighing two to three tons. All up, the gift was estimated to consist of about 30 000 kinds of articles, running to hundreds of thousands of individual items, with a total original value of £1 million (about $25 million at current values).[26]

With this mass of equipment, calculated as sufficient to last for six months of operations or twelve months at peacetime usage rates, were included armaments and radio gear. Amongst the former was gunnery equipment such as Vickers and Lewis .303 machine-guns, and bomb and drift sights, along with a full range of bombs and other aerial ammunition. Bomb-types available were five-ounce BIB (Baby incendiary), 20-pound HE (high explosive) Cooper bombs' and short, pear-shaped 112-pound HE bombs.[27]

Vickers guns were the only type of fixed armament fitted to the fuselages of SE5As, DH9s and DH9As. These were rigged to fire through the plane of rotation of the airscrew, with firing being controlled mechanically by

Constantinesco synchronising gear, named after its Romanian inventor. According to the account of a RAAF armament officer of the period:

> The CC gear, as it was commonly known, required very careful maintenance ... to ensure reasonable functioning, and had a most disconcerting disadvantage. The gun could be fired with safety only between fixed minimum and maximum revolutions (varying with the type of aircraft) ... As the engine revolutions increased in [a] dive, too much concentration by the pilot on the gunsight, and too little on the rev counter, caused the bullets to sever the airscrew blades when maximum revs were exceeded.

The Lewis gun was the only type of free gun in use, and was mounted either singly or in pairs on a Scarff ring which allowed fire to be directed through 360 degrees. Here, too, there were traps for the unwary: 'Omission to fit a deflector bag to catch the empty cartridge cases allowed a stream of hot cases to rain on the back of the pilot's neck when the gun was fired over the port side of the aircraft'![28]

The radio equipment received with the imperial gift included spark transmitters, continuous wave transmitters, crystal receivers and accessories.[29] The spark transmitters were used by aircraft for wireless telegraphy, while crystal receivers or Model TF valve receivers were used for ground stations. Transmitting equipment was notoriously unreliable, and, as the historian of the RAAF radio and radar organisation has recorded, pilots 'were not enthusiastic about its use'—one allegedly venting his frustration at the non-performance of a spark transmitter by jettisoning the set over the side of his aircraft! Despite the uncertain communications which was all that was possible with this wartime vintage gear, the gift equipment remained in use until 1928. By this stage much of the RAAF's radio equipment was being designed by and built under the direction of Flight Lieutenant Loris Balderson, who became the force's Signals Officer in 1923.[30]

Initial RAAF aircraft holdings

With delivery of the gift aircraft, Australia now had a large stock of aircraft with which to bring its air force into being. Indeed, some accounts point out that, when first formed, the infant AAF was in the extraordinary position of having more aircraft than officers and men on its personnel strength.[31] These accounts, however, do not agree on the size of the alleged surplus—with estimates of the number of aircraft on the strength of the Air Force on 31 March 1921 varying between 157 and 170.

In fact, the actual number of aircraft held by the AAF at its formation was no more than 153. This included only 127 gift aircraft, as one of the DH9As had already been lost in the disappearance of Captain Stutt and Sergeant Dalzell in September 1920 (see chapter 1). Added to it were 24 machines remaining of the 32 purchased by the Commonwealth in 1919, consisting of 13 Avros and only 11 Sopwith Pup Scouts (one having been struck off strength in April 1920). To these were added the Bristol Scout, survivor of the wartime CFS equipment, and the Vickers Vimy three-seat heavy bomber in which the Smith brothers carried out their epic England-to-Australia flight. This last

This photograph of a Westland Wapiti at No.3 Squadron, taken in 1932, provides some idea of aircraft armament of the period. This machine is fitted with 8½-pound bombs on its wing carrier. Also visible is the Vickers machine-gun fixed to the side of the engine cowling, and the Scarff ring around the rear cockpit on which a Lewis machine-gun could be fitted. Slung beneath the aircraft is the hook used in message picking-up (see next chapter). (*RAAF Historical*)

machine was presented to the Commonwealth in 1920 and passed across to the AAC, probably with the intention that it would be the nucleus of the four-plane Vimy squadron planned in 1919. Not included, although apparently still at Point Cook, was the BE2E (referred to as a BE2A)—by then regarded as obsolete and probably unflyable.[32] Also not included are six Fairey IIIDs, three-seat float seaplanes which the RAN ordered from England in 1920 for use with its warships. It was not until November 1921—more than seven months after the inauguration of the Air Force—that the six RAN planes arrived in Melbourne, at which time they were passed across to the Air Force.[33]

In any event, the notion of an extravagant surplus of machines over men was, in practical terms, a myth. The fact was that the bulk of the imperial gift aircraft were maintained in storage in the large wooden packing cases in which they had been shipped, and initially only a small number of each type was uncrated and assembled for use. These emerged from the cases in RAF camouflage of brown and green. Most kept these colours, although as machines underwent reconditioning they were coated with aluminium dope. The packing cases, resembling modern shipping containers, were not wasted. In July 1921 the Air Board approved of these being made available to personnel posted overseas on duty for storing furniture and personal effects.[34] Some were put to other purposes at Point Cook: several served as flight offices and stood beside the hangars on the aerodrome,[35] while others became garages for the married quarters.

One packing case was placed beside the road behind the orderly room at the gate in 1922 and housed the radio station. Recalls Charles Gates, a member of the radio section at this time:

> The case was sub-divided with an office at one end, and a receiving and operating room at the other. We secured a magnificent oregon pole—I suppose it must have been 50 feet [15 metres] high at least—and mounted this beside the hut to carry our aerials. We opened a radio channel between that old hut of ours and the navy office station at Victoria Barracks, and thereafter passed communications between Point Cook and Headquarters by radio.

By about 1925 the long distance capability of high frequency transmission was realised, and steps were taken in the RAAF for its utilisation. To quote Gates again:

> From then on we had a comfortable little hobby shop developing equipment of our own design. We started communicating with the new station at Richmond and shortly afterwards we set up another aeroplane case alongside the Point Cook–Laverton road just outside the station entrance.

It was from here that in 1926 communications were opened with the Signals Branch of the Air Ministry in London, and later with the other Air Force bases and detachments at Laverton, Bowen and Randwick also. These stations were formed into a RAAF radio network in 1928.[36]

RAAF records show that in September 1921 just two DH9As were serviceable, and only thirteen Avro 504s; a further fourteen Avros had been assembled but were then unserviceable. In July 1922 the Minister for Defence (Massy

A Fairey IIID A10–3, still in naval livery and bearing the naval designator 'ANA–3', takes to the water of Port Phillip Bay from the Point Cook jetty in 1922. This machine was the one used by Goble and McIntyre in their 1924 round-Australia flight (chapter 13). The buildings under construction in the background include the School of Aeronautics (*centre*) and Motor Garage (*left*). (*Cdre J.D. Goble*)

Greene) gave details to a questioner in Parliament which indicated that 101 of the gift machines were still being stored at that time.[37] The Minister's remarks also showed that 52 machines of all types had been assembled by then, though other sources show that the number actually serviceable on 1 July was only 25.[38] This situation was not markedly different two years later. On 30 June 1924, 90 of the gift planes were still in their storage crates; another 63 aircraft had been assembled for use, but of these only 17 were presently serviceable. According to records, this situation was slightly abnormal, however, in that the number of machines maintained for immediate use averaged 23.[39] Based on this evidence, it would seem that the RAAF normally operated less than 30 aircraft on a day-to-day basis in the first years of its existence. Not until the first squadrons were finally raised in 1925 were large numbers of the gift machines put into service.[40]

In August 1921 the Air Board adopted proposals from Squadron Leader Percy McBain, the Director of Equipment, for the introduction of a distinctive national system of aircraft numbering known as the 'A' series, employing the prefix A (for Australia) and a number denoting a particular aircraft type, followed by a second number signifying each separate machine of the type.[41] Under this system, the RAAF's DH9As—the number of which had been restored to 30 by the Air Board's decision to purchase a replacement for the one lost in 1920[42]—were identified as A1–1 to A1–30, these numbers being displayed on both sides of the fuselage of each aircraft. The designator A2 indicated the SE5A, A3 the Avro 504K, A4 the Sopwith Pup, and A6 the DH9. The Fairey IIIDs, which had arrived in Australia bearing the serial numbers ANA (for Australian Naval Aircraft) 1–6, became A10–1 to A10–6. Because the Smith brothers' Vickers Vimy was still expected to enter RAAF service this machine was actually numbered A5–1, but in the event it was transferred to the War Memorial instead.[43] The missing numbers in this sequence were apparently reserved for future acquisitions, but A11 was also allocated, to Felixstowe or Short F5 flying boats which it was planned to introduce but which became a casualty of financial stringencies imposed in 1922 (the number was reallocated to a later type).

In the case of the Avro 504 the number of machines initially allocated ran to only 47, owing to the fact that one of the Australian-purchased machines crashed at Point Cook only a week after the AAF's formation (see chapter 11) and was written off before the 'A' system was implemented. The purchase during 1922–23 of a further six aircraft of this type from a Sydney manufacturer (see chapter 9) brought the final number of planes bearing the A3 designator to 53.

Similarly with the SE5A, the number of aircraft with the A5 designator was increased by later acquisition. In this instance a single machine (A2–36) was purchased from the Aircraft Disposals Company in England in 1925.[44] The purpose of this addition was to obtain a two-seat training version with dual control, for converting pilots onto the single-seat Scouts. Some doubt evidently exists as to whether the aircraft which became A2–36 was actually the imported machine, or whether this latter aircraft simply replaced one of the RAAF's existing SE5As (possibly A2–2) which was modified locally.[45]

One other interesting addition to aircraft numbers was proposed in August 1925. This concerned the offer of Flying Officer Horace Miller, who had recently joined the RAAF as a technical officer in the aircraft repair section at Point Cook, to sell to the Air Force a DH9 bomber which he had recently bought from the Aircraft Disposals Company and privately imported. As Miller noted, the circumstances of a new recruit turning up in his own aircraft had caused considerable comment,[46] but, not put off at all by the unusual idea of the RAAF purchasing a warplane from one of its members, CAS (Williams) had recommended that the offer be accepted and the Air Board concurred at its meeting of 14 September. The Minister for Defence (Howse), however, decided otherwise.[47]

While Australia was well-supplied with spares for the aircraft which were part of the imperial gift, it was not so fortunate with the machines purchased for itself beforehand—except for the Avros, more of which had come with the gift. In the case of the Sopwith Scouts, which—like most wartime aircraft—had been mass produced to replace combat losses and were not built to last, severe difficulties were soon being experienced in maintaining them in flying condition. Just a year after the RAAF was inaugurated, the shortage of spares had become so acute with the remaining 11 Pups that the Air Board approved the cannibalisation of three machines (A4–5, A4–7, A4–11) to enable the other eight to be maintained.[48] This number was further reduced in August that year when Flight Lieutenant De La Rue crashed A4–8 into a Bessoneau hangar at Point Cook.[49] As one witness of the crash recalls, the pilot, renowned for his fiery Gallic temper, stepped out of the wreckage clouting and abusing the aeroplane.[50] The Pup was retired from RAAF service in 1925, by which time there were only six left.[51]

Although not beset with the same problem of insufficient spares for maintenance purposes, the RAAF's other aircraft types all suffered from deterioration due to age and were kept flying only through constant refurbishment. Air Marshal Sir George Jones recalls that when he became OC of the workshops at Point Cook in 1922, the main work of the aircraft and engine repair sections involved reconditioning the gift aircraft, some of which 'required almost complete rebuilding'.[52] The scale of the task was beyond the RAAF's resources alone, however, and from 1924 commercial companies carried out much of this work.[53] In all, 64 aircraft were reconditioned under contract—including four which underwent the process twice.[54]

By 1927, even reconditioning could not arrest the deterioration which was becoming evident in some of the aircraft. The problem was especially serious with the SE5A, which was the RAAF's sole fighter type even before the Sopwith Pup was withdrawn from service as a fighter-trainer. The SE5A had given good service and was a popular little aircraft. Its 200hp Wolseley Viper engine—the English version of the French Hispano-Suiza design built under licence—was an 8-cylinder water cooled type, with two monoblocs of four cylinders set at an angle of 90 degrees. This gave good reliability, though sometimes marred by stoppages in the fuel lines due to a system of flexible hose connections which was not impervious to petrol. Nonetheless, like the Sopwith Pup, the SE5A had not been made for an extended life, as was discovered after A2–25 crashed at Point Cook and some of its parts were

The scene in the Point Cook workshops during the early 1920s. Mechanics at left are working on what appears to be a DH9A, and a DH9, while a Fairey IIID is receiving attention in the background. (*RAAF Museum*)

subjected to testing.[55] In all likelihood this was the machine flown by Flight Lieutenant John Summers, which broke up after a vigorous display of aerobatics. A witness to this incident recalls that:

> on making a perfectly good normal landing, the fuselage began to disintegrate. Breaking apart midway between cockpit and empannage, with longerons broken and empannage dragged along by internal bracing wires, the luckless SE5A soon skidded to a halt. One wondered how tragic it might have been had the break-up occurred in the air, there being no parachutes worn in those days. A technical inspection of the aircraft showed that the moisture content of the spruce longerons had become so low as to render them brittle.[56]

On 3 August 1927 Squadron Leader Frederick Sandford, the RAAF's assistant director of technical services, telephoned the Munitions Supply Laboratories (MSL) at Maribyrnong with some disturbing news. An unused SE5A had been recently uncased from storage when, he said, evidence was found that this machine's timber components were in a possibly unserviceable condition. The machine allegedly concerned was A2–25, although—as noted above—this particular aircraft was actually the one crashed at Point Cook and was not freshly unpacked at all. Perhaps the RAAF was loath to admit that it had been (and likely still was) flying planes which were in an unairworthy state! The Air Board, Sandford said, had approved of the aircraft being sent to MSL for urgent examination and testing.[57]

In December 1927, the RAAF provided specimens to MSL of a further three SE5As (A2–9, A2–30, A2–33) about which it said 'there is some doubt as to the quality of the timber', along with specimens of wood from the airframes of three DH9s (A6–11, A6–12, A6–19) which were undergoing reconditioning for a second time.[58] While surviving records do not disclose the findings on the SE5As, it is known that on 14 January 1928 Sandford advised Wing Commander Malcolm Spicer, the RAF officer who had recently arrived as the

RAAF's Director of Supply and Research, that those for the DH9s had confirmed 'all the . . . timber is well below the required moisture content'; four of ten samples had failed a brittleness test while another had produced a doubtful result.[59]

When these findings were considered within RAAF Headquarters, it was suggested by Spicer that there was not yet cause for alarm over the condition of the DH9s. While the RAAF's aircraft were, he said, very near the critical point of strength, it had to be borne in mind that the specification for this type allowed for the strains imposed on machines when fully loaded with war equipment and required to perform evolutions in evading hostile planes:

> As these conditions do not appear likely to be reproduced in the near future, I think that it is only necessary to insist on a high standard of workmanship during reconditioning and there is no further need for anxiety . . . [It] is not possible to rebuild all existing equipment and I consider that the machines are adequate for their probable length of life.

While accepting this conclusion, Sandford asked whether—given the poor state of the timber in some machines—an order should not be issued prohibiting the performing of stunts in DH9 and DH9A aircraft. Sandford's superior as Director of Technical Services (Squadron Leader Grahame Dawson, another recent arrival from the RAF) noted that stunting of DH9As was not allowed in England, and agreed that some such restriction ought to apply in Australia. Accordingly a notice went out to COs at Point Cook, Richmond and Laverton on 22 February 1928 that both the DH9 and DH9A 'will not be spun'.[60]

With handling of aircraft having to be limited in this way, the days of much of the gift equipment were now obviously numbered. For the SE5A, however, there was not even a reprieve such as allowed the DH9s and 9As. On 20 March 1928 approval was given for all remaining machines of this type to be 'reduced to produce', including fourteen airframes which had never been assembled.[61] Next to go was the Avro 504K which was retired in June 1928. Although not part of the gift equipment, the Fairey IIID was also taken out of service around this time with the last two of these planes being approved for destruction in May 1929.

The DH9s were retired from service during 1929, by which time there were just three of them still serviceable. Quite apart from the problem of brittle timber in the final years, these planes had suffered from a troublesome engine. This was the 240hp Siddeley Puma, a modified version of the Beardmore-Halford-Pullinger (BHP) engine powering the DH9 when the type was first put into service during the war. The Puma was an orthodox 6-cylinder, water cooled, in-line type with two blocks of three cylinders each, described as having 'the appearance of an overgrown truck engine'. Claimed to have no serious vices and to be reliable in the air, this engine nonetheless gave considerable trouble with water leaks—so much so that one frustrated mechanic allegedly applied for a remuster from 'fitter aero engine' to 'plumber Puma'.[62]

The DH9As were steadily phased out of service during 1929–30, with ministerial approval for the write-off of the last eight still in use at Point Cook being given on 4 February 1930. The comparative longevity of this type was probably due to the fact that these mounted more reliable and powerful

American-manufactured 'Liberty' 400hp engines. A V-type, this had 12 individual cylinders with steel water jackets in two banks, set at the unusual angle of 45 degrees to lessen frontal area. Because this caused the unusual firing sequence of 45 or 75 degrees between any two consecutive cylinders in the firing order, it was equipped with battery instead of magneto ignition—the only such engine ever to be used by the RAAF.[63]

The departure from service of the bulk of the imperial gift provided one last piece of controversy. On 25 March 1929 CAS (Williams) recommended that quantities of the gift material—aircraft, engines and spare parts—should be sold off through the government's Contracts Board. This suggestion was strenuously opposed by Captain Jones, the Deputy Controller of Civilian Aviation, who was appalled by the prospect of worn and now dangerous equipment being sold to the general public:

> I can think of nothing that would hinder the development of civil aviation in Australia during the next few years to a greater extent than a decision to carry on with the proposed sale ... Even though a percentage of the material might be usefully employed it appears inevitable that a great deal of it would drift into the hands of inexperienced enthusiasts whose subsequent activities would be a constant source of worry to the Department.

He did not oppose the sale of three or four recently reconditioned DH9s to flying clubs, along with 14 Puma engines, 12 Rolls-Royce, 33 Viper engines, propellers and spare parts. He was, however, most adamant that five complete Avro 504s and another seven Avro fuselages, five SE5A airframes, six Clerget and Le Rhone engines, along with the airframes of Fairey IIIDs A10–4 and A10–5, should all be destroyed to prevent them falling into the wrong hands.[64]

Jones' view was strongly supported by Professor Henry Payne of Melbourne university, who wrote to the Minister for Defence (Glasgow) in his capacity as chairman of the Air Accidents Investigation Committee. Applauding the decision to put the aircraft concerned out of commission as 'an excellent move ... which should not be gone back upon', he nonetheless also deprecated the proposed sale and said:

> May I take the liberty of suggesting that these machines be totally destroyed rather than placed indiscriminately in the hands of the public. I make this suggestion because any accident arising must of necessity reflect against the credit of the Royal Australian Air Force.[65]

The matter was considered by the Defence Committee at a meeting on 21 May, at which it was decided to adopt the course proposed by Jones. An auction was accordingly held at No.1 AD, Laverton, on 10 July 1929, with over 700 lots for sale which realised approximately £2000.[66]

The question might well be asked what sort of Air Force, in terms of capability, was Australia operating during the 1920s. Did the RAAF's wartime aeroplanes, all wire-braced open-cockpit biplanes of wood and fabric construction, have any credibility as an element for national defence? Disregarding the meagre organisation with which the RAAF coped until 1925, it must be said that the combat aircraft which the service operated during its first decade were a less-than-ideal mixture of some of the most and least successful types spawned by the First World War.

The SE5A had been one of the new generation of fighters which restored Allied air superiority after the RFC reached its low ebb in 'Bloody April' of 1917. As such, it was regarded by many Australians who flew this type in Palestine and in France as the greatest fighter of the 1914–18 conflict. No fewer than sixteen members of No.2 Squadron, AFC, became aces flying the SE5A, and the majority of the 185 enemy aircraft claimed by this squadron were accounted for by this type.[67] Leslie Sutherland, a member of No.1 Squadron, AFC, who later flew the SE5A with the RAAF, wrote enthusiastically about its splendid handling capabilities—despite breaking his neck in one which he crashed in 1926. He declared that it 'climbed like a tree-kangaroo, and dived like a gannet'.[68] Adrian Cole, another ex-AFC pilot, expressed his delight the first time he re-entered the cockpit of a gift SE5A in June 1920 by recording in his log book: '*very much O.K.* (like old times again)'.[69] Armed with two .303 machine-guns (one mounted on top of the engine cowling and the other on the top wing), and capable of carrying a bomb-load of 100 pounds, the SE5A had been found to be an 'ideal machine, too, for ground strafing'.[70]

The DH9 and DH9A two-seat bombers dated from late in the war (both types were first flown in 1917), and both had the same armament and load capacity: one machine-gun forward and one aft, and a bomb-load of 460 pounds. They also had similar dimensions, with the DH9A having a larger wingspan (46 feet) than its predecessor (42 feet) and an extra 15 inches in height. There were strong differences in appearance, however, the DH9A having a square look about its engine area while the 9 had a more tapered nose with the cylinder block protruding above the engine cowling. Around 1923, the RAAF also modified the exhaust of the DH9, fitting it with a distinctive chimney-like pipe to carry engine fumes over the top wing and away from the aircraft's crew.

In terms of performance the advantage was with the DH9A, which perhaps determined that when the RAAF formed its first squadrons in 1925, the DH9 was allotted to Army cooperation duties while the 9A fulfilled the bomber role. Although originally designed as a high-speed, long-range daylight bomber for the IAF employed on the British strategic offensive against Germany late in 1918, the DH9 had initially given less than satisfactory performance. Not only was the design discovered to be inferior to the DH4 which it was intended to replace, but on arrival in France the type was plagued with severe serviceability problems with its BHP 230hp engine.[71] It also exhibited a flat gliding angle, which proved a trap for inexperienced pilots and produced a high crash rate.[72] A refit with the Napier Lion engine a few months before the armistice did much to retrieve the aircraft's name,[73] but despite this, the type's poor reputation was established. The RAF declared the DH9 obsolete in September 1921, and not all Dominion governments offered these machines under imperial gift arrangements accepted them.[74]

The only remaining aircraft type operated by the RAAF in its formative years, the Fairey IIID, had been designed as a general purpose machine for use as either a bomber or a spotter, on land or sea, and had first entered service with British forces in 1917. Its Rolls-Royce Eagle VIII engine—a 12-cylinder V-type, water-cooled and with a geared airscrew drive—was the ancestor of the famous Merlin which powered British aircraft of the Second World War.

The RAAF's first fighter was the SE5A. Pilots who had flown this type in wartime combat were enthusiastic about its handling capabilities, which were utilised by the RAAF in aerobatics exhibitions staged at public air displays. Business was evidently slow on the day this photograph was taken, judging from the airmen sleeping on the ground between the first two machines. (*Mr J.C. Miles*)

Although the type gave good performance, there were doubts about its suitability for the local defence environment. It was partly to test the engine and airframe, and the durability of the 3-ply floats and fabric in the climate of Australia's north that Goble and McIntyre made their epic 1924 round-Australia flight (see chapter 13).

The most relevant comparison that might be made, to consider whether the RAAF held any defence value for Australia apart from training, is with the aerial strength of Japan in the mid to late-1920s. The report on the air defence of Australia prepared in February 1920 by the first Air Board (see chapter 1) had, after all, noted Japan as Australia's most likely future adversary. This conclusion was repeated in a memorandum on air defence submitted by Williams on 21 April 1925, in which he specifically noted that Japan was expected to have the capacity to operate some 130 to 150 shipborne aircraft by 1928, of which two-thirds would probably be fighter types.[75]

The interesting thing to emerge from such a comparison is that the performance of Japanese aircraft up to the early 1930s was not markedly better than that of the types in Australia's possession. The reason for this was that Japan was slow to start building up her air forces before the early 1930s, relying almost exclusively on aircraft imported from Europe and America, or built under licence. Mitsubishi, for example, which produced major quantities of aircraft for both the army and navy in the period, held licences for British, US, German, and French designs of aircraft, aero-engines and many other components besides.[76]

Viewed from this perspective, and in the climate of international disarmament which prevailed after the Washington Treaties of 1922, the RAAF was not wholly without merit as a deterrent. At least the possession of even this modest force complicated the task of a potential enemy approaching Australian shores—as was demonstrated when a RAAF aircraft successfully surprised

Based on its wartime performance record, the DH9 was not especially favoured among Dominion governments offered this type under imperial gift arrangements. Nonetheless Australia received 28 of them, which were allotted to army cooperation duties. The discovery early in 1928 that the timber in many machines of this type was unacceptably brittle due to low moisture content led to restrictions being placed on the flying manoeuvres allowed to be performed in them, and they were finally withdrawn from RAAF service during 1929. Visible behind this machine are the canvas Bessoneau hangars which were in constant use for many years at Point Cook. (*Mr J.C. Miles*)

the visiting US Pacific Fleet in July 1925 (see chapter 13). If nothing else, this episode demonstrated that the quality of an air force was never simply a matter of equipment—the performance of aircraft depended, too, on the courage and skill of the men flying them.

First new types acquired

Apart from six locally built Avro 504s received in 1922–23, the RAAF had to make do for much of the 1920s with wartime vintage aircraft despite the steady deterioration and growing obsolescence of these machines. There were no orders for newer types placed until 1925, and when these acquisitions were delivered from their British makers the following year they comprised just nine aircraft: two DH60 Cirrus I Moths, one DH50A, and six Supermarine Seagull IIIs. These types were designated A7, A8 and A9 respectively. The cost of the Seagulls was found from the Navy's vote, so that the RAAF's first purchases of new types actually entailed just three non-combat machines.

Faced with the problem of aircraft replacement on a slender budget, priority was given to obtaining a more modern two-seat elementary trainer than the Avro 504. The two Moths which arrived in March 1926 for evaluation trials were fitted with the 60hp 4-cylinder Cirrus Mark I engine; though regarded as a 'bitza' because it was mainly designed around parts from the old wartime

Renault engine, it was the original light aero-engine of its type.[77] Both Moth machines were unfortunately short-lived—A7–1 crashed in May 1927 and A7–2 was converted to components a year later—but they proved to be just the first of this type which entered RAAF service.

In June 1928 the Avro 504 was officially replaced by deliveries of a further 34 DH60s which were distributed among the RAAF's three locations. These were the DH60X variant, fitted with improved Cirrus II 85hp and later Cirrus III 90hp engines; they were also the first RAAF machines fitted with slotted wings on the top mainplanes, a device which helped prevent stalling.[78] This first batch of aircraft was followed two years later by the Gipsy Moth (DH60G fitted with the Gipsy I 100hp engine), 98 of which were eventually obtained. The Cirrus Moth went out of RAAF service in 1935 but the Gipsy Moth continued until 1946, flying alongside the later development of the type, the DH82 Tiger Moth, which became the mainstay of the RAAF's elementary training effort during the Second World War.

An order for 20 DH82s was placed in November 1938, and these entered RAAF service between May and October 1939 with the designator A17.[79] The performance of the Tiger Moth was not markedly different to its predecessors, but it incorporated several improvements in having staggered and sweptback wings and an inverted engine, which allowed an improved forward view for the crew.[80] In addition to *ab initio* training, Moths were employed in search and rescue work, and were flown in the first public demonstrations of crop-dusting in Australia (see chapter 13). They were also used as amphibians: two Cirrus Moths fitted as floatplanes (A7–13 and A7–26) operated with the seaplane flight at Point Cook as conversion trainers for pilots progressing to the main seaplane types.

Like the first two A7 machines, the sole pre-war aircraft to carry the designator A8 also had only a short span of service with the RAAF. Received in May 1926, the DH50A was damaged in a crash at Cootamundra, NSW, in July 1929 and approved for conversion to spare parts. A four-seat transport with an enclosed passenger cabin, it was equipped with interchangeable wheel and duralumin float undercarriages. Although ostensibly intended 'for the use of the Governor-General',[81] the DH50A was apparently obtained with other uses in mind. Williams hints that he gained the approval of the Minister for Defence for the purchase by arguing its value as a VIP transport but notes that 'this one did not get much use as such'.[82]

A DH50 had been obtained in 1924 for the Controller of Civil Aviation (Brinsmead), who used it to carry out the first landplane flight around Australia while surveying air routes and landing grounds (see chapter 10). Williams had probably appreciated from this the suitability of the type for a particular task he had in mind, involving a difficult long-distance flight. Although powered by the same notorious 240hp Siddeley Puma as the DH9, the low compression of this engine had the attraction of permitting the use of ordinary motor spirit—often the only fuel available in remote regions. Williams made use of A8–1 immediately on its arrival to accomplish his historic journey to the South-West Pacific (see chapter 13), and followed this up by using it again on a 13 000 mile (21 000 kilometre) flight within Australia, undertaken in mid-1927 in company with two DH9s, to survey airfields for defence purposes.

By Williams' own account, the machine seems to have become the personal aeroplane of the CAS for much of its subsequent service: 'Having the DH50, which was not on the establishment of a unit, I used it for personal flying practice and occasionally for a weekend flight to Adelaide to see my parents. Being a cabin aircraft I sometimes took my wife, as well as a mechanic.'[83]

The acquisition of six Seagulls was related to requirements which derived from plans to support the Navy's program of charting the Australian east coast and the Great Barrier Reef. As a floatplane, the Fairey IIID had proved unsuitable for operations away from prepared entry and exit points to the water, so that the RAAF was obliged to obtain a boat amphibian for this work.[84] A separate requirement had also arisen to provide for the training of personnel for the seaplane carrier, HMAS *Albatross*, then under construction for the RAN. The Seagull was in fact the only British aircraft of this type then in service, but, as revealed by a statement by the Defence Minister (Howse) on 1 July 1926, it was anticipated 'that improved types would be available when the time arrives to order aircraft for the seaplane carrier'.[85] In fact, when *Albatross* was launched the Seagull was still the only aircraft the RAAF had to put on board it. Though used by the Royal Navy's Fleet Air Arm as a deck-landing spotter aircraft, throughout its Australian service the Seagull would never operate from a flight-deck but was lowered and retrieved over the ship's side.

In fact, however, the Seagull III hardly represented much improvement in technology over the Fairey. With dimensions almost identical to its predecessor, its speed was practically the same and its load carrying capacity in addition to the three-man crew was only 750 pounds better. But being a wooden-hulled amphibious flying boat, rather than a floatplane, it was able to handle heavier sea-states than the Fairey though its operations were still frequently restricted by bad weather.

Shortly after the six Seagulls on order were delivered in Melbourne in March–May 1926, this type was declared obsolete in the British service. In August the air liaison officer in London advised that a number of machines from the RAF reserves were up for disposal, three of which were previously unused. Purchase at scrap value—just £100 apiece instead of the price of nearly £6000 paid for each of the original planes—was approved the following month, and these three additional machines arrived in Australia during January 1927.[86]

As part of the upgrading of the RAAF's maritime capabilities, approval was given in June 1927 for the purchase of two Supermarine Southampton flying boats. These were the RAAF's first twin-engined aircraft (apart from the Vickers Vimy which never actually entered service), being powered by the same 450hp Napier Lion engine that was fitted singly in the Seagull III; this was a 12-cylinder water cooled engine of the broad arrow type, having three banks of four cylinders each. The Southampton was also the largest aircraft type operated by the RAAF before the Second World War, its wingspan of 75 feet being more than half as large again as the Seagull. In fact, when the two RAAF machines were shipped to Melbourne, arriving as deck cargo in January 1928, great difficulty was experienced in transporting the giant cases in which they were packed by road to Point Cook.[87]

The two machines—popularly known as 'Swamptons'—were trialed and put into service in a coastal reconnaissance flight at Point Cook. Re-numbered at this time too, these machines took over the A number formerly allocated to the Short F5s which had failed to materialise, S1158 and S1159 becoming A11–1 and A11–2 respectively. With a range of 930 miles, the Southamptons gave a capacity for long-range maritime patrol previously unknown in the RAAF, even though the big flying boats were only comparable to their predecessors in speed. The greater weight capacity of the Southamptons (14 600 pounds loaded) also facilitated their long-range capability by making possible the carriage of a crew of five, normally comprising two pilots who alternated during extended patrols, a bomb-aimer, a gunner and a radio operator. The cockpits were still all open and exposed to the elements, but at least the pilots—who sat in tandem behind the forward gunner—were provided with spray screens, and the radio cabin was enclosed within the hull beneath the wings.

The Southampton was a relatively modern type, having entered service with RAF squadrons only in August 1925. However, whereas the British service had progressed to the production of Southampton Mark IIs having duralumin hulls, the Australian machines retained the wooden hulls of the Mark I both for reasons of economy (being £4000 cheaper) and the local service's lack of maintenance experience with metal construction.[88] The two RAAF machines

Two Southampton flying boats reached Melbourne for the RAAF in January 1928. With a wingspan of 75 feet they were the largest aircraft in Australia at the time, and the huge crates in which the machines arrived presented major transportation difficulties to get them to Point Cook for assembly. (*RAAF Museum*)

The RAAF's two Supermarine Southamptons formed a long-range coastal patrol flight. Along with the seaplane trainers operated at No.1 FTS, Point Cook, they formed 'Seaplane Squadron' which remained in existence until absorbed within No.10 (General Reconnaissance) Squadron in July 1939. (*AVM E. Hey*)

remained in use for most of a decade, A11–1 being retired in 1937 and A11–2 in 1939. Used in many roles during the intervening years—including parachute training and instruction—they were a valuable long-distance transport on such tasks as conveying the panel interviewing applicants for the flying course at Point Cook on their travels around the States.[89]

In addition to these naval reconnaissance types, the RAAF received in the same period several experimental seaplanes—and landplanes also—built at the special section at Randwick under Wing Commander Lawrence Wackett. The story of these machines is told in chapter 9.

Replacement of combat types

With the impending retirement of the DH9s and DH9As—the RAAF's last remaining types of combat aircraft—a decision had to be made on a suitable type of replacement. Thinking in mid-1927 initially favoured the Armstrong Whitworth Atlas, of which six were approved for purchase in June (later increased to eight). The next month, however, CAS (Williams) expressed a preference for DH65 Hounds, and by the middle of 1928 the field of contenders had expanded to include the Supermarine Fairey IIIF and the Westland Wapiti. The Hound—being the cheapest of any of these—was the type selected for the RAAF on 3 July, and 32 were approved for purchase. In August CAS reviewed this decision and gained ministerial agreement to amend it to the Wapiti, principally on the grounds that the Hound was unlikely to enter service with the RAF, did not have upper and lower wings which were interchangeable, and did not offer a fixed price as was the case with the Wapiti. He also considered the Wapiti's composite wood and metal structure a suitable type to introduce metal construction into the RAAF and to begin the training of mechanics in the techniques involved in its maintenance.[90] The order finally placed in England was for 28 machines, taking into account the Wapiti's slightly higher price and the fact that only £150 000 had been allocated for the purchase by the government.

The Wapiti was a new two-seat type first flown during 1927. It was a general purpose machine, armed with one Vickers .303 machine-gun firing forward on fixed lines and a Lewis machine-gun in the aft cockpit, and also capable of carrying a 500 pound bomb-load. Destined for service against tribesmen on the frontiers of Britain's empire and to equip bomber squadrons of the Auxiliary Air Force (the RAF's equivalent of the Citizen Air Force), its designers had incorporated, for economy reasons, as many DH9A parts as possible.[91] Although never intended as a first-rate combat type, the Wapiti nonetheless had a very good power unit—the 9-cylinder Bristol Jupiter. This was the first geared radial engine type to be used in any numbers and gave the Wapiti a ceiling of 20 000 feet, a maximum speed of 133 miles per hour, and a range of 360 miles. Significantly, the chief Japanese fighters of the late 1930s were designed around developments of the same Jupiter engine, although with vastly different performance results.

Delivered to Australia between April and November 1929, the new machines were given the designator A5 originally allotted to the Vickers Vimy. The entry into service of the 28 Wapitis meant that operational strength of the RAAF during 1929 was 63 aircraft, with eight Bristol Bulldog fighters, nine Seagulls, two Southamptons and eighteen Moth trainers making up the balance. This was a small force, although probably no smaller than the effective strength of the RAAF in previous years. Until 1935 it was the Wapiti, supplemented by a token fighter element, which was the mainstay of Australia's aerial defence. The first nine Wapiti machines received were Mark I variants, fitted with a 420hp Bristol Jupiter VI engine and with wings and rear fuselage comprising a wooden frame covered by fabric, while later deliveries were Mark IIs, with an all-metal structure and a 460hp Jupiter VI engine. A further purchase of ten Wapitis was made in 1930–31, these being a modified Mark IIA version, featuring a split axle undercarriage and more powerful 550hp Jupiter VIII engine.

In mid-1937 the Air Board learnt that the RAF had six Mark IIA Wapitis available for disposal, and, while noting that it was not normal policy to purchase second-hand aircraft, decided to purchase these.[92] Although the Wapiti was well and truly obsolescent by this stage, the purchase was recommended as an expedient to keep an existing type in service, given Australia's inability at this time to obtain replacement aircraft from overseas suppliers and delays in deliveries of locally produced machines. The Minister for Defence (Thorby) approved the purchase of all six planes on offer, but it appears that delivery was taken in February 1938 of only five; certainly, the numbering of RAAF Wapitis did not extend beyond A5–43. Restoration to 'as new' condition of the second-hand machines, and their Jupiter engines, became the task of No.1 AD.[93] One of these aircraft (A5–42) was the last Wapiti in use by the RAAF, being finally converted to components in July 1944.

The Wapiti proved a very successful acquisition to the Air Force. The fact that its multi-purpose nature made it possible to limit the number of aircraft types needed to cover many operational roles suited the circumstances of a small service like the RAAF, which was not technically sophisticated and was expected to operate in a climate of economy. The Wapiti's reliable and hardy nature, if not particularly spectacular, was well suited to Australian conditions. Many RAAF members regarded the Wapiti as a quite remarkable aeroplane. One who flew it recalls: 'The Wapitis were great old machines. I remember the No.1 stoppage drill on the machine-gun involved leaning out of the cockpit and walloping it with a long-handled hammer.'[94]

Another also remembers nostalgically that: 'You could do incredible things in the Wapiti in the strong northerlies at Point Cook. I remember you could take off, and, without turning, fall back because the wind was so strong, and then land straight ahead again.'[95] In keeping with its description as a general purpose aircraft, the Wapiti fulfilled many roles in the RAAF: survey work, target towing, Army cooperation, communications and training.

At the same time as the RAAF was acquiring the Wapiti, it also received its small fighter element to replace the SE5A which had to be abruptly taken out of service in 1928. In line with the recommendation of Sir John Salmond that the RAAF should 'provide the Force with one or two units up to date in that important branch of Air Force work—air fighting',[96] the Air Board obtained approval in January 1929 for the purchase of six Bristol Bulldogs. After some haggling over price with the manufacturer, the RAAF order was increased by a further two machines the following May, by which stage delivery of the first batch had already been received.[97] These were formed into an un-numbered fighter 'squadron' located at Point Cook.

The Bulldog was a pugnacious-looking single-seat plane designed as a day and night fighter interceptor, built with a fabric-covered all-metal structure and fitted with a 490hp Jupiter VIIF engine. Influencing Australia's choice was the fact that the Bulldog had been selected by the RAF as its standard fighter; for three or four years after 1930 this type comprised about 70 per cent of Britain's fighter force for home defence, and it remained the most widely used RAF fighter until 1936.[98]

Although the Bulldog has become one of the most famous British aircraft, virtually a classic type of the period, it must be acknowledged that it reflected the serious deficiencies of this generation of design. Its armament of twin synchronised Vickers machine-guns, still essentially of rifle calibre, emphasised that thinking about aircraft firepower had advanced little since 1918.[99] In the years following the Great War, the only real progress in design and concept of the RAF's first-line fighters had been in aero-engines and a myriad of minor items relating to safety and reliability.[100] It was also inescapable that the Bulldog had been superseded almost immediately after its introduction in the RAF by superior Hawker Fury and Gloster Gauntlet types.

Regardless of its shortcomings, the Bulldog did keep the RAAF in touch with what were—at the time it was bought—the most up-to-date developments. The type also gave a biting edge to the otherwise second-rate capabilities of the RAAF's aircraft strength, and its dynamic aerobatic performance was quite purposely flaunted at air displays in the inter-war period. By 1935, however, the Bulldog was re-designated as a fighter-bomber type, which at least recognised that the RAAF had no first-line fighter aircraft in its inventory.

Despite the glimmer of modernity which the Bulldog represented during its first years, the fact remained that the RAAF was still far from being a technically advanced air force in terms of its equipment. Money was, predictably, the main determinant in this situation, as CAS was bluntly told in May 1930 when he attempted to do something about the deplorable state of the RAAF's naval cooperation element. Complaining that the RAAF detachment on board HMAS *Albatross* was 'operating with obsolete equipment of very poor performance which is fast wearing out . . . There are no reserves and the machine from this unit recently crashed cannot be replaced at present', Williams drew attention to an offer from the British company, Supermarine Aviation, to supply two metal boat-amphibians fitted with Armstrong-Siddeley Panther engines at a total cost of £22 916. He was, however, informed by the Minister (A.E. Green) that: 'In view of the financial position, this expenditure cannot be approved'.[101] It was the same story when, in December 1932, the Air Board sought a modern motor boat for crash rescue work at No.1 FTS. The existing Brook and Kelvin boats were now so unreliable and slow that a replacement was urgently required, but again the money—only £525 in this case—was refused.[102]

Major expansion

By 1933 the RAAF was in desperate need of new aircraft. As one technical specialist at No.1 AD later remarked:

> The Wapitis and Bulldogs had been totally disassembled and rebuilt so often that the only original evidence was the basic structural material, ie: square aluminium alloy tubing and jointing plates in the case of the Wapiti and rolled HT [high-tensile] steel sections which were the basis of the Bulldog design. The same applied to their Jupiter engines.[103]

This state of affairs had already been recognised by AMS (Cole), who in November prepared a submission to the Air Board pointing out that the RAAF's reserve of Wapitis had now been reduced to the stage where it was impossible to keep the number of aircraft in units up to their proper levels. He stated that enquiries had established that the aircraft best suited as a replacement was the Hawker Demon, as this would meet the RAAF's requirements in the bombing, Army cooperation and fighting roles. He therefore recommended that an order be placed immediately for eighteen machines of this type, with spare engines and all other necessary maintenance equipment.

The purchase of Demons was approved by the Minister for Defence (Pearce) on 15 November 1933.[104] Approval for the purchase of 24 Supermarine Seagull V amphibians was also given in January 1934, although the decision was not made public until May.[105] With the arrival of the first batches of these new types in 1935, the system of aircraft numbering was revised and what is known as the 'second A series' commenced.

The Demon, which took the designator A1, was a two-seat fabric-covered biplane with an all-metal structure. The first machines of this type arrived in Australia in March 1935, but even before these machines were received the government had decided to purchase more; in January 1935, the Minister for Defence (Parkhill) announced the placing of another 'substantial order'.[106] The quantity involved, which the Minister had refused to publicly specify, turned out to be 36 aircraft, with associated spare parts, tools and materials. All 54 of these planes were Mark I Demons, although they differed from those entering RAF service by having an improved power unit, the 600hp Rolls-Royce Kestrel V 12-cylinder V-type water cooled engine instead of the 584hp Kestrel VDR. The Kestrel V, the forerunner of the Merlin, was the first supercharged engine to enter RAAF service,[107] and with it RAAF Demons could attain a maximum speed of 182 miles per hour at 16 000 feet.

In 1936 a further ten Demons were ordered, these being a Mark II variant incorporating dual controls. This purchase was reportedly made necessary by the number of accidents experienced through the lack of operational trainers,[108] a deficiency eventually addressed in 1938 with the purchase of the RAAF's first Link Trainer—a ground-operated instrument flight simulator named for its American inventor, Edwin C. Link. It was deliveries of Demons, finally totalling 64, which gave the RAAF's aircraft strength such a dramatic boost in this period, lifting it from 86 at the start of 1936 to 141 a year later.

With the arrival of the Demon, the Wapiti was relegated to a training role although in reality the Demon was not that much of an improvement in type. While the rakish lines of the Demon gave it beauty, it suffered many of the same deficiencies as other British classic designs of this period. The Demon's armament was still just two .303 Vickers machine-guns firing forward, with another Lewis gun operated from the rear cockpit, and 'light bombs' carried under the wing. Actually a fighter variant of the Hawker Hart light day bomber, it had first seen RAF squadron service on an experimental basis in March 1931, before entering large-scale production in 1933. It was, therefore, in view of the rapid strides being made in this period, already an aging design even as it joined the RAAF. The prototype of the Hawker Hurricane took to the air in November 1935, followed by the prototype of the Supermarine Spitfire less than six months later; by mid-1936 the RAF had on order over 900 of these two types of eight-gun monoplane fighters which would take it through the first years of the Second World War.

More important for Australia was the fact that Japan's first low-wing monoplane fighters of stressed-skin construction—the Mitsubishi A5M and the Nakajima Ki-27—were also then entering service with that country's forces. The prototype of the A5M, the world's first carrier-based aircraft of this type, had already flown by the time the RAAF began receiving its Demons in 1935. Although still sporting a spatted undercarriage, an open cockpit and a pair of 7.7mm machine-guns just like a Sopwith Camel of 1918, the Bristol Jupiter powerplant of the A5M (or 'Claude' as it was dubbed in the West) nonetheless made it one of the world's most nimble aircraft—able to reach 279 miles per

The first flight of Hawker Demons (A1–5, A1–6, A1–7) on their way to No.3 Squadron at Richmond in 1935; the centre machine (A1–5) was flown by Squadron Leader (later AVM) A.L. Walters. Despite its rakish look and fearsome noise, the Demon was not much of an improvement over the fighter types of 1918. Even as the RAAF took delivery of the Demon, a new generation of designs which included the Spitfire and Hurricane were already in development and would render the Demon obsolete. (*Mrs J. Walters*)

hour and climb to 16 400 feet in less than six minutes. Against a potential opponent equipped with such machines the RAAF suddenly became a second-rate Air Force, years before the Mitsubishi Type OO—the famed A6M 'Zero' fighter—appeared on the Pacific scene.

The Seagull V amphibians which Australia acquired with the Demons had a more remarkable history—being, in fact, the first aircraft designed specifically for the RAAF. According to Williams, this type originated in a specification for a catapult-launched boat amphibian, drawn up in 1926 by himself and the Director of Technical Services, Wing Commander Herbert Harrison, when the RAAF found itself facing the prospect of putting the already obsolete Seagull III onto the new seaplane tender being acquired for the RAN.[109] While they had been unable to interest a British manufacturer at the time, in 1929 the Vickers Company came up with a design to fit the specification and indicated a willingness to build a prototype if Williams considered it suitable. Though Williams lacked any authority on behalf of the Australian government to give a commitment to purchase the aircraft if it was successful, development proceeded and the prototype was flown in June 1933.[110] So successful was the design, however, that the RAF subsequently took it into service too, renaming it the Walrus in August 1935.[111] The type went into large-scale production and saw extensive wartime use with British forces. Australia also bought more during the Second World War.

After the first two machines arrived in 1935, receiving the designator A2, the balance of the initial order for 24 Seagulls were delivered periodically up to July 1937. Because it was not confined to operating from a purpose-built carrier but was capable of being launched from the RAN's heavy cruisers, the type proved extremely valuable both for sea rescue work and reconnaissance, although it was also able to carry 760 pounds of bombs or depth charges under-wing. While modern in the sense that it was the first type operated by the RAAF to have an enclosed cabin for its three-man crew, the Seagull V was under-powered and pilots reported that it reached its claimed ceiling of 18 500 feet with great difficulty.[112]

Other aircraft were entering RAAF service at this time too, such as the Avro Cadet, the DH89 Dragon Rapide, and the Tugan Gannet. None of these were operational machines. The Avro 643 Cadet Mark II was a two-seat intermediate trainer, acquired to help prepare trainee pilots for the transition from elementary Moth machines to more advanced types. Developed from the Avro 626 Prefect in service with the RAF, the Cadet was produced exclusively for the RAAF and allotted the designator A6. Twelve of this type were ordered in May 1935, with the first six arriving in December and the remainder in 1936. Another ten were ordered in August 1936, followed by a further twelve in November 1937,[113] bringing the total of these machines eventually in use with the RAAF to 34. Fitted with a high performance engine for its era, and with a modified fuel system for prolonged periods of inverted flying, the Cadet proved an ideal machine for close formation aerobatics.

The purchase of the Dragon Rapides and Gannets was initially confined to just two machines, each of these two very similar twin-engine types. The Dragon Rapide was a popular civil passenger biplane, while the Gannet was a high-wing six-seat monoplane which had evolved from the Codock transport earlier built by Lawrence Wackett for Kingsford Smith. Both types were allocated to the survey and transport role. The first Dragon Rapide (A3–1) was received in August 1935 with the second machine (A3–2) following sometime later, certainly by May 1936, while the first two Gannets (A14–1 and A14–2, later re-numbered as A14–7 and A14–6 respectively) were delivered from the Tugan Aircraft Company in November 1935 and November 1936. No more Dragon Rapides were obtained, but during 1938 a further three Gannets were delivered by the Commonwealth Aircraft Corporation which had absorbed Tugan.

The decision to acquire the Demon and Seagull V proved to be just the start of an unprecedented purchasing program after the government announced to Parliament on 4 December 1934 a major expansion of the defence forces. Modernisation of the Air Force, combined with the minimum needs of the new general purpose squadrons to be raised under the expansion program, brought a requirement for a new twin-engine type suitable for coastal reconnaissance and bombing. The aircraft settled upon for the purpose was the Avro Anson GR1. The military variant of the twin-engine Avro 652 six-passenger commercial type, the Avro Anson GR1 (reputedly named after its maker, A.V. Roe and Son) had been developed to meet a British requirement for a land-based general reconnaissance machine for use by RAF Coastal Command. It had attracted considerable attention in the British press on entering squadron service from March 1936, as not only was it the first low-wing monoplane of the RAF's expansion program but it was also the first to employ a retractable undercarriage.[114] These two features also made the Anson a novelty in the

RAAF, despite the large number of turns entailed in manually cranking up and lowering the wheels. The introduction of an enclosed dorsal gun turret on the Anson was another first for the RAAF.

Crewed by three men (pilot, observer-navigator, and radio operator-air gunner), the Anson had a fabric-covered metal fuselage but wooden wings and tail. It was still only armed with one .303 machine-gun forward and another in the aft turret, and carried just 360 pounds of bombs or 500 pounds of depth charges. Two Armstrong Siddeley Cheetah IX 350hp engines gave it a maximum speed of 188 miles per hour at 7000 feet, a ceiling of 19 000 feet and an endurance of 790 miles. While impressive perhaps by the standards of the mid-1930s, the Anson, too, was obsolescent when war began.

Forty-eight Ansons were initially approved for the RAAF in November 1935, five of these being dual equipped planes for training purposes.[115] This order could not be filled immediately by the British manufacturers, who were fully committed meeting RAF requirements, and it was November 1936 before the first planes—diverted from an RAF contract—reached Melbourne. It was September 1938 before the last machines of the original 48 were delivered, by which time the RAAF had been obliged to order more. Plans to purchase Bristol Blenheims or Bolingbrokes to complete the equipment of the new squadrons being formed for coastal patrol work were delayed for various reasons, and by May 1938 a further ten Ansons had to be ordered as an expedient.[116] Later that year purchase of the Bristol Beaufort was approved, though none had entered service before the start of the Second World War, and in the event would be built in Australia rather than Britain (see chapter 15).

In the meantime more Ansons were added to the RAAF's strength, so that by the end of 1939 a total of 88 of these aircraft had been delivered. The result was that aircraft bearing the designator A4 had become the most numerous type in the RAAF's inventory on the outbreak of war. One of these machines (A4–41) was actually bought with £10 500 donated to the government by 90-year-old Sydney businessman and philanthropist Sir Thomas Buckland, who was concerned about the country's lack of defence preparations. After formally presenting the aircraft in September 1938 he was taken for a flight in it—rumoured to have been a condition of the gift.[117]

Still more new acquisitions were in hand as the decade of the 1930s drew to a close. In July 1939 the RAAF took delivery of the first Wirraways, a low-wing monoplane which was under large-scale local manufacture (see chapter 9), and by the outbreak of war on 3 September had received a total of seven of these machines. A two-seat trainer of good performance, this aircraft would become the subject of considerable controversy when, armed with twin .303 synchronised machine-guns in the forward fuselage and another single machine-gun in the rear cockpit, it was employed in a fighter role and pitted in combat against vastly superior Japanese types such as the Zero.

At the same time as the Wirraway was entering service, moves were underway to take delivery of nine Short S25 Sunderland Mark I flying-boats. In 1938 the RAAF expansion program included provision for a flying boat squadron of six planes to be based at Rathmines (NSW). The machines to equip this unit (No.10 Squadron) had not then been decided, and the Air Board was still obtaining details on a range of suitable types as options, among them the US

Avro Ansons pictured above Laverton. These aircraft were the first machines operated by the RAAF which had a retractable undercarriage. (*WgCdr D.J. French*)

Sikorsky S43 and Consolidated PBY (later known as the Catalina), and the British Lerwick and Sunderland designs. On the point of departing on exchange with the RAF in February 1939, Williams expressed his preference for the American PBY amphibian. However, when pressed in May for a recommendation as to the flying boat type to be ordered, the new CAS (Goble) declared in favour of the Sunderland. While agreeing that the Consolidated boat looked attractive on paper, Goble pointed out that the design had still to be tested as an amphibian and might require modifications which would delay production. On the other hand the Sunderland, though costing over £60 000, was 'a well proved type of boat' (being derived from the 'C' class Empire flying boat) and offered the RAAF the traditional advantage of operating a standard RAF type.[118] Later that month an order for nine boats was notified to London, but RAAF crews were still making preparations to ferry the first three of these aircraft to Australia when war began, and both men and boats were retained in England to serve alongside the RAF.

On the outbreak of the Second World War, the RAAF had 246 serviceable aircraft in Australia. Of this number, 164 or exactly two-thirds were counted as operational types although they were, in fact, Demons, Ansons, Seagulls and Wirraways; the remainder were training types: Wapitis, Avro Cadets and Moths. This hardly constituted an effective air defence force. Indeed, despite the appearance of further large additions in 1940 of types such as the Fairey Battle and the Lockheed Hudson, when it came to actually defending the Australian mainland from Japanese attack in February 1942 it was soon discovered that the RAAF still had—in the phrase of Sir George Jones—'hardly a feather to fly with'.[119]

This situation was essentially the culmination of the first two decades of the RAAF's existence, a period in which Australia never really developed a credible air defence. From 'living off the fat' of a free gift of surplus war equipment for seven years, the RAAF went to making do with often second-rate aircraft—and mostly in small quantities—in the interests of economy for another seven.

The birth of a new era: A20–3, the first production model of the Australian-built 'Wirraway' version of the NA–33. Although originally acquired purely as an advanced trainer, the type included all the most modern innovations such as retractable undercarriage and hydraulically operated flaps and undercarriage. When faced with difficulties with obtaining aircraft deliveries from traditional British suppliers, the RAAF decided to adopt a 'Wirraway' fighter version. (*AVM E. Hey*)

Once it came to undertaking a major expansion of the nation's aerial strength in the half-decade before 1939, Australia found itself caught in the dual binds of rapid obsolescence of types and an inability to obtain the best types coming off British production lines due to the urgent re-equipment of the RAF. Attempts to address this problem through local production were, regrettably, begun too late to have any significant impact.

7

Organisation and training

Under the organisation originally proposed for the RAAF in 1921 (see chapter 2), it was planned to raise six squadrons: two for corps reconnaissance (Army cooperation) duties, two fulfilling fighter roles, one unit of seaplanes (naval cooperation) and one of flying boats for coastal patrol. The four landplane units—one of each type to be established in Victoria and New South Wales—were all to have contained three flights of four aircraft, giving squadrons of only half the normal wartime strength; the two corps reconnaissance squadrons would each have two flights of DH9s and one of DH9As. The units having naval roles were each to have contained just two flights of three aircraft. As stated earlier, the basis of this organisation was laid at the beginning of 1922 when single flights of all except the flying-boat unit (No.6 Squadron) were raised at Point Cook.

When this provisional structure was abandoned in mid-1922, the RAAF was left with a single conglomerate squadron of six flights which bore no relation to any operational unit. In this organisation, 'A' and 'B' Flights were for training purposes, each being equipped with four Avro 504s and two Sopwith Pups; 'C' Flight contained four SE5As, 'D' Flight had four DH9s, and 'E' Flight initially three but eventually four DH9As; 'F' Flight was constituted with one Avro floatplane (A3–40) and one Fairey IIID (A10–3), with a further two of the latter type to be allocated later.[1]

With authorisation for the first squadrons to be raised in mid-1925, No.1 in Victoria and No.3 in New South Wales, these were organised along the lines planned four years earlier as units of three flights, each of four machines. However, to enable Army cooperation and air defence roles to be fulfilled by a single squadron in each State, instead of two as previously provided for, the new units were formed as 'composite' squadrons. This meant that each had an 'A' Flight of Army cooperation DH9s, a 'B' Flight of SE5A fighters, and a 'C' Flight of DH9A bombers.[2] As no authority was given for a seaplane unit to be formed at this time, the naval air element remained with No.1 FTS. This was reinforced by the arrival of two Southampton flying-boats in 1928 to form a coastal reconnaissance flight of 'Seaplane Squadron', which existed primarily to provide conversion training for pilots to fly seaplanes. This 'squadron' remained separate to No.101 (Fleet Cooperation) Flight which was raised in June 1926 to take part in the RAN survey of the Great Barrier Reef.

When the imperial gift equipment which the RAAF originally operated was replaced by the Wapiti general purpose machine at the end of the 1920s, the composite unit organisation was abandoned. Acquisition of eight Bristol Bulldog fighters in 1929 also saw the formation of 'Fighter Squadron', although this was not regarded as a unit in its own right but was placed at Point Cook where it was administered by No.1 FTS.

The 1930s saw a slow process of specialisation of the two main squadrons, with the Richmond-based unit adopting the title 'No.3 (Army Cooperation) Squadron' in July 1930.[3] No.1 Squadron became 'No.1 Single-Engined Bomber Squadron', then 'No.1 (Fighter Bomber) Squadron' in April 1936, and simply 'No.1 (Bomber) Squadron' in August 1937.[4] In December 1935, the six remaining Bulldogs of 'Fighter Squadron' had been designated as fighter-bomber aircraft and moved to Laverton as a flight of No.1 Squadron.

Following the expansion beginning in 1934, the RAAF's organisation became a mixture of units. By 1939 there were twelve squadrons in existence: one Bomber unit (No.1), one Army Cooperation unit (No.3), one Fleet Co-operation unit (No.9), five General Purpose units (Nos. 12, 21, 22, 23, 25—the last four all CAF), and four General Reconnaissance units (Nos. 2, 6, 10, 14). The basic unit organisation was still 12 aircraft, although the aircraft types varied. Nos. 1 and 3 Squadrons were both equipped with Demons, No.9 with Seagull V amphibians, the general purpose units with both Demons and Ansons, and the general reconnaissance units with Ansons solely (except for No.10 which was preparing to take delivery of Sunderland flying boats). As the British Air Staff noted of this organisation, although the general reconnaissance units were designed for coastal patrol in conjunction with naval forces and the general purpose squadrons to cooperate with Army fixed defences, it was obvious that in the event of Australia facing attack these units would bear 'the brunt of the early fighting as Bomber and Fighter Bombers'.[5]

Air defence roles

As indicated in the foregoing, from its inception the Air Force sought to introduce into its organisation a capability to perform air tasks unrelated to operations in strict support of either the naval or military forces. Such roles were most visible in the separate fighter squadrons proposed, a clear recognition that fighter-to-fighter combat or the interception of attacking bombers entailed a form of operations outside the province of Army or Navy commanders. At the same time, however, the inclusion of tactical bombers within the organisation of the Army cooperation squadrons to be formed indicated recognition of the need for a bombing capability over and beyond that required to support the operations of friendly ground forces. Since the DH9 was the assigned Army cooperation type, the DH9A flight within these units clearly was intended to have more of an offensive capacity. The consideration of preserving a mix of air defence and army support capabilities was uppermost in the concept of the 'composite' squadrons introduced in 1925.

From the first years of its existence, the RAAF had been keen to demonstrate the aerial defence aspect of its role—presumably to emphasise the rationale behind having a separate and independent air service within the nation's defence forces. This was clearly the purpose behind the mock combats staged over Melbourne in 1923 and a frequent feature of air shows around the country (see chapter 2). The fact remains, however, that a significant part of the training of the RAAF's fighter element was taken up with aerobatics practice in preparation for public appearances. With the retirement of the SE5A from the Air Force inventory in 1928, the specialist fighter role was preserved by the introduction of the Bristol Bulldog the following year. As stated elsewhere, there were only eight of these planes so that the combat capability they represented was minimal.

Despite the token nature of Fighter Squadron, within such a small force as the RAAF at that time there was undoubtedly a feeling that this was the elite unit of the Air Force. Records the biographer of Air Chief Marshal Sir Frederick Scherger, one of the best-known fliers of the Bulldog and then a flight commander in the squadron:

> The Bulldog right up until the commencement of hostilities was regarded in Australia as the peak of military aircraft . . . It foreshadowed exhilarating flying and the beginnings of a public-personality awareness which brought him [Scherger] recognition in many circles which later had an influence on his advancement.[6]

A former pilot of the unit reflects this outlook and also recalls the impact of Scherger:

> I was very pleased with my two years in Fighter Squadron. I liked that type of flying, which was more into the realm of air fighting. It usually required two aircraft to go up together and meet up above somewhere and see what we could do to each other. One outstanding personality was Scherger. He was an excellent pilot, and we had some excellent scraps in the air. I can't remember either of us ever coming out on top—I could never get him in my sights. One day we were circling around, turns getting gradually tighter and tighter. We were both on the same outlook, trying to do the same thing, until—at the same time—both aircraft stalled. We both spun out, missing each other by a few feet.[7]

This former member also recalls that aerobatics was probably the unit's main occupation, so that little had really changed from the days of the SE5A.

The incorporation of Fighter Squadron into No.1 Squadron in 1935 removed the specialist fighter from the RAAF's order-of-battle. Thereafter the Bulldogs operated alongside the Hawker Demon, another fighter-bomber type which was used for Army cooperation duties. Planning for the program of Air Force expansion underway from 1934 included provision for the formation of a new fighter squadron, however, despite some strong disapproval from the British Air Staff which produced a paper in 1937 noting that:

> under modern conditions fighters are the type least appropriate for Australian defence. There is no threat of shore-based air attack, no system of warning organization to facilitate interception exists or would appear to be possible, and a single fighter squadron constitutes little deterrent to sea-borne air attack in view of the size of Australia and the length of its coastline. It has been suggested that the

Bristol Bulldogs of 'Fighter Squadron' pictured at No.1 FTS, Point Cook, in August 1934. These highly versatile machines gave some teeth to the RAAF, which otherwise had only Wapitis as front-line aircraft. They soon became outdated, however, and the British Air Staff questioned whether Australia needed a fighter element at all. (*AVM E. Hey*)

> provision of this particular squadron has been determined primarily by the predeliction of Australian pilots for fighter types, and the necessity for stimulating interest in air development by pandering to this preference.[8]

This was, of course, an assessment which disregarded the advice contained in the 1928 Salmond Report regarding the importance of maintaining an element within the force to preserve and develop air fighting skills.

As for bombing, this capability was not initially exercised within the RAAF any more extensively than fighting skills. Mentions of bombing exercises in unit diaries are rare, one of the few being a practice undertaken on 1 June 1926 by two DH9s and two DH9As from 1 Squadron against a submarine hulk off Queenscliff. No direct hits were scored.[9]

A live bombing practice conducted at Point Cook with new Wapiti aircraft on 27 November 1929 was claimed to be the first time in the Air Force's history that the service had up-to-date equipment for training in air gunnery and bombing. On this occasion there were just three serials to the practice. In the first, two planes of No.3 Squadron, piloted by Flying Officer Sylvan Aurousseau (whose middle name of Hippolyte made two officers with this moniker in the one force) and Sergeant A.E. Brown, attacked a stationary land target with machine-gun fire. Following them came two more machines from 3 Squadron flown by Pilot Officers Gordon Rae and Jerrold Fleming, which made low bombing passes on land targets with each dropping forty 20-pound bombs. Finally, three machines of No.1 Squadron led by Flight Lieutenant Loris Balderson made passes from 10 000 feet against rafts moored on the bay about 2 kilometres south of the pier, each dropping three 112-pound HE bombs which sent spray 50 feet into the air.

By the early 1930s considerably more emphasis was being given to developing bombing skills. One evidence of this was the Armament Officers Bombing Trophy (a model of a Wapiti, now held in the RAAF Museum), which was competed for annually between 1932–36. Cessation of the competition did not reflect a reduction in training activity in bombing, so much as other factors. As one of the members of the 1931 Short Armament Officers Course (responsible for instituting the trophy) remarks: 'Like so many trophies, I think that it got forgotten'.[10]

Although practices undoubtedly continued, the skill these actually imparted was probably questionable. One then junior officer commissioned in 1937 complains that even at that time, with the RAAF in the process of expanding to a heightened level of defence preparedness, the standard of training was not realistic:

> We weren't training for war. We were not trained tactically, or mentally equipped for fighting. I fired more bullets in my first dogfight in the desert than I'd ever fired in four years before then. We didn't catch up to the standards of 1918 until 1942—we'd just forgotten about the lessons of the First World War.[11]

While this criticism was undoubtedly true, there were some in authority who recognised the problem and attempted to shake off the peacetime apathy. One of these people was then Squadron Leader Frank Bladin, who assumed command of No.1 Squadron in April 1934. He recounts that:

> When I was posted to command No.1 Squadron at Laverton, I discovered that the unit had not operated under field conditions away from its brick hangars and concrete tarmac since its inception some eight years previous ... Eventually I was able to alter all this, and to take the whole unit 300 miles away to a country airfield. The men lived in camp conditions and liked the novelty. I borrowed portion of a sheep station from a friend so that the pilots could carry out their bombing practice. My point is that nobody had seen the need for those changes.[12]

The place chosen by Bladin to give his unit experience in operating under active service conditions was Cootamundra (NSW). After an advance party was despatched by road on 17 November 1935 to establish the camp at the local showgrounds, the whole squadron moved by road to the new location and arrived on Saturday, 23 November, the same day as the squadron aircraft (two flights of Demons and a flight of Wapitis) flew in.[13] Remaining for two weeks, the unit returned to Laverton on 9 December.

Not just the unit gained from this change in routine. Bladin admits that he got the chance to better appreciate the qualities of some of his personnel, recalling three corporals in particular—Abicair, Rhyder and Relf—whom he dubbed his 'Three Musketeers'. Anticipating the CO's intention to detail working parties to the digging of drains and latrines the Sunday morning after arrival, they promptly volunteered themselves for Church parade in the township. Remarked Bladin: 'These three could smell danger a mile off'.[14] Certainly Abicair was to become one of the great characters of the Air Force, ending his career as a wing commander. Owing his entry to the RAAF to a brush with the law, allegedly involving the discharge of a revolver in a Melbourne cinema, he was even reputed to have been the driver of the notorious 1920s gangster, 'Squizzy' Taylor—a story which seems to have had its origins in the fact that Abicair worked as a mechanic at a large motor vehicle company from which Taylor occasionally hired cars.[15]

During the inter-war years RAAF squadrons almost never practised operating as a unit away from their permanent stations. To break this habit, the CO of No.1 Squadron (Squadron Leader F.M. Bladin) took his entire unit into camp at Cootamundra (NSW) in November 1935. For two weeks the squadron's Wapitis and Demons operated from the local showgrounds. (*J. Ryland papers, MS 5020, NLA*)

Although completely successful, the precedent which Bladin had set for disturbing the comfortable pattern of life on an established air station was not repeated by any of his successors until 1939. In June of that year Squadron Leader Patrick Heffernan took the squadron north to Deniliquin (NSW), where unit members were quartered in local hotels while carrying out navigational exercises for four days before returning to base.[16] Three months later Australia was at war.

Seconded officers

Soon after the RAAF's formation, Williams moved to obtain professionally trained junior officers from the other two services, principally to counter problems arising from the fact that very few of the wartime pilots forming the majority of the service's officer corps had any real grasp of administration (see chapter 12). Such a course had evidently been in his mind even before the Air Force was formally established, as he mentions in his autobiography a visit he paid to the Royal Military College at Duntroon in July 1920. The purpose of this visit was to talk with the commandant about obtaining graduates for the new air service, and he shrewdly took along Major Lawrence Wackett 'as an "exhibit" to the cadets of a graduate who had succeeded in the Flying Corps'.[17] Given the impracticality of subsequently establishing 'an Air Force College for Cadets' which he had at one stage hoped might be a possibility,[18] he was obliged to look to the naval and military colleges as sources of supply.

Under the scheme for which Williams won approval, it was arranged that young officers would be seconded to the Air Force for a period of years (two in the case of those coming from the Navy, three from the Army). Their first year was to be taken up with a flying course, after which they became officers on the general duties list. At the conclusion of their period of secondment, they could either transfer permanently or return to their original service. Either way, the scheme would result in the Air Force gaining a number of experienced officers with naval or military backgrounds and a corresponding increase in the number of officers in the other two services with a knowledge of Air Force work.

Thus, in January 1923 the RAAF's first *ab initio* flying course had among its twelve members five Army lieutenants (F.M. Bladin, U.E. Ewart, D.C. O'Brien-Mortimer, C.H. Simpson, and D.E.L. Wilson), all 1920 or 1921 graduates of Duntroon, and Lieutenant J.E. Hewitt, a 1919 graduate of the RANC at Jervis Bay. Their motives for taking this opportunity offered to them were no doubt varied, but a common reason—at least among the Army men—appears to have been an element of dissatisfaction with the career prospects offered by their original services.[19]

This initial intake was followed by another the next year with a further three Army lieutenants (W.J. Duncan, J.P.J. McCauley, G. Packer). On the 1925 course there were no fewer than five Army officers (Lieutenants V.H. Augenson, A.M. Charlesworth, P.M. Pitt, F.R.W. Scherger, E.W. Woodward) and three from the Navy (Lieutenants G.G. Carter, V.E. Kennedy, D. Ross). For the 1926 flying course at Point Cook, five Army officers (M.P. Allsopp, J.G. Cardale, C.W. Lord, O.W. Rennick, J. Waters) and two Navy officers (F.G. Crowther, G.A. Hall) were taken on secondment. In addition there were two Duntroon-trained officers—C.S. Wiggins and F.C. Ewen—who had already left the Army (in Ewen's case the New Zealand army) and entered the Air Force as flying officers.

A significant change which took place about this time was the taking of officers direct from the graduating class at Duntroon. The first instance of this practice had occurred in December 1924, when Squadron Leader William Anderson, acting AMP, and Squadron Leader Ray Brownell, visited the college two days before graduation to interview candidates for secondment. Two cadets applied, Sergeant Reg Pollard (a later CGS and that year's Sword of Honour winner), and Staff Cadet Fred Scherger (winner of the King's Medal).[20] Scherger was selected and commenced his attachment with the RAAF the next month; the other Army members of the 1925 flying course were all graduates of 1920 or 1923.

During 1925 agreement was reached with the Army that one member of that year's graduating class at Duntroon would go directly to the RAAF, this vacancy to be filled from among those cadets specialising in artillery.[21] Two cadets applied for duty with the RAAF, and it was Cardale who was selected; of the other four Army men seconded in 1926, one was a 1924 graduate and the others were all from the 1922 class. The practice thus begun was continued and expanded in subsequent years. Three members of the 1926 graduating class at RMC (Lieutenants F.J.B. Wight, L.J. Ryan, H.F.G. Banfield) went directly onto the flying course commencing at Point Cook in January 1927, to be joined by Waters—whose training the previous year had been deferred due to illness. There was one officer from the Navy (Lieutenant P. Bailhache) also on this course.

In fact, another important change had occurred during 1926, when four vacancies in the 1927 intake to Duntroon were allotted to candidates for the RAAF and filled by special selection by the Air Board.[22] The effect of this change was highly significant, as it meant that Duntroon now virtually operated as a joint Army-RAAF college. The ethos of the place and the content of the course, apart from the academic subjects on the civil syllabus, naturally remained military in bias. As one 1928 RAAF entrant observed:

> I think the only part of the Duntroon syllabus which would have set the RAAF cadets apart would have been the activities during the annual camp. The RAAF sent Wapiti aircraft from Point Cook and these landed on the gently sloping ground between the sports field on the Molonglo and the senior staff houses . . . We were given introductory flights and had to do such things as sketching imaginary battlefields and terrain around Lake George and Captains Flat.[23]

Nonetheless, the new arrangement meant that the Air Force was now able to nominate its own candidates at the outset of training, instead of having to make a choice in the final stages of a cadet's time at the college from among applicants who had initially fulfilled Army criteria to gain selection.

Trainees on the first post-war flying course at Point Cook check their instruction periods, 21 July 1923: (*left to right*) L. Brain, D.E.L. Wilson, C.H. Simpson, D.P. or A.D. Davidson. (*ACdre U.E. Ewart*)

The reason for this change lay in the dissatisfaction which Williams had felt late in 1925 with the standard being accepted by the Army for candidates seeking entry to Duntroon. He was 'not impressed' with cadets he saw at the college during a recent visit, and considered that the RAAF should 'consider the submission of a proposal for the establishment at Duntroon of what might be called an Air Force Wing'. The principal benefit of such a proposal was that the Air Force could expect to exercise some control over standards of entry to such a formation.[24]

The Director of Personnel and Training at RAAF Headquarters, Squadron Leader Adrian Cole, was in agreement with his chief on this matter. He arranged for Dr Richard Hosking, the civilian science instructor at No.1 FTS, to provide an analysis of the RMC syllabus to determine its precise relevance to Air Force requirements. Hosking was uniquely placed to perform this task, since before becoming the RAAF's sole education officer in January 1924 he had been Professor of Physics at Duntroon from the time the college opened in 1911 until retrenched during economies there in 1922. The conclusion reached by Hosking was that instruction at RMC offered little that was of direct relevance to officers coming to the Air Force.[25]

Following Hosking's analysis Cole asked him to prepare a detailed syllabus for a course specially adapted for Air Force needs, to provide a basis for estimating the educational standard which would have to be demanded at entry and for costing purposes. The syllabus which Hosking produced was arranged to fit in with the RMC course but would be capable of completion in just two years instead of four, as was presently the case at Duntroon. To implement his scheme, he recommended that the civil staff at the college would need to be increased by the appointment of lecturers for an Aeronautical Engineering department, headed by a professor of Aeronautical Science.[26]

Financial considerations remained the impediment to implementing this idea at this time, but evidence favouring its adoption continued to mount. In May 1926 the Army's Inspector-General (Chauvel) reported that Duntroon continued to attract a disappointing number of applicants and went on to remark that:

> I am of the opinion that insufficient publicity was given to the fact that candidates for permanent commissions in the Royal Australian Air Force would be trained at Duntroon. Judging by the number of applications received for training at Point Cook for *short service commissions*, this is a popular service, and the sooner aspiring aviators realize that the only roads to the permanent branches of the Air Force will be through Duntroon or Jervis Bay the better it will be for all three services.[27]

Perhaps resenting the fact that Army sought to trade on the RAAF link with Duntroon for recruiting purposes when the Air Force was itself unhappy with the benefits it derived from this arrangement, in February 1927 the Air Board pointed out to the secretary of the Defence Department that there was a negative aspect to the course Chauvel had argued. Some of the failure to attract RAAF candidates to Duntroon could be attributed to the fact that advertisements for the college indicated that cadets not accepted for the Air Force would be commissioned in the Army. The Air Board declared itself to be 'of the opinion that it will never get a satisfactory number of applicants for cadetships until it has its own Cadet College, or at least, its own course at one of the Colleges, and at which cadets will be trained to be Air Force officers and if found unsuitable will not be allotted to any other Service'.[28]

The Defence Committee duly considered this issue, and not surprisingly reaffirmed the object of ensuring that permanent commissions in the RAAF would be restricted to graduates of the naval and military colleges. It requested, however, that CAS submit a scheme for establishing an Air Wing at Duntroon along the lines of Cranwell. In May 1927 Williams asked Wing Commander Rolf Brown, acting AMP, for his views, and suggested that although the length of the Cranwell course was two years, the course at an Air Wing at Duntroon ought to be of three years duration so that the general education of RAAF cadets should not suffer in comparison with cadets from the Army.[29]

Brown was plainly not keen about the Air Wing proposal, though it would, he said, 'serve until we are in a position to establish our own College'. He was also concerned that establishment of the wing might make it more difficult to break away from Duntroon at a later stage. In any event little more came of the proposal at this time, and a year passed before CAS asked Goble, now back in the AMP's post after an absence overseas, to look at the question again in April 1928.[30]

Goble shared Brown's lack of enthusiasm for the scheme, pointing firstly to the difficulty that would be encountered in trying to modify the four-year course at Duntroon to meet the Air Force's requirements. He then went on to point to the avenue which had existed since 1920 for having Australian cadets trained at Cranwell itself (see chapter 4). The cost of having RAAF nominated cadets trained in England would be, he calculated, roughly half that at the Air Wing. And although these cadets would be required to follow their two years at Cranwell by serving three or four years with the RAF on secondment, meaning that they were not available to the RAAF for a total of five to six years, this was not markedly different from what the RAAF presently got or would get under the proposed arrangement. After four (or even three) years at Duntroon, newly commissioned officers 'would be of no real value for at least another two years' of training with their service. At least any officer trained with the RAF would come to the RAAF better experienced and with ready-made value as an instructor—all at considerably less cost.[31]

Williams did not like this criticism of a scheme which he plainly favoured, and on 19 June 1928 tersely pointed out to Goble that it was, 'of course, largely a matter of Government policy as to whether Air Force officers should be trained in Australia or Great Britain, apart altogether from cost which, however, will have a bearing on the decision'. In any event, CAS had an ace up his sleeve, because at this point the arrival was awaited of Sir John Salmond to make an inspection of the RAAF and Williams had hopes of finding a high-powered advocate for his pet proposal. As he now told Goble: 'I think we might hold this matter until Sir John Salmond arrives'.[32]

As Williams had hoped, the question of officer training was one of the aspects addressed by Salmond in the report he submitted following his visit.

While he considered that 'Ultimately, as the Force develops, it may be necessary to establish a separate cadet college solely for the purpose of training RAAF cadets', Salmond recommended that an Army cooperation squadron be established at Canberra in the meantime, to assist in joint operations training with the Army and specifically to provide flying training for RAAF cadets during their time at Duntroon. He also suggested forming a 'RAAF Wing' within the college itself, comprising annual intakes of six air cadets undertaking a three-year course. While Williams had got the endorsement he wanted, these ideas were opposed by the commandant of RMC and like much else of the Salmond Report were not acted upon.[33]

There was, of course, an inbuilt delay of four years before the first cadets selected by the Air Board under the new arrangements begun in 1927 would graduate from Duntroon and come to the Air Force—the first in 1930. In the meantime the practice continued of selecting final-year cadets and taking them into the RAAF directly from the RMC. Accordingly, immediately after graduating at the end of 1927, Lieutenant A.L. Walters went onto the flying course beginning in mid-1928, and in the same way four graduates of 1928 (A.G. Carr, V.E. Hancock, P.G. Heffernan, M.O. Watson) went onto the next course starting mid-1929.

Interestingly, some of these transfers demonstrated that the Army was not always tagging onto the RAAF's coat-tails as had earlier been suggested, and that opting for the Air Force was sometimes only the second preference of cadets. Walters, for example, later recounted that he decided to try for the RAAF when he missed out on being selected to join the engineers on graduation, which carried with it the opportunity of paid study for a university degree.[34] Hancock, who was the senior cadet of the college (the Battalion Sergeant Major) and winner of the Sword of Honour in his year, similarly recalls that he, too, had set his sights on the engineers and only decided to try for the RAAF when, three months before graduation, he learnt that no graduate would be selected for that corps in his year; 'Rather than specialising in Artillery to which I had been allocated, I expressed a desire to join the RAAF'.[35]

Four graduates of 1929 (J.J. Hutchinson, I.J. Lightfoot, C.E. Raymond, H.G. Scoullar) began flying training in January 1930, joined by a naval officer, Lieutenant G.J.I. Clarke. Only one officer, Lieutenant E.K.H. Klose from the Army, was on the flying course beginning in mid-1930, and—in something of a reversal of the now-established trend—he was actually a 1926 graduate of RMC.

In 1930, with the first batch of RAAF-selected cadets due to pass out from Duntroon at the end of the year, the situation of the college faced drastic change as a result of the Depression and the need for economies. Although not announced until later in the year, plans were afoot to close the Canberra campus and transfer the college to Sydney on a reduced scale—including a cut in the length of the course by eliminating much of its academic content. In light of the Army's intentions, on 20 June 1930 the Director of Training, Squadron Leader William Bostock, proposed that the RAAF cease sending cadets to RMC.

Bostock argued that one of the essentials for training youths for permanent commissions in the Air Force was early commencement of flying training in conjunction with instruction in basic Air Force subjects; it was also a great advantage if they were surrounded by an Air Force environment while undergoing training. One of the chief advantages of training air cadets at Duntroon had always been that 'the situation of the College is such that the above principle could, and would, when funds became available, be adopted', while the college also had the facilities for imparting a thorough grounding in general education. But, with the changes planned for the college by Army, he considered that these advantages would disappear—there was no suitable aerodrome at the new site, and if the basic educational curriculum was deleted 'then the course offers little common ground for the training of Army and Air Force Cadets'. He also believed that ending the existing practice of offering permanent commissions to direct entry officers might operate to the RAAF's advantage, by providing an additional inducement to officers on short service commissions who could now hope to achieve permanency. He therefore recommended that the Air Force stop sending cadets to RMC and in future appoint all general duties branch officers to short service commissions after they had completed their flying training.[36]

The thrust of Bostock's proposals was accepted by the Air Board, which on 23 June made a recommendation for discontinuance to the Minister. The Military Board quite readily accepted the decision, not least it seems because there was little space available for accommodating extra cadets at the college's new location in Sydney's Victoria Barracks. It therefore was anxious that the RAAF also 'take over' the eight air cadets who would still be mid-course when the college's transfer occurred at the end of the year. The Army was, however, not happy to see the association end entirely, and actually urged retention of the policy of secondment and appointment of Staff Corps officers to the RAAF.[37]

The Air Board agreed to remove all its cadets from RMC when the move took place, even though facing budgetary pressures of its own which were not helped by this entirely unexpected additional burden. The cadets were first offered the chance to accept appointments in the Commonwealth Public Service, or, if any preferred, nomination for short service commissions in the RAF in lieu of appointment in Australia, but in the event all elected to come to the RAAF. In December, therefore, apart from the four cadets of the graduating class who were due to join in any event (Lieutenants C.M. Blamey, W.L. Hely, A.G. Pither, H.E. Sharp), the Air Force also withdrew eight members of the three junior classes who were still only partially through the course (H.F. Boss-Walker, C.D. Candy, T.C. Curnow, W.H. Garing, R.A. Holmwood, I.D. McLachlan, A.C. Mills, A.M. Murdoch). This was plainly to the delight of those joining their designated service early, one of whom admits he had not enjoyed his time at Duntroon.[38]

The twelve ex-RMC officers and cadets of 1930 were all on the 1931 flying course. A second course beginning later the same year also had one ex-RMC officer, Lieutenant A.J. Dewar (graduate of 1928), as a member, along with four Navy officers (Sub-Lieutenants B.K. Phelan and A.G.F. Stewart, Midshipmen

In December 1930 the RAAF withdrew all its cadets in training at the Royal Military College, Duntroon, thus ending an association which had operated since 1924. In addition to four graduates who had completed their training and been commissioned as pilot officers—(*seated, from left*): Hely, Pither, Blamey, Sharp—there were eight cadets who also joined the next flying course (*from left*): Garing, Mills, Curnow, Candy, Holmwood, Boss-Walker, Murdoch. Cadet McLachlan is absent from the photo, having been admitted to hospital for knee surgery. (*Mrs E. Candy*)

G.A. Rattigan and G.E. Strangman). Although a further Navy officer (Lieutenant P.E. Carr) was attached for training the next year, this was effectively the end of the secondment scheme which had been in operation.

'A' flying courses

To develop and maintain the range of special skills needed to operate an efficient air service, the RAAF was faced with a steadily growing training requirement. As the core of ex-AFC and RAF personnel who originally formed the bulk of the Air Force progressively shrank in the post-war years, the RAAF was obliged to both replace and up-date the skills lost through wastage. For many years the requirement remained fairly spasmodic, due mainly to the small size of the force and the low pace of activity demanded by government. As late as 1930 Williams could observe as CAS that courses such as air gunnery, Army and Navy observers, and wireless operators, were only held from time to time as required.[39] Once the Air Force began expanding from 1934, however, this situation changed, with a range of training courses (air observers, signal trades, flight fitters, flight riggers, flight armourers, instrument makers, photographers) now necessarily conducted at No.1 FTS at regular short intervals.[40] In some areas, though, the need was constant from the start, and while technical staff could be recruited direct from civilian tradesmen, in the case of pilots there was no practical alternative—given the low level of such activity in civil aviation—to the Air Force conducting its own flying training courses.

As already indicated, the first course for new pilots which the RAAF held was in 1923. Seconded officers formed the bulk of the students on this and all subsequent courses until 1932, but from the beginning there was a mix of origins in evidence among the trainees. Thus, while six of the fourteen members of the first course were officers borrowed from Army and Navy, there was also one student already holding a RAAF commission (Flying Officer Thomas Swinbourne), and two airmen (LAC H.B. Hussey and AC1 M.G. Hawkesford) who were admitted under a scheme which would see them promoted to the rank of sergeant immediately on obtaining their pilot's wings. The five remaining students were cadets nominated by the Controller of Civil Aviation who were expected to take up employment with civil flying companies on completion of their training (see chapter 10).

This, then, became the pattern of what was known from 1925 as the 'A' Flying Course, to distinguish it from the shorter 'B' course instituted in that year for the training of pilots for the Citizen Air Force (see chapter 8). From 1925 the arrangement by which CCA-nominated cadets destined for civil industry were admitted came to an end, and their place was taken in 1926 by sixteen cadets to meet the Air Force's own needs following the formation of the RAAF's first permanent units. On completion of their training, these cadets were offered short service commissions or, in a few cases, permanent commissions. In addition there was still the usual quota of seconded officers, RAAF officers wishing to qualify as pilots, and airmen seeking to become NCO pilots. In some cases those admitted as cadets in later years were actually airmen who were aiming to join the commissioned ranks via the medium of flying training. From 1926, too, the practice began of sending a proportion of graduates from each course across to the RAF on short service commissions (see chapter 4).

The course itself was a mix of actual flying tuition and classroom instruction in a range of relevant subjects: mechanics, rigging, engines, airmanship, armaments, electrical, radio and navigation. In addition there were service subjects covered, such things as discipline, Air Force law, administration, and drill. Much of this instruction was given by serving officers on the staff of the FTS, though there was also a civilian science teacher in the person of Dr Richard Hosking. The length of the course varied over the years, beginning at eleven months, and shrinking to eight before returning to twelve months' duration in July 1929.

An important part of the course was cross-country flying, undertaken usually in the final month or so before graduation to confirm the classroom instruction in navigation. For the 1927 course, for example, cross-country exercises took some of the students to Echuca (Vic) and Deniliquin (NSW), others to Junee and Bong Bong (NSW), while still others flew to Richmond via Wangaratta (Vic) and Cootamundra.[41] Finishing the 1927 course was a cross-country exercise which ended for the 22 aircraft involved, described by one account as 'the largest formation seen to that date in Australian skies', at Ballarat: 'On arrival, their occupants were met by a fleet of limousines, were taken to Craig's Hotel, and forthwith, treated to a banquet in honor of the newly-graduated pilots!'[42]

The sights created by large numbers of planes proceeding to points around the countryside excited, naturally enough, considerable public attention. The appearance of eleven aircraft —a mix of DH9s, DH9As and Moths—at the western Victorian town of Nhill in March 1929 was enough to set the local paper talking about an aeroplane 'invasion':

> Eleven more planes will arrive tomorrow at various intervals during the morning and early afternoon and another squadron will also make the trip on Saturday making a total of 27 aeroplanes which is easily the largest number that has hitherto visited the town in such a short space of time.[43]

The 25 students of the 1929 course had been each required to make a similar flight to Deniliquin before the exercise to Nhill, meaning that they had covered 2250 kilometres. This led one Melbourne newspaper to report—not entirely accurately as we have seen—that:

> These performances have never been equalled by pupils in a flying training course. The longest flight undertaken previously was to Bendigo and back, when the machines were flown together in squadron formation. Solo cross-country flying is regarded as a very severe test for trainees.[44]

Notwithstanding that the students (and no doubt staff too) thought they were being subjected to an exacting test, in reality the standard aimed at was scarcely rigorous. As narrated in chapter 4, the deficiencies in the training of pilots passing through Point Cook—particularly in navigational skills—quickly showed up when FTS graduates were measured against their RAF contemporaries. In fact, the comparative standard remained low throughout the Air Force, with only few exceptions. On 25 April 1938 a large mixed formation of aircraft (Demons, Wapitis, Ansons, Moths) was returning south under Wing Commander Frank Bladin after the RAAF air display held at Richmond two days before. Although flown by experienced pilots, not cadets, when this formation encountered bad weather about Seymour (Vic) it was promptly scattered. The circumstances of the crash of one machine, and the forced landing of another, became the focus of special attention by the Air Accidents Investigation Committee which was then inquiring into another string of mishaps during an extended formation flight by No.1 Squadron the previous December (see chapter 11).[45]

In the straitened circumstances of the Air Force during the Depression a number of changes took place to the composition of the 'A' course. As previously described, the presence of seconded Army officers ended in 1932, although Navy still sent junior officers under this arrangement for a few more years. After 1932 the established practice of using the course to train NCO pilots also ceased. The last three airmen to receive such training—L.R. Ambrose, R.A. Sheppard and E.C. Sims—were, in fact, trained in response to a request from QANTAS for pilots who had additional experience as wireless operators. Recalls Williams:

> We had not yet trained the full number we required but I felt that if some of

> them wished to go to civil aviation they would probably find there a longer career than we could offer them in the Service. They would still be available in an emergency and their going would leave an opening in the Service for others for whom we had the facilities for training. I supported the proposal but the Air Member for Personnel did not like it.[46]

In effect, from the course beginning in mid-1932 the only avenue to commissioned rank in the Air Force was by completing the 'A' Flying Course.

This was not to say that the only students trained to fly at Point Cook were the RAAF's own cadets. Williams recalled being advised in December 1937 that the Minister for Defence had approved RAAF facilities at Point Cook being used to teach the Treasurer in the federal government, R.G. Casey, to fly. This unprecedented use of public facilities for the private benefit of an individual became a mild embarrassment when the son of the chairman of the Melbourne *Argus*, who was in a party of schoolboys visiting the base, happened to see Casey getting out of an aircraft (Avro Cadet A6–7) and mentioned this to his father, resulting in some unwanted publicity.[47]

Nonetheless, with the transition of the 'A' course to a system where it consisted solely of cadets under training, the FTS took on a different status within the Air Force. When the practice also began in 1932 of admitting two courses in each year, with one beginning in January and the next in July so that there were always two courses running concurrently, the FTS became to a certain extent the equivalent of the Army and Navy cadet colleges. Though the general educational content of the course was small compared to Duntroon, it was the RAAF's sole officer producing institution and consequently tended to take on the same aura. Recalls Air Vice-Marshal Frank Bladin, who was responsible as a wing commander for cadet training from the end of 1935 until 1938:

> I was quite disgruntled when the Chief of the Air Staff informed me that he intended to send me back to Point Cook. He apparently had in mind to establish, some day, a Cadet College but in the meantime I was to model standards on what I had experienced in my own cadet days at Duntroon Military College . . . I like to think that the organization I had built up was the forerunner of the RAAF Cadet College—in spirit and in its aim—if not in fact.[48]

The presence of so many Duntroon-trained men in the officer ranks of the Air Force had already helped shape the 'A' course before Bladin in 1935, with the introduction of a number of practices and traditions clearly borrowed from Duntroon. Among these were the appointment of cadet under-officers on the course beginning in 1932 (the first Senior Under-Officer being the future Air Vice-Marshal John Hobler), and the institution by 1934 of a Sword of Honour for the top graduate on each course.

One other practice attributable to the influence of ex-Duntroon officers was initiation ceremonies. These were in existence by 1931, with Sir Frederick Scherger recalled as one of those responsible for their introduction and their conduct.[49] In fact, staff members carried out the initiations until the time when concurrent courses provided a 'senior' course to initiate the new intake. The ceremony itself was a predictable blend of humorous pranks frequently verging on the dangerous, such as the shaving of pubic hair (with some nasty

At government direction the Federal Treasurer, R.G. Casey (later Lord Casey and Governor-General of Australia 1965–69), was taught to fly in 1937 using RAAF facilities at Point Cook. It fell to Squadron Leader F.R.W. Scherger (*at left*), later the RAAF's first air chief marshal, to instruct him. (*Mrs J. Gray*)

accidents reported) and being seated blindfolded and naked on blocks of ice. Hobler recalls that on his course the initiation was 'quite harsh': 'You knew you weren't going to die from it because they wouldn't be so bloody silly, but it went on and on, and was very frightening'.[50]

Excesses in the horse-play allegedly led to attempts to limit the scope of initiations, but accounts by cadets of later years indicate the ceremonies remained 'very solid' and unpleasant affairs, with initiates being given electric shocks, daubed with aircraft dope and branded (sometimes with the dope catching fire when not sufficiently dried), and (for those joining mid-year) being marched off the end of the pier into the icy winter waters of Port Phillip Bay.[51]

Along with the initiations there also arose the practice of hazing, although this appears not to have become the problem that it did at Duntroon. The fact is that there were those who bore the scars of this facet of cadet life at the RMC and resisted its full-blown appearance at FTS. Recalls Sir Frederick Scherger:

> When I became a flying instructor at Point Cook I had a seconded army officer passed on to me as a last chance (for him). Apparently I was regarded as a patient and tolerant instructor. It was interesting in that he was one of the few at Duntroon who had made my life difficult for a while. Yes—I could not resist making him very air sick and then insisting that he clean the aeroplane to my satisfaction.[52]

Nonetheless, a pale imitation of hazing did emerge in 'Cadet Squadron', stemming from the responsibility vested in the senior course for indoctrinating the new arrivals in service customs and traditions.[53] Consequently when one senior cadet used his service rifle to commit suicide in the wash room of one of the accommodation blocks during the second half of 1937, there was reportedly some concern at what might have been behind this occurrence. Recalls one member of the junior class at the time:

> There was a bit of an inquiry because he was known to have been into hazing—'bastardisation' is the modern word for it. This used to occur and there was a lot of it in those days, although generally there was nothing very serious happened. Anyhow, there were a lot of inquiries into whether one of the junior cadets might have helped him pull the trigger. I'm sure there was nothing to that suspicion. He was just a basically unstable type.[54]

Other junior cadets at that time agree that the individual concerned was erratic, one claiming to still have a suitcase which the senior cadet had shot through with bullets from a pistol after returning from weekend leave,[55] and another describing him as 'a bit of a brute' but adding that 'all the seniors were'.[56]

Point Cook in the late 1930s was clearly no pool of enlightenment, notwithstanding the recollection of one newcomer in mid-1935 that he was impressed by the diversity of other members of his course:

> The most surprising thing was the difference in character between the various

Sailing was a recreational activity introduced for air cadets at Point Cook when Wing Commander (later AVM) F.M. Bladin was OC of Cadet Squadron. The suggestion came from Flight Lieutenant Eric Douglas of Antarctic fame, who was a keen yachtsman. He selected the type of sailing dinghy used, saw to their construction, and supervised the training of the cadets. This scene is from September 1937. (*WgCdr D.J. French*)

Award of a Sword of Honour to the cadet graduating top of his course was a practice borrowed from Duntroon and inspired by RAAF officers who had formerly trained at the army college. Here the Governor-General, Lord Gowrie, makes one such presentation but neither the year nor recipient is identified. (*Mrs J. Gray*)

> individuals who had been selected for cadetships. There was a great variety of sizes, shapes, types, mentality, attitudes—you couldn't get a more different group of people. The Duntroon and naval cadets whom we dealt with and met in sport all looked the same to us. We could not see much of their differences, but you could see the differences of the Air Force group.[57]

There were, however, clear limits to the range of individuals which the RAAF was able or willing to accommodate. When presented in 1937 with a flamboyant personality like Francis James, the later controversial journalist who spent five years from 1969 as a prisoner in China, those limits were reached. Baffled by his reading tastes which included Liddell Hart and Trotsky, alarmed by his personal correspondence with British Cabinet ministers and permanent secretaries like Anthony Eden and Sir Robert Vansittart, and disdainful of his preference for symphony concerts and reading over participating in the initiation ceremony and 'healthy, manly games', staff members at the FTS quickly decided that James was unlikely to make an efficient officer—despite his apparent ability to handle flying and course study satisfactorily. He left Point Cook, only to join the RAF in 1940 where he served with considerable success.

His view of the RAAF was, naturally enough, coloured by his experience:

> The lasting impression I had of Point Cook and the RAAF was not of un-intellectualism but anti-intellectualism, the pressure for conformity. The only route to success in the Service in those days, given flying proficiency which *did* matter, was silence—at least the appearance of agreement, and ask no questions. Point Cook was worse than Duntroon in this respect—much, much worse—and it left its mark on the RAAF in the form of rigidity.[58]

James' differences with the FTS authorities were by no means a common experience and were not such as to dissuade the average 19 or 20 year-old from seeking admission. Competition for cadetships was usually intense, with applications in the early 1930s reaching 1200 for places which never exceeded a dozen; even at the peak of the pre-war expansion in 1938–39 the number of vacancies did not exceed 54. Hence, when the station commander at Richmond (Group Captain Cole) was detailed by Headquarters in 1937 to give a careers talk to Sydney school headmasters as part of the RAAF recruitment effort, he was not concerned with simple numbers but quality of applicants. Attempting to counter the concern that short service commissions provided little security of employment, he stressed that:

> The Board is anxious not to entice boys away from the security of professional careers. Where parents have money, or boys have the ability to earn scholarships, etc. we say *definitely* that it is better to discard the idea of the Air Force as a career. We do say, however, that the Air Force is definitely attractive to thousands of good type of public school boys.[59]

The selection process for admission to a flying course at Point Cook was, on all the evidence, a competitive one based solely on merit. The claim has been made, however, that political interference was sometimes brought to bear to disadvantage one candidate in favour of another less deserving. An alleged instance concerned the future Air Vice-Marshal Don Bennett, who entered No.1 FTS in July 1930. According to his English biographer, Bennett had first obtained a cadetship on the course commencing in January 1930, having topped the entrance examination. But his joining instructions were subsequently cancelled, allegedly because the politician uncle of a rejected applicant 'threatened trouble' for the Air Board 'so one cadet had to be dropped to make room. Who better than the lad from Toowoomba?'[60] While the credibility of this tale suffers from the unlikely claim that the Air Board, if facing such a predicament, would have chosen to forego its top prospect instead of another less promising applicant, there is more convincing evidence that, in at least one case, factors other than simple merit had influenced the selection board.

A few years after Bennett's reported experience, a stores hand at No.1 AD named Robert Hitchcock applied for a cadetship and succeeded in gaining selection in 1935. According to colleagues on the course and others, Hitchcock showed little aptitude for flying and ordinarily would have been expected to fail the course.[61] The fact was, however, that he was the eldest son of the mechanic who had perished when the 'Kookaburra' was lost in the Tanami Desert in 1929 (see chapter 10). A former room-mate at Laverton recalls

Hitchcock telling him of a verbal promise given to his widowed mother by the federal government at the time of the 'Kookaburra' tragedy, assuring her that her three sons would all be found employment in one or other of the defence services as soon as they reached the required age. (In fact, Hitchcock's two brothers also joined the RAAF during the Second World War. One of these, Wing Commander Harry Hitchcock, became post-war head of motor transport engineering in the service.)

It was this kindly pledge on the part of the government which brought young Robert Hitchcock into the RAAF in 1934, where he spent his first night in the service as a guest in the home of the CAS and his wife.[62] And it was this factor of sentiment which kept him on a flying course beyond the point where it was apparent that his training should not be continued. He graduated from No.1 FTS in July 1936.

Once passed into a unit, Hitchcock's inadequacies as a pilot soon became apparent to those made responsible for his advanced training. One of those, Flight Lieutenant Wilfred Compagnoni, recalls:

> After spending quite a while with him I realised that he would not make a pilot. It would have taken many hours and patience to make him even reasonably safe in an aeroplane. I went to my flight commander, Allan Walters, and suggested that his training be terminated. Cole [the station commander] told us that there was a direction that Hitchcock was to be taught to fly. Evidently it was a political decision. The Minister had directed that we were to persist with him.[63]

Despite further problems with Hitchcock's training, leading to a report being submitted to Headquarters that he was temperamentally unsuited to becoming an efficient pilot and a recommendation from the AMP (Air Vice-Marshal Nicholl) that his appointment be terminated,[64] the Air Board baulked from making the hard decision that should have been made. He was kept on, although his subsequent flying record remained poor. On 23 August 1938 he was involved in a collision of two Ansons of No.1 Squadron during take-off from Currie Aerodrome on King Island, but as only slight injuries to personnel resulted there was no significant action taken over this incident. The sad fact remains, however, that two years later Flight Lieutenant Hitchcock was the pilot of the Lockheed Hudson bomber A16–97 which inexplicably stalled on landing at Canberra and dived into low hills beside the airfield on 13 August 1940. All on board were killed in the crash, including the Minister for Air, two other Cabinet members, and the CGS (General Sir Brudenell White). The Tanami Desert might be said to have spawned two tragedies, both of which directly involved the RAAF.

Notions that the saga of Hitchcock was a reflection of the standards applying to FTS graduates generally would be completely unfounded. As even the RAF acknowledged, there might be deficiencies in ground training and such areas as tactics, but the flying capabilities of pilots trained at Point Cook were generally excellent. This point was made to one RAAF officer who called on the RAF commander (A.W. Tedder) at Cairo in 1939 and was told: 'I don't know how you Australians do your training, but I see the result and it is good'.[65] The future Lord Tedder and Marshal of the RAF might have been surprised at some of the methods used. One popular story concerns the day Flight Lieutenant John Balmer exited an aircraft in which he had been instructing and parachuted to earth, leaving the cadet pilot to fly the machine back to Point Cook. Newspapers reported that he had accidently fallen from the aircraft,[66] but the story circulated throughout the Air Force that he had jumped to provide an otherwise proficient student with the incentive to apply his skills properly! Other instructors employed less drastic methods, including Sergeant Francis Collopy whose outstanding work with trainee pilots was recognised with the award of the Air Force Medal in 1933.[67]

Sadly, there were many accidents during the 'A' Flying Courses, including one which might have had a profound effect on the keenness of a whole course of new trainees. As one then cadet entering Point Cook in July 1935 recalls:

> Our introduction to flying was slightly marred by a Wapiti with an instructor and a cadet on the course ahead of us taking off and hitting the H.T. [high tension] wires at the edge of the airfield resulting in a crash. The aircraft burst into flames, the instructor was thrown clear but the cadet was jammed in the cockpit by the fuel tank in the centre of the fuselage holding his legs and so he was killed. This all happened in our sight as we were marching down to the hangars at the time! Not very nice for a batch of new boys heading for their introduction to flying. As far as I remember, it didn't seem to adversely affect anyone.[68]

Accidents may have been an unavoidable facet of flying, but these were not so frequent at Point Cook as some imagined. One recently published claim that eleven out of fifteen cadets on an 'A' course in 1928 had been killed in crashes, mainly as a result of engine failures, is patent nonsense totally unsupported by the facts.[69] The highest number of deaths experienced on any flying course up to 1934 was only three, these occurring among the 30 members of the 1926 'A' course.

Specialist Navigation Course

In July 1938 six officers began the first and only Long Specialist Navigation Course held by the RAAF before the start of the Second World War. Lasting nine months, the course was instructed by Flight Lieutenants Bill Garing and Alister Murdoch, both of whom had completed specialist navigation training in England—Garing in 1935, Murdoch in 1937—and were now passing on that knowledge to officers who would conduct similar training in various squadrons.

A number of long navigation exercises were arranged as part of the training on the course, involving flights which were so unusual as to be regarded as newsworthy even at this late stage of public familiarity with aviation. The first of these was a 1900-kilometre night flight from Brisbane to Point Cook via Richmond, carried out in an Avro Anson (A4–35) on 4–5 September 1938. In reality, the purpose of the flight had been to convey a party of officers to attend the wedding of Flight Lieutenant Douglas Candy in Brisbane on 3 September, so that the conduct of the return leg as a test flight was practically incidental. It was also nearly fatal.

Leaving Archerfield aerodrome at 6.10 pm on 4 September with five personnel on board—Garing, Flight Lieutenant Frank Headlam, Flying Officer Geoff

Hartnell, and AC1s A. Gray and F.M. Rowson—the Anson touched down at Richmond four hours later.[70] Garing was in the pilot's seat, the other two officers (both course members) taking turns at navigating. Recalls Garing:

> The flight was pretty hairy. Landing at Richmond the only flare path was two red hurricane lamps. Of course, we knew from our navigating where the wind was blowing, and with the headlamps on the Anson I used to land on airfields with virtually no landing lights at all. The best we ever had were these smudgepot things, with a wick and a pot of kerosene—and you can imagine what they were like—but a string of these would be put out to mark the strip.[71]

About an hour later the plane took off and arrived over Melbourne after 2 am on 5 September during a heavy rainstorm. Garing complains that:

> Coming in to land at Point Cook, all the lights had been turned out. There was no flare path or anything and nothing had been done to light any of the buildings, although they knew we were coming alright. The storm was booming, a real rough night. The only way I could see to get into Point Cook was to go out to sea and come in over the tarmac and the line of old hangars down near the water, and set the aircraft at right angles to them. I did that, and dropped the aircraft down. Absolutely no light, just blackness and rain. The wheels touched and we rolled along the ground, when I got a gut feeling that something was wrong. I started to do a right-hand turn, but the wingtip of the Anson touched one of the trees along the road leading from the hospital to the headquarters and pulled us over onto one wheel.

A bolt for the exit followed in case the aircraft caught fire, which it did not. Despite this, there was extensive damage to the machine's undercarriage and mainplanes.[72]

As a matter of fact, Garing was mistaken about the lack of preparations for his flight's arrival. Both the duty pilot (or orderly officer) that night, Flying Officer Brian Eaton, and the officer responsible for laying the flare path, Flying Officer Ivan Podger, are adamant that a path was correctly laid, though in the conditions of high wind and poor visibility that morning Garing had obviously been unable to see it properly, if at all. Says Eaton: 'We laid the flare path fair up the middle of the landing ground, but he landed half-way up the flare path and went straight into the big trees near the hospital. Actually, I think he landed downwind.'[73]

Notwithstanding the drama of this initial attempt, the long-distance navigation exercises (including at night) continued. Two weeks later another Brisbane-to-Melbourne flight was carried out, this time during daylight hours. According to press accounts, an air sextant—a device described as 'new to Australian navigation'—was used on this trip and also on a second night flight which followed a week later. This latter occasion again involved an Anson with Garing in charge, and two students, Flying Officers Julius Cohen and Ralph Wiley, a wireless operator (Sergeant Johannesen) and a mechanic (AC1 H. May) on board.[74]

In mid-October further long-distance flights were planned to be made by two Ansons travelling from Melbourne to Adelaide by way of Broken Hill, on which students would be exercised in celestial methods of navigation. The first aircraft, carrying Garing and Flying Officers Arthur Watkins and John Daniell, was scheduled to depart Point Cook on 16 October and return the next day. The second plane, piloted by Murdoch and carrying Flying Officers Cohen and Wiley, was planned to follow the same route on 18–19 October.[75]

The next month, course members were taken on a twelve-day 10 800-kilometre flight around Australia which was hailed as 'the most ambitious training flight yet undertaken by the Royal Australian Air Force'.[76] Taking off from Point Cook, three Ansons each carried two students and two aircraftsmen—Headlam and Daniell, with AC1s Hoffman and M. Houghton in one; Watkins and Hartnell, with AC1s May and Gray in another; Wiley and Cohen, with AC1s Rowson and V.F. Sullivan in the third. Newspapers reported the announcement from RAAF Headquarters explaining that the objects of the flight were to: give the pupils experience in flying over the main air routes of Australia; apply their knowledge of navigation in flying over long distances; and make them familiar with the widely differing weather and flying conditions throughout the continent. Though flying in company, the planes were to travel at intervals so as to be out of sight of one another and thereby giving each crew the responsibility of navigating for themselves.[77]

From the time of first lift-off on 15 November the crews were confronted with the forces of nature, as recounted by one of the student officers in a letter home:

> We left Point Cook on Tuesday morning about six o'clock in winter sidcotts and decided to go to Richmond via Cootamundra. When we got to Sunshine [on the western outer edge of Melbourne] however the rain was so heavy and visibility so bad that we came back to Point Cook and landed . . . When it cleared a bit we took off again and had quite a good if very bumpy trip to Richmond. The extreme bumpiness combined with the first day's flying made me sick for the first time. We had some lunch at Richmond and then took off for Brisbane. There were dense bush fires all the way up the coast and the smoke was so thick that even though we could see the ground we had to fly blind while we were climbing and didn't get an horizon until we got to 12 000 feet. Actually I should say the smoke went up to about 13 000 feet . . . so we did a bit of coast crawling up to Coolangatta and then went inland . . . , running out of the smoke about a mile before we got to Archerfield.[78]

From Brisbane the flight headed further north, to Rockhampton, Townsville and Cairns, then west to Normanton, Cloncurry and Camooweal. From there it was on to the Northern Territory, to Daly Waters, Birdum and Darwin, before proceeding down the coast of Western Australia, through Wyndham, Halls Creek, Derby, Broome, Port Hedland, Onslow, Carnarvon and Geraldton, then direct to the RAAF base at Pearce, outside Perth. By 30 November the Ansons were safely back at Point Cook, their return coinciding with the arrival of two RAF Vickers Wellesleys at Melbourne (see chapter 4). The flight had reportedly been 'so free from trouble that the itinerary was altered to save a day'.[79]

The completion of this major undertaking was followed by a course in hydrographic surveying undertaken in Sydney with the Navy, before the students were back in the air once more on 13 December. This time the three Ansons (with Headlam in charge) proceeded to Normanton in Queensland, and from there came back to Point Cook in a series of flights which had them criss-crossing Central Australia—a flight lasting six days and taking them over another 7200 kilometres.[80] As one newspaper remarked: 'They will then be ready, early in January, for three weeks' continuous examinations, testing the work they have been doing during the last few months in the stiffest navigation course ever given in Australia.'[81]

For the successful students, the outcome of these examinations would be that they were eligible to apply for the Commonwealth's highest qualification in this field—a First Class Aircraft Navigator's Licence. So rare was this qualification that Garing actually gained the very first such licence only on 10 November 1938, while he was still engaged in instructing the course.[82]

In the event all the students on the RAAF course passed, three (Headlam, Hartnell and Wiley) with special distinction. Four of them were now expected to be posted to units, 'to instruct personnel in specialist navigation that will aid the sea reconnaissance duties of the RAAF', while the remaining two were to become instructors at Point Cook.[83] By 30 June 1939 Garing had been joined by three of his students as holders of first-class aircraft navigators' licences—Cohen, Hartnell and Watkins. In addition, one civilian pilot, Captain Gordon Taylor—who in June 1939 made a remarkable flight across the Indian Ocean to survey an alternative air route for the British and Australian governments as a defence precaution—also held a licence.[84] With this small number of specialist navigators Australia was, just two months later, to go to war.

Army cooperation

The first activity undertaken after the formation of the RAAF to demonstrate its capacity to work in conjunction with the Army was a 'cooperation shoot' with the 6-inch guns in the fixed defences at Queenscliff on 27 April 1922. This involved a DH9 (A6–6), flown by Squadron Leader William Anderson with Flight Lieutenant Adrian Cole as observer, carrying out spotting for the guns and reporting fall of shot by Type 57 radio to a ground receiving station inside the fort, manned by Flying Officer Hector Johnston, a corporal and a wireless mechanic. Despite the fact that the aircrew's spotting was admitted to be slow, due to cloud patches and lack of practice, and jamming of radio communications from a Melbourne wireless operator and a steamer sailing down Port Phillip Bay, the exercise was rated a success and drew a letter of appreciation from the CGS to the Air Board. For its part, RAAF Headquarters recognised the importance of this type of work, and expressed the hope that further and more advanced practices could be carried out.[85]

In fact, opportunities for this sort of cooperative training were not that abundant in the first years of the RAAF's existence, due to the lack of any Air Force squadrons in its meagre organisation, its geographical concentration solely in Victoria, and the fact that the Army was primarily a militia force whose training was limited to weekend and annual camps. Not that being restricted to one permanent base prevented Air Force planes and crews being deployed elsewhere. In November 1925, for example, two DH9 aircraft (apparently A6–1 and A6–26) were despatched to Western Australia to take part in exercises at an Army camp held at Rockingham, on the coast south of Fremantle.[86]

Nor, evidently, was there a disposition to deliberately exclude, or a reluctance to call upon, the expertise of the Air Force. When the CGS, Chauvel, held a staff tour at Canberra in September 1925, assembling virtually all the Army's general officers to consider exercise problems of a corps engaged in operations, an Air Force presence was included in the person of the CAS, Group Captain Williams. For a week Wiiliams mixed with 'a galaxy of serving senior officers of the old AIF', including Sir John Monash.[87]

By mid-1925, however, the RAAF had two operational units and had physically expanded to two States. Moreover, the organisation of the two composite squadrons which now existed both contained a flight equipped with DH9s specifically for Army cooperation duties. Barely two months after No.3 Squadron established its presence at Richmond in July 1925, it was called upon to carry out a cooperation exercise with an Army howitzer battery on the Holsworthy range. All the same, however, such activities were still not a common occurrence. No.1 Squadron did not perform its first artillery and infantry cooperation at Seymour until February 1926, and it was six months after its first task that No.3 Squadron was again involved in another shoot, observing the fall of shot on a sea target for the garrison artillery at South Head.

The geographic spread of squadrons at least facilitated the provision of aircraft for training in other States. In February 1926 two DH9s were sent from 1 Squadron to Adelaide to cooperate with the Army in South Australia for five days. It was, however, actually March 1928 before No.3 Squadron performed its first cooperation exercises in another State, in this case a single DH9A working with the 11th Mixed Brigade at Enoggera (Qld).[88]

Apart from observing for periodic artillery shoots, in most cases the form of 'cooperation' carried out was often very limited, consisting mainly of demonstrations of such things as signalling, artillery observation, low bombing and strafing, message dropping and picking-up, delivery of small supplies, and air-to-ground radio telegraphy. In effect the activities were little more than displays of the ways in which aircraft operated in support of ground troops rather than actual exercising, although for more than four weeks in February–March 1927 No.3 Squadron was involved with 1st Cavalry Division and 2nd Division at Liverpool and this must presumably have entailed something more advanced.

Probably of more practical value were various courses and attachments held for both Army and Air Force personnel. In December 1926, for example, an intelligence liaison course was conducted by No.1 Squadron at Point Cook for four Staff Corps officers, and a similar course for four officers was run by No.3 Squadron in July 1933. In September 1936 Captain W.N. Tinsley of the Staff Corps was attached to No.3 Squadron as a liaison officer; he gave some lectures in the unit but his period of exchange ended when he broke a leg 'whilst flying as passenger in a Demon'.[89] In October 1937 nineteen senior Army officers attended a three-day course of lectures at Richmond covering the various phases of Army cooperation, during which they were also given some experience of flying. During March 1938 one officer of No.3 Squadron was attached

In September 1925 a staff ride was held at Duntroon under the CGS, Lieutenant-General Sir Harry Chauvel (seated fifth from right). Virtually all the army's general officers were present, including Sir John Monash (on Chauvel's right). Group Captain Williams also attended as head of the RAAF; he stands second from the left. (*RAAF Museum*)

to the Tank Corps to gain experience with the Army on the ground, while an Army lieutenant was similarly attached to the squadron later in the year for six days to gain experience as an air liaison officer.

During the 1930s use of air support for Army ground exercises expanded slightly in scope, with some activities entailing more than routine demonstrations. In March 1934, for example, No.3 Squadron cooperated with the 14th Infantry Brigade at Liverpool using a Wapiti to carry out patrols until contacting ground forces, whereupon a low flying attack was mounted by three Moths. In similar vein was the cooperation performed by the same squadron with 16th Light Horse Machine Gun Regiment at Rutherford in February 1939, involving photography of transport vehicles in open formation and under cover, as a check on the effectiveness of camouflage, and low attacks against vehicles on roads in the open.

Cooperation between the Army and Air Force still often had about it an element of game-playing. For example, when the 8th Field Artillery Brigade went into bivouac at Laverton for its annual shoot late in 1931, No.1 Squadron celebrated the occasion with 'a spot of mimic warfare' between the artillery and the unit's newly-formed Aerodrome Defence Section (a similar section was also raised at No.3 Squadron from CAF personnel in July that same year):

> An aircraft of the Squadron kept the Artillery posted with the latest information concerning the 'enemy' by message dropping. The day being hot, Observers had no difficulty in locating the Artillery detachments at various Geelong Road hostelries. It is rumoured that the last message dropped savoured of Chinese warfare tactics, as it contained a demand for the cessation of hostilities owing to the approach of 'stand-down' and the fact that our machine-gunners wanted tea.[90]

Many activities, too, were still more of the nature of demonstrating uses to Army commanders who were rarely given the chance to practice the techniques involved. For instance, on 30 October 1932 a display was mounted at

A demonstration of message picking-up given during the air pageant at Richmond in November 1934. Wapiti A5–28 was piloted by Flying Officer W.G. Compagnoni, while the other machines were flown by Flight Lieutenant A.L. Walters and Flying Officer C.D. Candy. Much of the army cooperation done by the RAAF during the 1920s simply involved displays of this sort. (*GpCapt W.G. Compagnoni*)

Richmond by No.3 Squadron for 250 officers from a variety of Army units, purely to demonstrate the range of manoeuvres that aircraft could perform, including speed photography (involving photographs taken from the air being rushed to the photo section, developed and printed, then dropped to an Army commander from the air within half an hour). According to press accounts of the event: 'As each type of operation was in progress it was explained in detail'.[91]

Viewed in hindsight there also seems to have been slender value in some of the other activities undertaken, with one junior officer of the period recalling:

> Much of the army cooperation done at Liverpool seemed designed to get troops used to the frightening effects of aircraft. We used to dive on the troops in their trenches and pass just overhead. This would get them used to the noise—the Demon made an awful racket. It was good, and the Army appreciated it.[92]

This was, of course, by no means the full extent of the day-to-day activity going on in the squadron, which included a range of aspects relevant to the unit's designated role, among them drogue towing, signals, gunnery, bombing, photography and night flying, but the actual exercising of these skills in conjunction with Army units was another matter. Pilots gained very little at all from several hours of flying a Moth back and forth over Sydney Harbour while providing a passive target for 1st Anti-Aircraft Battery at Middle Head.

By 1937 the futility of such practices was beginning to be recognised. The minutes of the CAS Conference held at RAAF Headquarters on 8–9 April records Williams' concern that 'there was much that was of little training value, either to the Air Force or the Army unit' in the cooperation then being done. He noted that he intended to take up this question with the Army in discussing the next year's commitments.[93] Whether these representations were made remains unclear, but it is certain that a year later the Air Board was still concerned that much of the training it was being asked to provide by Army still held scant training value for the pilots involved. The CO of No.3 Squadron was encouraged to advise the local Army staff about any exercises they might be planning, to ensure that the activities required had sufficient merit to justify moving aircraft interstate.[94]

The fact was that the Military Board was also dissatisfied with the existing situation, which emerged in May 1938 when RAAF Headquarters received complaints regarding the failure of artillery spotting activities undertaken as part of air cooperation exercises conducted with the 3rd and 4th Divisions. After investigating the problem, the Air Board replied that the majority of the failures reported were the result of inadequate supervision and training of junior pilots participating from No.3 Squadron. The Army agreed to abandon economy measures which limited ranging fire being conducted by just one gun, admitting that this made observing fall of shot difficult for pilots, while the RAAF arranged to double the number of spotter aircraft used, so that inexperienced pilots carried out the actual observing of fire but under experienced supervision, and to send aircraft to Army camps as complete flights rather than singly.[95]

Regardless of such problems, by the late 1930s there was a clear intensification of the RAAF's involvement with militia training, itself taking on a new note of urgency at the time. Air Commodore Gordon Steege, a mid-1938 graduate of No.1 FTS who went to No.3 Squadron, recalls:

> There was plenty of work for the squadron to do. I remember there was a big combined exercise called Anzac I held around from Port Stephens in October 1938. The whole squadron moved to the Newcastle airfield at Broadmeadows. I can't say whether the exercise was a success or not, but the whole thing was run again a month later. We were involved in tactical reconnaissance and reporting of troop movements, that sort of thing, though we did do some fighter bomber work against incoming landing craft.
>
> We were doing a lot of army exercises at this time—directing for artillery on the range at Seymour in Victoria, also Army cooperation with the Riverina Horse at Wagga. I remember on that occasion our flight took up four Light Horse liaison officers to show them the effectiveness of their unit's camouflage, etc., and we gave them some aerobatics and made them thoroughly ill. They had their revenge though. A week later the Light Horse invited us RAAF pilots to go riding and gave us mounts which bolted back to camp at feeding time. We were set up nicely.[96]

By late 1939 No.3 Squadron, the RAAF's sole Army cooperation unit, was preparing to accompany the 6th Division of the Second AIF to Egypt on active service. Although the decision for its despatch was reversed for a time, by mid-1940 the squadron was finally on its way to the Middle East—where it saw action principally as a fighter unit.

Naval cooperation

Even before the RAAF was formed, experimentation with naval cooperation was being carried out using two Avro 504Ls (floatplane versions of the 504K) recived with the imperial gift equipment. As mentioned in chapter 1, in July 1920 one of these aircraft was taken aboard HMAS *Australia* with an officer and two mechanics from the AAC for trials. In September this machine and a new crew (Captain J.K Fryer-Smith and Lieutenant W.S. Walne) were transferred to the light cruiser *Melbourne* and taken on a cruise to north-eastern Australia and New Guinea. Test flights made over the next two months were not successful, however, as the plane was found to lack the power needed for tropical conditions, and by November the airframe was unserviceable through deterioration caused by five months of open storage on the cruiser's deck.[97]

Following these trials, Williams urged the Air Board to recommend to the Air Council that no further machines be allotted to the fleet until the Air Force had a type in service which was actually designed for the work.[98] When the Air Council considered the question on 15 March 1921, it acknowledged that the Avro type was not suitable for the purpose tried, but with new Fairey IIID seaplanes on the way it directed that a committee be formed to examine how the problems could be overcome. Membership of this body included the Navy's chief staff officer, Captain Claude Cumberlege (chairman), a representative of the Ship Construction Branch, and two Air Force officers: Squadron Leader McBain and Flight Lieutenant De La Rue.[99]

The problem this Seaplane Committee grappled with was how to fit a plane of the Fairey IIID's large dimensions and weight into the limited space available on warships of the size operated by the RAN. The committee's report in September 1921 could only suggest that host ships carry less coal and have their mainmast removed to make room. Various proposals to modify cruisers, or to convert auxiliary ships as seaplane carriers, were subsequently considered in 1922–23 but abandoned.[100]

Realisation that the Fairey was too big for carriage on warships, and no suitable vessel was available for conversion, led to attempts by Navy to obtain a smaller unarmed aircraft that would serve as ship's scout. As narrated in chapter 9, Navy's interest in this alternative led to proposals within RAAF Headquarters for a small flying boat meeting the requirement to be designed and built in Australia. Among those pressing this approach to the problem was Squadron Leader Lawrence Wackett, who urged on 3 January 1923 that the 'necessary technical personnel are available' to carry out the task.[101] Ultimately, however, the Navy was forced to the conclusion in June of that year that Australian warships could not operate aircraft.[102]

While these lengthy deliberations were taking place, the RAAF's maritime air element lay under-utilised at Point Cook. Some seaplane training was carried out but little else. In May 1922 De La Rue, OC of the Seaplane or 'F'

Flight, crashed the second of the two Avro 504L machines (A3–47) a few miles off Altona, causing irreparable damage to the aircraft which resulted in it being converted to spare parts.[103] According to the operator of the crash rescue boat at Point Cook, the pilot was brought ashore by some fishermen and was 'very upset at getting his corduroy breeches wet'.[104] The lack of activity for Seaplane Flight continued well into 1923, when the Parliamentary Joint Committee of Public Accounts found that:

> Although six Fairey seaplanes were purchased at a cost of £23 000, little use has been made of them. Four of the machines have been used, but word has recently been received from the British authorities that until certain alterations have been effected, these planes should not be flown.[105]

The naval cooperation role was not forgotten, however, least of all by Navy. In May 1923 a special branch of observers was instituted, which entailed selected navigators undergoing a three months course at Point Cook flying in the remaining Avro 504L trainer and the Fairey IIIDs.[106] The RAAF was not forgetting this aspect of its function either. When Flying Officer Arthur Hempel, having recently joined the RAAF from the RAF, was detailed to give a lecture at Point Cook on 'Range Finding in relation to Aircraft Co-operation with H.M. Ships and Coast Defences' in May 1923, an instruction was issued that 'As many "G" List Officers [i.e. pilots] as possible will attend'.[107]

The next month, June 1923, Hempel and Flying Officer Ernest Mustard joined the RAN winter cruise and were away until August; Hempel embarked in HMAS *Brisbane* while Mustard spent time in both HMAS *Adelaide* and the survey sloop HMAS *Geranium*. During the cruise, which took in the harbours and islands of North Queensland as well as Papua, New Guinea and the Solomon Islands, both officers compiled detailed descriptions of the places visited for possible RAAF use. With no aircraft embarked, however, there was no flying and little real call on the airmen's time.[108]

Although unable to operate from on board RAN cruisers, the Faireys were being used for survey purposes—most famously a round-Australia flight made by Wing Commander Goble, the acting CAS, and Flight Lieutenant Ivor McIntyre in April–May 1924 (see chapter 13). Even before this historic undertaking, Goble and McIntyre had teamed up in February for a route survey between Melbourne and Hobart in Fairey A10–2,[109] and other seaplane reconnaissances were carried out later, including one to Flinders Island in October 1925 involving two Faireys (A10–5: Flight Lieutenant Hempel (pilot) and Captain Edgar Johnston of Civil Aviation Branch; A10–6: Flight Lieutenant McIntyre (pilot) and LAC Miles).[110]

In 1924 also, a Fairey IIID was sent north to join the Navy in the mammoth task of surveying the Great Barrier Reef. This first trial, repeated the following year, foreshadowed a long-term commitment to supporting this important national project (see chapter 14), giving rise to the formation of No.101 (Fleet Cooperation) Flight in 1925 and prompting a decision to replace the Fairey floatplanes with Seagull III amphibians which were better suited to the work associated with the survey.

In 1925, however, the RAAF found itself suddenly confronted with an extraordinary situation. Early in June the Naval Board promulgated an order for the formation of a fleet air arm, a development which the Air Force strenuously opposed because this was seen as the first step towards eventual disbanding of the RAAF as a separate service. Williams would finally win this fight (see chapter 3), but in the meantime other events now intervened. During his address opening Parliament on 10 June, the Governor-General announced the government's plans to purchase a 6000-ton seaplane carrier named *Albatross*. Only when the contract was placed for this vessel in 1926 did the Air Board learn to its surprise that the Air Force was to provide the aircraft to go in this new ship, although there had been no prior discussion.[111]

The only planes the RAAF had available which were suitable for *Albatross* were the Seagull IIIs on order, although they had not been purchased for this purpose, merely for seaplane training and survey work. A specification for a new type of catapult-launched amphibian was accordingly prepared by the Air Staff in 1926 but failed to attract an aircraft manufacturer interested in building it; eventually this machine would enter RAAF service as the Supermarine Seagull V—but only after the *Albatross*, the ship from which it had been designed to operate, had been paid off from the RAN.

As it happened, no aircraft type was specified to the ship's designers, only that nine machines would be the maximum number required to be carried. In the absence of specific information, they assumed the Fairey seaplanes were the type intended for carriage and planning proceeded on this basis. Fortunately as it happened, the Seagull III's dimensions were very similar to the Fairey, and when the time came for aircraft to be placed on board and the RAAF's Seagulls were still the only relevant type available, it was found that these actually fitted. Despite the fact that embarkation of the Seagull III was an expedient which meant that only air operations of quite limited scope were possible, these planes were able to satisfy reconnaissance, gunnery, torpedo spotting and survey requirements provided sea conditions were favourable.

At the end of 1928, with the *Albatross* almost ready for commissioning, the Seagull IIIs of No.101 Flight vacated their temporary base at Bowen (Qld) and returned to Point Cook. Under a Cabinet decision in January 1928 not to persist with the fleet air arm concept, the Air Force was to provide the aircraft, pilots and maintenance personnel for the new carrier and the Navy would find the observers and radio telegraphists. While embarked in the carrier, however, the flight came under the operational control of the ship's commander.

On 21 February 1929, a month after HMAS *Albatross* was commissioned, the personnel of 101 Flight were taken on board. The normal RAAF complement on board was six officers (five pilots, one officer of the Stores and Accounting Branch) and 24 other ranks. Supplementing these numbers, the Navy provided five officer observers and six telegraphist air-gunners (TAGs) attached to the flight. Four days later six Seagull IIIs were hoisted aboard at Geelong. This was to provide the flight with five operational planes and one machine in reserve.[112]

No sooner had the ship joined the fleet than it became caught up in the drama over the disappearance of Kingsford Smith's aircraft *Southern Cross* in north-west Australia at the end of March (see chapter 10). Following this, *Albatross* began working-up exercises off the Australian coast, providing the reconnaissance element to the new County-class cruisers, *Australia* and

Canberra, which had also just joined the RAN. After taking part in combined exercises with two New Zealand cruisers in June, the carrier proceeded on a cruise of New Guinea waters in July–August with the Governor-General and Lady Stonehaven on board. In addition to the Seagulls, *Albatross* had the Wackett Widgeon II on board during the cruise, for the purpose of conducting tropical trials on this experimental type.

Bringing 1929 to a close, *Albatross* took part in a combined RAN–RAAF exercise in Port Phillip Bay in October. The point of this exercise was to test the carrier in making an air raid, along with assessing the efficiency of RAAF cooperation with Navy in repelling a seaborne air attack. According to reports on the exercise, the defence against the carrier attack was only successful because scouting Southamptons set off from Point Cook, without orders, some time before warning was actually received of approaching enemy aircraft. In fact, as noted by the CO of No.1 FTS, aerial patrols had failed to sight the approaching naval force. Strikes had then been mounted against these ships off Frankston, involving Moths (representing single-seat fighters) and Wapitis.[113] One RAAF pilot whose part in the scheme entailed simply flying over the Melbourne dock area probably summed up the feelings of many of those involved when he noted in his log-book that the exercise was 'A farce—nothing done or to see'.[114]

In December the command of No.101 Flight passed from Squadron Leader Hempel to Squadron Leader Victor Scriven, an RAF exchange officer. This marked the beginning of a period of difficult personal relations within the confined community on board the *Albatross*. The Flight Commander, a RAAF officer (J.E. Hewitt), has left a trenchantly critical description of Scriven's character,[115] and even those more charitable admit that he was not popular. Recalls Air Vice-Marshal Denis Creal: 'He didn't quite fit in exactly. He was frightfully "pukka", with a British way of doing things. And he always talked about "my servants" and that sort of thing, which didn't quite go down.'[116]

Scriven decided on attempting the recovery of an aircraft while the ship was still underway, a process which was recognised by all as full of risks. The attempt ended in a perilous episode which almost resulted in the loss of one of the carrier's cranes overboard, as well as a Seagull and its three-man crew.[117] Such occurrences and others described by Hewitt lend credence to the claim that Scriven eventually destroyed the morale of the flight, at one point declaring his entire flight unserviceable and no longer of any use for naval cooperation. The outcome of this action was that Scriven was relieved in February 1931 and command passed to Hewitt.[118]

With the exception just cited, most RAAF personnel who served in the *Albatross* describe it as a happy experience. Air Vice-Marshal Creal asserts that the arrangement 'worked just fine':

> The flight kept its own identity except it was in a navy ship. We wore our own uniform, that sort of thing. There was a bit of bother though, trying to get the two systems to link up. The Navy's disciplinary code is pretty strict and there were a few things you had to remember.

Others also recall the problem of matching up the different ways of doing things for two services working in a common environment. One former airman on the carrier states:

> When you were on the *Albatross* you had two sets of rules and regulations—you had the Navy and you had the Air Force. The Navy do everything at the double, and you never see it in the Air Force, a man doubling, not anywhere. As an Air Force chap you'd be meandering down the deck and somebody up on the bridge is singing out 'Hurry up, that man down there', and you'd just look up and toddle on![119]

From 1930 the pattern of activity for *Albatross* involved long cruises to various Australian states, the northern islands and New Zealand, along with participation in training with other RAN ships (usually the cruisers). In between No.101 Flight was disembarked and returned to Richmond. These activities presented flight members with plenty of challenge and excitement. On 15 February 1930 HMAS *Canberra* found one of the carrier's Seagulls down on the sea off the Tasmanian coast, having run out of fuel. Fortunately, it stopped and lowered a seaboat to deliver cans of petrol before proceeding on its way.[120] The next month, on 19 March, the flight suffered its first fatality when one of four aircraft aloft to observe fall of shot during gunnery practice by the cruisers at Norfolk Bay (Tas) went into a spin from 2000 feet and crashed. The RAAF pilot of this machine (A9–4), Pilot Officer Gordon Grant, and his naval observer were both injured; the radio operator, Leading Telegraphist D.O. McGowan, was killed. Recalls Creal:

> The telegraphist operating the radio set on board the Seagull wore head phones which had a fixed plug which could not be pulled off. When the aircraft was recovered it was found that poor McGowan had the cord around his neck and he'd been drowned. After that we put a flexible jack on this cord.[121]

Grant's spinal injuries also proved to be so serious as to ultimately cause him to be invalided out of the Air Force. The accident was believed to have been caused by the pilot having stalled the machine while trying to make a quick turn to facilitate photography.[122]

By 1931 the RAAF was facing major difficulties in maintaining enough Seagulls to equip *Albatross*.[123] This problem was addressed in a number of suggested solutions, including the fitting of four Wapiti aircraft with folding wings and floats, or procurement of new Fairey IIIF floatplanes.[124] Both schemes were rejected in the light of the emergence in 1929 of a new amphibian designed by the British Vickers company specifically to meet the RAAF specification. By 1933, however, with the effects of the Depression biting into the Australian defence vote, the decision was made to pay off *Albatross* into the Reserve Fleet. According to one NCO member of the RAAF detachment at the time, the decision to put the ship out of commission was not solely a matter of economy; Wing Commander Bill Bradley recalls that it actually followed an incident at Storm Bay (Tas) when the *Albatross* had pitched beyond the 35° roll which was permissible in rough weather, resulting in the breakage of much crockery and the movement of several aircraft in the hangars. This 'capsize' had raised a serious questionmark over the ship's stability.[125]

When the decision to pay the ship off took effect in April 1933, No.101 Flight returned to Richmond to await re-embarkation whenever the carrier might be recommissioned. This never happened, in fact, because although a catapult was fitted early in 1936 in anticipation of the ship being brought back into service when the first of the new Seagull V amphibians were delivered, and launch trials with a Seagull V were actually conducted in August, the need for a carrier had effectively passed. From 1936 the RAN's cruisers were each able to carry their own Seagull V amphibian—'thus negating the requirement for *Albatross*'.[126]

In the meantime, *Albatross* remained in Sydney harbour, where it continued to serve as an operating base for visiting seaplanes. Whenever required for this purpose, personnel were despatched from Richmond while the Navy provided operators for the ship's cranes. As well as providing refuelling and mooring facilities for seaplanes, minor repairs were carried out onboard.[127] In 1938 the *Albatross* was passed across to the Royal Navy in part payment for a new cruiser.

In fact, detachments of 101 Flight had begun operations from the cruisers as early as 1931. During September of that year a Seagull III was loaned to HMAS *Canberra* for trials during that cruiser's northern island deployment.

A Seagull III is hoisted aboard the navy's seaplane tender, HMAS *Albatross*, in 1929. The ship initially carried a RAAF detachment of six aircraft and 30 officers and men, but by 1931 serious difficulties were being experienced in keeping enough Seagulls serviceable to go on board. By the time a replacement type was received, the *Albatross* had been paid off into the Reserve. (*WgCdr J.H. McCorriston*)

This experiment was repeated in September the following year, when *Australia* took Seagull A9–2, a pilot (Squadron Leader Hewitt), and four airmen on a cruise of the islands in September–October 1932.[128] When the two cruisers proceeded to New Zealand on a winter cruise in 1933, both had Seagulls embarked which attracted considerable public interest during their visit to Auckland.[129]

Following the decommissioning of *Albatross*, the Naval Board arranged to have Seagulls with small RAAF detachments to fly and maintain them regularly embarked on both *Canberra* and *Australia*. This now became the principal form of naval cooperation for the Air Force throughout the last half of the decade of the 1930s, and was extended to all the RAN's cruisers. A revealing comment on the attitude taken by the Navy towards embarked RAAF personnel was the report by Flight Lieutenant Philip Graham on board the *Canberra* in November 1935, complaining that: 'Although cruiser detachments have been borne regularly . . . for some years, no provision has been made for the definite allotment of a cabin for the air force officer . . . [who] is the only specialist officer in the ship to whom a labelled cabin has not been allotted.'[130]

In 1935 the first Seagull Vs began reaching the RAN and replacing the older Seagull IIIs on board the cruisers. *Australia* received her new amphibian while in England in September, shortly before being sent to the Mediterranean during the Abyssinian crisis. It was at Malta on 26 March 1936 that this aircraft (A2–1) was involved in a near fatal accident when the sling broke while it was being hoisted on board after a flight. The observer on the plane, Lieutenant-Commander V.E. Kennedy, later recounted that he was in a hurry to get on board on this occasion and got a boat to take him to the ship, leaving the telegraphist to hook the aircraft on to the sling—a job which he normally did himself while standing on top of the aircraft centre section:

> The TAG hooked on and then climbed down into the interior of the aircraft. Had he (or I) stood in the normal position we must have been severely injured because the new slip parted when the crane jib was at maximum elevation, dropping the 'Seagull' bodily on top of a 4-inch A.A. [Anti-Aircraft] Gun. Luckily no one was severely hurt, but the Seagull was a mess.[131]

Also a mess was Flight Lieutenant James Alexander, who was also in the aircraft when it dropped, and recalls: 'The W/T [radio] operator threw up his hands so vigorously he hit himself and knocked himself out. I was standing next to the seat and scraped my shin to the bone.'[132] A replacement machine (A2–12) was sent out from England and reached the ship before it sailed to return to Australia.

In April 1936, No.101 Flight was re-equipped with new Seagull Vs and became 'No.5 (Fleet Cooperation) Squadron'. The unit still had the task of providing the pilots and maintenance personnel, as well as the aircraft, required by fleet units, but it now became heavily committed to the program of aerial survey of northern Australia in which the RAAF was involved.[133] It was a No.5 Squadron Seagull V (A2–15) temporarily stationed at Archerfield which was involved in a fatal accident on 28 November 1938. While assisting in a police search for a missing woman, the aircraft was following the course of the Albert River, south-east of Brisbane, when it collided with electricity cables

After HMAS *Albatross* was decommissioned, single Seagull IIIs and small RAAF detachments (usually an officer pilot and four airmen) were embarked in the Navy's cruisers to enable cooperation to continue. This arrangement worked well enough, although the limited space on ships this size meant that the aircraft had to be stowed where it was exposed to the weather. This Seagull (A9–8) embarked in HMAS *Canberra* was damaged during a gale en route to Hobart in April 1935. (*RAAF Historical*)

spanning the river near the Alberton Ferry. Killed in the crash and fire that followed were Pilot Officer Max Wiber, AC1s E.A. Everett and A.D. Milner, and a police constable on board.

Of course, quite apart from shipborne aircraft a considerable amount of naval cooperation was carried by the RAAF in other forms and by other squadrons. This might be providing aircraft to act as the target for warships during 'burst short' firing practices (involving live firing with rounds fused to explode well below the target aircraft). Or it might entail the sort of activity undertaken by No.1 Squadron in October 1933, with patrol and bombing exercises against RAN cruisers carried out from an advanced base at Yanakie (Vic). In February 1937, No.3 Squadron sent six Demons to Jervis Bay for exercises with the fleet. This formation divided into two at Bowral, and making use of cloud cover, closed on the ships to make simultaneous dive-bombing and torpedo attacks. One RAAF report of the exercise noted that 'It is considered that the Navy offered very little defence against this form of attack'.[134] In October of the same year the squadron engaged in exercises as the fleet sailed from Sydney to Jervis Bay, shadowing and attacking the ships during their progress. In April 1939 the RAAF was involved in the strategic exercise 'N.Z.3', described as the largest naval exercise in Australian waters since the First World War.

No less important than these activities was ground contacts between the two services, involving the attachment of naval liaison officers at RAAF bases and

Aircraft of the units based at Richmond lined up in 1935. The first two rows are Wapitis of No.3 (Army Cooperation) Squadron and the four machines at rear are new Seagull Vs of No.101 (Fleet Cooperation) Flight (renamed No.5 (Fleet Cooperation) Squadron in April 1936). From May 1935 No.3 Squadron began receiving deliveries of new Hawker Demons to replace the Wapiti. (*RAAF Museum*)

training of naval aircrews. Although the Air Board appeared initially reluctant to accept naval 'observers' in RAAF units, by July 1937 one such officer had begun a twelve-month posting at Richmond (later extended at the Air Board's request) and another such attachment was made at Laverton in July 1938. Aircrew training primarily took the form of an armament course run at Richmond for RAN observers and telegraphists. From an ad hoc arrangement undertaken in 1932 in response to a request from the commander of the Australian Squadron, Rear-Admiral R.C. Dalglish, this training had evolved by 1935 into a formal bombing and gunnery course and was conducted annually 'until May 1939 when the last one merged with war preparation'.[135]

8

The Citizen Air Force

From the time of General Legge's proposals for an air service to be part of Australia's peacetime defence structure, it was envisaged that such a force would have to be primarily established on a citizen basis. As the CGS had argued in September 1918, Australia could never afford to maintain a really large air service in peace-time: 'The Government will have to apply to the air service the principle which it has already applied to the land troops, *viz.*, supplement permanent forces by citizen forces'.[1]

This principle was accepted by the first Air Board established in January 1920, comprising Captain Nunn and Lieutenant-Colonel Goble from the Navy and Brigadier-General Blamey and Lieutenant-Colonel Williams from the Army. This temporary body agreed that:

> There appears to be no reason why the greater proportion of the *personnel* should not be organized on a 'Citizen Force' basis . . . The long period of continuous training which is necessary to qualify as an air pilot may make it difficult to obtain citizen officers. If this difficulty can be overcome, and it is believed that sufficient young men will offer for the service, there is no other insurmountable difficulty in the establishment of a Citizen Air Force.[2]

Although the permanent Air Board finally established later in 1920 also accepted the principle of strong citizen involvement in the new air force, which at that point had still to be officially raised, it adopted a somewhat more cautious approach. Under the scheme of organisation which Williams proposed in December 1920, it was planned to have an air force of seven squadrons, along with various administrative, supply and training units. Of all these, no more than three air squadrons (two corps reconnaissance, one torpedo-carrier) were to be manned by citizen force personnel, and in the case of the torpedo-carrier unit even that would need to have 'a fairly strong permanent nucleus'. In Williams' view:

> The forming of Citizen Force Units must necessarily at the commencement be looked upon as an experiment and the cost of such units and their value compared with those of permanent units after a period of trial. It may be found that the cost of such units is out of proportion to their value and that a smaller

> establishment of permanent units gives better results from the point of view of efficient defence. These are matters which can only be learned from experience.[3]

Outlining plans for creation of an Air Force Reserve, Williams' submission pointed out that this could be undertaken at an early opportunity because of the considerable number of former members of the AFC and RAF who were available and could be expected to enrol. It might, he said, also be possible to have reserve units organised in civil aerial transport companies.[4] The scheme, recommended by the Air Board on 22 December for the consideration of the Air Council, was 'approved generally' when considered by the Council the following day.[5]

Introducing an Air Defence Bill into Parliament on 8 April 1921, the Minister for Defence described a program for the new Air Force (see chapter 2) which provided for the creation of three squadrons in each of New South Wales and Victoria, one in each State to consist of CAF personnel. While the CAF component of this initial structure was to be 36 officers and 300 other ranks, as against a permanent establishment of 150 officers and 1000 other ranks, it was planned that new units formed in later years would be mainly citizen personnel.[6]

As explained elsewhere, this original program was very quickly modified in line with economies imposed on the Department of Defence. By July 1921, the Air Council was actually pointing out to the Board that under the approved policy for development of the Air Force it was contemplated that only the permanent personnel necessary for 'the Cadres of war units and the training and instruction of Citizen Force personnel' were to be maintained:

> The provision of complete permanent units (other than Naval Units) capable of taking the air immediately upon emergency occurring is not at present a part of that policy . . . The strength of permanent personnel must be determined solely upon its functions . . . and the numbers provided should be disposed and employed to attain at minimum cost the training of the maximum numbers of Citizen Force and Reserve personnel.[7]

Williams' earlier comments about the element of experimentation attached to plans for the CAF suggest he would have been less than pleased with this avowed intention of relying on an air force that was no more than a basis for future expansion. The concept, however, had the important support of British authorities. At the time the Air Council was issuing its stricture to the Board about keeping the number of permanent personnel to a minimum, the Air Staff in London commented warmly on Australia's happy situation at having 'ample warning of serious attack . . . [which enabled it to] raise the greater portion of its air force on a citizen army basis'.[8]

In November 1921 Williams' proposals for the formation of the Air Force Reserve came before the Air Council. He estimated that there were then about 500 trained pilots in Australia, a number which obviously represented a valuable reservoir for a service which itself numbered only 50 officers and 258 other ranks. The Council approved the formation of the Reserve on 10 November, but required that it 'should be made clear to applicants for enrolment . . . that no provision is made for the grant of clothing or any allowances and that annual training cannot be given'.[9] In the event no action was immediately forthcoming on the formal creation of the CAF or the Reserve. It was 1924 before provisional regulations were completed,[10] and another year before first steps were taken to bring either organisation into actual being.

Active CAF

The Citizen Air Force finally came into being on 19 June 1925, when the first appointments were made to the strength of the 'General Duties Branch—Citizen Force'. The number of such officers shown in the *Air Force List* of August 1925 (the first such published list for the RAAF) was only eleven—three flight lieutenants and eight flying officers. The next list produced in February 1927, however, included 42 names in the GD Branch (five flight lieutenants, sixteen flying officers, and 21 pilot officers), along with two in the Stores and Accounting Branch and one in the Medical Branch. Of those in the GD Branch, eighteen (four flight lieutenants, twelve flying officers, and two pilot officers) were shown with 19 June 1925 as their date of appointment; three other flying officers were shown as having been appointed less than a fortnight later, on 1 July 1925. Based on this evidence, it seems that the CAF began its existence with no more than 21 officers.

Immediately following these appointments, the first CAF units—Nos. 1 and 3 Squadrons—were raised on 1 July 1925 at Point Cook. No.3 Squadron subsequently transferred to Richmond three weeks later, and No.1 Squadron moved to Laverton in January 1928 (see chapter 5). Both were described as 'composite' general purpose squadrons, with approximately one-third of personnel being drawn from the PAF and the remainder from the CAF. The latter were distinguishable from their permanent force colleagues only by a dark blue, red and aqua triangular flash, the colour patch of the old AFC, worn at the top of the sleeves of their tunic. Both units comprised separate flights of fighters, day bombers, and Army cooperation aircraft—these roles being filled by SE5A, DH9A and DH9 types respectively—with the idea of providing a basis for quickly expanding each squadron to three, each flight providing the nucleus of a new squadron of that type. These two squadrons were, in fact, the RAAF's only operational units, thus representing a total defeat of the concept originally favoured by Williams in preparing his 1920 scheme for the new service.

Over the next four years the approved establishment of the CAF rose from a total of 276 (54 officers, 222 airmen) in the 1925–26 financial year to 325 (56 officers, 269 airmen) in 1927–28 and 1928–29.[11] As the officer strength shown in the *Air Force List* during 1927 was always within the approved ceiling, it can be assumed that the actual strength of the CAF as a whole was somewhat below the establishment level. Probably the CAF was no more than 300 strong throughout its existence in the late 1920s. In the same period, establishment levels for the PAF rose from 717 (89 officers, 628 airmen) in 1925–26 to 1000 (110 officers, 890 airmen) in 1928–29, with actual strengths rising from around 700 to 930 respectively (see Appendix 3). It can be seen, therefore, that the CAF generally comprised about one-quarter of the RAAF's total active strength during these years.

In accordance with a ruling of the Air Council in February 1923, the period of annual training required of CAF members was 25 days.[12] Under the regulations approved in 1924, seventeen days of this period were to be spent in camps of continuous training with the balance made up of whole day, half-day and night parades of minimum duration. Following the abandonment of the compulsory training provisions of the Defence Act in 1929, the CAF became an all-volunteer force and the prescribed training was varied. Accordingly, in the 1930s the training requirement—which remained at 25 days annually—might be fulfilled through attendance at either a sixteen-day camp and parades of a day or less duration or entirely through short parades without any camp.[13]

Training for CAF personnel got underway late in 1925, when No.3 Squadron held its first squadron parade. At the time of its transfer to Richmond in July the unit's only CAF members were officers, but on 1 October 90 compulsory trainees selected from various Army units were transferred to the squadron and taken on strength. Drill and technical training for these CAF airmen commenced on 17 October at the first unit parade.[14] Parades took place at Sydney's Victoria Barracks in Paddington, in a building formerly known as the 'Old Paymasters Hall' which the RAAF took over as a drill hall.[15] The first annual camp of continuous training for the squadron (both CAF and PAF) was held early in March 1926 at the west end of the Richmond aerodrome. Though hampered by bad weather a considerable amount of flying and ground instruction was carried out.[16]

The first annual camp at No.1 Squadron was held at the same time at Point Cook,[17] during which an intensive program of advanced training was undertaken:

> This included air gunnery using a Lewis camera gun, air to ground live gunnery, low bombing with 25-pound concrete bombs . . . co-operation with field artillery and army signallers, oblique and vertical air photography reconnaissance, and puff target practice; and, of course, much formation flying.[18]

It was, however, reportedly not until 22 July 1926 that the first night parade for this unit took place at the North Melbourne drill hall.[19] Thereafter ground training was provided at mid-week evening parades conducted throughout the year at the drill hall, with flying parades held at Point Cook on one Saturday each month.[20]

By the time No.1 Squadron held its third annual camp in March 1928, the unit was located at its new base at Laverton. Marking the conclusion of this camp, the squadron marched through the streets of Melbourne. The next year's effort had about it something of a carnival atmosphere, judging by various accounts. The unit diary records that:

> The Squadron (permanent and Citizen Force) went under canvas, the camp being erected on the vacant land between the School of Aeronautics and the Officer's Mess. A temporary kitchen was built and all cooking and catering under normal rationing conditions was carried out by Squadron personnel for approximately 250 airmen . . . Recreation tents were erected and some fine entertainments were run by the officers and airmen during the evenings.
>
> A few days before the end of the camp a sports days was held on the Station oval and some very good times for athletic events were recorded. Visitors were

The No.1 Squadron CAF camp at Laverton, viewed from the air in May 1929. These annual events in the unit calendar provided the most significant opportunity to conduct squadron training on a continuous basis, but the camps also came to have something of a carnival atmosphere which turned them into social events as well. (*RAAF Museum*)

> permitted to attend, afternoon tea being provided and a most enjoyable day was had by all.[21]

Not mentioned in this description was the appearance during the camp of a magazine called 'The Squadron Scout',[22] or the fact recalled by one participant that the occasion was 'enlivened' by the squadron turning out at Flemington racecourse on 18 March for the arrival of civil aviator Bert Hinkler, at the end of his famous solo flight from England.[23]

This, then, became the pattern of training for the CAF. Air Commodore Rex Taylor recalls that he was still studying at night for a diploma in engineering at Swinburne Technical College when he met up with a RAAF warrant officer recruiting for the Air Force in 1933. He put his name down, and was subsequently called to the North Melbourne drill hall which was CAF Headquarters:

> Within a couple of weeks I was a delighted member of the CAF. That meant going every Thursday night to the North Melbourne Drill Hall, and every second Saturday appearing at 1 pm at Laverton, where we spent the rest of that afternoon and all day Sunday until 4 pm . . . I was only an AC2—the lowest Air Force rank—but it was a wonderful way of converting my civilian skills into playing around with real live aeroplanes, and occasionally getting a flight in the backseat of a Wapiti or a Gipsy Moth. It was a real little Saturday afternoon club.[24]

The picture was much the same at Richmond. One member of 3 Squadron in the early 1930s, a PAF officer, recalls:

> At weekends the Citizen Air Force joined us, officers to be given flying training in Moths of which each flight had two. Airmen were given instruction in their trades which they received by being attached to airmen of the Permanent Force. Thus we worked right through weekends and had Mondays and Tuesdays off duty.[25]

A CAF pilot with this unit in 1934–35 also described weekend training at Richmond:

> We had a fair selection of non-commissioned airmen in the CAF. My gunner was a CAF bloke, and so too were many of the ground staff—the riggers and fitters—although I don't think they were at the stage where the system relied on them to maintain our aircraft. There would be between 30 and 50 airmen arrive with us by train. When we got off, we had to assemble the troops and march them down the road to Richmond. We'd spend from 10 or 11 am on Saturday until late Sunday doing training. There was no fixed routine, the work varied with weather and the number of aircraft that were serviceable for flying.[26]

Many others apart from Rex Taylor were struck by the atmosphere which pervaded the CAF as some sort of exclusive club. One who saw it as 'half way between an air force and a club' recalls the presence of First World War pilots like Thomas White (later Sir Thomas and a federal Cabinet minister) at the North Melbourne drill hall:

> I don't know what the arrangement was, whether they were just members of the Mess or there for the excitement. Their status with the RAAF was never clear, but they'd turn up at parades, and dining-in nights at Point Cook. And when an aerobatic team was required for an air show, often it would include some of these wartime pilots who were still classified as being fit to fly.[27]

Despite the impression which might be gained that the CAF operated rather as a social organisation than a serious part of the national defence structure, it was nonetheless true that citizen pilots were absolutely integral to the normal flying operations of Nos. 1 and 3 Squadrons. Indeed, many activities could not have been undertaken without them simply because the permanent cadre was too small. Thus, during the visit of the Duke and Duchess of York in 1927, the participation of large numbers of RAAF aircraft included in the program of events associated with the visit was only possible by the heavy utilisation of CAF personnel. The Air Force presence during the Opening of Parliament ceremonies at Canberra was bolstered by holding the annual training camp of No.3 Squadron in conjunction with this event (see chapter 13), and the role of CAF pilots during this period was graphically underscored by the death of Flying Officer Vincent Thornton in the tragic mid-air collision over Melbourne on 21 April, during the welcome to the royal couple (see chapter 11).

Similarly, when the South Australian section of the Australian Aero Club sought a RAAF presence at an air pageant it planned to hold to mark its opening in November 1927, the officer commanding No.1 Squadron, Flight Lieutenant Ray Brownell, was instructed by the Air Board to prepare to send nine machines across to Adelaide for this occasion. As Brownell's plans reveal, he was counting on the availability of CAF pilots to enable so many aircraft to take part, and in the event half the eight planes that were eventually despatched from Point Cook had citizen force pilots.[28]

Despite this vital connection between the PAF and the CAF—or really because of it—when Sir John Salmond delivered his findings on the RAAF in 1928, he was strongly critical of what he saw. The system of composite squadrons was, in his view, the antithesis of operational effectiveness and left the

RAAF 'totally unfit to undertake war operation'. The problem, as he saw it, was that 25 days was an insufficient period in which to properly train air crew for fighter or Army cooperation duties:

> It is an established fact that the efficient handling of modern war aircraft is attained only by the most constant practice, that the highest skill in the art of air fighting, the ability to navigate and bomb accurately and efficiency in co-operation with other arms, can be obtained only by constant training in a 'full-time', i.e., Permanent, unit.

For these reasons he recommended that a proportion of RAAF operational units should be maintained on a permanent basis. In his eyes, too, the concept of having composite squadrons as a basis for future expansion in an emergency presented such other problems that he considered it unsuitable to be retained for citizen force units:

> Arising from this triple composition [of flights for Army cooperation, bombing and fighter duties] are difficulties in maintenance within the squadron of three separate types and complications in training of personnel to three entirely dissimilar roles . . . I do not find the stability in these squadrons to justify this policy, well intended as it is, and I consider better value would be obtained in future by confining these and any future Citizen Air Force squadrons to one service type machine and one operational role only.[29]

Salmond's conclusions had obvious appeal to Williams, mirroring as they did the course he had wanted to take back in 1920. The onset of the Depression a year after the air marshal's visit frustrated any hope that funds might be made available to implement the recommendation for regular squadrons to be established, and not until expansion became government policy in 1934 did the time seem opportune to move in that direction. Then, however, Williams found himself confronted with the entrenched resistance of politicians for whom a citizen defence force was an article of faith. When the Air Board made provision in the 1935–36 Estimates for the formation of three general purpose squadrons manned entirely by PAF personnel, the Minister for Defence (Parkhill) wrote to CAS on 27 May 1935 to let him know of his disapproval:

> The underlying policy of our land defence forces is a Citizens' Force organisation based on a Permanent Force instructional nucleus and operational units are maintained with permanent personnel only when the work involved cannot be carried out satisfactorily by the Citizens' Forces. Unless there are sound reasons against it, I would prefer to see this policy followed by the Air Board in the organisation of new landplane units.[30]

Williams responded by warning the Minister that savings made by maintaining inefficient CAF units was a false economy which imperilled national defence. He therefore advised that the Air Board 'most strongly' recommended against further units being raised on a citizen force basis.[31]

Parkhill accepted this advice with some reluctance, and only after having turned to the Council of Defence and then the Air Ministry in London for support which was not forthcoming in either case. The course adopted in the short term was, however, for the CAF element of the two existing service squadrons to be withdrawn, and transferred to two new CAF units—Nos. 21

and 22 (Cadre) Squadrons—created at Laverton and Richmond on 20 April 1936. These new units still required a permanent core, and in the case of No.21 Squadron the proportion of PAF to CAF personnel was quite unchanged: seven officers and 43 airmen (PAF) to 20 officers and 106 airmen (CAF).[32] The real difference was that Nos. 1 and 3 Squadrons now became fully permanent—No.1 as a bomber and No.3 as an Army cooperation unit—although both were initially short of personnel and equipment.

On its formation No.21 Squadron was equipped with four Hawker Demons, two Wapitis and three DH60 Gipsy Moths. The existence of its permanent cadre, staffing the headquarters and one of three flights, enabled the unit to operate on other than a weekend basis, with the result that elements of the squadron participated in many of the activities which fell to the RAAF as a whole. Aircraft took part in aerial searches for missing aviators, in the North Australia Survey, and in air pageants staged for the general public. At Christmas 1936 No.21 Squadron also took over the routine task of daily meteorological flights from No.1 Squadron while that unit went on annual leave, and the next Christmas carried out bushfire patrols.[33]

On 1 July 1936 both Nos. 21 and 22 Squadrons had dropped 'Cadre' from their titles and took the name of the capital city on which they were based, becoming now No.21 (City of Melbourne) Squadron and No.22 (City of Sydney) Squadron; a third such unit, No.23 (City of Perth) Squadron, was formed in May 1937 (its story is related primarily in chapter 5). At the end of 1936 and during early 1937 the two squadrons received new aircraft. No.21 Squadron took delivery of its first Avro Anson (A4–1) in December 1936, and the following March No.22 Squadron received its first machine of this type. On 19 January 1937 No.21 Squadron also took temporary custody from No.1 Squadron of the RAAF's remaining Bulldogs pending the formation of a new squadron to be numbered No.2.[34] These last-mentioned machines were eagerly utilised by squadron members; during their time with the unit, one (A12–8) was crashed into the sea at Point Cook and written off. One CAF officer who got to fly these planes enthused over their aerobatic capability: 'To play around big cumulus clouds in a Bulldog was a delightful way of spending an afternoon'.[35]

Sadly, it seems that much of the training carried out was perfunctory or of little real relevance to developing skills. One member of No.21 Squadron recalls being asked by the permanent officer commanding the unit in 1938 about which parts of the training syllabus he had yet to complete: 'At that stage I hadn't done any photography, and he asked "Did you have a camera with you on that day?" As a matter of fact I did. "Alright", he said, "you've finished your camera work"'.[36]

The aura of an exclusive social club still clung to the CAF. The annual training camp for No.21 Squadron, conducted at Easter 1937 (26–29 March), was keenly watched by the Governor of Victoria, Lord Huntingfield. He inspected the work of the unit during the camp, was taken for a flight in Anson A4–7, then called a parade 'and expressed his interest in, and appreciation of, the work of the Citizen Air Force, and he confirmed this the next afternoon when he telephoned the Commanding Officer', Flight Lieutenant Robert Dalton.[37] Huntingfield's interest did not end here, however, and on 1 July he

Aircraft of No.21 (City of Melbourne) Squadron lined up at Laverton in 1938; in the foreground are Avro Ansons, with three Bristol Bulldogs next, followed by Hawker Demons. (*Mr P.C. Eaton*)

became honorary air commodore of the squadron.[38] Not to be outdone, the first honorary air commodore of No.22 Squadron from 1 August 1937 was the Governor of New South Wales, Lord Wakehurst.

In November No.21 Squadron was involved in escorting the Victorian governor to Canberra to visit the Governor-General, Lord Gowrie. This exercise entailed the unit's seven Ansons and three Demons flying in formation to Canberra, sitting on the unsealed aerodrome which later became Fairbairn airbase while the CO, now Squadron Leader Charles ('Moth') Eaton, and Squadron Leader Fred Thomas (one of the pilots and an honorary aide) joined the viceregal party at lunch, then flying back to Laverton.[39] The trip was notable only for the fact that one of the Demon pilots, Pilot Officer Ian Campbell, tipped his machine (A1–9) on its nose on landing at Canberra.[40] Later Huntingfield was instrumental in establishing an affiliation between No.21 Squadron and No.600 (City of London) Squadron in the Auxiliary Air Force of the RAF, announced in August 1938,[41] and in having the crest of his coat of arms incorporated in the unit badge authorised for the squadron in July 1939.[42]

The Active CAF undoubtedly had about it a distinctive cachet, combining the glamour of flying with a certain social standing. In 1938 it attracted even the eldest son of Prime Minister Joseph Lyons as a recruit to No.21 Squadron;[43] enlisted as an air cadet on 26 December 1938, young Desmond Lyons was appointed to the CAF as a pilot officer on 3 March 1940. Many of those initially drawn to service with the CAF subsequently made the transition to careers in the permanent force. Men such as Rex Taylor previously mentioned. And John Miles, who joined the CAF as a radio operator mechanic in 1926 before becoming a cadet on the 1929 'A' flying course, and later became well-known as a test pilot.[44] Yet another was Bill Garing, who joined the CAF in 1927 as a wireless operator/air gunner with No.1 Squadron; he recalls:

> I enjoyed the CAF so much, and I began to get the feeling for flying. Eventually I went to see the CO, then Squadron Leader Johnnie Summers, and explained to him that I thought I'd like to join the Permanent Air Force. I must be very frank. He looked at me and said, 'The only way to really become a firm and long-time

> commissioned officer in the RAAF is to go to Duntroon. I hate Duntrooners, but it's the only way.'[45]

Garing followed this advice and won a distinguished record during the Second World War, but like many others he remembers his start was with the Citizen Air Force.

The true test of the value of the CAF as an effective element in the nation's defence came, of course, with the Second World War. It is inescapable that the rapid expansion of the Air Force on the outbreak of war in Europe was only achievable because of the reservoir of partly trained personnel then available in the Active CAF and the Reserve. There can be no doubting, either, the ability of the bulk of this personnel, a fact to which the high rank and other distinctions achieved fully attests. Yet one doubt must be retained, reflecting on the effectiveness of the citizen force concept rather than the personnel concerned. It is the doubt expressed by a pre-war regular, who later said:

> The training given to the mostly professional men who came up for flying training was rather more sparse than we got as short service commissioned officers. I could not help but notice that these officers, although some were granted fairly senior rank when war actually broke out and they took their positions in the various units, the lack of training became very apparent. They were far too senior for their experience in the service and I think that in some ways it showed up rather badly.[46]

'B' flying courses

The formal creation of the CAF in mid-1925 brought with it a requirement to train members of the new organisation as pilots. Plainly, there were too few fliers in the RAAF's permanent element to provide sufficient personnel for flying duties as the citizen scheme progressed. Air Force authorities recognised, however, that the course through which new pilots for the PAF passed was unsuitable for members of the citizen forces. Few men could afford the time away from occupations which provided their livelihood to undertake a lengthy course, and it was considered impracticable to conduct effective pilot training during normal CAF weekend parades and an annual camp.

A condensed course was accordingly devised, designated the 'B' Course to distinguish it from the longer one already conducted and which now became known as the 'A' Course. This short course concentrated on providing flying and aircraft instruction, with just the bare basics of Air Force life thrown in, and was intended to be followed up with advanced flying instruction every second weekend and non-flying training at normal CAF parade nights. The course was also designed particularly to accommodate university students, most of whom were liable for compulsory training under the Defence Act anyway, during the period between December and March when they were free of lectures. Apart from the glamour of learning to fly, the prospect of earning pay of 10 shillings a day during the long vacation was a considerable attraction to students.

During the latter half of 1925, recruiting for the new flying course was conducted at Sydney and Melbourne universities. As these were the centres in proximity to the newly-formed CAF units, it was impractical for candidates from other States to be seriously considered. After interviewing and medically checking 200 applicants, a total of 30 were accepted—17 from Victoria and 13 from New South Wales.[47] These were required to undertake to serve in the RAAF for eight years, although four years could be spent on the Reserve.

Those selected for the first 'B' flying course reported on 14 December 1925. Most travelled by train to Laverton, and were met there by the adjutant of No.1 FTS, Flying Officer Eric Woodward, who supervised their movement to Point Cook in Crossley tenders. Also on hand for their reception was Flying Officer Frank Bladin, who had been given overall responsibility for running the course. A few of the cadets arrived by car, one of these being Wilfred Brookes (later to distinguish himself as CO of No.78 Fighter Wing in New Guinea in 1944), who had been pursuing an interest in aviation since purchasing an Anzani aircraft engine at the auctions conducted at the CFS in 1919 (see chapter 6). He has stated: 'I remember I had a small car, which was a great advantage. It was quite a pleasant day, so I motored down to Laverton. The Aircraft Depot was still being built then, and the site was only paddocks and a train stop.'[48]

Another member of the course, Leslie Robertson, records that the cadets were quartered, two to a room, in dusty huts of corrugated iron supplied with cold running water. At parades the course was placed in a separate flight behind the main body, under the strict eye of Sergeant Major Charles Oliver. Having formerly served with an hussar regiment in India and the South African mounted police, Oliver had the responsibility of 'imparting extra spit and polish' not usually encountered in CAF units and proved also to be 'a general adviser, not to say friend, to the cadets'. For aviation training the course was split into two half-flights, with one group carrying out flying in the morning and ground instruction in the afternoon, while the other worked on the reverse arrangement. Robertson writes:

> A primitive air traffic control system existed. There was no tower and no radio contact with the aircraft. An officer, wearing a red armband, was daily appointed to patrol the tarmac and generally keep an eye on . . . aircraft movements. The only real control consisted of a coloured flag designating a right or left hand turn to be taken after take-off depending on the wind. There were, of course, a crash ambulance and fire tender always standing by under orders from the officer.[49]

After the first week, each cadet had logged over four hours flying; most went solo after 11 or 12 hours of dual instruction. The solo flights were initially done in two-seater Avro 504K trainers, for which cadets were introduced to 'Aircraftsman Sandbag'—a 120-pound sack of sand placed in the empty seat to trim the aircraft in the absence of a passenger's weight.[50] After several solo flights the cadets were transferred to service aircraft. To quote Robertson again:

> Due to the condensed course it was deemed impracticable to put each cadet through all of the three types; thus the trainees were split into three groups, according to their ability. Those showing evident dash and devilment went to the SE5A fighters. Normally a single-seater, a specially modified training version was used having room for dual instruction. The 'staid' types with good landing records

Once cadets on the first 'B' flying course in 1925 went solo, they began training on the types of service aircraft to which they seemed best suited by temperament and ability. Cadet W.D. Brookes went onto SE5A fighters, and is shown here at the controls of A2–11. (*Sir Wilfred Brookes*)

> took on the heavy DH9A. The slightly less expert were left with the DH9, a delightfully amiable light bomber.[51]

Brookes was among those allocated to fighters. He recalls:

> That Christmas they converted one of the SE5As into a twin seater, called the 'Flying Pig'. The wing loading was higher than other aircraft then flying, and when you pulled off the power—such as it was—the plane just fell like a stone. First I was on 504Ks and I got off after five-and-a-half hours dual. When I had between 15 and 20 hours in 504Ks, I was then converted to SE5As. I think I was the first or second pilot actually to be converted in the 'Pig'. Before that, people had to go up from the Avro solo.[52] [For more on the 'Pig', see chapter 6.]

Inevitably there were incidents associated with the flying training. Robertson remembers the cadet who 'became expert at hooking the landing skid under the top wire of the aerodrome fence in his initial landing attempts, thus failing to encourage his colleague next in line for solo'. Several cadets also experienced forced landings in Avros, including one at Laverton which nearly ended in disaster. Recalls Robertson:

> At that time the Laverton aerodrome was under construction with frequent blasting out of rock formations. Not knowing this, the cadet made an excellent forced landing in what appeared to be a suitable space only to be shocked by finding himself surrounded by explosions hurling rocks in all directions.[53]

Some incidents had an amusing side, as in the case of the instructor who took off with a cadet in a DH9 only to have his joystick come adrift from its socket.

'Despite the muffled sounds in the rubber intercom the pupil failed to comprehend until the instructor beat him about the helmet with the stick. The aircraft, by then in a steep climb, was very quickly put into a more comfortable attitude.' In similar vein was the antics of one instructor, a wartime Sopwith Camel pilot, who took a cadet on a dawn flight for what he termed 'contact chasing':

> For ten minutes the Avro averaged 25 feet height as they flew through farmyards and dodged outhouses much to the bemusement of the occupants . . . and the cadet! Finally, climbing to 1000 feet, the instructor rolled into a vertical dive above a stricken aircraftsman on the tarmac. On pulling out at 100 feet the tight turn caused the cadet to feel his forehead creeping down to his chin with forearms welded to wooden legs. The feeling passed off as horizontal flight was resumed. The aircraftsman passed out and the pilot was 'seen' by his CO![54]

Some of the incidents were distinctly less than amusing, such as one involving:

> a regular who in a SE5A made contact with an Avro carrying a cadet with an instructor during a landing approach. Fortunately the trainer landed safely although with six inches missing from the starboard upper rear main spar whilst the SE pancaked with its undercarriage dangling in a dribble.[55]

On 12 March 1926 Cadet W.S. Hamilton crashed a DH9 (A6–11) in truly spectacular fashion while making a solo flight. He had taxied out to the western side of the airfield to take off into the wind past the line of Bessoneau hangars. When ready for take-off, he opened up the throttle and watched over the left side of the cockpit as the machine gathered speed. By not looking over the right-hand side also, he did not notice that his sudden over-handling of the throttle had caused his machine's nose to swing so that he was now headed directly for the open end of the second last hangar. Robertson remembers that the DH9 emerged at the other end of the structure 'stripped clean of the upstanding exhaust pipe and the fuel pump windmills, to say nothing of all the wings and the centre-section'![56] The crash was witnessed by Brookes too, who recalls:

> It was a fluke that it was morning tea time and the troops had just left their jobs inside the hangar and were sitting outside in the sun with their backs to the wall, drinking tea. Hamilton was actually just lifting off the ground when he disappeared inside, smashing three or four machines as well as his own.[57]

Francis Bladin, the adjutant, also recounts that he was rushed to the scene from his office in a motorbike with sidecar to find that the ten mechanics on the ground had escaped with only one suffering superficial injuries, and the pilot still lying shocked in the cockpit under a pile of debris. He discovered, however, that there had nearly been two other casualties:

> I met these two as I got out of the side-car, standing still shaking in a state of shock under the nose of the wingless aircraft protruding through the canvas back wall. One was my warrant officer, Hugh Cully, his face covered with small blood spots where it had been punctured by flying splinters. The other was Sergeant Best, the orderly sergeant that day. The two of them had been on their way to carry out the daily inspection of barrack rooms. Cully could not speak at all, and Best, who suffered from an asthmatic larynx, could only wheeze 'My Gawd'.

The chaotic scene after Cadet Hamilton accidentally flew DH9 A6–11 inside a Bessoneau hangar at Point Cook on 12 March 1926. Hamilton has just been assisted uninjured from the cockpit; a mechanic who sustained minor injuries is lying on a stretcher out of frame at left; a DH9A, its back broken in the crash, is visible inside the hangar. (*Sir Wilfred Brookes*)

In all, thirteen men had escaped injury or death by the narrowest of margins. Bladin records that a few days later Cadet Hamilton attempted a second take-off, this time to the south over the sea:

> He mishandled the petrol valve and the engine malfunctioned when he was about a hundred feet off the ground. At the time I was lecturing to a rather disinterested and sleepy sea of faces on Air Force Law . . . the sea of faces changed to a sea of backsides as my audience rushed to the windows in time to see the aircraft splash into the sea. We had to cease that pilot's training, as a third time might have been unlucky.[58]

On the day that Hamilton suffered his first mishap, the city coroner was already conducting an inquest into an earlier fatal crash on the course. Cadet Alfred Greenwood, a 19-year-old dental student at Melbourne University, had been on solo practice in an SE5A (A2–28) when, at about 10.30 am on 19 February, he cut off the engine to glide in for a landing. Still at 150–200 feet when he realised he had misjudged his approach height, the pilot tried to restart the engine and go around for another attempt. The engine failed to pick up again, however, and he was left with a choice of landing in the sea, along the beach, or trying to regain the aerodrome. Deciding on the latter course, he made a right-hand turn which the machine had almost completed when its nose was seen to drop. Entering a dive and then a spin, the aircraft struck the ground immediately afterwards, killing Greenwood instantly.[59]

Robertson comments that:

> The not infrequent landing mishaps together with the one fatality brought the Course some unwelcome publicity when a notorious Melbourne weekly with some heavy headlines announced the activities of the 'Australian Flying Corpse'. The details showed that there must have been a 'mole' on the Station. An enterprising

> aircraft hand general had, indeed, been making a little on the side by his reports.[60]

The RAAF had reason to be sensitive about the accident rate of the 'B' Course. At its 'passing out' on 30 March 1926, the CO of No.1 FTS, Squadron Leader Adrian Cole, referred in his final address to the course's unfortunate record in this regard. There had been no fewer than 24 crashes—involving 14 Avros, five DH9s, three DH9As, and two SE5As—and it was, he said, amazing that the majority had not caused serious injuries to personnel: 'either "Young Australia" is made of India-rubber or else you have gained so much experience in the art of crashing that you can do so moderately safely'. Cole calculated that the replacement value of the crashed aircraft alone was about £30 000, and when hidden expenses were included the final cost to the Air Force was more likely in the vicinity of £50 000—a sum approximating $1.3 million at current values.[61]

On 25 May 1926 the 26 cadets who had successfully completed their training were detailed for flying duties with Nos. 1 and 3 Squadrons. They retained their cadet rank, however, pending gazettal of the appointments. When this occurred in June, 20 became pilot officers and the rest were made sergeant pilots, with effect from 1 April.[62] Robertson writes: 'The Pilot Officers obtained the necessary silver topped black canes and had the narrow gold band added below the golden crown and eagle on their uniform cuffs'.[63] Notwithstanding this rise in status, all now embarked on advanced training.

In December 1926, a second 'B' course began training. This course was similarly constituted, although among its number was AC2 Fred Thomas (a later Lord Mayor of Melbourne), who was already a member of No.1 Squadron.[64] It was unfortunately similar to its predecessor, too, in that there was again a fatal crash once the solo phase of training was reached. On 14 February 1927 Cadet Alexander Dix was making a landing approach in an SE5A (A2–19) when, like Greenwood the year before, he stalled the aircraft (that is, lost sufficient speed to maintain flight) and crashed, killing himself and completely destroying the machine.[65] Nine members of this course were appointed pilot officers with effect from 25 March 1927,[66] among them Raymond Garrett (later to become president of the Victorian Legislative Council and a knight).

A shift in this pattern of these courses occurred following Sir John Salmond's 1928 inspection and report on the state of the RAAF. Among the air marshal's recommendations was a suggestion that the length of the 'A' Flying Course, at that time about eight months, was too short to adequately cover the training curriculum and needed to be extended to twelve months' duration. Extending the 'A' course, however, would make it impossible to also conduct 'B' courses at No.1 FTS, and Salmond proposed that these should in future be run by the two CAF squadrons themselves.

> If, as seems possible, the number of university candidates falls short of requirements, the advantage of conducting Citizen Air Force training at Squadrons becomes apparent, as it will then be practicable to select for training other volunteers in the same district . . . The scheme will [also] have the advantage of training these officers in the State to which they belong, and will obviate the

necessity and expense of transferring CAF Officers from other States to Point Cook, Victoria.[67]

Salmond's recommendation was in fact taken up in December 1929, with nine cadets starting a 'B' course at Laverton and another eight at Richmond.[68] These were, however, the only such courses conducted for some time, due to economies which hit the service during the Depression. Not until 1932 were 'B' courses resumed, with the first one of these in Victoria being run back at Point Cook; it was 1933 before another 'B' course was conducted at Laverton by No.1 Squadron.[69] Group Captain John Graham, who began his CAF flying career with the first of the revived courses in December 1932, recalls that among those attending with him were John ('Sam') Balmer (killed in 1944 on his last operational mission as CO of No.467 Squadron (RAAF) serving with the RAF), Pat Fairbairn (brother of James Fairbairn, MHR who later became Minister for Air), and John Lerew (later RAAF commander at Rabaul in 1942).[70] At the same time, a 'B' Course was being held at Richmond for the first time, among the trainees there being Clarrie Glasscock (killed over New Britain in 1943 as CO of No.30 Squadron).

As Salmond had foreseen, the holding of courses at Richmond widened their appeal among Sydney candidates. For the 'B' Course commencing on 4 December 1933, there were reportedly several hundred applicants for the twelve vacancies available.[71] According to Tom Lawrence, a later controller-general of the Department of Productivity and one of the members of this course, those selected included other than just the usual university students: 'We were a mixed bag. One fellow was a wholesale green-grocer, another a farmer. Then there was a lawyer, a doctor, a veterinary scientist.'[72] And one of the regulars tasked with instructing them recalls:

> The CAF types were pretty good. A large percentage of them were Sydney university students and they were a very good type of person . . . so it was interesting having to train them. They came down every second weekend for training and turned out to be very good all-round, a better sort in general to the Air Force type.[73]

Considerable prestige also became attached to obtaining selection for these courses. On his first official visit to Richmond on 29 March 1935, the new Governor of New South Wales (Brigadier-General Sir Alexander Hore-Ruthven, VC) inspected the graduating 'B' Course members, which included Peter Jeffrey (later to command No.3 Squadron in Syria).[74]

These arrangements quickly came in for further change when Nos. 1 and 3 Squadrons became wholly PAF units and the CAF were separated into 'twenties' squadrons manned by small permanent cadres. These new units in their turn now became responsible for training their own pilots. Although both No.21 (City of Melbourne) Squadron and No.22 (City of Sydney) Squadron were established in April 1936, it was not until the following year that 'B' Courses were finally got underway. No.21 Squadron had an intake of 12 cadets under instruction during the continuous training camp over Christmas 1937 and New Year 1938;[75] six of these received their Pilot's Flying Badge on 7 June 1938,[76] followed by another two, including David Colquhoun (later to achieve prominence in the post-war RAAF), on 22 June 1938.[77] No.22 Squadron received its first intake of nine cadets on 13 December 1937, and seven of these completed their training in August 1938.[78]

The formation of No.23 (City of Perth) Squadron at Laverton only in May 1937 and delays associated with its planned move to Pearce (see chapter 5), meant that it was not until October 1938 that the first CAF trainees were enlisted on a 'B' Course with this unit. From over 100 applicants,[79] just eight were selected as air cadets on this course, and by the time the first month of training was completed on 6 November the number had been reduced to six.[80] Those remaining continued to receive air and ground training at fortnightly weekend camps of instruction, until graduated and posted to the squadron on 2 September 1939. As circumstances would have it, these six were to become the only batch of pupils to receive their initial flying training at Pearce before the start of war.

For a proportion of graduates of the 'B' course, their attainment of this flying qualification did not necessarily mean that they were confined to serving in the CAF. Up to 1934 there were no fewer than six instances where graduates of various 'B' courses had been permitted to subsequently upgrade their qualifications to 'A' standard by undergoing the longer course.[81] The officer trainees in this position kept their CAF title as honorary rank, but were paid as cadets; in at least one case, the flying and general service experience of the officer concerned meant that he was not required to take the first term's work.[82] The purpose of this arrangement in four cases appears to have been to enable the person concerned to apply for a short service commission in the RAF, but for one officer it lead to a permanent RAAF commission (which in the event proved to be not so permanent when he was transferred to the Reserve during the retrenchments of the Depression).

With the commencement of the RAAF's expansion from 1934, however, and a growing need for more pilots in the PAF as this program gathered pace, it was an increasingly common practice for CAF-trained pilots to be offered short service commissions in the PAF. Accordingly, such figures as John Balmer, John Lerew, and Peter Jeffrey made the transition from the CAF to the regular force before the outbreak of war totally transformed the RAAF scene.

Air Force Reserve

The Air Force Reserve was conceived as a means of retaining the services of personnel who possessed skills of value to the RAAF in a defence emergency. Its members were expected to comprise time-expired members of the Active CAF, former members of the PAF, and pilots in civil aviation who were required by the terms of their engagement to serve in the Reserve.

Because of the anticipated source of its personnel, the development of the Reserve naturally lagged behind the growth of the Active CAF and took shape only very slowly. When included for the first time in the *Air Force List* of August 1927, just three officers—all general duties branch—were shown to be members: Flying Officer Arthur Long, and Pilot Officers Robert Griffin and Donald Mack. Long was among those originally appointed to the CAF on its formation, and transferred to the Reserve on 1 July 1927; Griffin and Mack had completed the first 'B' Flying Course conducted at Point Cook, and were transferred from the Active CAF on 1 July 1926 and 1 August 1927 respectively.

By September 1934, however, the Reserve had a strength of 87 officers, not including two officers holding honorary commissions (Air Commodore Sir Charles Kingsford Smith and Flight Lieutenant Charles Ulm). Three of this number were flying officers in the stores and accounting branch, three were doctors in the medical branch, two were chaplains, and the rest were general duties pilots. Of the 79 officers in the last-mentioned category, there were two wing commanders, a squadron leader, 19 flight lieutenants, 45 flying officers, and 12 pilot officers. Reflecting the severe impact of the Depression, nearly three-quarters of all these officer appointments to the Reserve had been made during or after 1930, and the vast majority were officers transferred from the PAF or the Active CAF. In August 1939, on the eve of the Second World War, 158 officers were now on the Reserve: 137 of these were general duties personnel, while there was still three in the equipment branch, 10 in the medical branch, and eight chaplains.

According to a 1934 pamphlet outlining conditions of entry to the Reserve,[83] airmen also formed part of the Reserve. Surviving records do not indicate the extent of any such enlistments so it is not possible to give a complete figure for the overall strength of the Reserve at any one time. The 1934 pamphlet does at least give a clear idea of how the Air Force Reserve was structured and what was required of its members at that time. According to this document, the majority of personnel in its categories were expected to have prior service in

A scene at No.22 (City of Sydney) Squadron, Richmond, in April 1938. By this time the increased allocation of funds devoted to expanding the RAAF had ensured the provision of a high standard of modern facilities at permanent air stations. (*RAAF Historical*)

The Governor-General, Lord Gowrie, inspects Pearce (WA) in 1940, at which time the construction work underway at the base is still very much in evidence in the background. The station was first occupied in March 1938 by No.23 (City of Perth) Squadron, renumbered No.25 Squadron in January 1939. The unit was ultimately intended to have been a CAF unit with a permanent cadre, but its first six CAF pilots were not commissioned until the day before the Second World War began. By this time Pearce was also home to a PAF unit, No.14 Squadron. (*RAAF Museum*)

either the RAAF or the RAF, or be regularly engaged in civil aviation. In the case of pilots, technical officers and airmen, however, enlistment was also open to qualified men without previous Air Force experience, although these were not issued with any uniform.

Enrolment in the Reserve was for an initial period of four years; provided they remained medically fit, members might be re-enrolled for further terms of four years until they reached a retiring age of 40 (higher for ranks of squadron leader and above, and airmen other than pilots). They were liable to be called up for war service in defence of Australia or a Commonwealth-controlled territory, 'but will not, without their consent, be required to undergo training outside the Commonwealth or a Territory'. Unless called up for service they were unpaid.

Actual demands made on members were small. They were required to notify the Air Board of their address annually, and undertake levels of training sufficient to retain the skills of value to the Air Force. In the case of pilots this was at least twelve hours' solo flying on approved aircraft types; for other reservists it was 'such period of appropriate voluntary training as the Air Board approves'. Even these minimal levels might not be enforced:

> The prescribed training may, with the authority of the Air Board, be waived in any year in which it is not practicable to carry out such training. The Air Board also may exempt a reservist from the prescribed annual training on such reservist producing satisfactory evidence that, by the nature of his employment, his flying or airforce efficiency has been maintained at a satisfactory standard.

For members of the PAF and Active CAF who were required by regulations to render service in the Reserve, failure to carry out prescribed training or to remain efficient risked being liable to pay £20 'for each year by which his service in the Reserve falls short of four years'.

Despite the obligations which membership of the Reserve supposedly entailed, the reality of the arrangement was apparently somewhat different. In 1933 one Sydney newspaper summed up the situation in the following terms:

> It is doubtful if the trained and tried pilots of any other country are as cavalierly treated by their Governments as the men comprising the Reserve of Officers of the RAAF. At a cost of perhaps £2000 apiece they have been trained as military pilots, and on completion of their service term have been placed in the reserve, where they stagnate. Definite orders forbid any commanding officer allowing a reservist to fly RAAF equipment. He may wear his jaunty dark-blue uniform with its gold 'wings' at social functions, but the fact remains, he is 'a ground bird'.[84]

Even allowing for journalistic licence, it seems clear that membership of the Air Force Reserve could be for some an experience at once as frustrating as it might be rewarding.

One of those for whom the Reserve did not mean this was Garnet Malley. Having finished the war as a fighter ace and awarded the MC and AFC, he had joined the CAF as a flight lieutenant in June 1925 and was promoted to squadron leader before being transferred to the Reserve in 1930. In 1931, however, he went to China, first becoming aviation adviser to the Kwangtung government until 1936 and then aviation adviser to Generalissimo Chiang Kai-shek. Williams was no doubt surprised when, as CAS, he received an urgent request through British diplomatic channels in China early in 1937 to promote Malley to honorary rank of wing commander in the RAAF Reserve. As Malley belatedly explained to CAS more than a year later, the reason for this arose from his appointment 'together with an Italian General to reorganise the [Chinese] Air Council':

> There were to be five full Generals including the Italian General. It was necessary, from the British Ambassador's point of view, that I should be on this Aeronautical Commission. On account of my position here, the Ambassador (Sir Hughe Knatchbull-Hugessen) felt that it was most advisable that my rank should be raised . . . in keeping with my associates on this Commission.[85]

Malley learnt of his promotion late in February 1937, news which prompted the personal congratulations of Madame Chiang Kai-shek.[86] Few members of the Reserve found their time as fulfilling as Malley, however, and for most the experience simply marked the end of their active association with the RAAF—at least until the outbreak of the Second World War.

RAAF Wireless Reserve

In September 1928 the Defence Department received an offer from the Wireless Institute of Australia for the services of its entire organisation to be placed at the department's disposal as a defence communications reserve. The prime mover behind this offer was the institute's federal president, Kingsley Love, a 33-year-old electrical engineer. Having served in France with No.4 Squadron, AFC, during the war, he had become a flying officer on the formation of the CAF. It was evidently during his time in camp at Point Cook in March 1928 that he first tested the concept behind the Wireless Reserve, by organising the transmission of a defence message around Australia in Morse-code using the network of amateur radio operators.[87]

No action had been taken by the Air Board on Love's approach before another opportunity arose the next year to provide a practical demonstration of the utility of the amateur network at the time of the *Kookaburra* tragedy in the Tanami desert (see chapter 10). When three DH9As were despatched from Laverton on 17 April 1929 to join two RAAF planes of this type already searching for the missing aviators Hitchcock and Anderson, the leader of this flight, Flying Officer Leo Ryan, carried in his machine a short-wave AT2 transmitter. Ryan was briefed that arrangements had been made with Love in Adelaide to monitor radio messages from the RAAF party, and relay them through another amateur in Melbourne to pass on to Air Force Headquarters.[88] The aircraft equipment proved unsuccessful, however, and the RAAF party preferred to use the Post Office telegraph system to communicate with the Air Board.[89]

Distrust of radio as a reliable communications medium, a view widely held throughout the RAAF, was evidently the cause of the Air Board's delay in availing itself of the Wireless Institute's offer. From press reports, it appears that a direction from the Minister for Defence (Glasgow) in the wake of the *Kookaburra* effort was necessary before the institute's offer was finally accepted.[90] The Minister's instructions were complied with, and the 'RAAF Reserve, Wireless Section' was set up to utilise the services of owners of licensed experimental radio stations. The purposes of the new organisation were: to train radio operators in peace who would be available to the Air Force in an emergency; to provide channels of communications for RAAF aircraft operating at distances from their bases; to form a group of meteorological reporting centres; and to provide information on the condition of landing grounds and seaplane alighting sites.[91]

The Wireless Reserve remained in existence and, as intended, proved extremely useful when RAAF aircraft were deployed on long distance flights which took them beyond the eastern states. When three Wapitis from No.1 Squadron and three Bulldogs from Fighter Squadron at No.1 FTS were sent on a training flight to Western Australia in October 1932, contact was made with members of the Wireless Reserve at Adelaide and Perth and these stations were 'exercised wherever possible'.[92] During an identical flight to the West in October of the following year, the same procedure was again followed.[93] While these contacts with the reserve were evidently quite satisfactory, elsewhere the organisation was not so healthy. Indeed, it gave an appearance of being close to moribund.

When Flight Lieutenant Douglas Wilson led a RAAF formation on an annual training flight to Tasmania in March 1933, the Wireless Reserve in that State was also utilised and tested whenever possible. Wilson reported that, in fact, the reserve network actually played a very active role during the flight's tour, to the extent that his party seldom used its own ground radio installation because it was found to be much easier to use reserve stations. He commented, however, on the serious problems found to exist in the organisation, particularly in respect to morale. Although operators had displayed the greatest enthusiasm, there was general dissatisfaction that the Wireless Reserve seemed disorganised, ignored and forgotten. Members complained that they lacked any definite status, effectively had nothing to do, rarely received notification of RAAF flights taking place (or got this notice too late to play any supporting role), were issued with out-of-date lists of operating signals along with details of procedures which, in some cases, were incorrect.[94]

Wilson urged that measures be taken to address these deficiencies, including the issue of up-dated operating information and some form of identification recognising membership in the reserve. He proposed that one Sunday a month an aircraft fly from Laverton to work with reserve operators for an hour. This would maintain contact, give them practice in procedure, test wave lengths, and generally give them something useful to do.

These comments on the Wireless Reserve's sorry state had immediate and important results, with a decision being taken later that month to abolish the existing loose system of organisation and replace it with one designed for more effective control. Under the new arrangements, selected radio operators would be actually enlisted in the Wireless Reserve and, although not provided with uniform, would be issued with an official badge. The reserve network was now also removed from bands used by radio amateurs, with training being conducted on frequencies allotted for RAAF use. An organisation was adopted having 'guard stations' in each State capital city being responsible for training sections of four or five stations; these sections were to be practised two nights a week, and available for duty at other times when needed.[95] A list of station callsigns which is undated but was probably issued during 1934 shows that on this revised basis the reserve had some 86 members, and that, while strongest in Victoria (31) and New South Wales (24), its organisation extended to all mainland States and territories.[96]

Also bringing the reserve structure under more effective control was a decision to give it a headquarters and to place this within the Directorate of Organisation and Staff Duties at RAAF Headquarters. A CAF signals officer, Pilot Officer Robert Cunningham from the Melbourne suburb of Malvern, was appointed OC and Director of Communications, with a CAF Reserve pilot officer, Vaughan Marshall of Kew, as his deputy director.[97] These two officers remained associated with the reserve in succeeding years, and in August 1939 were both still in the Directorate of Signals—Cunningham as a flight lieutenant and Marshall as a flying officer.

The ultimate usefulness of this organisation is difficult to gauge. Wing Commander Charles Gates, whose long involvement in the RAAF's radio organisation culminated in his appointment as Director of Signals in 1942 and later Deputy Director of Radio Services (Signals), recalls that:

> Obviously the dream was that the RAAF would have a ready supply of useful wireless operators should the need arise. When the need did arise, it was well met by the very fine output of our own Signals School ... I can't say that I knew many people during the war who had been members of the Wireless Reserve. A few good officers come to mind: Bob Cunningham, Arthur Tinkler, and John Traill. Notable was Bill Gronow who, as a Wing Commander, did a splendid job running our technical development and maintenance administration within the Directorate of Signals. Of course, there was Vaughan Marshall, whom [Group Captain Carn] Wiggins selected as his successor when he became Chief Signals Officer under Bostock in RAAF Command. It was a most contentious decision and one on which Wiggins himself had some doubts ... before deciding.[98]

On the basis of this evidence, it is probably reasonable to conclude that the RAAF Wireless Reserve had fulfilled the purpose for which it was intended.

Australian Air League

Mention should be made here of the Australian Air League, for, although not in any way a formal part of the RAAF's structure, this body was nonetheless a significant 'feeder' of personnel for the service in much the same way as was the CAF and the Reserve. The inspiration for the organisation came in 1933 from a chance remark by 12-year-old Keith Robey, who, when asked by his father, George, why he had not joined the Boy Scouts, replied that he would prefer to join a movement that could teach him to fly.[99] Thinking on this remark and finding that no such group existed, George Robey set about forming one. Beginning in 1934, the new organisation quickly grew. By mid-1935 nine branches had been formed in Sydney and several regional centres,[100] and at the end of its first year of existence the league could boast some 4000 members.[101]

As George Robey (now the league's honorary general secretary) explained in the Sydney press, the purpose of the league was to foster air-mindedness in Australian youth. It would, he said, 'provide a common meeting ground and create an atmosphere rich in the traditions of aviation which should be of incalculable value in the training of our future aviators'.[102] The league's Commissioner for Training, Leslie Pittendrigh (who had more reason than most people to appreciate the benefits of aviation—see chapter 10), further explained that the new organisation was based 'on the same lines as the Boy Scout movement', with three sections for cadets aged 9–16 years, 16–21, and over 21.[103]

The new body was not organised with a direct or overt defence purpose in mind, nor did it operate with any form of direct sponsorship from the Air Force. This placed it in a different position to the Air Training Corps, established in 1941 with the aim of instilling discipline and a knowledge of the RAAF in young Australians, and developing qualities of leadership, self-reliance and initiative. Nonetheless, the relevance of the Air League to defence was quickly recognised. When William Hughes, who had been Australia's prime minister during the First World War and was now Federal Minister for Health, opened the league's hall at North Sydney on 15 June 1935, he declared that aviation was 'vital to removal of the handicap of distances or to defence of Australia'. His remarks were echoed by Sir Charles Kingsford Smith, who expressed a belief that 'If air-mindedness was inculcated in young Australians, other nations would think hard before attacking Australia'.[104]

On the outbreak of war, the league placed its entire organisation at the government's disposal in the hope that it would be taken over for war training. The expectation that this would happen produced a surge in membership, from around 6000 to around 10 000, 'as young men saw the League as a stepping stone to the RAAF'.[105] Absorption of the league into the formal RAAF structure did not occur, however, and the organisation soon found itself facing major problems. An editorial in the *Journal* in mid-1942 noted that one of the biggest difficulties then confronting the league was the staffing shortage created by the departure of over 1000 officers and cadets on active service.[106] By early 1943 the wing president noted that: 'The call to Arms of so many of our Officers and Senior Cadets has depleted our ranks to such a great extent that it has made it almost impossible to carry on. Our original plan of training has, more or less, been reduced to a minimum.'[107]

Despite the lack of official encouragement, there was clear evidence of the league's significance to the RAAF. Much of the material appearing in wartime issues of the *Air League Journal*—the official organ of the New South Wales wing—came directly from the RAAF Directorate of Public Relations, giving it the appearance of an Air Force propaganda organ. The pages of the *Journal* also provided news of the league's many enlistees in the RAAF.[108] To show this aspect it is only necessary to cite a few personal examples, beginning with Keith Robey, who joined up as soon as his now-widowed mother would give her consent and served as an RAAF pilot in the South-West Pacific theatre.[109]

Another example was that of Wallace Stewart, who in November 1937 had been one of the first league cadets to win a 'Flying Scholarship' financing a cadet's pilot training through the Aero Club to gain an 'A' licence; promptly joining the RAAF, Stewart was killed in an aircraft crash in June 1940.[110] Yet another was Ray Blackie of North Bondi, who had joined the Air League at the beginning of 1939 as a senior cadet and rose within the organisation to become an instructor and adjutant. He also won a league flying scholarship before enlisting in the RAAF in 1940. After completing initial training and elementary flying training, he went to Canada for advanced flying and there gained his 'wings'. Arriving in England in 1941 he flew Hurricanes and Spitfires in defensive operations and on raids over northern France. Returning to Australia early in 1943 and serving with the RAAF, Sergeant Pilot Ray Blackie was killed in an aircraft accident in April of that year.[111] In the stories of Keith Robey, Wallace Stewart and Ray Blackie might be read the experiences of a great many others like them, illustrating the part which the Australian Air League also played in the story of the RAAF.

9

Local development and production

While Australia was establishing the beginnings of its military air arm before the First World War, there had been talk of setting up the means to locally produce aircraft for defence purposes. Departing for the 1911 Imperial Conference in London, the Defence Minister, Senator George Pearce, spoke of the government's plans to purchase two aircraft from Britain 'almost immediately' for the creation of an aerial corps, followed by the creation of 'a factory where aeroplanes can be built'.[1] Despite this initial intention, and the arguments of some who continued to urge local manufacture of overseas designs under licence,[2] it soon transpired that the government was not disposed to move beyond direct importation of the country's defence needs in aircraft. Although a Bristol-type Boxkite was constructed at Point Cook in 1915 to serve as a second basic trainer, this foray into local construction was an expedient adopted solely because British suppliers were unable to meet Australian requirements, due to the priority being accorded Britain's own war needs.[3] It was not resorted to again.

Plans for post-war local manufacture

In the last year of the war the question of local manufacture received serious and detailed consideration, following proposals by the CGS (Major-General Legge) for the creation of an air corps operating 200 Australian-built warplanes (see chapter 1). After Cabinet endorsed the scheme in April, the CGS was directed to take the necessary steps to arrange for local construction of the machines required. Plans were already underway at this time for a central military arsenal to be built at Canberra to manufacture the peacetime needs of the defence forces in a range of equipment, including aircraft.[4] But it was recognised that this scheme was some way off realisation, and in the meantime Legge established a three-man advisory body called the Aeroplane Construction Committee, charged with ascertaining precisely what was the nation's capacity to build aircraft with its existing means of production.[5]

In September a second advisory body was formed to advise on the 'types and patterns of engines and other air equipment to be manufactured in the Commonwealth'. This new committee unequivocally recommended adoption of the DH9A aircraft and the American-made Liberty engine for the construction program, thus forming an identical view with the Aeroplane Construction Committee which a fortnight earlier had expressed its view to Legge that the Liberty engine should be the power plant adopted for the locally built machines (which by implication would be DH9As).[6]

In the contraction of government spending which came with the end of the war, planning for a program of local aircraft manufacture was thrown into doubt. The scheme for a federal arsenal in the Federal Capital Territory stalled and was eventually abandoned in mid-1919. Thinking had, in any event, moved from favouring a single centralised facility in a secure inland locality to a more dispersed concept of utilising private industry in major population centres, based on the existing government munitions factories at Maribyrnong, outside Melbourne. Accordingly, the functions of the former Arsenal Branch of the Defence Department were transferred to a Board of Factory Administration in 1920. A year later, in August 1921, all matters of defence supply were vested in the Munitions Supply Board (MSB), a statutory body under the chairmanship of the Controller-General of Munitions Supply, A.E. Leighton.[7]

As part of these new arrangements, it was proposed that the new MSB would undertake the production—among other things—of 50 aircraft and 12 000 bombs. Australia had, in fact, already acquired by the end of the war most of the machinery and equipment for an aircraft factory, along with personnel trained in airframe and aero-engine manufacture, but the necessary approval to actually get the construction program underway was not forthcoming.[8] The reason for this was the government policy for naval, military and civil aviation discussed in Cabinet on 7 September 1920. While restating its desire to stimulate the building of aircraft in Australia, the government now preferred to encourage civil industry in this area and foreshadowed a later statement by Prime Minister William Hughes regarding the form of government assistance that would be given to firms undertaking manufacture.[9] However, when Hughes gave his speech to Parliament on defence policy two days later, declaring himself 'a fanatic in my belief in aviation' and announcing the allocation of £500 000 for building up an efficient air force with another £100 000 devoted to civil aviation,[10] he said nothing about the form of any such incentives. The promised statement was still undelivered a year later.

Although the new MSB sought the establishment of an aircraft factory in 1922, it found the government unwilling to fund such a facility. The Minister for Defence at this time, Walter Massy Greene, expressed to the board his confidence that a commercial industry would set itself up with the encouragement of a bounty: 'The best means to encourage local manufacture appears to me to be the passage by Parliament of a Bill authorising the payment of a bounty to be based on the size, capacity and power of the machine'. The MSB thought the Minister's views to be nonsense, pointing out that the small size of the domestic market meant there was no prospect of any commercial enterprise making a go of aircraft manufacture for many years, with or without a bounty or tariff. Disregarding this advice, Massy Greene advised the board in October 1922 that Cabinet had approved his proposal.[11]

While the new Air Force which came into being in the midst of these developments had an obvious interest in the question, its early preoccupation with the task of getting the service functioning smoothly precluded it having much input to debate. RAAF involvement did not extend beyond encouraging the study of aerodynamics by presenting the University of Melbourne with a BE2 aircraft, including spares and materials, in August 1921.[12] In November of the same year the Air Board considered a request from Professor Henry Payne, Dean of the Faculty of Engineering at the university, for a grant towards the cost of a small aerodynamics laboratory. A former AFC officer on the staff of the university had obtained a wind tunnel in England at low cost, and Payne asked the Defence Department for half the £6000 needed to erect the laboratory and instal the tunnel. Williams supported the proposal, considering that the laboratory would be valuable to the RAAF 'in connection with experiments and designs for war machines and in the training of its Technical Officers'. In the event the Assistant Minister for Defence (Major-General Sir Granville Ryrie) approved the allocation of just £1500 from RAAF funds in December.[13]

AA&ECo. machines for the RAAF

As a consequence of the policy adopted by the government, the first venture of significance taken in manufacturing military aircraft in Australia actually came from a private company. At the end of the war, a number of civil enterprises had begun air operations across Australia, mostly run by wartime pilots and offering passenger services which amounted to little more than joy flights. Among these, however, was the Australian Aircraft & Engineering Co. Ltd. of Sydney, which began as a partnership between two former AFC men, Nigel Love and W.J. Warneford, and the chief engineer of the British aircraft manufacturers A.V. Roe & Co. Ltd. (Harry Broadsmith), who had secured the Australian agency for Avro machines. This company imported four Avro 504s and began charter operations and joy flights from leased grazing paddocks near Cooks River at Mascot early in 1920. Parts sufficient to assemble 20 more 504 machines were also imported at the end of 1919 and work began on assembling these at Mascot the following February.[14]

Like many early private aviation concerns, the AA&ECo. was undercapitalised and soon found itself struggling. On 18 April 1921 Love (the managing director) approached the Minister for Defence, Senator Pearce, with a proposal that his company supply the Air Force with Avro 504s manufactured locally from Australian materials. Pearce directed the RAAF's Director of Equipment (McBain) to obtain definite proposals from the company, and later that month Love was invited to spell out what he had in mind.

In his submission, Love stated that over the previous twelve months the company's engineer (Broadsmith) had obtained results from exhaustive tests of local timbers showing that Australian-grown materials were quite suitable for use in aircraft construction, even though other component parts would still need to be imported. He proposed now that the Commonwealth place an order with AA&ECo. for 24 Avro 504Ks at a cost of £1150 per machine, to be fitted with Clerget engines supplied by the Defence Department from stocks received as part of the imperial gift. The first machine would be delivered in six months and the remainder over a further eighteen month period.

Reporting to the Air Board, McBain expressed his doubts that the company could fulfil the promised delivery schedule, considering that Love was 'rather optimistic' about overcoming difficulties. He accordingly recommended seeking advice from the Arsenal Branch (meaning the Board of Factory Administration) regarding the claims made for Australian raw materials and the inspection procedures that would be necessary to ensure quality control; if a favourable response was received here, he said, an order for six aircraft might be placed purely 'to encourage the manufacture of Aircraft in Australia'. The Board observed in discussion that the Air Force already had a 300 per cent reserve of Avro machines, but, noting that 'this type is not likely to go out of date as a training machine in the near future', it endorsed the suggestion that six machines be ordered, subject to confirmation of the proposal's soundness.[15]

When the Board of Factory Administration looked at the proposal on 11 May, it decided the scheme was feasible although the Arsenal Branch was not in a position to do the necessary inspection work. It noted Love's admission in a letter to the Air Board on 19 April that his company was in an unstable financial position and reflected on the fact that AA&ECo. would not be in any better health on completing the proposed contract. But, it concluded, if later the Air Council decided the time was ripe for encouraging such a manufacturing enterprise, then Love's proposal seemed 'a fit object for consideration' and affirmed that it could arrange the inspection work required.[16]

Further consultations followed about providing an adequate inspection service but eventually, on 1 July, the Air Board reaffirmed its suggestion that six machines be ordered from AA&ECo., the first two utilising local timber but imported fittings and the remainder using Australian timber and fittings made locally from imported materials. Six days later the Air Council confirmed this proposal, recommending to the government that six machines be ordered at a total cost of £7100 and that, in response to a request from the company, one-third of that amount might be paid in advance subject to satisfactory guarantees. The Air Board was to provide an inspector of woodwork and the services of an inspector of metal work were to be obtained; all charges were to be against Air Force votes. Most importantly, the Council concluded by stating that it desired:

> to point out that a single order and the provision of the inspectional staff necessary to its completion cannot from the point of view of policy and economy be satisfactory. The Council would urge the advantage of maintaining steadfastly any policy once approved of supporting local manufacture.

Whether or not the government specifically accepted the latter injunction, Cabinet approved the order for AA&ECo. on 19 July.[17]

As McBain had anticipated, the delivery schedule envisaged by the company proved unrealistic in the face of the difficulties of the undertaking. Problems were encountered in obtaining mountain ash and white ash from the Snowy River area near Cooma (NSW), and in preparing wood for aviation uses.[18] Not until nearly a year later were the first two aircraft completed. On 15 June 1922 Dame Mary Hughes, wife of the prime minister, was at Mascot to cut a ribbon attached to the tail of the first machine, numbered A3–48, and send it on its maiden flight,[19] after which it was officially accepted by Squadron Leader Frank Barnwell of the RAAF. The following month A3–49 was also finished, and, after acceptance trials on 19 July, both aircraft were flown from Mascot to Point Cook by Love and Barnwell.[20] The remaining four machines followed at intervals until the last was delivered in March 1923.

The fourth machine in the run of six (A3–51) arrived at Point Cook in anything but a flyable condition, having been crashed during its acceptance tests. When Flying Officer Loris Balderson took the plane up at Mascot on 22 November 1922, he took with him AC1 Eric Douglas as front-seat passenger. The aircraft had reached an altitude of about 1500 feet when Douglas signalled that there was a problem with the throttle control, and Balderson decided to land to have this fault rectified. As he later explained to RAAF Headquarters:

> As I approached the aerodrome, I noticed that I was a little short of the landing ground and endeavoured to open up the engine to make good the error, the engine did not respond to the throttle and fine adjustment controls which caused the machine to strike the fence bounding the aerodrome.[21]

Witnesses to the crash provided a more colourful description of the event, recalling that Douglas foresaw the impending collision and stood up in his seat moments before the Avro demolished six panels of fence: 'the engine was thrust back into the front cockpit; the plane momentarily rose—the pilot had

The first Avro 504K built by the Australian Aircraft & Engineering Company on order for the RAAF (A3–48) is prepared for take off on its maiden flight after being formally handed over at Mascot on 15 June 1922. The machine is piloted by Squadron Leader F.S. Barnwell, a noted British aircraft designer of the First World War who served in the RAAF in 1921–23. (*E.W. Searle collection, NLA*)

switched on his engine—and then subsided in a disintegrated mass of wreckage, none of it more than two feet above the ground. A truly comprehensive crash.' Douglas's quick reaction had saved his life, although he suffered severe cuts to the forehead from broken wires and the crash had propelled him forward so that he now squatted astride the crankshaft, unconscious and bleeding. Stumbling from the wreckage, the pilot noticed his companion's plight and reached into his pockets to produce a small camera, crying away rescuers rushing to Douglas' assistance 'until I take this ruddy picture'. Given the trouble he went to, it was a shame that Balderson's own unsteady state resulted in the spoiling of these photographs.[22]

As if this crash was not bad enough for Avro A3–51, it was placed in a hangar at Mascot awaiting a decision as to whether it would be rebuilt by AA&ECo. or sent to Point Cook for repair. While in storage, a storm destroyed the hangar on 3 January 1923, forcing the removal of the aircraft outside where it lay exposed to the elements. AA&ECo., still short of suitable timber for spars with which to complete the last machine of the RAAF order, approached the Defence Department and was able to purchase the top planes for £40. The hapless aircraft was then despatched by rail to No.1 AD, where it arrived on 23 January. After reconditioning, A3–51 eventually gave five years of service before being scrapped.[23]

The slow rate of delivery and associated problems could not detract from the fact that local manufacture had been successfully accomplished. There were, however, further problems with these locally built aircraft. The Australian timbers used instead of imported spruce not only meant that they weighed 36 kg (or 6.5 per cent) more than the British-built models,[24] but they were also found to have wings which were excessively strong. Inspection of the wreckage of A3–51 revealed that 'with a similar crash with the Standard Avro Wings, the latter would have been crumpled up to matchwood . . . which points to the Wings being far too strong and unnecessarily heavy . . . there is much room for more economical design'. Responding to the inspector's findings, the technical officer at RAAF Headquarters, Squadron Leader Lawrence Wackett, told the Second Air Member of the Air Board on 12 February 1923 that:

> When it was decided to build these Avros with Australian timber it was kept in mind that interchangeability of parts was essential. The resulting increase in strength was unavoidable and while it is recognised that it is not an economical design, it would really be less economical to interfere with the interchangeability.

The increased strength, Wackett declared, 'will undoubtedly increase the confidence of the flying personnel'.[25]

Even before the completion of this contract, AA&ECo. was looking for more such work. At the hand-over ceremony for the first RAAF machine, Love had taken the opportunity to stress in front of the prime minister that the company's capital expenditure for plant, and the cost of experimentation with local materials, had been very heavy 'and could not be borne without further orders for machines or financial assistance in some form or other'.[26] On 15 September 1922 the Minister for Defence (Massy Greene) met with Love and several New South Wales members of the House of Representatives. Love now wanted assistance measures to enable manufacturing efforts to extend into the production of scouts and bomber aircraft, as well as another order for 504s for the Air Force and a contract to carry out reconditioning work on gift aircraft then in storage at Point Cook.

When Cabinet considered these proposals on 17 October it declined to place further orders with AA&ECo.[27] The result was that in March 1923, shortly after the last of the six 504s built for the RAAF was formally handed over, Love announced that the company would go into voluntary liquidation. Australia's first significant foray into meeting its own needs for defence aircraft had thus fizzled out without providing any lasting benefit.

With the demise of AA&ECo. there was something of a loss of confidence in the government's original hope that commercial interests would provide a basis for an aircraft industry. In July 1923 the Parliamentary Public Accounts Committee presented the report of its investigation into government expenditure on air services in which it concluded that:

> the best results will be obtained by setting up, in conjunction with a research establishment, a parent Government factory, well stocked with the necessary raw materials pending their ultimate local production, which would be responsible for experimental and developmental work, and for the organization of engineering works throughout the country, so that in time of emergency the whole strength of Australia's engineering capacity could be readily turned to the efficient production of Defence requirements.[28]

Taking up the spirit of the committee's report, in October 1923 the Air Board received a memorandum from the Second Air Member, Squadron Leader William Anderson, outlining 'a proposal for Government Assistance to facilitate the establishment of an Aircraft factory capable of producing in Australia, the requirements of the Royal Australian Air Force and Civil Aviation Coys'.[29] The push from this quarter was probably the inspiration for a joint meeting of the Air Board and the MSB which took place in November 1924 to consider the question of local manufacture. These deliberations concluded that the best that might be then aimed for was encouragement of private firms to undertake production of aircraft components, with erection of completed planes or engines being carried out in some government-controlled assembling shop.[30]

The finding of the joint board's meeting was to prove an unexpected impediment in mid-1925, when the Larkin Aircraft Supply Company (LASCo.) of Melbourne offered to undertake the manufacture of aircraft provided the government was prepared to give it an initial order for the RAAF. The number of planes sought was not large, only between three and eight depending on type, and Williams as CAS was very attracted to the proposal. He noted that the company had acquired considerable experience through fulfilling Defence contracts for reconditioning imperial gift machines, and their work had been satisfactory. As the DH9 was the type the Air Force had least of, he indicated that five of these would not be in any way in excess of the RAAF's needs.[31]

When LASCo.'s offer was referred to the MSB on 30 July, however, that body based its response on the previous joint board conclusion. It considered Defence Department demand to be too uncertain to justify the investment involved, and therefore urged that the proposal 'be not entertained'. Williams was horrified, and advised the secretary of the department on 28 October that:

> [Assembly] at a central Air Force Depot, or some other Government establishment ... cannot be said to be the best step towards establishing the aircraft industry in Australia. Before that industry is satisfactorily established, we must have a factory or factories capable as nearly as possible of producing the machine from start to finish ... The proposal of the Larkin Aircraft Supply Company gets us much nearer to the desired goal.

He accordingly recommended an order for six aircraft be placed with LASCo., and that £20 000 be provided for this purpose. Having heard nothing by late December, he again drew the secretary's attention to his recommendation but without any action being taken on it.[32]

The reason for the rejection of Williams' advice is perhaps explained by the response received by the De Havilland Aircraft Company when it approached the government in mid-1926 about undertaking local manufacture itself. The Minister for Defence (Howse) told the company that no assurances of aircraft orders for the RAAF could be given for either the 1926–27 financial year, or even the next. This did not dissuade the company from a second approach in March 1927, indicating that it was prepared to 'lay down a properly equipped factory' if it received an order for 20 Army cooperation aircraft to cost between £55 000 and £60 000 in the first year, and fifteen such machines in subsequent years. Williams realised that this proposal was nothing less than a demand for a De Havilland's monopoly on the supply of the RAAF's aircraft requirements, and, as it happened: 'We don't want 20 Army co-operation type of landplanes next year or the year after I think, but in any case we cannot yet decide on a type and when we do it will be a question of calling for tenders ... [It] would then be up to DeHavillands to compete with others in price.'

The reply the company received from the Defence Department accordingly stressed that the government could not 'commit itself to the support of an Aircraft firm in such a way as would afford that firm a monopoly of Government orders'. Moreover, the policy of conforming to RAF aircraft types as closely as possible, subject to local conditions, meant it was 'probable that the requirements of the Defence Department in aircraft will not be confined to the designs of one Aircraft firm alone.'[33]

Barnwell and Wackett

Even as the AA&ECo. experiment came to an end in 1923, the basis for the next step in the process was already present on the Australian scene in the person of Frank Barnwell, the RAAF squadron leader who had taken delivery of Love's first 504 destined for the Air Force. Barnwell was no ordinary air officer. He was, in fact, an eminent figure in the British aircraft industry, having been the Bristol Company's chief designer and engineer since 1915. In May 1921 the Air Board had received an unsolicited offer of his services, after he approached the High Commission in London about employment opportunities in Australia. According to one account, he had quarrelled with his firm in England and was keen to come to Australia 'for a holiday and to allow time for a reversal in his firm's policy'.[34]

When Barnwell's availability was put before the Air Board, his offer was at once accepted. Williams reported that Barnwell was 'recognised in England as one of the leading designers there and . . . probably the only one of these . . . who is a qualified pilot and flies his own machines'. As the man responsible for the Bristol Fighter—'which, for all round usefulness, can probably be said to be the finest machine produced during the war' and was still the RAF's standard aircraft—Barnwell's services would be, Williams considered, 'an invaluable acquisition to this country in whatever capacity he may be serving'.

The Board appeared to take very little convincing in adopting a recommendation that he be offered temporary appointment at Headquarters as a squadron leader (technical), and urged this course on the Air Council. If the Council approved, it proposed the offer be cabled through the High Commission, informing Barnwell 'at the same time that the Commonwealth Government does not contemplate the manufacture of aircraft in Australia in the near future'.[35] The Air Council, which considered the matter on 24 May, was no less keen to see Barnwell's services utilised.[36]

Despite the warning sounded at the time of his appointment about the absence of Australian plans to embark on aircraft manufacture, Barnwell was evidently keen to become involved in design work in his new post. His urging on this score was most likely behind the proposal put to the Air Board in September 1922 for technical officers at Headquarters to be given the task of designing a replacement for the RAAF's trainers. Squadron Leader Percy McBain, the Director of Equipment, argued that:

> The Technical Branch of the RAAF have officers on strength who have had many years' experience in designing British aircraft, and it is . . . certain that a more efficient training machine than the Avro 504K can be designed. It is also certain that such a machine can be manufactured in Australia at a cost which will be approximately half that of the Avro.

The response of Board members to this idea was no more than lukewarm. While the importance of local manufacture from Australian materials was recognised, it remained approved policy that RAAF equipment should generally be the same as that of the RAF, and expenditure of funds on experiments which duplicated work done in England was not justified. The most that the Board seemed willing to do was allow the Technical Branch 'to undertake the drawings of a training machine . . . provided that it does not interfere with the normal work of the organisation now established'.[37]

Despite this disappointment, Barnwell was not dissuaded from trying to get some form of local design work adopted. When discussions arose out of the RAN's need for a seaplane that could operate from its cruisers, it appeared he had his chance. Even before Barnwell actually left England early in 1922, he was tasked from Melbourne with collecting and bringing with him 'fullest information' about the types of maritime aircraft available and the equipment needed to operate them. His involvement in this capacity ensured that he had continuing input to the debate following his arrival.[38] After Navy proposed in December 1922 that a small single-seat seaplane be obtained for scouting purposes instead of the large and heavy Fairey IIID then in service, Barnwell drafted a minute suggesting that the most complete answer to Navy's stated requirements was a small flying boat specially designed for the purpose. In conveying this idea to the Naval Board, it was stated:

> This machine could be designed by the Air Board Technical Staff and built of Australian timbers, except ply-wood, in our own workshops at Point Cook, by our own personnel. The cost of material would be approximately £500. The engine recommended is the 250 H.P. Siddeley Puma of which we have a number in reserve.[39]

Before the Air Board had an answer to this proposal from the Navy, Barnwell's connection with the RAAF had come to an end. By now thoroughly frustrated with the Defence Department's failure to properly use his talents, he resigned his commission and returned to England to resume his post at the Bristol works where he remained until 1938. By all accounts he considered his period in Australia to have been a waste of time,[40] but there was one outcome which perhaps gave him consolation. At Headquarters he had encountered a kindred spirit in Squadron Leader Wackett. Also a technical enthusiast, Wackett was a wartime graduate of the Royal Military College, Duntroon (and then the only such qualified officer serving with the Air Force). He had served with the flying corps in the Middle East and France and returned to Australia after being awarded both the DFC and AFC. Aged just 25 in 1921, he was now technical staff officer involved with controlling the gift equipment in storage.

Late in 1920 Wackett had been planning to go into private business and was attempting to obtain his release from the permanent forces to start up an air mail service in Queensland using flying boats. With promised backing from business sources, he had been recruiting former wartime colleagues with technical qualifications for the new company, indicating that his firm proposed 'to build and design aircraft as orders are received and hope[d] also to tender for Government orders'.[41] The reality of the post-war aviation industry had soon become plain, however, and by mid-1921 Wackett confided to Walter Nicholson, a wartime colleague to whom he had offered a job, that he knew of nothing 'doing in the civil aviation line and personally I have given up all idea of it.'[42]

He may have recognised that the climate was not then favourable for the sort of enterprise he had initially contemplated, but Wackett was far from ready to bury his hopes entirely. To further his ideas of becoming a qualified aeronautical engineer, he gained admission in 1922 to the University of Melbourne which arranged a special course enabling him to undertake subjects necessary to complete a Bachelor of Science degree in a single year of study.[43] Having successfully completed this course and gained his degree, Wackett turned his attention to aircraft design and here, of course, found in Barnwell the ideal person to extend his knowledge in this field. Wackett later wrote that when Barnwell 'discovered that I was the only person in Australia really interested in his work . . . [he was] willing to give his time to teaching me . . . I was receiving what amounted to a post-graduate course in aeronautical engineering from an instructor who was, at the time, the foremost exponent of the subject'.[44]

Tutored by Barnwell, Wackett soon wished to try his hand at building aircraft of his own design. The opportunity to do so arose during the first half of 1923 when proposals were made for the establishment of an experimental workshop for producing components needed by the RAAF. The suggestion was by no means wholly Wackett's, as the idea had been floated by the Air Staff in

Britain as early as February 1921 that the testing of materials such as glue or fabric could be more usefully undertaken on a local basis.[45] Moreover, in April 1923 the Air Board had considered a submission from the Second Air Member (Anderson) that there was enough research and experimental work required for the RAAF to warrant a special workshop being formed to handle these tasks.[46]

On 23 June the Air Council approved in principle the establishment of such a workshop, provided this was on a small scale and could be funded within existing Estimates. The Council accepted that there was essential work for the section to do and that 'it is more economical to detach from No.1 Station [Point Cook] the personnel required than to attempt the work there', but seemed equally disposed to countenance the idea because a suitable building was available free of charge at the Harness Factory, in the inner Melbourne suburb of Clifton Hill, and 'the necessary machinery is available or can be obtained free of cost'.[47]

Subsequent to this approval being obtained, the proposal was modified to provide for the experimental station being established inside Victoria Barracks. Shortly after this change was made, Wackett learnt of a workshop in Sydney which was being offered for sale by the Defence Department. Located in the eastern suburb of Randwick, in Avoca Street, this was the former factory of Father A.J. Shaw which the Navy had taken over in 1916 for producing radio apparatus, and which had then been taken over by the Repatriation department for training returned servicemen. In November the Air Board decided to recommend taking over this facility to house the new RAAF workshop, and the Minister (Bowden) approved.[48]

Wackett recalled that he then suggested to Headquarters that he be allowed to build a machine of his own design, arguing that this would initiate aircraft design and construction in Australia to the RAAF's benefit in time of war: 'They would not support the suggestion, but agreed that if I could get funds from some other source they would approve.'[49] Here Wackett was successful in winning support from a somewhat unexpected quarter. Colonel Brinsmead, the Controller of Civil Aviation, agreed to provide the £2500 Wackett estimated he needed to design and build a seaplane in Australia. Clearly inspiring Wackett's choice of a flying boat as his first undertaking was the fact that the Navy had, up until the time that the Air Council actually approved his scheme, still been looking at Barnwell's proposal for a locally designed and constructed small seaplane that could be embarked on warships.[50]

The RAAF Experimental Section

The new workshop, known as the RAAF Experimental Section, was officially formed on 14 January 1924.[51] Though Wackett was appointed the first officer commanding, for the first month of its existence it operated under Flight Lieutenant Charles Harman. Wackett maintained that he was originally allowed just three NCOs and some six other men from Melbourne to form a nucleus staff,[52] but about the time of his arrival the personnel of the section had already grown to 21 in number. The disciplinary NCO was Sergeant-Major Les Carter, one of the original mechanics of the pre-war CFS, as was Flight

Sergeant Ted Shorland, and as well there was Corporal Arthur Styles, four LACs and thirteen airmen.[53] Within a few months the unit's strength had increased to about 30, with the addition of skilled tradesmen from the now defunct AA&ECo. Among those who joined from this company was Broadsmith, who became Wackett's assistant.

The utility of the section for the RAAF's operations generally became very quickly apparent. Quite apart from anything produced by the workshop, it was the only RAAF presence in New South Wales until an air station was established at Richmond and a squadron permanently based there in mid-1925. Thus, when three DH9As were despatched from Point Cook to Brisbane for that city's centenary celebrations in 1924, and all crashed at various points around New South Wales, a party from the section was on hand at the Sydney wharves to receive the disassembled remains of one of the machines shipped back from where it had crashed near Maclean on the Clarence River.[54] Similarly, in April 1924, when Goble and McIntyre embarked on their round-Australia flight and had the pilot's compass explode shortly after departing Point Cook, they were able to call on the Experimental Section to make essential repairs that got them on their way again.[55]

Even after the Richmond station became operational, the section at Randwick provided the RAAF with an invaluable recruiting shopfront in Australia's largest city. Bill Bradley recalls that in December 1926, as a 23-year-old who had recently lost his job at a Sydney marine engineering firm, he was walking past the section when he saw a sign saying 'Airmen wanted'. He walked in off the street and, after a medical examination, a written test, a trade test, and an interview with Flying Officer Bill Palstra (deputy director of personnel services at RAAF Headquarters), found himself on a train south that night, headed for Laverton.[56]

Before work on the seaplane project—begun in March 1924—had progressed very far, the talents of the Experimental Section team were demonstrated in another capacity. In July the New South Wales section of the Australian Aero Club announced a competition for ultra-light aeroplanes to be held at the Richmond aerodrome in November, with large cash prizes donated by Lebbeus Hordern, the Defence Department, several companies, and the club. Wrote Wackett:

> I decided to make an aeroplane of this type, at my own expense, with the assistance of some of my enthusiastic mechanics. I was to provide the design and money and the others volunteered to contribute their labour. We secured use of a private garage and worked evenings and week-ends.[57]

The machine produced was a Parasol monoplane pusher named the 'Warbler', powered by a 1490cc two-cylinder engine which was entirely Wackett's own design. As it happened, another three aircraft were commissioned from Harry Broadsmith—now working at the section following the collapse of AA&ECo.—for entry in the competition. Using the living-room of his home as a workshop, and with the assistance of three former employees now also working at the Randwick Section, he completed two of these machines in 13 weeks of working evenings and week-ends.[58]

The Warbler became the centre of attention when it appeared at Richmond by virtue of the fact that both machine and engine were 'home-made'. Initially, however, its performance was disappointing. Flight Lieutenant Ray Brownell, who had been seconded to the Civil Aviation Branch for the competition to give refresher flying training to entrants and to test-fly the machines, was distinctly less than impressed:

> In giving this machine its initial test I ran the whole length of . . . [the] aerodrome and finally managed to lift the plane off the ground just before the boundary fence, but it refused to climb more than about twenty feet and I had no option but to head for a gap in between two trees and just managed to get through by tilting my wings. Needless to say I pretty quickly flew round and landed, and regardless of rank and seniority I told the Squadron Leader exactly what I thought of his experimental machine![59]

Despite Brownell's lack of enthusiasm the press proclaimed Wackett the hero of the Richmond meeting, it being considered that he had made the first definite contribution to Australian aircraft construction with his self-designed engine—for which he received an award. But it was Broadsmith's designs which took the most prizemoney.[60] As both Wackett and Broadsmith were with the Experimental Section, their success has been described as 'a RAAF triumph',[61] but the reality reportedly was that relations between the two—already unhappy before the competition—worsened as a result, and Broadsmith soon afterwards felt constrained to return to England.[62]

The Widgeon I

Six months later the seaplane commissioned as the section's first design task was approaching completion. It was a small five-seat flying boat (length 30 feet, span 40 feet), powered by a single 230hp Armstrong Siddeley Puma engine. Following a detailed study he had made of the properties of local timbers,[63] Wackett constructed his machine with a hull of Queensland maple and cudgery. Appropriately, he gave it the name 'Widgeon', after a bird (also called the Pink-eared Duck) which was native to the lakes, streams and swamps of continental Australia. As the finishing touches were being made, he wrote to the Air Board in April 1925 pointing out how closely construction of the machine had met his design specifications and extolling the Widgeon's likely utility in a service role such as coastal patrol. Now very confident that his aircraft would give superior performance to comparable types on the market, and knowing that the department was then in the process of ordering Seagull IIIs from England,[64] he suggested that 'no action be taken with the purchase of any marine aircraft until the "Widgeon" has been seen on test'.

Although he was Wackett's sponsor, Brinsmead took a more cautious view in notes he prepared for the Air Council on this proposal. Testing was an essential and inevitable part of developing any new aircraft type and it would be 'a considerable credit' to Wackett if his machine did not need any modifications after it was trialled. All this would take time, Brinsmead wrote:

> and unless units necessary to operate aircraft from the proposed seaplane carrier

Wackett's 'Widgeon I' is transported by Army tender through Sydney streets on 1 July 1925 to the camp at La Perouse, on Botany Bay, in preparation for its first test flights. (*RAAF Musuem*)

> are formed and equipped at an early date they will not have reached the necessary standard of training to enable them to go aboard the carrier when it is completed. I am definitely of the opinion, therefore, that whether the Widgeon is successful or not, the units referred to above cannot wait for it.[65]

On 1 July 1925 the aircraft was transported from Randwick by Army tender and housed in a canvas hangar at a little cove at nearby La Perouse, on the northern shore of Botany Bay, in readiness for its test flights the following week. Instead of RAAF markings the Widgeon bore the registration G-AEKB, which did not conform with the usual prefix of Australian civil aircraft (G-AU) but derived from the initials of the Minister for Defence, E.K. Bowden.[66] A RAAF contingent was despatched from Point Cook for the event, including airmen to guard the aircraft at night. One member of this party, AC1 Bert Gardiner, recalls that the only visits were from awed Aborigines who inhabited the beach.[67] Other personnel accompanied the crash launch which was prudently on hand as a safety measure.

After launching and flotation tests on 7 July, the Widgeon was to take to the air the next day. A throng of more than 500 spectators turned up in bleak and cold weather to witness the trials, among them CAS and others interested in aviation including Sir Keith Smith. Taxi runs were made in the morning, during which a sandbank was hit at speed but fortunately without damage to the aircraft, but with rain threatening and whitecap waves on the bay, it seemed doubtful the trials could be completed. The aircraft had been brought ashore and christened by Wackett's wife with the traditional bottle of champagne when, about 4 pm, it was decided to proceed in attempting a take-off.

The Widgeon was floated out again, with Wackett as pilot, and Brinsmead, Sergeant Harry Becker (foreman at the section) and AC Cropp as passengers.

'Widgeon I' is readied for its maiden flight in a sheltered cove at La Perouse on 7 July 1925. The aircraft's registration markings, derived from the initials of the Minister for Defence, E.K. Bowden, are clearly visible. The day's tests ended in disaster, but the machine was repaired and successfully flown five months later. (*RAAF Museum*)

In a cloud of foam she headed southward out into the bay towards Kurnell, 'her nose beginning to lift into flight, when a long roller heaved under her, and she dived into the trough and immediately stood on her head'.[68] Anxious moments followed as the crash launch sped to the scene, but none of those aboard were injured apart from Brinsmead who received a cut hand; the aircraft, however, had its bow staved-in, caused by the impact of striking the wave following the breaker which upset it.

Considerable criticism flowed in the press following this initial failure, with some alleging that money wasted on Wackett's experimentation was retarding, not hastening, the development of Australian aviation. Despite this agitation which might have threatened the future of the unit, Wackett was given permission to persevere with the Widgeon. The aircraft was repaired and modified: different timber was used in the hull, the position of the petrol tanks was changed, and alterations were made to the nose to reduce the amount of water shipped inside. On 3 December the Widgeon successfully flew, again with Wackett at the controls, taking off from Botany Bay and making several circuits of the bay lasting ten minutes.[69]

Service trials were undertaken in April 1926. To avoid the unpredictable swells near the heads, base camp was set up further around Botany Bay, at Sandringham beach just south of Dolls Point. In a report on the trials tabled in Federal Parliament on 27 May 1926 by the Minister for Defence (now Major-General Sir Neville Howse, VC), it was stated that the Air Board considered the Widgeon's performance to be very good and to meet all requirements.[70] It was subsequently modified as an amphibian by the fitting of a retractable wheeled undercarriage, given a 300hp Nimbus engine, and other changes incorporated. In October the Widgeon made its first flight as an amphibian, taking off from Mascot and landing on Sydney harbour near the Cockatoo Island naval dockyard.[71] The following February, after trials at Mascot, the aircraft was flown to Cockatoo Island for technical inspection by dockyard engineers. Wackett developed a proposal to build Widgeons for defence purposes at the naval dockyard and submitted it to the director, Robert Farquhar, but nothing came of this idea.[72]

Although its development had been funded by the CCA, on 28 June 1927 the Widgeon was flown by Wackett, accompanied by a mechanic, Corporal Morley, on a direct non-stop overland flight to Point Cook, where it was handed over to the Flying Training School for training seaplane pilots.[73] The aircraft came in for considerable use in this role, but its condition fairly quickly deteriorated and it proved difficult to coax off the water. It then remained for a long time in the hangar; according to Charles Gates, then an NCO on the station:

> There was one pretty serious defect. The wing spars were Monel metal, the wing ribs pressed out of brass or something of the sort. The sea air caused electrolysis between the two and corroded them. No-one wanted to fly the damned thing . . . [so] it was put in the corner and left there.[74]

In the hangar the Widgeon was the target for some waggish humour from Flight Sergeant Theodore Holst of the motor boat crew (known as 'Matey' by virtue of having previously served in the RAN until 1920). Bill Symons, then a station member at Point Cook, recalled that:

> He cut out one of those parrots that swing back and forth [and put it in the Widgeon's boxkite tail structure], and underneath they made a nest and put a great big egg in the nest . . . underneath they painted a board 'Wackettus Widgeonus—Made in Australia' . . . Wackett walked around one day . . . and they forgot to take the parrot out of the tailplane and the egg, nest and board. When he saw it, I thought he was going through the roof of the hangar . . . [75]

It being considered not worth the expense of a complete overhaul, in October 1929 the aircraft was approved for disposal and broken up.[76]

With the relative success of the Widgeon behind him, Wackett's position became more secure. The staff of the section, already over 60 in February 1926, expanded to 80,[77] and Wackett was given temporary rank of wing commander on 1 July. In 1 November 1926 an engine and rigging section was formed within the workshop under Flight Lieutenant Arthur Murphy.[78] The section now undertook quite a lot of engineering work, fitting new aluminium seats to enable the wearing of parachutes in aircraft,[79] and reconditioning the aging engines of the imperial gift aircraft still used by the RAAF. Since this was work shared with a number of engineering firms (some of whom relied heavily on their RAAF contracts to stay in business) and was not in the least experimental, naturally there was grumbling in some quarters that the section was operating in competition with private industry.[80]

The 'Widgeon I' at Point Cook, where it languished in a hangar corner until approved for destruction in October 1929. The airmen pictured with it are T.C.H. Holst, Kelly, G.L. Grendon, Stocker, and Kersley. (*SqnLdr G.L. Grendon*)

Widgeon II

Approval was also obtained to build a more advanced amphibian, again funded by CCA. Called Widgeon II, it was completed in December 1927. Its dimensions were little different to the first machine (length 30 feet; span 43 feet), and it had a hull of Tasmanian blackwood and northern rivers white beech, a propellor of Queensland maple, and wing struts of hoop pine.[81] It was fitted with an Armstrong Siddeley 440hp Jaguar air-cooled radial engine, giving an estimated maximum speed of 130 mph. Widgeon II made its first trial flight at Mascot on 21 February 1928, and the next month it was entered in the New South Wales aerial derby, minus its wingtip floats, and came third.[82] It was then sent to Point Cook for testing, Wackett himself flying it there and back accompanied by Corporal Morley and AC W.K. Tomkins.

In late April Prime Minister Stanley Bruce announced that the Widgeon would be sent to Singapore the following month, to meet and accompany a RAF flight of four Southampton flying boats making a show-the-flag tour of Australia. Wackett was selected to fly his machine, with Flying Officer Harold Owen (an engineer and also a pilot) and Tomkins as his crew.[83] Departing Richmond on 1 May, the Widgeon flew as a landplane over the inland route to Darwin, maintaining radio contact with RAAF Headquarters the whole way.[84] Back at Randwick, Sergeant James Reynolds, a radio operator borrowed from the Richmond station, prepared to maintain late-night watch for the next leg to Singapore.[85]

By this stage Wackett had discovered some of the frailties of his machine: rough landings on unmade airfields repeatedly caused cracking of the Monel metal wheel rims; during trials on Darwin Harbour it was found that the low density of air in the tropics made it impossible to get the aircraft off the water, and the prolonged runs involved in these attempts produced cracking in the wooden propellor through high-speed contact with spray.[86] To overcome the loss of power, Wackett decided to shed weight by removing the radio and returning it and the operator to Sydney by ship, but the Air Board, obviously worried about safety, directed Wackett not to leave for Koepang without permission, and on 13 May a government decision was made to cancel the flight to Singapore altogether.[87] Instead Wackett was ordered to Broome, Western Australia, to await the visiting Southamptons.

With the arrival of the RAF machines, the Widgeon accompanied them as they flew down the west coast, finally separating from them at Adelaide and proceeding directly to Richmond, thereby completing a 14 500-kilometre flight which was the first around Australia in a locally designed aircraft. On 1 July the Widgeon arrived back at Mascot and was soon in pieces at Randwick undergoing overhaul, Wackett evidently expecting to be called upon to accompany the flying boats on their return leg to Singapore.[88] This did not happen, however, and later that year the Widgeon was passed to Point Cook where it was flown without an A number but with RAAF rondels (unlike the Widgeon I which had neither).

In July 1929 the Widgeon was loaded on the Navy's seaplane tender HMAS *Albatross*, for tropic trials during a cruise of New Guinea waters with the Governor-General on board. The machine proved less able to handle the tropical conditions than the Seagull amphibians already used for naval cooperation work. John McCorriston (then a member of 101 Flight) recalls that the unit CO, Flight Lieutenant Arthur ('George') Hempel, took the Widgeon out for take-off but could not get it airborne from the glass-smooth water, needing some chop to the surface to provide lift: 'After taxiing a couple of miles, he came back and said to us, "Put it away, it's neither a speedboat nor an aeroplane". As far as I remember, it only got the one try-out.'[89]

The ship arrived at Portarlington, on the Bellarine Peninsula east of Geelong, in late October, and the Widgeon was unloaded. Flight Lieutenant Joe Hewitt was assigned to fly the aircraft back to Point Cook. After a mechanically troubled flight which caused him first to alight on Corio Bay, then return to *Albatross*, Hewitt finally arrived at his destination next day and handed the machine over to Flight Lieutenant Freddie Briggs of Seaplane Squadron. Briggs, a recent transfer from the RAF, asked Hewitt how he liked the Widgeon; 'You can have it', was the unenthusiastic reply.[90]

Others who had the opportunity to fly the Widgeon at this time were equally unimpressed. Two members of the Seaplane Squadron, Flying Officer David Ross and Sergeant George Grendon, had the aircraft up for a flight on one occasion when a problem was suddenly experienced with the controls. Grendon recalls:

> It got into a sort of incipient right spin, wanting to go one way all the time which meant we had to go in a circle, a flat spin. We lost about two and a half thousand feet, but luckily we had been at about 6000 feet. We didn't think it was going to come out. Dave was fighting like hell, doing everything, but she was out of the pilot's control.[91]

Back on the ground a check was made of the rigging, and for pulley damage, but this disclosed no reason for what had happened, leading the two crewmen to consider that there was an inherent design fault.

Just a few months later the Widgeon was destroyed when it crashed into Port Phillip Bay, killing Flight Lieutenant Briggs, Captain the Hon. Hugh ('Puck') Grosvenor (aide-de-camp to the Governor of South Australia) and a mechanic, LAC D.C. Ewen. Grosvenor, a popular figure in local aviation circles, was planning a record-breaking dash from Australia to England with Briggs as his relief pilot, and it was to gain experience for this flight—to have been made in an amphibian—that the Widgeon was taken up on 6 January 1930. They were flying at 400 feet about a mile off Point Cook when the aircraft was seen to suddenly nose over in flight and dive into the sea with terrific impact, striking almost vertically and immediately plunging to the bottom. The wreckage was located in 30 feet of water and salvage barges lifted the wreck to the surface the next day, but without finding any bodies still in it. Brigg's wife and five-year-old daughter were already on their way to England, expecting to meet him there in March after the flight with Grosvenor.[92]

The reason for the crash was essentially indeterminable, though suspicion remained strong that Grosvenor (officially a civilian passenger on the flight) had control of the machine at the time, rather than Briggs who was a senior instructor on flying boats and an expert in this branch of airmanship. Whether or not this suspicion was warranted, the RAAF came in for some strong criticism as a result of the loss of Widgeon II. When the Commonwealth auditor-general sent to the Air Board early in 1930 a report from its chief inspector criticising the 'loss of Government property' in the crash due to 'lax departmental methods', this drew a very sharp reply.

The Board pointed out that any disobedience of orders which might have occurred 'rests with a person who is not now present to give his reasons . . . and it is useless pursuing the matter further'. It also rejected any suggestion that the Widgeon had been 'aerodynamically inefficient', declaring itself unaware of any qualifications possessed by the audit inspector to comment on matters which had been examined in two courts of inquiry by experts having the first hand evidence before them. It considered deplorable the fact that the Board needed to reply to 'a statement affecting policy, discipline, training and administration' made by a person having no competence in these areas, and suggested the inspector should 'confine himself to matters of an Audit nature'.[93]

Irrespective of the Board's blunt reminder of the limits of the auditor-general's role, the criticisms of the audit report had clearly hit a very raw nerve—as could be judged by the tone of the reply. In fact, the Board was aware of suspicions about the soundness of the Widgeon as a factor in the crash. In January 1930 the court of inquiry headed by Squadron Leader Allan Hepburn had reported its doubts as to whether the aircraft's design 'was carried out in accordance with ordinary engineering standards of safety', and suggested that, although this aspect was outside the court's scope, another inquiry should be instituted. The Board debated whether to submit the

Widgeon II design drawings to the British Air Ministry for checking, but decided against this on the grounds of cost and the belief that no useful purpose would be served.[94]

With another Wackett machine—Warrigal II—about to be produced by the Experimental Section, however, the Board decided to order a check of the designer's figures to satisfy itself that the new aircraft was airworthy before flying tests began. The report obtained in May from Squadron Leader Herbert Harrison, director of technical services, showed that there was nothing in Wackett's stress calculations to provide a reason why the Warrigal machine should not be flown. In disposing of the notion that the Widgeon had been 'aerodynamically inefficient', the Board was thus able to tell the auditor-general: 'The fact is that the Board in having the figures for "Warrigal" II checked have found that Wing Commander Wackett has been working on a factor of safety which is more conservative than that which is now accepted by the Air Ministry in England.'[95] But this avoided mentioning why examination of the design data for Warrigal II might have been thought necessary in the first place.

Warrigals I & II

A year before the Widgeon crash, the first of two machines given the name 'Warrigal' (Aboriginal for 'wild' or 'untamed') had been produced by the section at Randwick. In 1927 specifications were issued for two new types wanted for the RAAF, one an Army cooperation aircraft to replace the aging DH9s, and the other an advanced trainer. Harry Broadsmith, returning to Sydney in March that year with plans that Avro might commence manufacturing in Sydney,[96] had submitted a design for the first type of machine; this reached the mock-up stage before being rejected by the Air Board in favour of purchasing the Westland Wapiti direct from England.[97] Meanwhile, Wackett had begun work on satisfying the second requirement with a two-seat biplane which might prove a suitable replacement for the Avro 504s then approaching the end of their useful life. When ministerial approval for Wackett to proceed with his design was given by Sir Neville Howse, it had been foreshadowed that two machines would probably be necessary, the second aircraft—a development of the first—to be used in seaplane training but with the capacity for conversion to service use as a fighter if required.[98]

The Warrigal I was completed late in 1928. As with its predecessors, extensive use was made of Australian materials, especially timbers. Tasmanian blackwood, white beech and hoop pine were used in fine strips, fastened together with bronze rivets. Ribs for braces were moulded from strips of Australian brass founded at the ammunition works at Maribyrnong, and a nose cone was cast locally from stainless steel and phosphor bronze to resist water corrosion. The power unit was, however, again imported—this time a 200hp Armstrong Siddeley Lynx radial engine.[99]

After testing at Richmond, the machine was flown to Point Cook by Wackett on 29 January 1929 in preparation for its formal hand-over at No.1 FTS two days later when it was christened by the wife of CAS.[100] The Warrigal I was, by available accounts, not well received by the RAAF pilots called on to test-fly it. Air Commodore 'Paddy' Heffernan recalls:

> From what I heard later, I learnt that it was most unpopular, being under-powered and overweight. It also had most peculiar spinning habits. On one occasion it was being flown by Squadron Leader Charles Eaton with Sergeant Eric Douglas as passenger and was deliberately placed in a spin. After several turns Eaton attempted recovery, but nothing happened. As the aircraft was still in an uncontrolled spin at 2000 feet, he called on Douglas to abandon ship. Douglas undid his belt and stood up and this action apparently varied the air flow over the tail surfaces and the spin stopped at about 1000 feet. As the spin had stopped, Douglas did not jump, but two very frightened pilots got out of the aircraft after landing.[101]

Perhaps it was just as well that on 19 September, less than nine months after its delivery, it was written off in a landing accident at Laverton while being flown by Flying Officer Andrew Gerrand.[102]

Following close behind the first machine, was the improved Warrigal II. Powered by a 450hp Armstrong Siddeley Jaguar engine giving a maximum speed of 140 mph, this aircraft was fitted with interchangeable wheel and float

Following its dedication by Chaplain F.G. Hughes, Mrs Ettie Williams, wife of CAS, christens Wackett's 'Warrigal I' by breaking a bottle of champagne over its propeller boss. The ceremony, performed on 31 January 1929 in front of 60 guests, marked the formal hand-over of the Warrigal to No.1 FTS. Afterwards Wackett took Mrs Williams on a flight, and then Squadron Leader Bostock gave an aerial display in the new machine. (*Mrs J. Stewart*)

undercarriages. But even before construction was completed, the aircraft became bound up in a heated wrangle over the experimental section and its future. The basis of this controversy lay in remarks in Sir John Salmond's 1928 report on the RAAF (see chapter 4), which queried the point of the Randwick section's existence and the cost of maintaining it (£25 000 in 1927–1928). Though he considered the skilled work done in the section to have been 'of a very high order and . . . ably directed', Salmond observed:

> If the object . . . is to prove that it is possible to build aircraft entirely or almost entirely from materials produced in Australia, then that object has been attained. If the object after having built them was to put them to the trade for production, I can find no authority in any correspondence for such a policy. That, however, appears to me to be the only justification, and . . . [unless] the Government is prepared to adopt some such policy . . . at considerable and recurring financial cost . . . I recommend that its functions as a factory cease, and that the reconditioning work be done elsewhere.[103]

Although Wackett did not consider Salmond's judgement to have been at all impartial, accusing him of having been an agent of British aircraft manufacturing interests keen to ensure that Australia continued to buy the RAAF's needs from the old country,[104] the recommendation of the report could not easily be ignored. It was inescapable that the experimental section had been an expensive proposition, and the benefits of its output of one-off aircraft rather debatable. Wackett estimated that the Widgeon I had cost £7180 to produce, and the Widgeon II £9329,[105] but the auditor-general calculated the real costs at £14 360 and £18 478 respectively;[106] in comparison, the first six Seagull III amphibians bought from Britain in 1925–26 had only cost around £6000 apiece.

In September 1929 the auditor-general began querying the costs of the two Warrigal machines, pointing out that no specific ministerial authority appeared to have been obtained for the second aircraft. After pressing the matter for some months, the auditor was finally told the total cost of the Warrigal I had been £10 573, but that no complete figure for the Warrigal II was yet possible.[107] All this took place before the crash of the Widgeon II had even raised other awkward questions about the Wackett machines.

The Air Board had considered closing down the Randwick section during 1929. In December that year the secretary of the Defence Department was told that this step had been initially delayed by a proposal from the Commonwealth Shipping Board that a nucleus of the section should be transferred to Cockatoo Island, but that now the Air Board had decided to close down the section as soon as work on the Warrigal II was finished; it was considered a waste of time, effort and money to cease immediately when the machine would be completed in a few weeks. The Minister for Defence in the newly elected Labor government, Albert Green, approved this course—and gave belated sanction to Warrigal II's construction—on 2 January 1930.

Not everyone on the Board remained supportive of this approach, especially when, after the auditor's comments concerning the Widgeon crash had been received, Wackett reported early in February that the Warrigal was still six weeks off completion. The AMP (Goble), in arguing for a check of Wackett's design figures for the new Warrigal, flatly declared that if this check could not be obtained economically in Australia then 'serious consideration ought to be given to stopping construction of the aircraft, and cutting the loss'. As Moth trainers were already giving satisfactory service in the RAAF, he could see little justification for producing more Warrigals even if the type was a success. Williams countered by pointing out that, in deciding on the development of Warrigal II, 'the Board had in mind not so much its introduction as a training machine as its adoption ... for use in the Army Co-operation Squadron recommended by Sir John Salmond'.[108] Nonetheless, even Williams, as Wackett's best ally, could not protect him indefinitely. The Experimental Section ceased operating on 2 April, and two weeks later the CAS directed Wackett to send the Warrigal II as early as possible by road to Richmond, where it was to be held 'pending further instructions from these Headquarters'.

Early the next month the Board decided that preliminary trials of the aircraft should be carried out at Richmond, with full tests following at No.1 FTS, Point Cook. On 15 May Wackett was instructed to supervise the machine's erection at No.3 Squadron, Richmond, for the first proving trials. More than six weeks later, however, No.3 Squadron advised Melbourne that no tests had taken place, and the unit complained that it was impossible to get a decision from Wing Commander Wackett. In fact, this was just one of the unit's difficulties in dealing with the Experimental Section's head.

At this time Wackett had resigned from the RAAF and was winding up the Randwick section, transferring his operations to the Cockatoo Island naval dockyard in a civilian capacity. John Miles, then a flying officer, recalls that he was sent to collect some tools and machinery from Wackett's workshop:

> He had a fair amount of equipment out there, lathes and things like that, which we needed badly at Richmond. I got out there, carrying a letter of authority and having advised Wack we'd be picking the stuff up, and he refused to let anyone have it at all. I was ordered off the place and, being a junior officer, I just went away. The next day they went out again, after more correspondence or phone calls with Wack, and it had all gone—gone to Cockatoo Island. He took the lot. Old Wack was quite odd at that time.[109]

No.3 Squadron sought and gained permission from the Air Board to go ahead with tests on its own. Carried out by Flying Officer 'Bertie' Simms, these were completed by 18 July with Simms reporting that the Warrigal was 'a good general purpose machine and could be used successfully for dual instruction, army co-operation etc.'[110] While awaiting transfer to No.1 FTS for service trials the opportunity was taken by many curious and eager pilots at Richmond to take the aircraft on flights, and on one of these it was crashed and sustained damage requiring repairs at the Cockatoo Island dockyard. On 12 September, however, it was flown to Point Cook via Junee by Flying Officer Allan Walters and Corporal H.J. Eddison. In August 1931 Warrigal underwent more test flights before being passed across to Fighter Squadron the following month for evaluation trials. These had not been completed when the aircraft sustained considerable damage in a mishap on 6 October. Flying Officer Simms was making a normal landing when he heard a loud crack as the undercarriage collapsed.

This aircraft is usually described in publications as the first of the Warrigal machines produced by L.J. Wackett at the RAAF Experimental Section, but Wackett's own papers confirm that it was actually the 'Warrigal II'. It is shown here at Richmond, where it underwent initial flight testing in July 1930. (*Mr J.C. Miles*)

After repairs were made, the machine was ready for service trials which began in January 1932. In these tests, involving carriage of a full load of petrol and bombs, the performance of the Warrigal was found to suffer markedly. The aircraft was unstable, heavy on the controls, exhibited inadequate rudder control, and was erratic in aerobatics; in nearly all aspects it compared poorly with the Wapiti. In September approval was given to fit the float undercarriage designed for the Warrigal and for the aircraft to be issued to Seaplane Squadron for trials 'with a view to its use for instruction and bay patrols if found suitable'. Two months later it was reported that, although the aircraft had flown only six hours in the seaplane role, involving only take-offs and landings but no service trials, it was clear that the undercarriage was unsuitable. Apart from this weakness, however, it had been successful as a seaplane and was considered to be valuable for patrols over Port Phillip Bay when seas were too rough for the RAAF's Moth seaplanes.

By March 1933, the CO of No.1 FTS (Wing Commander De La Rue) was recommending the disposal of the Warrigal. It had, he said, been unserviceable for a considerable time, and it was doubtful if any modification to the undercarriage would ever prove efficient. The AMS (Cole) supported this recommendation to the Air Board, arguing that, as the Warrigal's performance was so far below that of the Wapiti or any new types ordered, it was not economical for the RAAF to continue to develop or operate it. Since the aircraft (as an experimental machine) could not be sold to a civil operator, he suggested accepting an offer from the principal of the Melbourne Technical College to use it for instruction in aircraft design. This course was approved by the Minister for Defence (Pearce) on 6 July 1933, and six days later the Air Board gave instructions for the removal of the engine and the stripping of all standard instruments and accessories. The Jaguar engine from the Warrigal II, along with the Lynx engine from Warrigal I which had been in store since February 1930, were sold to New England Aviation Ltd. for £350.[111]

Thus the story of the RAAF Experimental Section, and the machines produced there, came to a close. What had been achieved? As events would show, the Randwick unit had been a useful—probably essential—training ground for the man who would ultimately take in hand the vital work commenced in connexion with local aircraft production in the years immediately preceding the Second World War. But the four aircraft designed by Wackett, while often innovative, had cost the RAAF far more than the price of comparable types available from overseas. Was this justified? In the opinion of Sir John McCauley, a later CAS, Wackett's efforts may have kept alive some knowledge of aircraft design in Australia, but his machines were below the level of technology of the time. It was Wackett's driving personality, he claimed, which produced the Randwick section, but there had been nobody able to control him.[112]

Other RAAF members of the period share this general assessment of Wackett's influence and achievement, although their views are the recollections of junior officers entering the service at the time that Wackett resigned his permanent commission and left. As Heffernan remarks, 'With all due respect to the brilliant producer of these early aircraft, I regret that none of them made any impact, apart from an unfavourable one, on any of the pilots who flew them'. Air Commodore Bill Garing, who still recalls as 'too long' the four-and-a-half unnerving minutes flying time he acquired in the Warrigal II, observes that, 'In many respects Wackett was perhaps ahead of his time but he never seemed to get a very happy marriage of ideas . . . Mind you, his Experimental Section at Randwick didn't do a lot of good but it didn't do any harm—it got Australians thinking about flying.'[113]

Post-Randwick developments

The winding up of Wackett's Experimental Section may have ended, at least temporarily, the development of new aircraft types under direct RAAF sponsorship, but it did not terminate local aircraft production meeting Air Force needs. Long before the decision was taken to close the Randwick unit, action had been put in hand for the construction of aircraft for the RAAF at alternative sites around Australia.

After acquiring two DH60 Moths from England in 1926 to ascertain this type's suitability in a training role, the RAAF placed orders for more from January 1928. The initial machines were shipped from England in a knocked-down state and assembled at the De Havilland company's depot at Essendon, outside Melbourne. While the importing of airframes continued, subsequent orders also provided for a large number to be built at Geelong, Victoria, by LASCo. and sub-contracting companies.

Since 1925 LASCo. had maintained a keen look-out for Defence work. Recalling that in July 1927 the company had responded to RAAF specifications for Army cooperation and light bombing aircraft by submitting a mock-up fuselage, in February 1928 its managing director, Herbert ('Jimmy') Larkin, followed up by sending drawings of Avro-designed machines that met the specifications. Was the RAAF still interested in purchasing aircraft of this type, he asked, and was there any prospect of the type then being developed at Randwick (the Warrigal) being placed in production locally?[114] In March 1929, the Minister for Defence (Glasgow) announced that LASCo. had been contracted to supply 32 Moths over three years.[115]

In addition to the LASCo. machines, other Moths were built elsewhere for the RAAF. After eight copies of a metal fuselage version were imported from England and completed at De Havilland's Mascot works in January 1930, the Air Board was keen to obtain a comparison of the costs of further importing these machines or continuing the construction of the wooden version in Australia. In September 1932 the Board decided to place an order with the seaplane repair section now operating under Wackett at Cockatoo Island, occupying the drawing offices and electrical shop at the naval dockyard.[116] Acquisition of a single machine was proposed, purely 'to try the Section out' as to whether it could produce a Moth in wood for £350 to £400 as Wackett claimed. Approving the proposal, the Minister for Defence (Pearce) stressed the trial nature of the purchase.[117] The aircraft (A7–55) was completed and first flew about April 1933, but as no further orders followed it would seem that the dockyard's aircraft section was not as competitive as expected.

On the basis of this trial, the Air Board approached the MSB in February 1933 about a further six metal-framed aircraft being supplied from the Ordnance Factory at Maribyrnong. Previously, the RAAF had made little use of the government munitions factories, apart from the production of a few components and items such as special aircraft dopes and paints. In March the following year, however, an order was placed with the MSB for the metal Moths at a unit cost of £430. When these machines were completed in March 1936 MSB claimed a price of £729 each, resulting in protracted haggling which ended with the Air Board agreeing to pay £586 per airframe. The wings for these aircraft were sub-contracted elsewhere, with final assembly done at Laverton by the RAAF.[118] Five more Moths were rebuilt—apparently from wrecked airframes—at the De Havilland workshops at Mascot.

Interestingly, even these modest arrangements to meet basic training requirements were opposed by Air Marshal Sir John Salmond during his 1928 inspection visit. Williams later recalled that Salmond was critical of the RAAF's plan to adopt the De Havilland Moth as a preliminary trainer over the Avro 504, which was still the machine fulfilling this role in the RAF. The two spent an afternoon arguing the merits of the RAAF's decision, though in the event the British air force followed the Australian lead just a few years later.[119] Since the DH60 was a British aircraft, like the Avro, Salmond's objections could hardly be construed as identical to those alleged by Wackett to have been behind his opposition to the Experimental Section, but this was seemingly another example of British resistance to any displays of independence on the part of Australian defence authorities.

The closure of the Randwick unit was also not the end of the RAAF's connection with aircraft designed by Wackett. While working at the Cockatoo

Dockyard, he was commissioned in 1933 by Sir Charles Kingsford Smith to build an airliner suitable for use on a trans-Tasman service the latter hoped to establish. The result was a high wing cabin monoplane seating a pilot and six passengers. Called the 'Codock' (so-named for its place of origin), it was powered by twin Napier Javelin engines. Tested in April 1934, the machine was not a success; according to Harry Purvis, who worked on it, 'Napiers made some of the best engines in the world but the Javelin was one that constantly gave trouble and these were no exception'.[120]

Wackett worked on improving the basic design of the Codock and, when the dockyard's aircraft section was closed down in 1934, he left to set up the Tugan Aircraft Company where he began construction of the LJW7 (Lawrence James Wackett's seventh design) 'Gannet'. It was now fitted with Gipsy VI engines which were far more reliable—which was just as well as the aircraft lacked single-engine performance. Wackett recounted that through his friendship with Williams, the CAS, the RAAF agreed to purchase the Gannet 'if it passed the necessary tests for a certificate of air-worthiness'.[121] After the first completed machine was successfully tested at Mascot in October 1935,[122] the RAAF received its first Gannet the next month. A second machine was purchased by the RAAF before Tugan Aircraft was bought by the Commonwealth Aircraft Corporation (CAC) and production of a further three machines of this type carried out by this company—the first of these (A14–3) being the first aircraft delivered from CAC's factory at Fishermens Bend, Melbourne, in February 1938.[123]

Production of the Wirraway

With the establishment of CAC Australia had been finally able to achieve the goal of meeting defence requirements in aircraft through local manufacture. The scheme resulting in CAC's formation had its genesis in negotiations between the government and certain companies from October 1935, although the precise circumstances behind these contacts are obscure. According to accounts, the first steps were taken as early as August 1935 at the instance of Mr Essington Lewis, then managing director of Broken Hill Pty. Co. Ltd. (BHP), who succeeded in convincing the government to convene a conference of interested parties to discuss commencing the production of aircraft and engines.[124]

Whether the meeting was Lewis' idea, or came from W.S. Robinson (joint managing director of Broken Hill Associated Smelters (BHS)), L.J. Hartnett (managing director of the car company General Motors-Holden (GMH)), or the Minister for Defence, Archdale Parkhill, appears to be the subject of conflicting claims. According to one historian, there is evidence that Parkhill was far from 'being simply directed' by the other claimant architects of the scheme, but had, shortly after assuming the defence portfolio in late 1934, become convinced that a local aircraft factory was a necessity. He was alarmed by the delay being experienced by Australia in getting defence orders for aircraft filled from her traditional British suppliers, although he initially looked to a different solution to that eventually adopted. In 1935 he explored the

The LJW7 or 'Gannet' was developed during Wackett's period with the Tugan Aircraft Company based on the 'Codock' airliner he designed for Kingsford Smith in 1933. The first machine of this type joined the RAAF in November 1935. The aircraft shown here is possibly A14–3, which was the first plane produced by the Commonwealth Aircraft Corporation after it absorbed Tugan; it was also the first Australian aircraft to undertake a flight outside Australia, being used to fly CAS to Singapore in February 1938. (*J. Ryland papers, NLA, MS 5020*)

possibility of persuading a British aircraft company to set up a factory in Australia to meet local needs, but failed to find a taker.[125]

The outcome of this lack of British interest was that in early 1936 a syndicate was formed by Essington Lewis comprising BHP, BHS and GMH. Announcing the move on 15 February, Lewis stated the intention to obtain the services of Lawrence Wackett to take charge of the operation. According to Sir Richard Williams (CAS at the time), he had personally recommended Wackett for this task when contacted for his opinion on another 'certain officer then employed in the Service'.[126] Four days later Wackett departed Australia to investigate the aircraft industry overseas, in preparation for setting up the local plant. He was joined by two other technical experts with strong RAAF backgrounds similar to his own: Squadron Leaders Arthur Murphy, then OC of the No.1 AD workshops, and Herbert Harrison, formerly director of Technical Services before transferring to the Citizen Air Force Reserve in May 1935.

The news that Australia was preparing to go its own way was received with consternation in London. The British board of trade belatedly recognised the commercial disadvantage of a major Dominion establishing its own aircraft industry, and at a meeting of officials on 20 February declared itself as 'strongly opposed'. The Air Ministry, however, was more relaxed, considering that provided the types manufactured were British, the only objection that mattered was the fact that American interests were included in the Australian syndicate in the form of GMH. If foreign participation was allowed, this would preclude continuing Australian access to British technical secrets. London's qualified approval of the Australian plan was, therefore, conditional on extinction of the American shareholding in the concern.[127]

British opposition to GMH placed Australia in a difficult position. The company was important to the scheme's success in view of its experience in body-building engineering, but, as one historian has pointed out:

> from the United Kingdom viewpoint it was just this experience in the motor car industry which posed a danger to British interests. Thus the argument . . . [about] American access to British defence secrets was simply a 'screen' to guard against 'the more serious danger . . . '[128]

An accommodation of the professed British concern was made by reducing the size of GMH's holding in the new consortium, and the Dominions Office was advised in March that GMH was reducing its interest to just 10 per cent. This left the way open to a UK aircraft firm to join the syndicate, but apart from a brief flirtation with the Hawker-Siddeley group nothing came of this attempt. Eventually other partners were found to dilute GMH's profile, initially Imperial Chemical Industries (ICI) and later Orient Steam Navigation Co., but there was no question of dispensing with GMH altogether because the corporation needed the services of that company's managing director, Lawrence Hartnett.[129]

While the issue of American involvement had been enough to bring UK-Australian relations under strain, worse was to follow when the results of Wackett's technical mission became known. Having visited Britain, several European countries and the United States, the mission returned with a recommendation for a non-British design. This was the North American Aviation Company's NA-16, a two-seater general purpose, all-metal low-wing monoplane with welded steel tube fuselage, powered by a Pratt and Whitney Wasp 9-cylinder air-cooled radial supercharged geared engine.

Predictably, this preference of Wackett's technical mission resulted in uproar. Throughout October–November 1936 great pressure was exerted on the Australian government by Britain to reverse the decision, with claims being made that Australia was destroying the basis of defence cooperation. The Lyons government defended its action, with the Minister for Defence (Parkhill) issuing a formal statement in January 1937. This justified the choice of the NA-16 on the grounds of urgency, and the fact that the absence of a British design which incorporated the required modern technology, was beyond experimental stage, or not too complex for CAC to cut its teeth on, meant there was simply no alternative.[130] To help mend bridges, Lyons also asked Hartnett to go to London 'on his behalf, in effect . . . as the equivalent of a Minister Plenipotentiary', to explain the Australian decision more fully.[131] Probably the prime minister hoped to defuse some of the pressure which was building by showing Air Ministry critics that the man representing the GMH interests considered so objectionable on security grounds was actually a 'Vauxhall-trained Englishman'.[132]

Notwithstanding the British reaction, on 7 January 1937 a letter was sent to CAC advising that, subject to agreement on prices and specific terms, the government was prepared to place an order for 40 'locally manufactured standard NA-16 type' aircraft and ten spare engines, at a cost of around £320 000 and with delivery spread over two years. The letter made clear that the Commonwealth was fully conscious of the enormity of the step that was thus being taken:

> You are aware that it has been the policy to use only approved types of British aircraft and engines for the equipment of units of the Royal Australian Air Force and for this reason the Government has given serious consideration to your proposals which involve the introduction of foreign, in preference to British, types of aircraft and engines. Owing, however, to the abnormal position now being experienced in the British aircraft industry, it is agreed that the proposals submitted by your Company offer the most satisfactory means of establishing the industry expeditiously and on a sound basis.

CAC acknowledged the advice of the government's intention on 12 January. At this stage, it said, the prototype NA-16 was expected to be available to the RAAF for testing about July.[133]

In September Wackett wrote to Parkhill's successor as Defence Minister (Thorby) to propose that the Commonwealth consider varying its initial order. He explained that, even at the time the Aviation Commission selected the NA-16 in 1936 as the most advanced development available in an aircraft of its type, it was known that certain improvements to this machine were in hand, including conversion to a retractable undercarriage which the Air Board was keen to have incorporated in its machines. Consequently, Wackett had made a second visit to North American Aviation in February 1937 to obtain the licence for the NA-16 type with improved undercarriage. While there, he recounted, he had discovered that the United States army had since become interested in this aircraft too, as the basis for a more advanced 'Basic Combat' type. The army's interest had resulted in various manufacturers being invited to submit machines in a competition in May 1937, to decide which was the best. This contest had been won by the NA-33, an improved NA-16, and North American had since received an order from the United States Army Air Corps for 150 of this aircraft.[134]

Wackett pointed out that CAC now held the licences for both the NA-16 and the NA-33, and samples of the two planes had been ordered by the corporation to give the Air Board the opportunity of evaluating them. The NA-16 was already in Australia and had been flown, while the NA-33 had been shipped and was expected to arrive soon. He now proposed that the Air Board consider confirming its initial order for 40 machines, but for the more advanced NA-33 version 'which has been specially designed as a military type as distinct from the advanced trainer type of the original NA-16'. This new type, he assured Thorby, contained 'all the most modern developments found on the most advanced military aircraft in any part of the world'—including retractable undercarriage, geared and supercharged engine of increased output, hydraulic operation of flaps and undercarriage, constant speed three-bladed propeller, internal bomb-gear, rotating seat for the gunner, prone bombing facilities, electrical installation for radio, and a blind flying practice hood.

The cost of the NA-33 was £8098 as against the original price for the NA-16 of £7200 (£7550 for the version with retractable undercarriage), but the difference was that the more advanced version could 'be considered as an effective weapon for air defence'. Some consideration had already been given to fitting it for a military role, and Wackett believed that the excellent flying qualities of the type, along with its range, could be retained even while burdening it with both front and rear guns, ammunition, and a 500-pound bomb-load.

At the same time as the Air Board was examining this new proposal from Wackett, it was also giving consideration to the question of selecting an aircraft type to meet the RAAF need for new *ab initio* trainers. The Board planned to buy up to six of the chosen type from an English manufacturer, then have CAC build a further twelve a year under a licensing arrangement. In May 1937 the Board told CAC that it had two types, the Percival Vega Gull and the Miles Hawk, under study 'and it is likely that one of these two types will be selected'. When CAC suggested that Wackett, who was then in the United States, should visit England to investigate the manufacturing requirements of whichever of these two types was selected, the Board agreed that there might be advantage in having him ascertain the licensing terms.[135]

In June, however, the Air Liaison Office in London advised RAAF Headquarters that the Vega Gull was not classed as an aerobatic type and that the Miles Hawk had been fitted with a Gipsy Major engine and was now called the Miles Magister. The ALO also reported that Wackett's investigations had led him to strongly recommend the Magister, because of its satisfactory aerobatics capability and greatly improved construction compared to the older Hawk. The Air Board asked at this point about the possibility of acquiring a single Magister for test purposes, and in July raised the question of terms involved in obtaining rights for local manufacture.

By April 1938 the Air Board had rethought its approach to the trainer issue. When asked by the ALO where this matter rested, RAAF Headquarters advised that it was not intending to buy more than the single Miles Magister already received, and that the idea of manufacturing this type locally under licence would not be proceeded with. As a result of this decision the Magister which arrived in Australia in February 1938 and was numbered A15–1 became an orphan aircraft in the service. Although flown in public displays given by the Air Force at Flemington and Richmond in April that year, where it was demonstrated alongside the NA-16 and NA-33, it had no useful function. Later that year it was evaluated in connexion with the trainer design being developed by CAC, which first flew as the Wackett Trainer in October 1939, but by July 1940 it was scrapped.

The Air Board's response to Wackett's proposal to acquire the NA-33 as a military machine rather than simply a trainer came in March 1938. By this stage it had been decided to name the aircraft type the 'Wirraway' (Aboriginal for 'challenge'). In a minute from the secretary of the Air Board (Coleman) to the secretary of the Defence Department (Shedden) it was argued that while the advanced version of the Wirraway was certainly more costly than the version originally looked at, the difference was 'only a little more . . . [and] is definitely covered by the value of additional items to be supplied'. The Board 'has felt all along that the prices quoted for local manufacture are most

Test-flights on the NA-16 began at Laverton on 3 September 1937 and were completed on 15 September, when officials of the Commonwealth Aircraft Corporation (CAC) visited the base. The tests were carried out by Squadron Leader F.R.W. Scherger (*second from left*), pictured with (*from left*): Mr A.G. Brown (CAC secretary), WgCdr A.W. Murphy (OC Workshops, No.1 AD), SqnLdr H.C. Harrison (retired RAAF director of technical services), and Messrs T.W. Air, Knight and Smith of CAC. (*Mrs J. Gray*)

reasonable', bearing in the mind that the industry was just being established. Considering also the prices paid for British aircraft, CAC's price was 'by no means excessive'. Based on the formula usually applied by the MSB for Air Force orders met in its factories (RAF rates plus 33.3 per cent), the cost of locally producing the Demon—'a machine of much inferior performance to the 'Wirraway''—would be £7500, and this was without the many modern features in the Wirraway. The Board was 'definitely of opinion that the prices quoted are reasonable' and recommended that they be accepted without further investigation.[136]

The NA-33 did not reach Australia until November 1937. Both it and the NA-16 were subjected to rigorous testing by the RAAF, leading in January 1938 to the Air Board's acceptance of CAC claims that the 33 was much superior in performance to the 16 and should be the type put into production. By this stage production had already commenced in one department of CAC's factory and was expected to be in full swing within three months. The corporation was therefore pressing for completion of contract details and negotiation of more definite terms than it presently had.[137] Before approving the Air Board's recommendation, however, the Minister asked to see all the evidence available to the Board regarding the reasonableness of the prices quoted by CAC. It was after seeing comparisons of the cost of importing the inferior Demon from Britain, and taking into account the benefits of local manufacture, that on 5 May he gave formal authority for an order to be placed with CAC for 40 NA-33s at a cost of £8098 each, and ten spare engines at £2600 each. The requirement for CAC to steadily increase the level of Australian materials and workmanship involved in meeting this order was, however, to be specifically written into the contract.[138]

The NA-16 and NA-33 were eventually taken onto strength as A20–1 and A20–2 respectively. As the prototype for the production program the latter aircraft was a prize piece of machinery to be jealously guarded. Requests from pilots wishing to fly it were numerous, but only granted after a period of dual instruction. One airman-pilot who got the chance received the nasty scare of a forced landing which might have wrecked this most valuable plane:

> Unfortunately I mishandled the engine controls on take-off and it misfired and lost power just as I was passing over the aerodrome fence. I was too low, and had insufficient time to determine the cause of the trouble, and immediately decided to close the throttle and land the aircraft. My landing was successful with only damage to the tail wheel tyre. The aircraft was towed to the aerodrome by the rescue truck, and the engine was found to perform perfectly . . . While the commanding officer would have realised the reason for the engine misfiring, it was never discussed, nor as far as I know, was it recorded. He would have been more concerned than me, as the aircraft was the most important aircraft in the RAAF.[139]

The first Australian-built Wirraway (A20–3) was test flown from Fishermen's Bend on 27 March 1939 by Hubert Boss-Walker, a former flight lieutenant who had transferred to the RAAF Reserve the previous July. This machine was not actually delivered to the RAAF until 1 August, by which stage the Air Force had already received A20–4 and A20–5 three weeks earlier. The next three aircraft off the production line were delivered at the end of August. Taken with A20–2, which was the prototype for the Australian version actually built, the RAAF had seven Wirraways on its strength at the outbreak of war.[140]

Apart from the production facilities established at Fishermens Bend, another important step towards establishing a viable local aircraft industry was also laid in 1937, with moves being taken towards providing the research effort necessarily associated with maintaining such an industry. On 31 July 1937 H.E. Wimperis, recently retired as the British Air Ministry's director of scientific research at Farnborough, reached Australia as the Commonwealth government's adviser on inaugurating aeronautical research. On hand to welcome his ship when he arrived at Sydney were three Demons from No.3 Squadron which provided an escort.[141]

After visiting airports, university engineering faculties and other engineering establishments including the munitions supply establishments at Maribyrnong, as well as RAAF stations at Richmond, Laverton, Point Cook and Pearce, Wimperis produced a report in December which called for the creation of certain facilities for design testing purposes, along with a committee on aeronautical research to continue providing expert advice to the government.

Agreeing with those who argued Australia needed the means to conduct research leading to the production of aircraft adapted to local conditions, he recommended construction of an aeronautics laboratory equipped with two wind tunnels, to be located at Fishermens Bend where it would be adjacent to CAC; the setting up of a department of aeronautics at a university, preferably Sydney, to promote the training of aeronautical engineers; and an engines test house to be built at the RAAF station at Laverton.[142]

The Wimperis scheme was approved in 1938 but progressed only slowly; the proposed laboratory was not even commenced until August 1939. Nonetheless, taken with a British initiative to re-establish itself in the local aircraft-production scene after the debacle over the Wirraway (see chapter 15), the measures taken by 1939–40 had finally provided Australia with the basis for meeting an important part of aircraft needs for defence purposes from its own resources.

10

Civil aviation

In the years immediately following 1918, thinking about the shape and nature of Australia's air defences was dominated by the perceived connection between military and civil aviation. From the outset the debate on this matter revolved around the question of resources. As General Legge argued as CGS, it was 'altogether to the interest of the Government to assist civil flying, because Australia is a country which can never afford to maintain a really large air service in peace-time'. For this reason he developed plans for an air corps meeting the Army's needs primarily on a citizen force basis, and urged also the formation of local aero clubs. These, he was convinced, 'are to be encouraged because they, in turn, would stimulate the construction of 'planes for private or commercial purposes, and any 'plane that can carry passengers can carry bombs'.[1]

The general conviction that civil aviation could make a vital contribution towards strengthening national defence was reflected in the arrangements put in place by the government to foster and control the post-war growth of commercial air activities in Australia. It was decided during 1920 to create a Civil Aviation Branch (CAB) within the Defence Department. On 16 December that year, 37-year-old Horace Brinsmead was appointed Controller of Civil Aviation (CCA) and member of the Air Council.[2]

Williams, then the Army's director of air services, later made clear that Brinsmead was not his choice. Asked at the time by the Minister for Defence to give a verbal recommendation on the applicants for the position, he had suggested the name of 'one with extensive flying experience, including command of a squadron'.[3] Brinsmead was not a pilot, Williams sniffed, nor had he had anything to do with aircraft operations. This was apparently true, despite later claims to the contrary,[4] but he had served on the AFC staff at AIF Headquarters, London, where he displayed such marked administrative ability that he became Staff Officer for Aviation in September 1918 and was promoted lieutenant-colonel. The Minister, evidently valuing administrative above flying skills, ignored Williams' advice.

The first task confronting Brinsmead was to create his branch, which he considered needed a minimum of three officers to start work.[5] For a non-pilot controller, it was obviously desirable that his principal subordinates should be knowledgeable aviators. Accordingly, the officers appointed on 10 February 1921 were all wartime pilots,[6] and were actually paid at a level equivalent to captains in the Air Force.[7] Of these, Captain Ernest Jones was still serving in the AAC, resigning as Equipment Officer at the CFS to take up his new appointment as Superintendent of Civilian Flying Operations.[8] To assist him in determining the medical fitness of pilots to hold a licence, the Air Council gave approval for using the services of the doctor at Point Cook, Captain Arthur Lawrence, as a temporary measure.[9]

Early work of the Civil Aviation Branch

Creating a network of commercial air services involved a number of essential preliminaries, such as surveying air routes between the principal cities and towns, and establishing airfields. The carrying forward of this work assumed priority in the branch's program, but a problem immediately existed in that the CAB had few material resources available to it. Brinsmead attempted to overcome this difficulty by asking the Air Force in March 1921 to issue him with an Avro 504 from its stocks, along with a camera and other photographic equipment.[10]

The Air Board was not sympathetic to this request. While it had no objection to making a free gift to CCA of any equipment which was obsolete or surplus to service needs, it wanted to retain as a reserve all the Avros it had which were not actually in use. The Board suggested that CCA 'to take steps to obtain a commercial machine more suited to the special requirements of his Branch', but in the meantime it was prepared to loan an aircraft for six to twelve months. The camera requested would be similarly loaned, until the Air Force's units were established, and to meet CAB photographic requirements the Board agreed that material 'for developing and printing . . . can be made available and the work carried out by Air Force personnel'.

Brinsmead was plainly disappointed by this response. On 24 March he wrote back to the Air Board to say that, while he recognised the importance of his branch having an aircraft of commercial type available for its use, he had only been trying to economise by asking for a machine that might be surplus to the Air Force's establishment. If the Board could not oblige except on a loan or repayment basis, then he would certainly 'take the necessary steps' while accepting the interim arrangement offered.

Notwithstanding the Air Board's attitude on this occasion, Brinsmead persisted in trying to use RAAF resources to get things moving in the civil sphere. As narrated previously (see chapters 1 and 2), from early 1921 he had been promoting the idea of using the RAAF to initiate an experimental airmail service—strongly resisted by Williams. A year later, in March 1922, the suggestion arose that the Air Force should provide an aircraft from its stocks to enable an aerial medical service to be begun at Charleville (Qld) for residents in outback Australia—a scheme which predated the Flying Doctor Service by six years. The proposal arrived in the Defence Department in letters addressed to Williams and Brinsmead, dated 9 February 1922 and signed by H.B. Davis, honorary secretary of the Queensland section of the Australian Aero Club. It had been suggested, Davis said (by whom he did not state), that he should approach the department:

> with a view to obtaining, either as a free gift or at a low figure, one of the Aeroplanes given to Australia by Great Britain. I think I am right in saying that these machines are hardly likely to be used by the RAAF for a very considerable time, and as they are likely to perish . . . they might be better employed on such a service.

In reply, both Williams and Brinsmead told Davis that they would be happy to discuss the proposal during a forthcoming visit to Melbourne by the president of the Queensland section of the Aero Club, but Brinsmead went further and stated that he had put a suggestion for a gift to be made of a DH9 to the Air Board for comment. Williams, however, did not consider it 'advisable to distribute the aircraft in our possession in this way', and his attitude was supported by the Assistant Minister, Sir Granville Ryrie, who decided to refuse the Aero Club's request.[11]

The following November Brinsmead reported to the Air Council that Queensland & Northern Territory Aerial Services (QANTAS) had applied for a government subsidy to establish an aerial service between the western Queensland towns of Charleville and Cloncurry. In the absence of sufficient CAB funds for this purpose, Brinsmead said he 'could only suggest for the consideration of the Council that three of the DH9 machines that were presented by the British to the Australian Government should be handed over free of charge' to QANTAS. Adopting this step would, he maintained, be doing no more than putting into effect the original intention of the Air Ministry that the aircraft presented to the Dominions and colonies would be 'available for the development of Civil Aviation in the countries concerned'.[12]

Again, however, Brinsmead was defeated in his bid to get his hands on any of the imperial gift planes, as the Air Council decided it was 'unable to recommend the depletion of Air Force Reserves by the handing over . . . of three DH9 machines as suggested'. The following June, with QANTAS now experiencing serious delays in obtaining the Vickers Vulcan aircraft from England which it had ordered to operate on the new route, a second effort was made to buy or borrow two RAAF DH9s; on Brinsmead's advice the company even telegrammed the prime minister seeking his intervention, but still to no avail.[13]

Later there was some relaxation of the policy regarding loans of RAAF aircraft, though only in cases involving finite periods. In August–September 1924 the CAB was able to arrange the loan of two DH9s (A6–4 and A6–5) to Australian Aerial Services to overcome a situation where the company was unable to maintain its mail services due to the unserviceability of all its aircraft. Given temporary civil registration, A6–4 and A6–5 were returned to the Air Force in December that year.[14]

The loan of an Air Force machine to the CAB at least allowed a start to be made in May 1921 on inspecting air routes and selecting sites for aerodromes and emergency landing grounds. On 16 November, however, the Superintendent of Aerodromes, Captain Edgar Johnston, crashed the Avro in northern New South Wales while surveying a route between Melbourne and Brisbane accompanied by Corporal George Gottschalk, a mechanic borrowed from the Air Force. Experiencing engine trouble along the Clarence River, he made a forced landing about 25 kilometres from Stanthorpe (Qld) in which the Avro was badly damaged. Johnston returned to Melbourne by train, leaving Gottschalk to dismantle the aircraft for return to Point Cook.[15]

Fortunately, the CAB's own aircraft—a Bristol Tourer, a conversion of the wartime Bristol F2B Fighter—was available by this time.[16] Registered as G-AUCA, this was housed at the RAAF station at Point Cook. It was this machine that Brinsmead used in late November to attend the inauguration of the Geraldton–Derby air mail service for which he had fought so hard. Flying the plane during the trip was Captain Jones, with Gottschalk again borrowed from the Air Force as mechanic.

The Bristol Tourer remained in service with the CAB until 1923. In March of that year it flew as escort to a Vulcan airliner finally being delivered to QANTAS for use on the Charleville–Cloncurry route. The Vulcan arrived in Melbourne in February and underwent assembly at the RAAF base before departing for Longreach, Queensland, on 14 March to give demonstration flights prior to acceptance from the manufacturer. The Tourer crashed on arrival at Bourke (NSW), and was so severely damaged that on 12 April a board of survey was convened at Point Cook, comprising Flight Lieutenant William Bostock (president), Flying Officer George Mackinolty, and a CCA officer, to make a recommendation on its disposal.[17]

With the loss of the Tourer the CAB was once again dependent on the Air Force for aerial transport. Brinsmead and his staff now used Avro A3–49, which was the RAAF Headquarters communication aircraft housed at the Shaw-Ross Aviation Company's aerodrome at Fishermans Bend, Port Melbourne. This machine was allotted to enable pilots on the staff to obtain flying practice, and was available to CAB whenever not actually in use by the Air Force. During 1923–24 it was used by Civil Aviation officers on 54 separate occasions, all at no charge.[18]

Eventually, in mid-1924, four new replacements for the Bristol Tourer arrived in Australia for the CAB.[19] Among the new machines was a DH50A, built by the De Havilland company in England specifically for use as a commercial airliner (it had a cabin seating four passengers). On 7 August Brinsmead left Melbourne in this aircraft with Jones at the controls on a round-Australia flight. The aircraft made a westerly circuit of the continent, arriving back at Melbourne 22 days later. Coming just a few months after the flight by Goble and McIntyre, this was practically as great an achievement as the RAAF trip but received far less acclaim.

Despite now having aircraft for its own use, the CAB often still relied on personnel or machines borrowed from the Air Force. For example, in October 1926, when Johnston took CAB's DH37 (G-AUAA) on a flight to Normanton in north Queensland which was notable for entailing the longest one-day flight then made in Australia (Longreach to Melbourne), he was accompanied by LAC J.J. Davies of the RAAF.[20] Sometimes the arrangement worked in

This Bristol Tourer, a converted wartime fighter, was the first aircraft to belong to the Civil Aviation Branch. Prior to this the Branch used an Avro 504 on loan from the Air Force. Registered in July 1921, the Tourer was operated from the airbase at Point Cook until crashed in March 1923, whereupon the Branch again became dependent on the RAAF for more than a year. (*ACdre U.E. Ewart*)

reverse, as on 22 June 1925 when Flight Lieutenant Bostock took G-AUAA on a 30-minute practice flight from Point Cook with Squadron Leader Roy Marsden as passenger.[21] And between May 1931 and May 1932 the DH50A was loaned to the RAAF for temporary service (though retaining its civil registration as VH-UAB).[22]

In 1928, when Johnston and the senior aircraft inspector, Jack Howard, planned a survey flight to New Guinea, taking in the north-east coastline of Australia as well, they arranged the loan of RAAF Seagull III amphibian A9–3, complete with a mechanic, Corporal Robert Souter. Though Johnston was a flight lieutenant in the CAF, in addition to his official position, he had no experience on the type and needed some instruction from Squadron Leader Hempel at Point Cook before the flight could begin. Leaving from Point Cook on 26 June, the aircraft and party was away for nearly four months, returning on 19 October.[23]

Notwithstanding the efforts of CAB in the first years of its existence, the fact remains that a far greater proportion of air route surveys in this period were performed by the RAAF (see chapter 2). With more resources of men and machines, the Air Force was better placed to advance this work and—given its importance to operations for defence purposes—laying out safe routes was no less high on the RAAF's early list of priorities. The CAB consequently looked to other means for improving the operating climate for commercial air activity.

RAAF training of civil pilots

One initiative already taken before the creation of the CAB had clearly established the link that was to exist between military and civil flying. As early as

December 1919 a circular was issued by the Defence Department offering refresher flying training at the CFS to former AFC and RAF pilots wishing to undergo a course at their own expense. The purpose of this was stated as being 'to assist qualified pilots taking up commercial flying as an enterprise, and to afford them opportunities to regain their old efficiency without risk to the public'.[24]

Nine pilots had taken advantage of this offer during 1920, and it was decided to continue this arrangement with CCA assuming the responsibility for recommending to the Air Board those to be admitted.[25] A further fourteen attended in the RAAF's first year of existence (1921–22),[26] followed by another five in 1923.[27] Refresher training for civil pilots was still being conducted at Point Cook as late as 1926, with Raymond Parer completing a course in August, shortly before starting an air service on the New Guinea goldfields, and two other pilots, Lester Brain from QANTAS and Arthur Evans, receiving similar training in November.[28]

Useful though such training was, Brinsmead very quickly recognised that there were underlying problems facing civil aviation in Australia. In July 1921 he had received a report from his Superintendent of Aircraft stating that inspection of the 60 civilian planes in the country had revealed that 'very few indeed' had been built later than 1918, and only three landplanes and an equal number of seaplanes had any value for defence purposes.[29] Brinsmead included these findings in a confidential assessment of the state of the industry to the Air Council, adding that, with three possible exceptions, every one of the 41 firms then operating air services was of doubtful financial viability. Although the situation was deplorable, he did not see it as entirely hopeless, considering that 'civil aviation has not yet really commenced in Australia'. What did concern him was that:

> I am doubtful if there will be sufficient reliable pilots in Australia to maintain these Services after Air Force requirements are satisfied, and they have taken the best; this difficulty will be a very real one in a year or two unless steps are now taken to make provision for training pilots in sufficient numbers to satisfy the probable requirements.[30]

With this in mind, he had already spoken to Williams to suggest that the RAAF should annually take a number of pupils who had trained in aero clubs and bring these to 'the standard laid down for a Pilot's Flying Certificate for machines for the purposes of public transport'. Though these pilots would be for civilian employment, they would be enrolled in the RAAF Reserve or the CAF and thus available for call-up to meet defence requirements.

This scheme held attraction for the Air Force and was accordingly taken up in conjunction with its plans to begin conducting *ab initio* flying courses from the start of 1923. Among the trainees embarking on the first 11-month course at the end of January were five cadets nominated by the CAB: Lester Brain, Robert Doyle, and Dudley, Alan, and Louis Davidson (who were not related).[31] Only three of the five graduated to take civil flying jobs—Brain becoming a nationally known figure in the Australian aviation scene. Alan and Louis

Davidson were discharged before completing the course, although Alan was later granted a short service commission in the RAAF, becoming a pilot officer in 1924.[32]

Even before this first course was underway, Brinsmead had attempted to build on acceptance of his scheme but here he encountered resistance. After AA&ECo. approached the Minister for Defence (Massy Greene) in April 1922 seeking financial assistance to keep the firm going (see previous chapter), Brinsmead was prompted to suggest that the Department purchase an aircraft which the company was in the process of manufacturing, and place an order for another two of the same type. This was the six-passenger airliner which AA&ECo. had begun developing in October 1921. Designed by Harry Broadsmith and named the B1 (some accounts refer to it as the B6), the aircraft was powered originally by a Liberty engine until CAB provided a Rolls-Royce Eagle VIII engine for installation. Brinsmead suggested that the three aircraft bought from the company should be used at No.1 FTS to give special training to the civil cadets on the flying course. The opportunity to fly heavy commercial machines for a few hours prior to graduation would, he said, improve their employment prospects, and this form of subsidy should also 'ensure a successful future to at least one firm that may be the nucleus of a thriving industry in the future.'[33]

RAAF Headquarters flatly opposed such an arrangement however. Williams clearly regarded the proposal as another bid to divert RAAF resources from their proper defence purposes, and he was not alone in this view. Squadron Leader William Anderson, the acting DPT, declared:'It does not appear to me to be one of the functions of the RAAF to teach pilots to fly purely commercial machines such as are never likely to be in use in the Air Force.' Training of civil pilots by the RAAF had only been justified by the fact that these pilots would subsequently serve in the Air Force Reserve, and the flying practice proposed by CCA did not add to their fitness for this role.

The reply from the Air Board accordingly rejected Brinsmead's scheme as 'uneconomical and unsound if it could be avoided'. Apart from the considerable costs involved in maintaining a flight of three B1s at Point Cook, it was argued that the training given to civil cadets by the Air Force would go beyond anything ordinarily required of a civil pilot and that the B1 was not very different to the DH9 and DH9A machines on which cadets were to be trained in any event. The Minister sought further details of the estimated costs in July, and on being told these worked out at a total in excess of £20 000, he approved the purchase of only one machine for CAB.

Although the RAAF won this round, this was evidently not the end of the fight over the principle involved in the B1 purchase. In September 1923 the new Minister for Defence (Bowden) announced that two English Electric Wren aircraft had been ordered for CAB—for use in training at Point Cook. This order was cancelled three months later when the manufacturer increased the quoted price, but two DH53 Humming Bird machines were substituted and delivered during 1924.[34]

Meanwhile, in November 1923, applications were called for civil aviation vacancies on the second course commencing in January 1924. The three successful candidates, who all graduated at the end of 1924, were Basil Daish,

Arthur Evans and Arthur Affleck. On this occasion, however, only Daish came direct from civilian life, as the other two were already enlisted as airmen in the Air Force. Evans later received a short service RAAF commission, becoming a pilot officer in January 1928. For the 1925 intake, four CCA nominees were selected and all were RAAF airmen. However, as this course was starting it was anounced that the training of civil aviation cadets would be discontinued. Consequently, after graduating, the four—George Rice-Oxley, George McCausland, Les Trist, and Frank Drayton—all stayed with the Air Force.[35]

The aero club connexion

During the latter half of 1924, interest began to be taken locally in the light aeroplane clubs appearing in England. These clubs were considered valuable adjuncts to the air defence of Britain, attracting support and assistance from the Air Ministry, and the belief arose that such clubs could have similar utility in Australia. A meeting was held in December of that year to reform the Victorian section of the Australian Aero Club, at which two serving RAAF officers were elected as office bearers: Flight Lieutenant Harry Cobby as one of two honorary secretaries, and Squadron Leader Henry Wrigley as a committeeman.[36]

By November 1925 approaches to the Minister for Defence (Sir Neville Howse) had produced government approval for assistance to be provided to aero clubs, with equipment to be loaned for the training of pilots and a cash bonus of £20 to clubs for each new pilot they trained to 'A' class standard. In July–August 1926 six DH60 Moth aircraft ordered by the CAB arrived from England, and during the latter month they were issued as loan equipment to get the Victorian and New South Wales sections of the Australian Aero Club operating under the new arrangements.[37]

The success of this movement in getting off the ground was not a universal cause for satisfaction, especially within RAAF Headquarters where the government's enthusiasm for the aero clubs was seen as creating additional strain on scarce defence funds. Indeed, there were senior Air Force figures who doubted that civil aviation as a whole did much for defence at all. Wing Commander Goble gave clear expression to such a view as acting CAS in 1924, when he went on record as saying that much commercial flying simply consisted of 'flying from one place to another'.[38] This attitude for once placed him in agreement with Williams who, following his return to Australia, sounded a caution in June 1925 about the value of flying clubs for defence. While he generally agreed with statements attributed to Britain's CAS (Trenchard) that 'if a large number of machines and pilots can do useful commercial work in their civil capacity they will be of inestimable value as a Reserve for War', he was at pains to point out that persons trained under the Australian aero club scheme would still require 'some months of further training in RAAF schools before they were fitted for the RAAF Reserve or for the Citizen Air Force and before they would be of use for war'. On these grounds Williams argued for any available funds to be put into building up the Air Force rather than sponsoring aero clubs.[39]

Despite reservations at this level, over the next few years there was a rapid proliferation of aero clubs across the Commonwealth and it was perhaps predictable that many of these drew their instructional staff from former Air Force pilots. For example, on its formation the Victorian section obtained the services of Flight Lieutenant Ernest ('Pard') Mustard as its flying instructor, and the South Australian section employed Flight Lieutenant Ivor McIntyre when it formed in November 1927. Several clubs were also given retired RAAF aircraft to assist their activities; in October 1929 the New South Wales and Queensland sections each received a DH9 (A6–9 and A6–21 respectively), and the South Australian section received A6–19 the following January.[40]

In later years, especially following the onset of the Depression in 1930, the government continued to provide support for the clubs and, indeed, for owners and operators of civil aircraft generally. Small quantities of obsolescent aircraft and aero engines no longer required for Air Force purposes were still issued to clubs on loan, but in addition the rental rate charged for accommodating private aircraft in government hangars, including at RAAF bases, was reduced to a level only one-third of that charged in pre-Depression days.[41]

This was by no means the end of Air Force involvement with the club scheme. In March 1930 the RAAF's Director of Training (Squadron Leader Bill Bostock) was sent through Victoria, New South Wales and Queensland examining civil air instructors and students, with the chief instructor at Point Cook (Squadron Leader John Summers) doing the same in South and Western Australia. These tours were carried out at the request of CAB in an effort to maintain the standard of flying tuition imparted by the clubs.[42]

The aero club assistance scheme undoubtedly increased the opportunities for young men to gain private pilot licences. There were, however, authoritative voices who were dissatisfied with the pace of expansion in this area and urged the value of citizen training on an even greater scale. Sir Edward Ellington, Inspector-General of the RAF, included a section in his 1938 report on the RAAF which dealt with the 'relation of civil aviation and aero clubs to air defence'. Repeating Sir John Salmond's advice that the 'closest possible co-operation' between the two air branches of the Defence Department was imperative,[43] he gave as his view that 'more co-operation between Civil Aviation and the Royal Australian Air Force is desirable':

> There is nothing which an air transport pilot is required to know which is not also essential before . . . the pilot of a bombing or general reconnaissance aircraft can be considered fully trained. Both are required to navigate over long distances reliably and punctually, and both must have an intimate knowledge of weather conditions and the various aids to the navigation and operation of aircraft, in order to ensure the safety of their crews and passengers.[44]

Ellington recommended several practical cooperative measures, including one suggestion that commercial pilots should first gain experience as members of the RAAF (to free companies of the cost burden of early training, and to ensure that 'civil pilots are fit for service in war by reason of their early training in armament duties'), and another that civil and Air Force pilots should be exchanged for short periods.[45] But the key proposal he made was for the adoption of a scheme organised along the lines of Britain's Volunteer Air

Flight Lieutenant Ivor McIntyre—famous for his part in the 1924 round-Australia flight—became flying instructor for the South Australian section of the Australian Aero Club when it formed in November 1927. He is pictured here during the club's opening day on 5 December. Barely three months later he crashed the club's Moth trainer giving an aerobatics display at Parafield, and died the next day, 12 March 1928. (*Mr C.H.G. Gardiner*)

Force Reserve, which used flying clubs and civil transport companies to train pilots both for reserve service and as a preliminary to those undertaking service training in the RAF. Ellington considered such a scheme would allow the FTS course at Point Cook to be reduced to nine months, including the addition of a course on navigation.[46]

While the Air Board agreed with proposals to give civil pilots experience in the RAAF, and for periodic exchanges of pilots, it was incensed by remarks about the state of cooperation between the civil and military branches. The Board snorted that it could not believe Ellington 'intended to infer that the desire to so co-operate is lacking'. On the contrary, the Board claimed, it had 'always looked upon Civil Aviation in its every form as an asset to Air Defence, and with this in view has endeavoured to assist its development to the fullest possible extent'.[47]

So far as the volunteer reserve proposal was concerned, the Board was not impressed and pointed out that its implementation would mean dispensing with a proven training course already available at Point Cook, in favour of having this done outside the RAAF 'where neither the aircraft, other instructional equipment, nor the necessary instructors are now to be found.' The Board also considered that nine months was 'too short a period for the basic training of those who are to be appointed to the permanent Air Force'.[48]

The CAB, on the other hand, was of the view that the 'introduction of the Civil Air Guard system [meaning the volunteer reserve] into Australia would present no great difficulty' provided the government was prepared to countenance the expenditure involved. They considered that 'All the nine approved aero clubs now functioning are fully qualified to form sections of the Air Guard and to provide the necessary flying training though assistance to increase aircraft fleets might be necessary if substantial numbers of members of the Guard are to be trained'.[49] They were more concerned by Ellington's other proposals, believing it entirely wrong to restrict airline operators to employing only RAAF-trained pilots, and pointing to the need for Air Force pilots to be trained in the 'special duties that are the lot of a second pilot on a modern air liner' before exchanges took place.[50]

In responding to the two sets of advice, the government issued a sharp reminder to the Air Board that it 'places great emphasis on the value of civil flying and aviation as an adjunct to Air Defence' and that its civil aviation policy 'has been directed to the creation of airmindedness and civil resources'. While seizing the opportunity to rebuke the Air Board on this point, the government went on to reveal that it was not actually prepared 'at present to make any greater provision for the Defence Vote for additional facilities for citizen training', so that Ellington's suggestion was, in reality, a dead issue from the outset.[51]

Notwithstanding the enthusiasm of Ellington for even greater utilisation of aero clubs and the government's professed commitment towards this same approach, there were now—just a year away from the start of the Second World War—significant reservations evident about the underlying assumption that aero clubs were a defence asset. More than a decade after Williams made his sceptical remarks there were others who now shared his belief that an 'A' class licence obtained through an aero club had little value in defence terms. A year after Ellington's visit one aviation magazine was editorialising on the fact that aero clubs attracted the wrong sort of pupil, usually only the well-off. The people who learnt to fly in the clubs were not chosen for their aptitude or promising ability, physical qualities, or value as defence reservists, but purely by a sort of means test. As one historian remarked of such criticism: 'The inference is clear: there was little direct defence value to be gained from fostering this sort of training in the interwar years'.[52]

Price of progress

Contrary to Brinsmead's initial high hopes, aviation did not quickly establish itself as a flourishing large-scale enterprise in Australia. Scheduled services in the 1920s were few, undertaken mostly by operators who still needed government subsidies to survive. Much of the civil flying done during this period involved demonstrations and record-breaking attempts aimed at catching the public imagination. The Air Force played a supporting role in many of these pioneer flights, providing facilities and maintenance at its bases for visiting overseas fliers such as Alan Cobham, Jean Batten (see chapter 4), Captain W. Lancaster and Mrs Keith Miller, and local aviators, Charles Kingsford Smith and Charles Ulm. Assistance was also given to others, like Major Hereward de Havilland, Francis Chichester, Amy Johnson and James Mollison.[53]

The association between the Air Force and some of these folk heroes of the air assumed special significance. After making the first solo England–Australia flight in February of 1928, Bert Hinkler was made an honorary squadron leader in the RAAF Reserve and awarded the Air Force Cross.[54] After their historic trans-Pacific flight from California to Brisbane in 1928, Kingsford Smith and Ulm were similarly awarded the AFC and appointed to honorary commissions in the RAAF—'Smithy' with the rank of squadron leader, Ulm as flight lieutenant.[55] An acclaimed east-west crossing of the Atlantic in June 1930 earned Kingsford Smith a promotion to wing commander, and after a record-breaking solo flight from England to Australia in October of the same year he received the further step to air commodore.[56]

The publicity associated with these various flights was in some respects a fillip for the Air Force, yet there was a downside to this picture, to be seen in the appearance of Kingsford Smith and Ulm wearing RAAF uniform in New Zealand after their trans-Tasman flight in September 1928. The appropriateness of civilians representing an Australian service in this way seems to have been nowhere questioned. It was noteworthy, too, that in being made air commodore, Kingsford Smith had been granted the same rank as the most senior officer in the RAAF—the CAS. Williams was later to complain about this situation, remarking that: 'The question of granting such commissions was not referred to the Air Board by the Government, and I considered it wrong in principle . . . these men had never been members of the RAAF and other appropriate honours should have been awarded to them.'[57] Understandably, Williams considered that the practice detracted from his own status as head of a 'fighting Service', and, in the absence of civilians who distinguished themselves on land or sea being appointed to honorary rank as generals or admirals, he had a point.

Moreover, the Air Force chief complained that his service 'did not even see' Hinkler, which was certainly not the case with Kingsford Smith and Ulm. When making the first non-stop flight across Australia in August 1928, their famous *Southern Cross* machine actually departed from Point Cook, just as Richmond airbase was the starting point for the first trans-Tasman crossing the following month. An estimated 25 000 people jammed the Richmond–Windsor road for their return flight from New Zealand in October, and when the aerodrome power supply failed shortly before the midnight landing 500 motorists were organised to line up their cars to illuminate the runway with headlights.[58] Similarly the failed 1935 Tasman flight which doomed Kingsford Smith's airline also originated and ended at Richmond. The government subsequently purchased the *Southern Cross* for preservation purposes, and, when the aircraft made its last flight from Mascot to Richmond on 18 July 1935 for official presentation and storage until arrangements were made for its public display, an escort of RAAF Demons flew in close formation around it.[59]

Assistance to civilian airmen sometimes embroiled the RAAF in controversy, as was the case with the attempt by three New Zealanders—John Moncrieff, George Hood, and J.L. Kight—to make a crossing of the Tasman nearly eight months before Kingsford Smith successfully did so. This party planned to make the crossing in a small Ryan monoplane which they had shipped from the US and landed at Melbourne, but in September 1927 the Australian Prime Minister, S.M. Bruce, effectively banned the attempt on the grounds of the aircraft's unsuitability. In a cable to his New Zealand counterpart, however, Bruce undertook to extend every courtesy to any licensed Dominion pilots flying an aircraft authorised by the New Zealand government, so that when Prime Minister Coates replied on 31 October that his aviation authorities regarded the Ryan as capable of making the trip the Australian government had little choice but to accede.[60]

In the last days of December CAS was advised by Bruce that the RAAF was to provide assistance for the forthcoming flight. Hangar accommodation along with labour to help with assembling the plane was to be made available at Point Cook, and accommodation for the three pilots. Facilities were also to be extended at Richmond after the assembled Ryan was flown there to prepare for the final departure from Mascot for New Zealand.[61] Controversy continued to surround arrangements for the flight, until finally, on 10 January 1928, the

Australia's noted aviators C.E. Kingsford Smith (*left*) and C.T.P. Ulm received honorary commissions in the RAAF in recognition of their achievements. Kingsford Smith ended up with the rank of air commodore, the same as that held by the CAS, leading to complaints about the propriety of such appointments. (*NLA*)

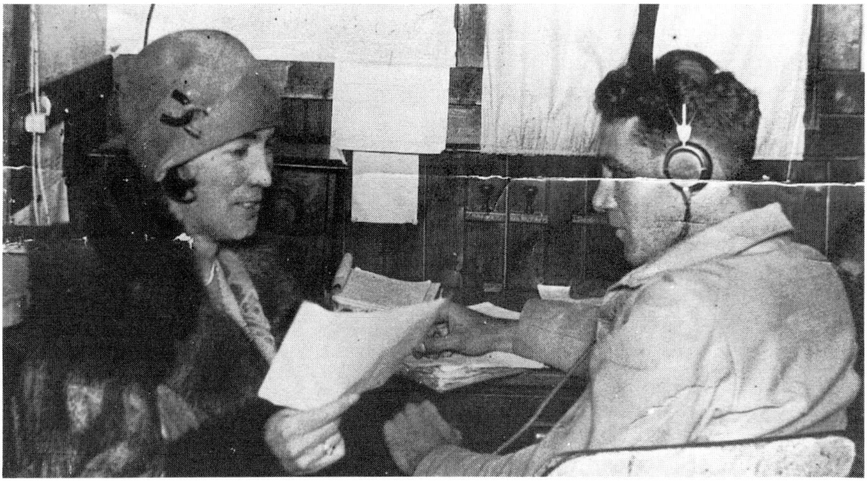

During the return flight of the 'Southern Cross' from New Zealand in October 1928, Mrs Mary Ulm follows her husband's progress from radio reports received at the RAAF base at Richmond. The radio operator with her is AC1 J.E. Reynolds. (*Mrs M. Bennett*)

Ryan took off from Richmond (Mascot was too small for the aircraft to take off with full load) with Moncrieff and Hood at the controls. Nothing was heard of the aircraft or crew again. It was, as one account noted, 'a lonely monument to a flight which should have been banned'.[62]

Occurrences such as this threw into question the whole notion that a pilot in an aircraft of whatever description was still an asset to national defence. In fact, very much the opposite lesson was beginning to be discovered. When civilian aviators attempted flights in machines which were inadequately prepared or unsuitable for the purpose undertaken, the Air Force frequently found itself involved in mounting demanding and usually expensive searches.

In the last days of March 1929, for example, Kingsford Smith set out from the RAAF base at Richmond in *Southern Cross* on a record-breaking flight to England. When he lost radio contact with the ground in bad weather over north-west Australia, this touched off a major national effort to locate and rescue the downed fliers. Responding to public clamour for a search, the Minister for Defence stated on 1 April that the RAAF would certainly be involved if this proved necessary. Two days later, the Air Board directed Wing Commander Lawrence Wackett to leave from Richmond in his Widgeon II amphibian to assist in search operations.[63]

At this time the Widgeon was being readied for trials on board the Navy's seaplane carrier HMAS *Albatross* and had been undergoing modifications in the Mascot workshops. The aircraft needed to be test-flown before embarking on such a major trip, and while the machine was taken to Richmond by road Wackett looked for a let-up in bad weather over Sydney for the necessary opportunity to take off. Local weather conditions were only part of Wackett's problems, however, as heavy rains across northern Australia had badly affected

The Ryan monoplane in which New Zealanders J. Moncrieff and G. Hood planned to cross the Tasman Sea, pictured at Richmond airbase prior to the ill-fated flight in January 1928. The Australian government's attempts to prevent the flight were over-ridden by New Zealand pressure. (*Mr J.C. Miles*)

landing grounds along his route. When advised of this Wackett, recalling his difficulties with outback airstrips on the Widgeon's round-Australia flight in May–June 1928 (see chapter 9), doubted his chances of getting through to the West. His message to the Air Board on 4 April giving expression to his concern drew only a response that he should be ready to depart as soon as possible.[64]

Persisting with his objections, Wackett signalled again that bad aerodromes and a lack of fuel and oil at certain airfields compelled embarking on a trip of 6500 kilometres via southern Western Australia. This, he suggested, made it impracticable to render adequate assistance in reasonable time, and he proposed that use of the Widgeon should be abandoned in favour of chartering aircraft already available in the West. This representation drew only an admonition from the Air Board to get on with the task given to him; abandonment could not be considered—primarily, it appears, because the Minister for Defence (Glasgow) had directed otherwise.

The next day Wackett reported that he had taken off from Richmond fully loaded but had turned back because he was dissatisfied with the Widgeon's performance. The modifications which had been carried out seemed to have impaired the plane's handling, and as he was not confident about attempting the task he was not prepared to go on. The Board, clearly suspecting him of obduracy, called for more details of his reasons, for the Minister's information; it added that he seemed to be advising the press of his intentions before telling RAAF Headquarters, and this situation was to stop. Wackett denied that he was communicating with the press—the fact was that as journalists had access to Richmond they could see and overhear everything for themselves—but he made clear that he considered he was being unreasonably pressured to depart in an aircraft which still needed to be properly tested following its modification. His final word on this matter was to tell the Board that he was quite prepared to 'make way for anyone else who may be able to get better results'.

Having been unable to bully Wackett against his better judgement, the Minister had no choice but to take the somewhat embarrassing course of issuing a public statement on the reasons why the RAAF was unable to participate in the search for *Southern Cross*. In the event, this was not the end of the planned RAAF involvement. Following an approach for assistance from the West Australian premier, Prime Minister S.M. Bruce decided on 11 April 1929 that the *Albatross*, carrying its RAAF contingent of six Seagull amphibian aircraft, should sail from Sydney to join in the search of northern waters and rivers. The ship steamed out of harbour the next day, but had got no further than Port Stephens when word was received that *Southern Cross* had been sighted where it had made a forced-landing on mud flats in the Glenelg River estuary, between Wyndham and Derby in Western Australia. Though the *Albatross* was recalled, there was still work for the RAAF to do arising from this incident.

On 7 April a Westland Widgeon light aircraft called *Kookaburra* had left Sydney to join in the search effort. Piloted by Keith Anderson, a former partner of Kingsford Smith, and with Bob Hitchcock as mechanic, this machine reached Alice Springs but then went missing after taking off for Wyndham on 10 April. By the following day, Brinsmead was beginning to feel some concern regarding the whereabouts of *Kookaburra* and made enquiries which indicated the machine's occupants were probably in trouble. The next day a full conference was held in the Defence Department, at which the Minister gave approval for two DH9As to be despatched to Alice Springs, there to await further orders about conducting a search.

The man chosen to lead the two RAAF aircraft was tall, lean 34-year-old Flight Lieutenant Charles ('Moth') Eaton. On the same day as the Minister's approval was given, Eaton left Laverton in A1–1 with Corporal P. Sullivan, in company with A1–7 crewed by Flying Officer Andrew Gerrand and Sergeant John Campbell. Four days later, on the recommendation of the AMP (in the absence of CAS in Brisbane), the Minister approved a further three DH9As being sent. These machines—A1–5 (Flying Officer Leo Ryan, Corporal Doug Endean), A1–20 (Sergeant Eric Douglas, LAC W.J. Smith) and A1–28 (Flying Officer Max Allen, Corporal G.W. Allen)—set off from Laverton on 17 April under Ryan's command and joined Eaton at Alice Springs two days later.[65]

In a diary which he maintained of the operation, Douglas recorded that:

> We put on board the usual cross-country kit of emergency rations, water bottles, hand tools, tying down gear, a few engine spares, some materials for air frame repairs and a first aid outfit. In addition, we stowed blankets, a service rifle and ammunition, a spare landing wheel and a spare propeller.[66]

A stock of aircraft spares was also sent to Alice Springs in the care of Corporal Cottee, arriving on 23 April.

After initially flying a program of patrol searches from Alice Springs, Eaton decided to shift the focus of his operations further west. Sending Flying Officers Ryan and Allen on to Tennant Creek on 20 April, to link up with a QANTAS airliner, *Atalanta* (a DH50 under Lester Brain), coming from Brisbane to join the search, he planned to take the other three RAAF machines via Tennant Creek to Wave Hill station with the intention of using that place as an advanced base. Unfortunately Allen's machine (A1–28) became unserviceable on the 20th, and the next day the engine of Eaton's own machine suddenly seized approaching Tennant Creek, about 13 kilometres south of the town. Douglas first noticed white smoke coming from Eaton's aircraft, and when he flew closer observed molten aluminium dropping from the engine. Quickly he signalled Eaton, indicating a site alongside the overland telegraph for attempting a crash landing. According to Eaton's diary:

> I had no alternative but to land in small rolling hills along the telegraph wire. It was a good crash, the trees stripping my wings, pulled up and I just flopped to the ground. The pilot of the other aircraft, seeing that I was not injured, went on and landed at the Tennant Creek telegraph line station. The mechanic, who was with me, and I started to walk into Tennant Creek, where we were met by a duckboard with . . . the linesmen of the Tennant Creek telegraph station.[67]

The aircraft was a write-off and, after removing valuable instruments, it was burned on orders from the Air Board.

While Eaton was thus preoccupied, Brain sent news that he had discovered *Kookaburra* that day, located in the Tanami Desert about 130 kilometres east-south-east of Wave Hill. A body was sighted lying in the shadow of one wing, but no sign of the other occupant had been observed. Leaving Flying Officer Allen, Sullivan, Smith and AC1 Allen at Tennant Creek to work on A1–28, Eaton took the remaining three DH9As to Newcastle Waters on the next day to confer with Brain. During this discussion, the idea of landing in the vicinity of the crash-site was dismissed; instead it was decided that a ground party was needed and this was best despatched from Wave Hill. The four aircraft then departed for Wave Hill, flying over the forlorn resting place of *Kookaburra* on the way.

During 23 April arrangements were made to assemble the ground party. Another airliner, *Canberra* (a DH61 under Captain Les Holden), now also arrived at Wave Hill but as its services were not thought to be required it immediately departed again, along with *Atalanta*. At this point Eaton reportedly sought permission from the Air Board to parachute himself into the site in the hope that the second airman from *Kookaburra* might still be alive, but the Board advised against this action unless evidence was obtained that justified the risk.[68] The next day the ground party, consisting of Eaton, Douglas, the pastoral supervisor for the Australian Investment Company, and three Aborigines, set off. Guided by the aircraft which flew out daily in their support, the party pushed on by vehicle and then horseback towards *Kookaburra*. The journey was a nightmare, with Eaton recording that: 'As the horses pushed through the scrub, myriads of flies rose and settled on horses and men alike. They would not be brushed off and we were just covered in a sticky mess.'[69]

On 27 April the ground party— by now in a pitiful state itself—reached the *Kookaburra* and established that there was little that could be done to lessen the tragedy. The body of the second crewman was located and both bodies were buried where they lay: Hitchcock by the aircraft and Anderson some 400 metres to the south-west. The Widgeon was also inspected in an attempt to discover the reason for its forced landing. It was found that Anderson had maintained a rough diary on the fabric of the tail plane; Eaton cut this section out to take back to the Air Board (it is now in the Western Australian Museum). From the diary's notes it was learnt that the damage had been done by a loosened push-rod causing a cylinder to cut out, and although the required adjustment had been made, the two men were then unable to clear a large enough runway in the scrub before being overcome by heat and thirst. It was discovered that the aircraft was in reasonable condition, and Douglas was soon able to get the engine running, but there was no question of attempting to fly the machine out now. After taking some photographs the party made their way back to Wave Hill, arriving there on 1 May.

Over the next few days work was carried out on all aircraft to prepare them for the return flight. By 7 May all seemed in readiness, but on attempting to depart it was found that A1–20 had engine problems. The permission of the Air Board was obtained the following day to write the aircraft off, the expense of rectification being considered to be unwarranted, and A1–5 and A1–7 set off carrying Eaton, Douglas, Gerrand and Campbell; Ryan and Endean were left behind to destroy A1–20, then make their way to Darwin by rail. On the arrival of the two DH9As at Tennant Creek on 8 May, it was found that A1–28 was still plagued with engine trouble. Accordingly, A1–5 and A1–7 went on to Alice Springs the next day, and from there Eaton obtained approval from Air Force Headquarters to destroy A1–28 too, it being considered that fixing its problems would incur unnecessary public expense on an aircraft practically at the end of its useful life. Flying Officer Allen was instructed to salvage all valuable instruments and parts from the troublesome machine, then bring Sullivan, Smith and AC1 Allen on to Alice Springs to arrange their passage, and that of Corporal Cottee, back to Melbourne.

When the two remaining DH9as under Eaton arrived back at Laverton on 12 May, this brought to an end the first phase of the RAAF's involvement with this tragic episode. While most Australians were fully conscious of the disaster which had consumed Hitchcock and Anderson, the public remained for the most part unaware of the fact that the search and attempted rescue which was undertaken had been at a cost of three Air Force machines.

Even as the RAAF contingent had been preparing to depart from Wave Hill,

The DH61 airliner 'Canberra' and three of the RAAF DH9As engaged in the search for the 'Kookaburra', pictured at Wave Hill Station (NT) on 23 April 1929. The leader of the Air Force party, Flight Lieutenant C. Eaton, and the pilot of 'Canberra', Captain Les Holden, confer beneath the propeller of the airliner. This search was to result in the loss of three of the five RAAF aircraft involved. (*RAAF Historical*)

Flight Lieutenant C. Eaton (*right*) and Sergeant G.E. Douglas stand beside the grave of Lieutenant K.V. Anderson, one of the two civilian aviators who perished when the 'Kookaburra' force-landed in the Tanami Desert north-west of Alice Springs (NT) in 1929. After the downed machine was located from the air on 21 April, Eaton led a ground party into the area to locate and bury the bodies. (*RAAF Historical*)

the possibility arose of a second ground party being sent in by the Department of Home Affairs to recover the bodies. The Air Board pressed for details of this planned expedition and the cooperation which might be sought from the RAAF party already on the scene, being concerned not to delay the return of Eaton's aircraft any longer than necessary because of the deterioration observed in them. In the climatic conditions of Central Australia, the machines were already requiring spare parts, and laminations in the airscrews were found to be splitting.

All in all, Eaton said in his report, the aircraft and engines had performed 'as well as could be expected' but they were unsuited to such operations, constantly tending to boil whenever they flew at low altitudes. If a second ground party was to be sent in for the bodies, he urged that two of the new Wapitis entering RAAF service should be used. Of particular interest was the fact that one of the aircraft (A1–5) had been specially fitted with a short-wave transmitter designed and built in the wireless section at No.1 Squadron, and carried as an experiment to provide the contingent with communications. A regular radio watch was mounted at Laverton to listen for messages from this set but unfortunately, Eaton reported, the equipment had not been a success and was dismantled for the return journey. Another radio transmitter had been hastily built and installed in the airliner *Canberra* by Sergeant Oliver Studley at No.3 Squadron, specially to prepare this aircraft to join in the search for *Southern Cross.*[70]

In passing Eaton's report to the secretary of the Defence Department, the Air Board stated its desire to:

> bring to notice the fact that this work was carried out by the personnel concerned with machines and equipment which cannot be said to be suitable for the work and which the Board has for some years looked upon merely as training equipment suitable for use over aerodromes or on organised air routes.

The search, the Board said, had entailed nearly 240 hours flying time 'under most trying conditions, principally over country where a forced landing meant certain crash', and it went on to point out 'the continuous series of small difficulties that had to be overcome in order to keep machines serviceable and carry out the work and ... the amount of work required to be done by individual members involving long hours and little sleep'. Another tribute to the RAAF airmen who had shown such professionalism in their work was contained in a separate report submitted by Lester Brain, who wrote:

> Knowing the old machines with which they are equipped, the class and amount of country over which they have flown, the severe flying conditions and the spirit displayed by them in their work, I should like to commend the Air Force party engaged on the search to the greatest praise and to your special attention.[71]

In May, a three-man committee was appointed by the government to inquire into matters connected with the whole episode, and in particular to investigate claims which had arisen that the disappearance of *Southern Cross* had in fact been a pre-arranged publicity stunt which had misfired. In its report, this Air Inquiry Committee also made a point of commending 'the splendid efforts of the various Air Force pilots who were dispatched'.[72] The suggestions that some special recognition might be accorded the RAAF party were eventually acted on when Eaton was awarded the Air Force Cross in 1931 for his part in the *Kookaburra* search, although by this time he had been involved in another two such episodes.[73]

Even this was not the end of the affair so far as the Air Force was concerned. During April 1929 the government decided to arrange for the return of the remains of Anderson and Hitchcock for proper burial. Recovery of the bodies proceeded, without further call on the RAAF, and Hitchcock was buried in Perth on 3 July in a relatively modest ceremony attended by his family and relatives. Anderson's funeral in Sydney three days later was an enormous public event, and the RAAF was involved in providing pallbearers, as well as aircraft for a flypast during which they dropped wreaths in tribute.[74] The dead aviators may have been civilians having no connexion with the Air Force, but for this occasion all thoughts of what the episode had cost the service were forgotten. Ten years later, a legacy of the *Kookaburra* incident was to present the RAAF with further tragedy (see chapter 7).

Despite some important lessons afforded by the loss of *Kookaburra* about the risks associated with flying in Australia's remote regions, the events in the

Tanami Desert were by no means one of a kind. In late 1930 another aviator was reported missing in the vicinity of Alice Springs. This was Errol Coote, pilot for a well-equipped expedition hunting Lasseter's fabled gold-reef, who was marooned when his plane suffered a broken propeller on landing 8 kilometres south-west of Ayers Rock on 29 October. He was found alive and uninjured nine days later, though not before three RAAF DH60G Moths had already left Point Cook on 8 November to start searching for him. These aircraft, fitted with extra fuel tanks to give them endurance of about six hours, were again under Charles Eaton's command.[75]

The very next month, on 31 December, Eaton received orders to prepare for another rescue mission into Central Australia. On this occasion a Moth piloted by Leslie Pittendrigh, and carrying a mining expert for the Central Australian Gold Exploration Company named Hamre, had disappeared eleven days earlier while travelling from Alice Springs to Ilbilla, 350 kilometres to the west. Eaton departed from Point Cook in A7–34 on 1 January 1931, accompanied once more by Andrew Gerrand flying A7–37. Arriving on 3 January, Eaton deduced that the missing aircraft had probably come down north of the McDonnell Ranges, and proceeded to Hermannsburg to conduct the search from there.[76]

After eight patrols over subsequent days had produced no result, and Gerrand's machine was put out of action on 6 January after hitting a concealed stump while taxiing, Eaton called for further assistance. Two more DH60G Moths (A7–41, Flying Officer Robert Dalton, and A7–42, Flying Officer Arthur Evans) were despatched on 7 January, and arrived to join the search two days later. Success came the very next day, 10 January, after a 'T' marker was sighted on the ground near Dashwood Creek, and Pittendrigh and Hamre were located, apparently uninjured.

When the missing aviators were found, food and water was dropped to them by parachute along with a message telling them to stay put until collected by vehicle. The RAAF aircraft then withdrew to Alice Springs, where Eaton made arrangements for two cars to head for Dashwood Creek. When this expedition set off at 9.30 pm that same night, it contained Eaton and the Government Resident in one car, and a doctor and a driver in the other. Before leaving, he arranged for Evans and Dalton to fly out again next day to drop more food supplies to Pittendrigh and Hamre.

The stranded aviators were reached by 2.30 pm the next day, when it was found that the two aircraft had not only made the food-drop, as instructed, but one had actually landed in a clearing nearby. Evans had done so to ascertain if, after 21 days in the desert in temperatures averaging more than 43° centigrade, the condition of the men was desperate; satisfied that they appeared normal but weak, he had taken off again. An hour later another ground party had reached them, and after Eaton and the cars made their appearance soon afterwards the whole lot set off for Alice Springs around 3.30 pm, arriving at their destination at 7.00 am on 12 January.

On 15 January Evans and Dalton took off on a flight to look for the Moth which Pittendrigh and Hamre had abandoned. Eaton did not accompany them this day, as he was suffering badly from conjunctivitis due to dust, and in the event bad visibility caused Evans and Dalton to return. A second attempt was

Flight Lieutenant C. Eaton pictured with Aboriginal children during yet another rescue mission into Central Australia in November 1930, this time in search of a missing aviator flying for an expedition hunting Lasseter's fabled gold reef. (*Mr P.C. Eaton*)

made the next day, the missing machine being found in a clearing about 16 kilometres from the easterly point of Haasts Bluff. Things now began to go wrong for the RAAF airmen. Dalton suffered a broken connecting rod and made a forced landing 24 kilometres from Alice Springs. He recorded that he noticed his engine problem when his oil pressure suddenly dropped to zero, and 'at the same time the engine lost its revs. and pieces of the engine were flying past the cockpit'. Gerrand also force-landed on a creek-bed near the telegraph station, about 5 kilometres from the town. After despatching a car to Dalton, Eaton collected Gerrand and found the cause of his problem, enabling him to take off again and land at the Alice Springs aerodrome.

Eaton spent some time carrying out instructions he had received from the Air Board to inspect a proposed site for a new aerodrome some 11 kilometres south of Alice Springs, before being ordered back to Point Cook with Gerrand. Evans and Dalton were instructed to follow when the latter's aircraft was again serviceable. Apart from the Air Force Cross awarded to him at this time, as a legacy from the 1929 *Kookaburra* search, Eaton received another recognition of his several involvements in search operations in this region, with Lake Eaton (about 40 kilometres north-east of Ilbilla) being named in his honour.

In December 1932 yet another appeal for RAAF assistance was received, this time involving a downed aircraft in Western Australia. The appeal came in a telegram to the Minister for Defence from Perth solicitor Frank Leake, who advised that a three-man party had left Forrest by air to locate a gold find near the Livesey Range, supposed to be Lasseter's reef. On board the WA Airways machine were the pilot, H.F. Baker, prospector Patrick Whelan, and a mining engineer named Norman Stuckey. As the Minister, Sir George Pearce, was absent from Melbourne (he was actually in his home state of Western Australia), the Secretary of the Defence Department arranged with the Acting CAS

(Goble) to have two aircraft readied in case it was considered necessary to send them. In the meantime it was suggested to Major Norman Brearley, the manager of WA Airways, that the company use its own aircraft for the search.[77]

On receipt of the initial advice of a possible requirement, the RAAF had begun readying two alternative search teams of Wapitis and Moths—it being unknown on the available information which would be more suitable. Crews were kept on 30 minutes standby to leave until more advice was finally received on 28 December of the actual requirement. It was now learnt that the missing machine was a DH50 which had been damaged in a forced landing on a salt lake some 400 kilometres north of Forrest. The aircraft had been repaired to the point where it was flyable with a reduced load, so that when it had taken off again to return to Forrest it had been necessary to leave behind one of the occupants (Whelan) on the salt lake.

Brearley now advised that he had made several drops of water by air to Whelan but most of it had been lost through bursting on impact. More needed to be delivered to save Whelan's life, and he suggested Wapitis could do this by parachute. Officials in the Defence Department were not impressed with the idea of despatching aircraft 1800 kilometres to deliver water, believing that Brearley could surely improvise a suitable kit to accomplish this task from his own resources. Nonetheless, it was considered 'better to act with mercy than with judgement' in such circumstances, and the Acting Minister (Senator McLachlan) authorised the immediate despatch of two Air Force planes.

The aircraft sent were Wapitis from No.1 Squadron: A5–35, Flying Officer Frank Wright, LAC W. Thompson (W/T operator); and A5–32, Flying Officer John Miles, Sergeant John Campbell (fitter aero), and Sergeant William Cameron (carpenter rigger). These left Laverton on 29 December and arrived at Forrest the following day. In the next few days both aircraft flew sorties to the lake to drop supplies to Whelan. Finally, on 5 January, they escorted the repaired DH50 back out to the lake, where Baker safely touched down and lifted Whelan out.

Back at Forrest, participants in the rescue celebrated this successful outcome with a dinner at the WA Airways' hostel before the Air Force machines left to return to Laverton, little imagining the consequences which were to flow from the whole episode. On his return Flying Officer Wright reported events which had occurred on the arrival of the RAAF party at Forrest. A conference had taken place between the Air Force officers, Brearley, and representatives of the gold prospecting company which had Brearley's aircraft on charter at the time it went missing. Wright and Miles were considerably disquieted when confronted with the attitude of the others that 'now the Air Force had arrived their own responsibility in the rescue of Whelan was finished'. The company representatives left for Perth on the evening of 31 December, and Brearley the following morning. Before the company officials departed, it had been necessary for the RAAF party to prevent them taking away tinned foods brought to Forrest for supplying Whelan.

In summing up the situation to the secretary on 19 January 1933, the Air Board declared its opinion that 'Flying Officer Wright and the personnel under his command deserve much commendation for the efficient manner in which the Air Force part of the operations was performed', but that the situation

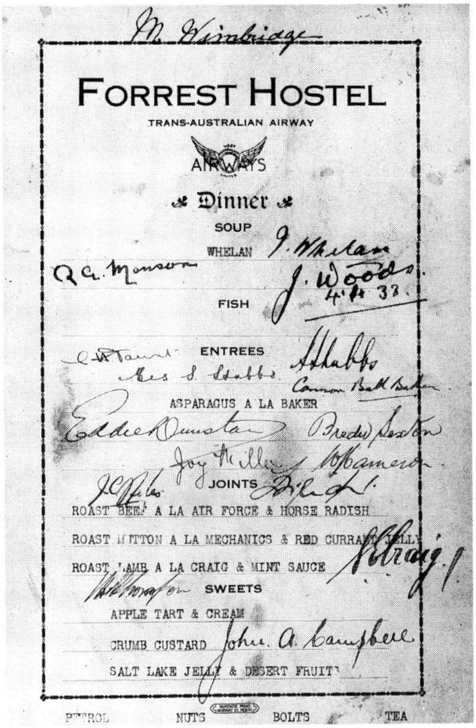

FORREST HOSTEL

TRANS-AUSTRALIAN AIRWAY

AIRWAYS

Dinner

SOUP

WHELAN

FISH

ENTREES

ASPARAGUS A LA BAKER

JOINTS

ROAST BEEF A LA AIR FORCE & HORSE RADISH

ROAST MUTTON A LA MECHANICS & RED CURRANT JELLY

ROAST LAMB A LA CRAIG & MINT SAUCE

SWEETS

APPLE TART & CREAM

CRUMB CUSTARD

SALT LAKE JELLY & DESERT FRUIT

PETROL NUTS BOLTS TEA

In December 1932 a prospector named Whelan was stranded on a dry salt lake north of Forrest (WA) following the forced-landing of a DH50 on private charter from West Australian Airways. Two Wapitis were sent from No.1 Squadron to assist in dropping water and supplies to keep Whelan alive until he could be lifted out again. The success of this mission was celebrated with a dinner back at Forrest, but meanwhile Defence Department authorities began examining ways of discouraging reckless ventures which made such rescues necessary. (*Mr J.C. Miles*)

reflected 'little credit on Major Brearley and his organisation generally'. The Board expressed its irritation with the fact that, having despatched the DH50 in the first place—in great secrecy because of the flight's purpose—and although his company had suitable aircraft on hand, Brearley had adopted an attitude that the responsibility, risk, and cost of search operations was the government's. The prospecting company's action of attempting to spirit away by train to Perth the foodstuffs assembled for Whelan appeared particularly reprehensible, as did Brearley's action in sending a collect telegram to the Department costing over £3 demanding forms enabling him to bill the Air Force for the cost of hangar accommodation for the RAAF machines engaged in the search, and food and accommodation for the personnel.

Pointing out that this was 'the fourth occasion on which Air Force expeditions have been despatched to Central Australia to the rescue of single engined civil aircraft flying alone, ill-equipped and with inadequate ground organization', the Board went on to suggest a tightening of the civil aviation regulations to 'prevent repetitions of unjustifiable expense to the taxpayer and risk to Air Force personnel and equipment'. The Board considered the government should put its foot down and seek recovery of the cost of the Whelan rescue—calculated at nearly £400—from WA Airways. This recommendation was endorsed by Cabinet, subject to the Crown Solicitor being consulted.

Clearly anticipating just such a development, Brearley had meanwhile been writing to Pearce to lobby against his company being billed, arguing that responsibility rested with the gold company but that, in any event, taxpayers ought to bear the cost of risks involved in pioneering and developing the country. Fortunately for him, advice from the Crown Solicitor's office was provided to the Defence Department in February 1933 indicating that, as neither company had expressly asked for RAAF aircraft, nor accepted liability (either expressly or implied), the costs incurred in the Whelan operation were not recoverable from either.

While accepting this advice the Minister still wanted the companies asked what portion of the expenses they were prepared to reimburse the government, and a letter to this effect went out on 3 March 1933. On 16 March, Leake (who now emerged as the solicitor for Lasseter's Reef Gold Mining Company) bluntly responded that his company was under no liability to the Defence Department. When WA Airways answered in similar terms a few weeks later, the Defence Department asked if it was still expected to pay the £39 which the airline had charged for accommodation and meals. Brearley replied that if the Department did not proceed with its claim, his company would forego its charges.

While this correspondence effectively brought the Whelan episode to a close, matters no longer stood entirely where they had before. The CAB had been looking at the form of a public statement to be made, declaring that search operations entailed in supporting private profit ventures which went wrong were not the responsibility of the Commonwealth, and would be undertaken only at the request and cost of the organisers of such ventures. Johnston (as Acting CCA) observed that even if liability in such instances could not actually be enforced, as in the case of the Whelan operation, a statement to this effect would still be 'good evidence of the Government's future intentions'. The statement was given to the press, as drafted, by Senator Pearce on 24 January 1933.

The practical effect of the government's changed attitude was demonstrated in 1935, when the question arose of a RAAF search being undertaken in New Guinea. A considerable number of ex-Air Force personalities had become engaged in civil aviation in the Territory—Ern ('Pard') Mustar (he had dropped the final 'd' from his name), 'Jerry' Pentland, George Fonteneau, Les Trist and George Gottschalk among them—and on this occasion it was Stuart Campbell, of Antarctic fame (see chapter 14), who had gone missing. Flying a Short S16/1 Scion, a small high-wing monoplane amphibian being used on aerial survey work, Campbell and a passenger disappeared in the vast, flat swamps of the Fly River country while on a flight from Daru to Kiunga. While Campbell's employer, Papuan Concessions, pressed several boats into the search, it also asked the government for assistance in the form of a RAAF

Southampton flying-boat (A11–1) then operating in the Gulf of Carpentaria, out of Normanton. Significantly, the company offered to pay all out-of-pocket expenses and the request was granted.[78]

The Southampton arrived at Daru on 26 July, captained by Squadron Leader 'George' Hempel and crewed by Pilot Officer Lance Burt, Corporal Percy Neale and LACs Clark and O'Donnell. The next morning the big flying-boat took off and headed up the Fly. The missing Scion was quite soon spotted, moored to the jetty at Madiri Plantation, and the RAAF aircraft made a landing on the river to ascertain the plight of Campbell and Howden. This was soon established not to be desperate, the little amphibian having a cracked crankcase, and Hempel went back to Daru to report the situation. But as it turned out, this did not leave the Southampton free to return to Australia. As one published account narrates:

> During a landing, the hull of the big boat had struck a reef of coral; repairs were under way when it was discovered that one of the Eagle engines needed replacing. It was five weeks before the repairs were completed and the new engine installed, and Hempel then flew to Port Moresby. He commenced survey work along the Papuan coast, investigating possible sites for aerodromes at Kikori, Kerema and Yule Island, but the new engine proved to be defective and the Southampton remained in Port Moresby until additional parts were delivered. It was not until 27th September that . . . Hempel and his crew were able to lift the flying-boat off the brilliant blue waters of Port Moresby harbour on their flight to Thursday Island, from where they flew around Australia, following the coastline.[79]

Fortunately for the company, in this case the department was happy to waive reimbursement for what had proved such a long (and costly) operation. The Defence Minister considered that in carrying out the search the Southampton had explored country not covered by previous RAAF flights and later it had performed valuable reconnaissance:

> The information and experience thus gained will be of considerable future value to the Air Force from a strategic as well as a survey point of view, and it also enabled a report to be submitted to the Civil Aviation Branch on the possibility of obtaining landing grounds at Daru, and on the coast of New Guinea. In view of these circumstances, I have approved that no claim should be made against Papuan Concessions.[80]

Civil aviation takes off

While the greater number of pilots licences granted following the formation of aero clubs may have increased the RAAF's problems in such matters as searches for lost aviators in light planes, there was also an upsurge in the activity of regular air transport services from 1929 which complicated the RAAF's existence. Inevitably, the RAAF was called upon whenever emergencies arose here too, as in March 1931 when the Australian National Airways (ANA) Avro 10 airliner *Southern Cloud* went missing on a flight between Sydney and Melbourne. In what was Australia's first major civil aviation disaster, a massive search operation was joined by RAAF aircraft. Some 23 flights were undertaken in Victoria by No.1 Squadron, and two Wapitis from

Richmond were ordered to check reported sightings in the Kiandra–Tumbarumba area of New South Wales (where the missing aircraft was eventually found 27 years later).[81]

On 19 October 1934 *Miss Hobart*, a DH86 airliner of Holyman Airways, vanished in the vicinity of Wilson's Promontory with 11 passengers and crew on board while on a flight across Bass Strait to Tasmania. A year later another Holyman machine of this type, the *Loina*, disappeared on this same route when it crashed into the sea off Flinders Island on 2 October 1935, with the loss of five lives.[82] The RAAF was again caught up in these two disasters, both with the Southamptons from Point Cook (A11–1 and A11–2) making air searches and Squadron Leaders Ellis Wackett and Herbert Harrison (the serving and recently retired directors of technical services at RAAF Headquarters) joining a conference of experts formed after the second incident to investigate the cause of the mishaps. After yet a third accident to one of their DH86s, which suffered a forced landing on Hunter Island in Bass Strait on 13 December 1935, Holymans was obliged to ask for the loan of a RAAF Dragon Rapide (A3–1) to keep their services operating until replacement aircraft could be obtained; this request was approved by the Defence Department on 23 December.[83]

Holyman Airways' misfortunes touched the Air Force in a very direct way, because the airline actually operated from Laverton during 1934–35—with Defence Department approval—and the company's DH84 and DH86 aircraft were a familiar sight both there and at Point Cook. Air Commodore Rex Taylor, an airman in this period, recalls that: 'The memory of the DH86 type is firmly impressed upon my mind for I was fined ten shillings by my superior officer for flying on a test flight in *Loina* with Clarrie Scott [an ex-RAAF NCO pilot flying for Holyman], without . . . permission'.[84]

In February 1937, an Airlines of Australia (AOA) three-engined Stinson airliner also disappeared 40 minutes after taking off from Brisbane's Archerfield aerodrome bound for Sydney. Again a massive search was mounted, lasting seven days and involving some 40 RAAF, commercial and private aircraft, but it was a Queensland bushman, following his own hunch on foot, who found the wreckage and two survivors of what became known as the Lamington Plateau diasaster.[85]

Less sensational but even more important were other developments associated with events affecting the airline industry in this period. Although governments remained committed to an approach which might be loosely described as holding that what was good for civil aviation was good for national defence, it began to be more evident that the reverse linkage did not necessarily apply. In fact, when Sir Neville Howse retired as Minister for Defence in March 1927, he told the prime minister that had he retained that portfolio he would have recommended the separation of civil aviation from the defence administration:

> Beyond the fact that Civil Pilots would in case of war require less training to fit them for Air Force duties than persons without such experience and that the ground organizations for operations of Civil air routes built up in time of peace would be of great value in war—there is little if anything in the development of Civil Aviation directly connected with Defence.[86]

The DH86 airliner 'Miss Hobart' pictured at Point Cook just a few days before its disappearance in Bass Strait in October 1934. The company operating the aircraft, Holyman Airways, conducted its commercial services to Tasmania from Laverton airbase during 1934–35. (*AVM E. Hey*)

In a period of rapid expansion in commercial air travel, it was evidently beginning to be felt that civil aviation had outgrown its links with the Defence Department.

By mid-1930 this pressure had intensified to the point where the question arose of whether the CAB should be transferred to the Department of Transport. Asked for its views on the proposal, the Air Board replied on 23 June that it could see 'no adequate reason for transferring the Civil Aviation Branch, but can see many reasons why it should not be so transferred'. The role which the Defence Department would assume in war regarding civil aviation was, it argued, a sufficient reason for leaving matters as they were, especially as the organisation already established was functioning 'quite as satisfactorily' within Defence as it could anywhere else.[87]

The Air Board admitted that some duplication between itself and the CAB then existed, but it noted that a subcommittee of the Defence Committee was presently examining this very question and urged that any decision on the proposed transfer should be deferred until the sub-committee's report was available. In fact, the Air Board claimed, the close relationship between civil and military aviation provided scope for economies in administration not possible in any other department, while transferring the CAB would make impossible the close cooperation which was so important to the work of both CCA and the Board.

The present proposal emanated, according to the Air Board, from press agitation over alleged 'military' control over civil aviation. Such claims were considered quite groundless, as in fact the CAB was not controlled in any way by any section of the defence forces. On the other hand, civil aviation had in the past benefited greatly from being under the umbrella of the Defence Department, in such things as obtaining assistance for aero clubs with air pageants; making equipment available from Air Force stocks; loaning of aircraft to civil operators, and allowing their use of RAAF facilities; the release of personnel to take up civil aviation posts; and the use of RAAF aircraft for surveying civil air routes. In the event, a decision was taken to retain the branch within Defence, although a significant change was made with RAAF-controlled landing grounds being passed across to the CAB (see chapter 5).

Despite changes in the relationship between civil and military aviation which occurred at this time, in other important respects the links remained no less strong. In the matter of staffing, in particular, the RAAF continued to be a source of experienced personnel. In May 1929 a former commanding officer of Point Cook, Squadron Leader Alan Murray Jones took up duty as superintendent of flying operations. He proved himself an 'able administrator' but resigned in July 1931 to become the De Havilland company's Australian manager.[88] He was succeeded in August by Flight Lieutenant David Ross, who resigned from the Air Force to take up the appointment.[89] Also among the CAB's staff of 40 in 1936 was to be found Ted Shorland, an original member of the CFS staff of 1913.[90]

When, in April 1936, the branch was reconstituted as the four-man Civil Aviation Board headed by a controller-general, another prominent RAAF identity assumed a senior position in the new body. This was Wing Commander Harry Cobby, who had joined civil aviation the previous year and now sat on the Board as controller of operations.[91] He continued to have regular contact with his former Air Force colleagues, because in addition to matters concerning air transport services, licensing of personnel, issuing of certificates of airworthiness and aircraft inspection, and wireless and meteorological facilities, his duties included aero clubs and liaison with the RAAF.[92]

The connection was retained in other ways, too. Australia's interests at the International Committee for Aerial Navigation and later the International Air Navigation Commission were represented from 1927 by officers from the RAAF's liaison office at the Air Ministry.[93] Proposals in 1935 for the establishment of an 'Empire Air Mail Service' (see chapter 4) also involved the RAAF in reconnoitring the route between Darwin and Sydney.

The expertise of Air Force specialists was also utilised when the government finally took steps to provide direction finding and communication aids over the nation's commercial air routes. No provision of any kind had been made for such equipment before 1934, but pressures to address this need arose following the disappearance of the *Miss Hobart* just weeeks after it had been assembled at the Point Cook RAAF base and made its first flight on the Bass Strait route on 1 October 1934.[94]

Valuable experiments had already been performed in October 1933 when Flight Lieutenant Carn Wiggins led a team of RAAF and RAN personnel to Western Junction aerodrome (near Launceston, Tas), where they set up a portable naval direction-finding (DF) apparatus. In conjunction with the Tasmanian Aero Club's Gipsy Moth which had been fitted with a radio transmitter, tests were then made for several days to find the best position to instal permanent DF equipment before the party moved on to perform similar tests at Brighton aerodrome (Hobart).[95] In response to the problems experienced by airliners crossing Bass Strait, it was now decided to locate the RAAF's mobile radio station, used during the 1934 Centenary air race, at Western Junction for an indefinite period.

Because the CAB had no radio organisation of its own, the Western Junction facility continued to be operated by the RAAF on the arrangement that the Air Force operators were carrying out a research program while providing a radio communication service to scheduled airline traffic. The station came on air early in March 1935, operating from a tender initially manned by Sergeant James Reddrop and AC W.H. Murden (replaced in November 1936 by Flight Sergeant Henry Bain and Sergeant Nicholls).[96] The service provided by this first RAAF 'aeradio' facility proved so successful that a Marconi DF unit was also installed and handed over to the RAAF at the end of July of that year.[97]

An aeradio station was also established at Darwin in January 1936, covering the Darwin–Koepang section of the Brisbane–Singapore air route. Providing similar services to the station at Western Junction, this too was operated by the RAAF for the Civil Aviation Board.[98] At Richmond, watch was maintained on high-frequencies and the direction-finding station there exclusively for the C class flying boats operating the regular Empire Air Mail services to New Zealand, until the CAB established its own facilities late in 1938.[99]

Meanwhile, in March 1935, the government sent Squadron Leader Wiggins, the RAAF's director of signals, overseas on behalf of CAB and the Air Board. Visiting the United States, Britain and Europe, he studied 'the whole question of radio communication and navigation as applied to commercial aviation throughout the world, with the object of advising as to the nature of the radio organisation that should be set up for use on Australian air routes'.[100] Following Wiggins' return in December 1935 and receipt of his report, it was decided to test and (if successful) instal, a navigational beacon system on certain internal air routes. In May 1937 Wiggins was attached to the CAB and took charge of a new section established in the Operations Branch to administer the civil aviation radio program.[101]

As the new beacon system was slowly acquired, the RAAF found it was still expected to play a role. When the first medium wave beacon was installed at Essendon at the end of 1936, the tests on this unit were carried out by an Air Force officer using a DH Moth, which was the only aircraft type available though not really suitable for the purpose. Following the Lamington Plateau disaster in February 1937, the CAB immediately moved to acquire high frequency beacons. Unfortunately, problems now arose over obtaining a modern aircraft needed to test the new installations, with the Finance Member of the CAB, Major Melville Langslow (who also served in the same capacity on the Air Board), arguing that, for reasons of economy, use should be made of a RAAF Anson.[102] Unconvinced by technical advice that an Anson was unsuitable and would need costly modifications taking six months to complete, Langslow resisted proposals to charter a civil airliner until the possibility of converting an Anson for the job was reviewed.

The situation which had arisen, where the latest RAAF warplane was unsuitable for testing systems serving modern commercial airliners, was actually an interesting reflection on the dramatic shift which had occurred in the relationship between civil and military aviation since 1921. As one writer very fairly points out:

> In the mid-1930s, when the RAAF was only thinking about modern aircraft, and Wirraway production in Australia was still several years off, mechanics in civil aviation were making themselves familiar with the latest American technology—high-powered radial engines, stressed-skin metal construction, and retractable undercarriages—on aircraft entering commercial service ... RAAF mechanics did not begin to handle aeroplanes of equivalent technology to the DC-2 [which started commercial service in Australia early in 1936] until 1940, when the airforce started to take delivery of the Lockheed Hudson medium bomber, a derivative of the Lockheed 14 which had been operating commercially in Australia since mid-1938. So the crews of the commercial airlines gained experience of high-performance modern aircraft many years before their defence colleagues, and they eventually assisted and trained air force personnel.[103]

The existence of this difference was brought home sharply to one RAAF officer at about this time, while returning from a visit to Western Australia in a Demon:

> Coming home I found that at about 12 000 feet there was a good tail wind, and boy-oh-boy, did the countryside run past us. It must have been at 100 miles per hour. The thing that shocked me was that at that height, which was the optimum to get the wind, what should blithely pass me—not at high speed, but without too much trouble—but a DC2 of one of the commercial operators. I mentioned this to some of the senior officers at a director's meeting at RAAF Headquarters, and no-one appeared concerned. The explanation was that the Demon was fitted as a fighting machine and the difference was to be expected.[104]

As a result of the RAAF's technical situation in 1938, and the action of Langslow, testing of the beacons was held up and these were still not operational by 25 October 1938, when a DC2 airliner *Kyeema*, flying the regular service between Melbourne and Adelaide, overshot Essendon airport in low cloud and crashed into the crest of The Basin at Mount Dandenong with the deaths of all the 14 passengers and four crew members on board.

Impact of the *Kyeema* crash

The *Kyeema* crash entailed the worst loss of life of any aircraft accident in Australia so far, and in the face a press and public outcry an open inquiry into the circumstances was conducted by the Air Accidents Investigation Committee for the first time. For this inquiry, too, the committee's normal composition—Colonel Thomas Murdoch (chairman), Group Captain Eric Harrison (the Defence Department's director of aeronautical inspection), and Mr David Ross (of the Civil Aviation Board)—was augmented by two additional members: Mr Edmund Herring, QC, and the nationally recognised aviator Captain Gordon Taylor. After three days of open hearings at the Exhibition Buildings in Melbourne between 28–30 November 1938, the committee submitted its report on 5 December.

The findings of the inquiry pointed to several contributing causes for what had happened, principally relating to the airline company's failure to observe procedures. But the committee's most severe criticism was reserved for the CAB, which was accused of tardiness and maladministration. For this situation the report found the controller-general personally to blame, but it also singled out Langslow. Criticising the Finance Member's attitude that 'it was his duty at all costs to keep expenditure down to the lowest possible point', the committee directly attributed delays over installation of beacon equipment to Langslow's 'insistence on the limitation of expenditure':

> This may be regarded as meritorious in a Finance Member, but there comes a time when the saving of a few pounds is much less important than the expedition that may effect the saving of lives . . . We think the Finance Member misconceived his position and acted rather as a representative of the Treasury than as a Member of a Board which was responsible for pursuing an active policy.[105]

Notwithstanding the damning observations regarding his conduct contained in the report, Langslow was able to ride out the storm which could conceivably have cost him his seat on the Air Board as well as the CAB. Following the government's study of the Air Accident Investigation Committee's findings, the Minister for Defence (Thorby) announced the appointment of an interdepartmental committee comprising members from Defence, the Postmaster-General's Department, Commonwealth Railways and the CAB, specifically to investigate the CAB's administration.[106] The report of this IDC, chaired by Mr Robert Lawson, chief engineer of the Postmaster-General's Department, was delivered on 30 December 1938 but given no publicity at the time. Its conclusion, that the problems of the civil aviation administration were 'a Board failure for which the blame must be attributable to all,'[107] stood in mitigation of Langslow's personal responsibility and enabled him to preserve his position.

When the air component of the Defence Department was formed into a separate ministry in November 1939, it was Langslow who was appointed secretary and first permanent head of the new Department of Air—a post he retained for the next 12 years. From this eminent position he was not inclined to forget those who had so strongly criticised him in the *Kyeema* inquiry. The only member of that committee who came within his official purview was Harrison, then popularly regarded as the 'Father of the RAAF',[108] who had retired from the Air Force in March 1938 but retained his departmental appointment in a civilian capacity. Langslow's attempts to make life difficult for this officer were a continuing source of bitterness and friction within the department, reportedly extending to CAS, now Air Vice-Marshal Goble, who felt it necessary from time to time to protect Harrison from these attacks.[109]

While Langslow was able to avoid the damaging effects of the outcry over the *Kyeema* disaster, this was not the case for the CAB as a whole. Even before the committee's report was made, the Minister for Defence had decided that the board should be removed from the Defence Department and constituted as a separate Department of Civil Aviation with a director-general as its permanent

An Airlines of Australia publicity still featuring the DC2 airliner 'Kyeema' which crashed in October 1938 with the worst loss of life in any air accident up to that time. This disaster produced an inquiry which was savagely critical of the Finance Member of the Civil Aviation Board and the Air Board, and produced personal antagonism against Group Captain E. Harrison—who many regarded as the 'Father of the RAAF'. (*NLA*)

head. This change was gazetted on 24 November 1938, and provided an effective means of demoting the then controller-general, Edgar Johnston. A statement from the prime minister announced that Johnston 'will no longer be in complete control [of civil aviation], but will have charge of one of three branches of the Department'.[110]

It was a odd twist of fate that the sequence of events which resulted in Johnston's ousting —that is, an air crash followed by a public inquiry—paralleled exactly the circumstances associated with Williams' removal from the RAAF's top post at the very same time (see chapter four). It was noteworthy, too, that the inquiries associated with the *Kyeema* disaster were regarded for years afterwards as the classic example of failed public administration. In September 1940, Lawson (now a member of the Air Board in his capacity as Director-General of Supply and Production) was prompted to draw the case of the CAB in 1938 to the attention of certain members of the Air Board, suggesting that in the matter of delegations the Board could then 'be charged with remissness in the same way and for the same reasons as the Civil Aviation Board was charged'.[111]

11

Safety record

No sooner had the Australian Air Force been inaugurated than the point was dramatically made that the business of the new service involved mastering a risky and unforgiving medium. With the loss of Stutt and Dalzell in September 1920 (see chapter 1) still a recent tragic memory, the point was given new emphasis by the first fatal crash in seven years of flying at Point Cook just one week after the AAF came into formal being. On 6 April 1921 an Avro 504K (H3021) stalled on a training flight and spun 500 feet nose-first into the ground about a half-mile north of the hangars. The pilot, Flying Officer James Fryer-Smith, received a fracture to the base of the skull and severe cuts to the face and body, while his passenger, air mechanic Corporal Bertie Whicker, suffered injuries from which he died in Caufield Military Hospital that evening.[1]

Such occurrences raised important questions for the new service regarding compensation for personnel injured or killed in the course of their duties. The case of Stutt and Dalzell was still generating correspondence six months later regarding the amount of compensation due to their dependents. In Fryer-Smith's case, although he recovered from his injuries he was not classed as medically fit to return to flying duty. In September 1921 the Air Council expressed the view that his case 'is one showing the need for the establishment of a pension scheme for the Royal Australian Air Force'. Ordinarily, Fryer-Smith's appointment would have been terminated, but it was recognised that under existing regulations he was entitled to very little compensation. In the circumstances, alternative employment as an equipment officer had been found for him, but, as the Council recognised, similar cases were bound to occur and it could not always be expected that other employment would continue to be found 'as an act of grace'.[2]

In the event it was four years before the RAAF suffered an accident of similar severity to Fryer-Smith's. On 25 March 1925 another No.1 FTS Avro (A3–28) on a training flight stalled while approaching to land and spun into a paddock at the rear of the Point Cook hangars.[3] The pupil pilot, Flying Officer Alan Charlesworth, newly seconded from the Army, sustained multiple injuries from which he recovered to resume his career and ultimately attain the rank of air vice-marshal. The instructor, Flying Officer Stewart Mailer, was killed however, becoming the first RAAF officer fatality.[4]

In contrast to the RAAF's good accident record in its first five years of existence, 1926 was a horror year in which seven members were killed in five separate accidents. The first of these, occurring at Canberra on 11 February (as described later in this chapter), cost two lives. A little over a week later, the first of a series of mishaps to No.1 FTS aircraft took place. On 19 February a CAF trainee died in a single-seat SE5A when he overshot a landing and crashed (see chapter 8). Another pupil pilot, Cadet Aubrey Percival, was killed during a solo training flight on 3 June when Avro A3–27 dived into the ground at Altona North; the cause of this crash was not definitely ascertained but reports by witnesses suggest Percival had stalled the aircraft.[5] Cadet Thomas Watson, another pupil pilot, and his instructor, Flying Officer William Holtham, were killed 3 kilometres north of Manor Station, near Werribee, on 1 July when their DH9 (A6–13) stalled during forced landing practice, crashed, and caught fire. Sergeant Ernest Ebell, yet another trainee, died on 5 September when he encountered storms while returning from a cross-country flight test in A6–17; attempting a genuine forced landing near Geelong, he crashed the machine and was killed.

While fatalities attracted most attention and shaped public perceptions of Air Force efficiency, the number of such occurrences provided a very misleading yardstick. RAAF statistics of accidents covering 1921–37 reveals the existence of a complex situation. For instance, in this 18-year period there were 37 fatal Air Force crashes resulting in the deaths of 56 occupants (two of them civilians) and injury to another 16. But, as the figures show, the number of fatal accidents was only a tiny proportion of all accidents occurring in any one year. For example, the twelve fatal accidents during 1932–37 comprised only 3 per cent of 420 mishaps to aircraft recorded in this period; the majority of this total (more than three-quarters) involved only minor damage, and just 24 were of such severity as to result in aircraft being written-off.[6]

A mounting toll

After the high number of deaths in 1926, the following year proved to be no better for the Air Force. Four more aircraft accidents resulted in another seven fatalities, equalling the previous year's grim harvest. On 14 February a training crash at Point Cook resulted in the death of a CAF student pilot (named Dix—see chapter 8); on 21 April two DH9s taking part in a flypast marking the arrival at Melbourne of the Duke and Duchess of York collided in mid-air, resulting in the deaths of their four occupants; less than three weeks later another SE5A crashed in a flypast for the duke of York at Canberra, again killing the occupant; finally, on 27 October a student pilot on an air navigation exercise, Cadet Clarence Seach, was killed after he stalled DH9 A6–3 while attempting a forced landing at Healesville (Vic).[7] Given that two of these incidents happened in unusually public circumstances, 1927 was actually more disastrous for the RAAF's image even though the actual loss of life was no worse than in 1926.

The Air Force had already come in for considerable public criticism after the sudden upsurge in casualties the previous year, resulting in the Minister for Defence instituting a court of inquiry into the accidents question. The report of this body had appeared in September 1927, in which an attempt was made to educate the public about the true facts of the issue, principally that the prime cause of all air accidents was error of judgement by pilots:

> In nearly every case ... in which the cause of an accident is traceable to error of judgement on the part of the pilot, it was found that the mistake was apparently an involuntary stall near the ground, and every fatal accident in the RAAF can be attributed only to this cause.[8]

Notwithstanding this disclosure, Air Force crashes—however minor—were still treated to sensationalist headlines. RAAF authorities responded by instructing pilots involved in mishaps of any nature to give no details whatever to press representatives, and a notice appeared in the Victorian *Police Gazette* forbidding constables from informing newspapers about the occurrence of RAAF accidents, a practice which often saw journalists arriving at crash-scenes before Air Force investigators.[9]

Into this strained atmosphere was injected the tragedies of April–May 1927. The first of these occurred when the RAAF put some 30 aircraft into the air to provide an escort for the Duke and Duchess of York from the time they disembarked from HMS *Renown* until the procession pulled through the gates of Government House. Shortly after 4 pm the seven DH9s in V-formation led by Flight Lieutenant George Jones, then at 1000 feet directly over St. Kilda Road, entered a shallow dive which was the customary form of aerial salute.

Jones later recounted how he recovered from the dive and looked back to check that members of his formation were in their allotted positions, only to see that 'the two near machines on my starboard quarter were falling apart, evidently after a collision'. As Jones watched, one of the machines entered a vertical dive and disappeared through the roof of a building some 300 metres west of the Government House gates (a Post Office garage in Sturt Street, South Melbourne). The sheet of flame which erupted through the hole told its own story. The other machine commenced a flat spin, the pilot clearly struggling to regain control with a badly damaged port wing. Eventually this aircraft crashed into a road (Dodds Street) about 50 metres east of where the other machine had hit.[10]

The collision was clearly observed by airmen in other machines. One recounted seeing the DH9 on the outside end of the starboard arm of the V-formation—that is, A6–26, piloted by Flying Officer Vincent Thornton and carrying Sergeant Herbert Hay (photographer) as passenger—hold the dive slightly longer than the rest so that when this aircraft commenced to climb it came up underneath the machine next to it in formation (A6–5, crewed by Flight Lieutenant Robert Dines and Corporal James Ramsden). The nose and port wing root of the lower machine was seen to strike the other plane halfway down the fuselage from below, almost cutting it in half. The two machines were locked together momentarily, then separated and commenced their plunge earthwards.[11]

The collision also had innumerable witnesses on the ground, including the

Duke of York. Having completed an inspection of an Air Force guard of honour drawn up in the grounds of Government House, he was just turning to speak to the guard commander (Squadron Leader Harry Cobby) when over Cobby's shoulder he saw the planes collide. 'Two of your buggers gone', airmen in the ranks heard the duke say.[12] Also in the inspection party was CAS, who later described being alerted to the fact that something was wrong by a sudden unsteadiness in the ranks of the guard. Turning to see what was taking their attention he now also saw the DH9s falling from the sky: 'I expected to see four men with parachutes leave them. The aircraft were so low that they must jump at once if the parachutes were to have room to open and break their fall and I was almost shouting 'Jump'. But nothing of the sort happened . . .' Williams would later ascertain that none of the four men were wearing parachutes because 'air crews did not like them, they were uncomfortable to wear'.[13]

The effect of this diasaster on the RAAF pilots participating in other formations was one of deep shock. George Grendon, travelling in a formation of DH9As behind Jones' DH9s, recalls that he tapped the pilot of his machine (William Gardner) on the shoulder to point out the drama taking place:

> He was already gazing at it. We climbed straight up and went back to Point Cook, landing in the middle of the drome. We didn't speak—Gardner was absolutely silenced, couldn't even taxi the aircraft. Thornton had been a particular friend of his. At that stage I hadn't been in the RAAF long enough to make close friends, but I could imagine how hard it hit others.[14]

The Duke of York inspects a RAAF guard of honour in the grounds of Government House, Melbourne, on 21 April 1927. The Duke (*back to camera*) has turned to speak to the Governor-General, Lord Stonehaven, and CAS, Group Captain Williams, with the guard commander, Squadron Leader A.H. Cobby, on his right. Moments later the Duke turned to Cobby, only to observe two machines in the RAAF formation overhead collide and plummet to the ground. (*RAAF Museum*)

Not all the pilots were so shocked as to be incapable of further flying. According to one newspaper account, following 'the unwritten law of the British and Australian Air Forces' other airmen flung themselves into an impromptu display of aerobatics.[15] Actually, the aerobatics display was a pre-arranged affair which owed nothing to any notion of pilots 'keeping their tail up' after a flying tragedy. Recalls Sir Wilfred Brookes:

> During the day I flew A2–13 [an SE5A]. I went low up St. Kilda Road doing sharp turns around right-angled roads. I wasn't even aware of the catastrophe until I landed, and then all hell broke loose. Nobody wanted to take the responsibility for giving permission . . . The *Age* newspaper got my number while I was under the wires. The first edition had come out with headlines 'What a wonderful sight', that sort of thing—then it cut off. After that the RAAF could do no right. Ensuing editions called for the pilot of A2–13 to be court-martialed. Had they left it at that I probably would have been. But they made the mistake of keeping up the pressure until the Air Board said, 'Who is running the Air Force—us or the Melbourne press?'[16]

Apart from the agitation of newspapers RAAF Headquarters received a number of letters about the aerobatics display, with several members of the public complaining at the disregard of safety entailed in what they had witnessed.[17]

A fortnight after the Melbourne tragedy, on 9 May, disaster struck again during ceremonies in Canberra to mark the opening of Parliament House. At a mass military review in the afternoon, after the duke had formally inaugurated the new legislative building that morning, an SE5A (A2–24) was seen to inexplicably peel off from one of the formations passing overhead at 500 feet. The machine's stalling turn developed into a dive and it ploughed into a small rise some 600 metres directly in front of Parliament House. Landing on its nose and the port wing with a terrific crash, the aircraft turned over amid a cloud of dust which rose 15 metres high.[18]

The pilot of the aircraft was actually the adjutant of No.1 FTS, Flying Officer Francis Ewen, who was taken to hospital with serious injuries, including breaks to both arms and one leg, and chest injuries. He died at 7 pm that same evening, without being able to speak about what had happened. As Ewen had been unable to provide any information about his accident, the cause remained a mystery. The ensuing coronial inquiry could only return a verdict of accidental death, the coroner observing that it 'was one of those inexplicable events that cannot be accounted for . . . '.[19] It only remained for the Air Force contingent in camp to attend to the melancholy business of burying Ewen before departing for their respective bases. The funeral took place at St. Johns Church on 11 May, with a procession comprising several hundred men from each of the three services led by the RAAF Band. Wreaths were sent by the duke and duchess (who by this time probably dreaded the prospect of any more flypasts in their honour), the prime minister, the Commonwealth and New Zealand governments, the speaker of the House of Representatives and the president of the Senate, and the RAAF. During the burial service three Air Force planes circled the churchyard and dropped wreaths by the graveside in a final tribute to Ewen.[20]

In the week before the ceremonies at Canberra on 9 May 1927, the RAAF contingent suffered an embarrassing number of accidents to its planes—though fortunately without serious injury to personnel. This SE5A appears to have escaped significant damage in its mishap, described as having occurred at south-east Canberra. Machine A2–24 appeared similarly undamaged after it overturned while making a landing and was later flown by Flying Officer F.C. Ewen during the military review before the Duke of York. Ewen was fatally injured when his aircraft inexplicably fell out of formation and dived into the ground in front of Parliament House. (*Mr A.W. Skimin*)

With the necessary obsequies performed, it was time to weigh the damage. The RAAF's public image had undoubtedly taken a battering, especially with reports in the press of a succession of aircraft accidents at Canberra in the week leading up to the actual opening ceremony. There had been at least four forced landings,[21] two of them by the same pilot on consecutive days.[22] Fortunately none of these mishaps had resulted in major injury to any personnel, but some aircraft had been put out of action.[23] According to one RAAF member who was there, among the machines involved in the mishaps was the SE5A flown by Ewen at the time of his fatal crash:

> That aircraft was flown up to Canberra by Don Carroll, a sergeant-pilot who was the son of a Country Party senator from Western Australia. When he landed it at Canberra, the thing went over on its back. It was tipped up the right way again, and no damage was said to have been done. But Ewen was killed in it the next day.[24]

The week after the Canberra ceremonies it would be claimed that the final tally was actually ten crashes.[25]

Inevitably there was concern at this none-too-glowing record. By the end of the week of Ewen's accident, the Minister for Defence (Glasgow) had announced the holding of a departmental inquiry 'into every aspect of the Royal Australian Air Force', and threatened that this 'may be followed by more searching investigation if preliminary reports justify such a step'.[26] The following week, Prime Minister Bruce announced the appointment of an Air Accidents Investigation Committee (AAIC) 'to inquire into the mishaps which have recently occurred in connection with aviation'. The five members of this panel were Professor Henry Payne of Melbourne university (chairman); Squadron

Leader Eric Harrison of the RAAF; Captain Ernest Jones of CAB; Mr Marcus Bell, superintendent of the Defence Laboratories; and Lieutenant-Colonel H.B.L. Gipps, chief inspector of MSB.[27] Harrison owed his place of the committee to his being the RAAF's Assistant Director of Technical Services—a position with special responsibility for aircraft inspection and an area in which he had been specially trained since 1919. Another RAAF officer, Flight Lieutenant Thomas Swinbourne, was appointed as secretary to the committee, which came into formal being on 26 May 1927.

Williams expressed some impatience and irritation over this action on the government's part. As he made clear in private correspondence with his British counterpart later that year, he saw this as caving in to agitation by the press. He had also suggested to the Minister for Defence that it evinced a lack of confidence in the Air Board, although he was assured that this was not so and that the committee was appointed purely for political reasons. Nonetheless, he told the British CAS, he saw the AAIC as 'a possible source of trouble':

> There are only two members with any flying experience and the Committee, with a lack of expert knowledge goes over the ground which we, in the ordinary way, cover much more expeditiously with some knowledge of the subject. However, on the other hand, apart from wasting a good deal of valuable time of officers, it has so far, I think, not done any harm.[28]

Williams was probably being candid about his reasons for objecting to the formation of the AAIC, but there were actually sound reasons for establishing such an independent watchdog. Whether or not CAS liked to admit it, he knew that the RAAF's own court of inquiry convened at Point Cook on 22 April, the day after the South Melbourne tragedy, had been a hasty and slipshod affair. Its investigation of this accident had concluded that the cause lay in A6–26 having lost its station in the formation coming out of the dive, and in trying to regain position having overtaken and crashed into A6–5; there was, however, 'no evidence of negligence or contributory negligence on the part of any one concerned'. CAS could see that the inquiry had not done its job properly, and on 30 April he sent the report back to AMP with the complaint that 'the Court has not exerted itself to any extent to get complete information with regard to this accident' and pointing to discrepancies in evidence which had not been checked or examined to obtain a resolution. On 3 May, the CO at No.1 FTS was directed to reopen the inquiry to obtain further evidence.[29]

In the revised findings which resulted, considerable emphasis was attached to the fact the Thornton was a CAF pilot who had not been able to give his full time to formation practice conducted in preparation for the event on 21 April. In fact, he had participated in only one of four rehearsals. As was recognised within RAAF Headquarters, Thornton's previous experience with No.4 Squadron, AFC, during the war meant that his failure to attend practice runs for such a simple manouevre as an aerial salute was unlikely to have been a vital factor in what occurred. Williams therefore can not have been convinced that the truth of matters had yet emerged, but at this point the Air Board deferred approving the court's findings pending the separate investigation being carried out by the AAIC. It nonetheless rejected the suggestion that Thornton's lack of practice of the diving manouevre had contributed to the accident.[30]

Williams had been shrewd enough to realise the inadequacy of the RAAF inquiry, yet in his concern to get to the bottom of the episode he can hardly have been prepared for what he would have found if he had kept pressing. The fact was that a number of the pilots participating in the aerial display of 21 April had actually been drinking in the Mess before take-off. Normally the use of alcohol when flying was prohibited, but Sir Wilfred Brookes recalls that there had been a lot of talk by wartime officers about the march of the AIF through London in 1919. On that occasion AFC pilots (many of them drinking beforehand) gave a memorable—some said truly frightening—aerial display, during which aircraft were brought right down into the streets of the British capital.[31] This was plainly the inspiration behind the similar display given over Melbourne and some pilots had undoubtedly been affected by drinking beforehand. Says Brookes: 'Dines was a tee-totaller, but Thornton in my opinion had had a good deal to drink. I had a few glasses myself, but not as many as Thornton. It was, of course, Thornton who flew into Dines.'[32]

Given Williams' own abstemious habits, there was an obvious disinclination on anybody's part to bring these facts to the notice of CAS, let alone to agencies outside the service. For that reason, the investigations of the AAIC were hardly welcome either. In the event, the new body got no nearer the truth than did Williams. Like the RAAF court, it concluded in its report dated 13 July that A6–26 had been to blame for the mishap, due to an error of judgement on Thornton's part regarding where his machine was in relation to A6–5. The committee also thought that Thornton's desire to assist the photographer he was carrying may have caused him to lose sight of A6–5, but that in any event the formation salute had been insufficiently rehearsed. When the AAIC's report was passed to it for comment in August the Air Board disagreed with all these judgements, especially the finding that Thornton as the pilot of A6–26 had been to blame:

> There is no evidence, nor is such evidence available, to show what might have happened in the air, and the accident may not have been caused by an error of judgement in the ordinary accepted meaning of that statement.
>
> Further, to place the responsibility for the accident and for the deaths which followed on an individual in this way achieves no good purpose, and it is suggested that the Committee might like to review the matter in this light.[33]

One can only speculate whether the Air Board would have taken this view had the alcohol factor been known.

In any event, the AAIC's role as a permanent body to study accident causes and suggest preventative measures would itself come in for some questioning in later years. In April 1933 a voluntary committee of aeronautical engineers criticised the AAIC to the government, alleging insufficient enquiry, faulty conclusions and especially unfair reflections on the ability of deceased pilots. It recommended that the membership of the AAIC—already reduced to three members in February 1931 as an economy measure—should be completely changed to include experts drawn from specific professional and scientific fields. Despite this strong attack, the AAIC survived essentially unchanged. Harrison, by now a wing commander and soon to become group captain, retained his role on the committee.[34]

Pilot error

As indicated in the previous section of this chapter, it was recognised that the most common cause of all accidents suffered by the RAAF involved some misjudgement or failing on the part of the pilot controlling the aircraft involved. Nowhere was this point underscored more poignantly than in the sad succession of young air cadets whose funeral processions provided photo reports for the Melbourne newspapers.[35] The fact was that almost half of all accidents in the RAAF occurred in training units, where the inexperience of trainees combined with the unforgiving nature of the air medium provided a deadly mix.

Trainees were by no means the only ones prone to making mistakes, however. Even pilots with long hours of flying experience were guilty of misjudgement, as often as not brought on through over-confidence. As one pilot recalls:

> I realised only too well about the dangers of over-confidence when I became a flying instructor. There were three danger points: the day you went solo, when you reckoned you were as hot as Richthofen; the day you got your wings; and when you got to about 500 hours and reckoned you could fly any bloody thing. If you survived those three periods, there was a reasonable chance you'd be an old, bold pilot![36]

Illustrating this piece of wisdom was a crash at Richmond on 25 February 1929, when Moth A7–12 failed to clear trees near the aerodrome boundary; the pilot, Flight Lieutenant C.T. Anderson of the RAF sustained multiple injuries from which he recovered, but his passenger, LAC Colless, was killed.

Also among the Air Force's losses in this period was Flying Officer Charles Blamey, son of Major-General (later Field Marshal Sir) Thomas Blamey. A Duntroon graduate who began his career in the RAAF at the end of 1930, young Blamey was the pilot of Moth A7–41 which stalled on approach for a practice forced landing at Richmond on 6 December 1932. The aircraft crashed, killing Blamey and causing minor injuries to his passenger, AC1 J.V. Plumb. Later press reports claimed inaccurately that the wreckage of this aircraft was put up for public sale, and was purchased by a RAAF airman, Sergeant Wilfred Brown of No.3 Squadron, who rebuilt it during 1933 as a private machine. Brown was killed in his plane on 10 February 1934, when it nose-dived into Cooks River while taking off from Mascot. With him on the fatal flight was a CAF airman, Sergeant John Bundy. Brown's machine was, in fact, an ex-RAAF plane, but it appears it was actually the former A7–28 which had been crashed at Mascot on 2 April 1932.[37] On that occasion the pilot, Squadron Leader George Jones, had failed to clear the boundary fence after taking off downwind, resulting in considerable damage to the Moth and extensive minor injuries to Jones and his passenger, Pilot Officer Doug Candy.

Occasionally, incidents of this sort had the potential for quite sensational consequences, as on 16 February 1926 when DH9 A6–19 flown by Flight Lieutenant Vernon Burgess struck a tree and crashed while taking off from Flinders Naval depot. As it happened, Burgess had with him as passenger none other than the Governor-General, Lord Stonehaven. Fortunately neither His Excellency nor Burgess was injured, and the aircraft was not seriously damaged.[38]

The wreckage of the Moth aircraft in which Flying Officer C.M. Blamey was killed in a crash at Richmond on 6 December 1932. Blamey, a 1930 Duntroon graduate on secondment from the Army, was the son of Major-General (later Field Marshal) Sir Thomas Blamey. His death emphasised that not just raw trainees made mistakes and that the air was an unforgiving medium. (*Mrs J. Stewart*)

When an accident did not actually result in dire consequences, it was frequently possible to see the funny side of such events. One such mishap was the indignity suffered by a seconded RAN officer, Flying Officer Vincent ('Bones') Kennedy, while attempting to land an Avro 504 after a solo flight at Point Cook in 1925. Finding his approach too short, he had choked his engine attempting to restart it to go around again; he suddenly found himself heading for a crash landing on tennis courts near the married quarters—where a game was in progress. There was no crash, however, as Kennedy's machine became snagged in high-tension power lines which ran beside the road.[39] The Avro was suspended above the ground, nicely balanced at an angle of 45 degrees and slowly swinging backwards and forwards. After the power was cut, a tall step ladder rescued the pilot from this precarious position, but getting the plane down was another matter. Recalls Ulex Ewart, an officer on the base at the time:

> The problem of rescuing the Avro was a technical job to which I was not privy, but I heard that it involved positioning a crash tender (a four wheeled trailer with a flat top used for transporting aircraft) and calling in a linesman to cut the wires and later to restore services.[40]

Ewart also remembers an unorthodox landing which he made when he was assistant adjutant at Point Cook in 1924. On this occasion he took a Sopwith Pup to fly the unexpended portion of a payroll to the Shaw-Ross aerodrome at Port Melbourne, where an officer from RAAF Headquarters was to be on hand to receive it. Cutting his landing approach a little too fine, the Pup's wheels hit a ditch hidden in long grass at the edge of the airfield and the machine immediately flipped over onto its back:

> So severe was it that my seat belt broke and I was catapulted out of the cockpit. I can remember the ground coming rapidly upwards towards me and instinctively

> put my hands over my head. I landed on these, turned a forward somersault and landed upright on my feet. Intent on delivering the pay to the officer I saw running towards me, I walked to meet him fumbling in the pocket of my uniform jacket for the pay bag. The officer was Flying Officer De La Rue, who was bubbling over with excitement and, pointing to the Pup lying on its back, shouted 'Look, look! Look what you've done'. I looked. And then, only then, did he accept the pay bag from me.[41]

Indignity was also the lot of Air Cadet Ernest Knox-Knight, who had the misfortune to run into a cow on 8 September 1926. Knox-Knight was making a landing at Geelong in DH9 A6–16 when, as he was about 2 feet from the ground, one of about 20 cows which had been running away from his approach suddenly turned about:

> I immediately opened the throttle, but failed to clear the cow & broke the port undercarriage strut. Not knowing how much damage I had done, I decided to land there instead of coming home & 'pancaked' from about six feet, whereupon the undercarriage collapsed.

The DH9 was hauled back to Point Cook by road, while the cow became the subject of a compensation claim finally settled for £10.[42] Knox-Knight was known as 'Mr Ox-Knight' for some time afterwards.[43]

An incident with grimly humorous overtones concerned a member of No.3 Squadron, Flying Officer Leslie Sutherland, who was part of a RAAF presence at Canberra during the weekend of 22–23 October 1926 for the annual sports and gymkhana held at the Royal Military College, Duntroon. Sutherland was flying SE5A A2–26, a machine which he regarded as his personal and favourite 'mount':

> I groomed and cared for that hussy in *de luxe* style. I spent hours and hours of my own time with her; and quite a lot of what would have been cigarette-money was lavished on her appearance—varnish, metal polish, and little knick-knacks to improve her 'turn-out'. Of course, I was very proud of her. But what did she do for all this loving care and attention? She bit the hand that fed her. Actually, it was more serious than that, for she broke my neck.[44]

By most other accounts Sutherland's injury was hardly the aeroplane's fault. The mishap occurred when he left Duntroon to return to Richmond on 24 October. Taking off from the college playing fields, Sutherland clipped the top of a fence and crashed, upside down, in trees by the Molonglo River.[45] He was, however, uninjured to this point. First rescuer on the scene was Brigadier-General Walter McNicoll, who had been visiting Duntroon for the sports weekend (his son was then in the senior class of cadets).[46] Sutherland assured the general that he was alright, and with the words 'I think I'll get out now' released his seat belt, promptly falling on his head onto the ground and cracking his spinal column.[47] The permanent stiff neck which resulted from this injury ultimately led to his discharge from the RAAF—as he later remarked philosphically of his allegedly ungrateful steed, A2–26: 'Still, I recovered, and she did not. So there is some justice.'[48]

Structural failures

Because the RAAF was obliged to continue flying First World War vintage aircraft for most of the first decade of its existence, it might seem reasonable to suspect that a factor in a number of the accidents suffered up to 1930 was the aircraft themselves. But examination of the evidence provides no justification for an image of Air Force planes dropping from the skies through sheer old age, having been flown beyond their frail limits. Obsolescence of an aircraft type did not, of course, automatically imply unairworthiness. Provided airframes remained structurally sound, and engines and other systems received adequate maintenance, then safety was not unduly imperilled by the continued operation of imperial gift machines by the RAAF. Of course, had the Air Force continued to operate SE5A aircraft after evidence was found of deterioration in their airframes in 1927–28, and had a ban not also been imposed then on stunting in DH9s (see chapter 6), then clearly a spate of accidents involving mid-air break-ups would have been a distinct possibility. But, in fact, instances of mishaps due to this cause were extremely rare.

Astonishingly, the first crash recognised to have been due to structural failure involved not one of the RAAF's antique machines but its very latest type, the Bristol Bulldog fighter. When machine A12–4 broke up in mid-air over Point Cook on 15 May 1930, it had been in service only two weeks and had less than 20 hours flying time. On this occasion Pilot Officer Gordon Rae had been detailed to practice inverted aerobatics for 30 minutes, and after completing several stunts decided to attempt an 'inverted loop' at about 4000 feet, a manoeuvre in which a circle is described starting with a dive instead of a climb.

On his first attempt Rae got more than halfway round the loop and was almost vertical and climbing before the machine stalled and fell out. He tried a second time, this time using extra acceleration gained by keeping his machine in the vertical dive for a few seconds before taking it onto its back. Having got the aircraft onto its back, Rae heard a loud crack and suddenly lost control as the plane began to spin down from about 2000 feet, still in an inverted position. The pilot bailed out of the plummeting machine which crashed and was completely destroyed. Rae's emergency parachute descent made him the first RAAF pilot in the exclusive Caterpillar Club, the membership of which was confined to those who owed their survival to a British Irvin parachute.[49]

From eye witness accounts that the wings had split and folded forward, it was concluded that the wings of A12–4 had collapsed due to the port landing wires breaking under excessive tensile strain.[50] Notwithstanding the fact that the Bulldog had been successfully looped in Britain, the Air Board banned pilots from attempting outside loops, and also 'bunts' (half of an outside loop). It was reportedly discovered only later that the RAF had experienced an identical mishap but had failed to tell the RAAF.[51]

Problems were also experienced with the RAAF's other modern replacement for the old gift machines, the Westland Wapiti. Part of the problem experienced here arose from attempts to aerodynamically stretch this general purpose aircraft beyond the role for which it was designed. Ideas that it might be used to emulate the dive-bombing heroes of a popular US movie called *Hell Divers* proved particularly misguided. Although it was found that this aircraft could be held in a near vertical attitude in a dive, the plain fact was that it was never built to withstand the stresses of such usage. The fuel tank was liable to cave in under the pressure generated, and in one reported incident a pilot was almost asphyxiated with benzol when the blade of the small propellor driving the generator flew off and punctured the fuel tank; 'One very groggy pilot had to be assisted from his machine!'[52]

Plans were laid for a dive-bombing competition to be held within No.1 Squadron, using the unit's Wapiti planes, and a handsome statuette was even produced as a trophy. This was reportedly the handiwork of Sergeant R.J. Parker, who was a clever sculptor in addition to his trade as a carpenter rigger.[53] Parker had also completed a flying course though his skills in this field of endeavour were reportedly less impressive, and not long afterwards he was the pilot involved in an accident while practising for the competition. This was probably the mishap at Point Cook on 9 October 1934 which was witnessed by Warrant Officer A.J. Skinner:

> I was walking down the hangar area path with Squadron Leader George Banting [an RAF exchange officer], and we were having a chat about our respective jobs when out of a clear blue sky we heard a plane approaching from the direction of Laverton. As I remember it, George said 'What the b - - - - - hell are these people going to do now?' and just then the plane gave us the answer. It was a Wapiti and came down in a practice dive-bombing sighting. They pulled out of that dive and climbed about 1000 feet before coming in on another dive; but they never pulled out of that one and went straight into the ground. We were both running towards the crash, and the first thing I saw was the engine rolling around the field like a wheel, having snapped off on impact. Then the dust began to settle and we saw two figures pull themselves out of the dust and start to run in circles around what was left of the plane, which was practically flat to the ground—all except the tail empannage. Fortunately it never caught fire, and there were nil injuries to the crew (the bomber had a scratch on his nose). It was a miracle.[54]

The crash was reportedly due to a wing strut coming out of its socket in the dive,[55] an occurrence which was enough to cause cancellation of the unit dive-bombing contest. The handsome trophy remained unawarded and is on display in the RAAF Museum.

Notwithstanding this experience, Wapitis were still used as dive-bombers and public displays of the aircraft in this role continued to be given as late as 1934. A reminder of the limits to the Wapiti's versatility came on 16 April 1935, however, when A5–31 crashed at Werribee while being test-flown from No.1 AD. What was believed to have been structural failure sent the machine plummeting from about 4000 feet and the pilot, Flying Officer Clive Edgerton, was killed.[56]

Even the Hawker Demons which entered service with the RAAF from 1935 were discovered to be not as robust as their modern status led some to expect. On 14 May 1937 Pilot Officer Donald Ashton-Shorter of No.3 Squadron was demonstrating the aerobatic capability of Demon A1–40 to members of the militia unit at Townsville (Qld). Fortunately he was not carrying the customary observer on this flight. Taking his machine to 3000 feet, Ashton-Shorter commenced a near-vertical dive towards the Ross Plains aerodrome. At between

250 and 500 feet he began to pull the aircraft out of its dive, but the port upper mainplane broke away and folded back, followed by the lower mainplanes. The machine continued on into the ground, killing him instantly.[57]

The only other instance attributed to structural failure found in RAAF records was a crash near Windsor (NSW) on 16 October 1938, which again involved a relatively new aircraft type, the Avro Anson. On this occasion machine A4–8 of No.22 Squadron was being flown by a 19-year-old CAF pupil pilot, Cadet John Fletcher, on a height test. The aircraft appeared to power dive from 6000 feet, during which there was an explosion and the starboard wing tore off at approximately 4000 feet. Wreckage was spread over three miles around McGrath's Hill, near Windsor. At an inquest in October, the coroner criticised RAAF inspection procedures and commented also on his belief that A4–8 had not been in good order before the crash. Both criticisms were rejected by the CO at Richmond, Group Captain De La Rue.[58]

Bad weather accidents

Inevitably, there was a proportion of mishaps which were directly attributable to adverse weather encountered in flight and the fact that aircraft of the period were ill-equipped to cope with such conditions. 'Cloud flying' instruments available to pilots in the RAAF's early planes consisted of little more than a compass and a small curved spirit level that indicated whether the wings were in a horizontal plane.

The perils of flying in poor conditions were graphically brought home by an incident which occurred when six DH9s were sent to Sydney during the visit of the British special naval squadron in 1924 (see chapter 2). For the flight, resort was made to the usual navigational aid of following the railway line. On this occasion, however, bad weather encountered after passing Goulburn (NSW) forced the RAAF machines down from a safe altitude of 2000 feet to only 100 feet to keep within sight of the train tracks. Approaching Moss Vale it was practically impossible to see ahead. The biographer of one member of this formation recounts that:

> Suddenly, less than twenty yards ahead, he saw the railway track disappear into a tunnel in a steep hillside! A stuffed cloud! It seemed obvious that all the machines would be piled on to the mountain-side. Mustar pulled back his control lever and zoomed almost vertically into the mist. The . . . following machines, with a split second's extra warning, did likewise.

The loss of all six aircraft and crews had been averted by a split second.[59]

Others who became caught in low cloud or fog, or fierce storm activity, were less fortunate. As mentioned previously, the accident which claimed the life of Sergeant Ebell, a pupil pilot, in September 1926 was attributable to this cause. Another trainee, CAF Cadet Murray, was killed on 21 April 1935 when he flew his Wapiti (A5–18) into a hill at Gordon, near Ballarat, in bad weather.

Cloud nearly claimed the crews of two Demons of No.3 Squadron early in 1937. On 3 February, A1–3 (Pilot Officer Donald Ashton-Shorter, pilot; Sergeant William Gould) and A1–8 (Pilot Officer Gerald Buscombe, pilot;

Sergeant A.A.B. Slight) took off from Laverton to fly to Hobart (Tas) where they were to carry out Army cooperation work, photography, and participate in annual Regatta celebrations. Off the Victorian coast they entered cloud which extended all the way across Bass Strait and shrouded northern Tasmania too. The two planes were to have refuelled at Waratah, south of Wynward, but this destination was missed in the murk and they overflew it. Now short of fuel, both machines faced the prospect of making blind landings. On board A1–8, Buscombe later recalled that: 'I said it looked like we would have to go over the side, but he [Gould] said he had lost his parachute'.[60]

The crew of A1–3 were fortunate enough to glimpse a road through a break in the fog, enabling them to touch down west of Waratah with only slight damage. Buscombe and Gould in A1–8 were not so lucky, however, and crashed near Mount Cleveland about 10 kilometres from the town. Both were uninjured in the crash, but they were now stranded in remote country and in need of shelter from the weather. Meanwhile a search effort was mounted, and the crash-site was located by Anson A4–1 piloted by Flight Lieutenant Gordon Rae, who dropped a flask of soup to the pair on the ground before turning away to report his find. A ground party reached the airmen a couple of days later and brought them out; after salvageable parts of A1–8 were recovered, the wreck was burnt. The remains of this machine were recovered in 1977 and completely rebuilt at the RAAF Museum, where it remains on display.

On 10 August 1938, an Avro Anson (A4–29) of No.2 Squadron was flying in low cloud when it hit a hill near Dromana (Vic) called Arthur's Seat. Four men on board were killed (Sergeant John Gillespie, pilot; Pilot Officer Robert Symonds, navigator; AC1 Kenneth McKerrow, radio operator; and AC1 Robert Mawson, engineer), but miraculously AC1 James Glover—a passenger sitting in the rear gunner's compartment—survived with only minor injuries. The aircraft had been one of five Ansons engaged on a navigational exercise from Laverton, and the flight had been recalled due to the deteriorating weather at the time of the accident. In the official RAAF investigation, however, the cause was attributed to human error rather than the dense fog and light rain conditions encountered.[61]

Throwing into doubt the fairness of such official pronouncements was the experience five days later of Pilot Officer John Balmer, who made a forced landing in Anson A4–36 near Whitfield (Vic) during which a wing was torn off the aircraft. Cause of this mishap was recognised to have been the formation of ice on the wings, and it was revealed that the Anson was not fitted with any of the standard modern equipment for flying in adverse weather conditions, despite the fact that this relatively new British type cost over £7000. As noted later in this chapter, this and subsequent accidents to Ansons raised important questions about the operational capability of this aircraft type.

Acts of indiscipline

A regrettable fact concerning some accidents was that they were due to the action of pilots flying their aircraft in a manner which contravened orders or failed to observe safe operating procedures, sometimes while engaging in acts of devilment that carried a high risk of going wrong. While discussion of this aspect risks maligning individuals, particularly persons whose death in accidents prevented them from explaining their actions to subsequent inquiries, it is necessary to give due weight to indiscipline as a factor in the RAAF's overall safety record.

In its early years, the RAAF was heavily influenced by the ethos of the air services in the First World War. In this conflict Australian pilots had won a reputation for the way in which they met the harsh demands of air combat, acquiring in the eyes of some almost mythical status. One British officer, Colonel Louis Strange, who commanded a wing in France which included No.4 Squadron, AFC, wrote a post-war account strongly promoting this image. Though his idea of Australia was rather quaint, owing more to the American Wild West, he applauded the individual initiative of Australians as a quality 'inherited from ancestors who had been cattlemen, sheep-ranchers, poachers, trappers, outriders, overland post and transport drivers . . . [They] were equally at home when cutting out cattle at a round-up or shooting the rapids in a canoe'. As Strange told it, the Australian squadrons made him think of 'the sheriff and his posse going out after bushrangers'.[62]

Unfortunately there were wartime pilots in the RAAF who accepted their cowboy image at face value and attempted to live up to its romantic ethos. The biographer of First World War air ace 'Gerry' Pentland records something of the high jinks which this officer got up to in the RAAF's early years, in the course of such activities as artillery spotting for the Army. These missions were normally carried out by two pilots who took it in turns to fly and act as observer:

> The procedure could become boring to even the keenest of recruits and it was particularly irksome to a high-spirited pair like Pentland and [A.W.] Vigers when they found themselves teamed together, so on a couple of occasions they relieved the monotony by dispensing with the landing routine and changed places in midair.[63]

Officially, of course, conduct of this sort was not countenanced. Thus when Pentland and Vigers were spotted one day doing their mid-air change-over they were 'hauled up on the carpet'. Similarly, when Flying Officer Charles Matheson's flouting of take-off procedures resulted in the crashing of Sopwith Pup A4–4 early in 1925, his breach of safety did not escape the wrath of officialdom. After running up the engine, Matheson did not bother with taxiing to the lee side of the airfield but opened the throttle to full bore and began a take-off along the tarmac beside the hangars. Recalls one who witnessed his fall from grace:

> Just as he became airborne, the port wing of the Pup struck a guy rope of one of the hangars. The Pup flipped over into a half roll over the top of and a few feet above the hangar and landed right side up on the road the other side with a resounding crash. It took some cutting of wires and other impedimenta to get Charlie out, when it was found that his sole injury was a broken leg. Of course that marked the end of his career in the RAAF. When fully recovered from a spell in hospital Charlie's commission was gracefully terminated.[64]

On 31 January 1930 the Melbourne *Argus* reported court martial proceedings taken against two officers of No.1 Squadron—Flying Officer Andrew Gerrand and Pilot Officer Gordon Rae—charged with carrying out aerobatics at low altitude. Rae was found not guilty, and Gerrand received only a reprimand.

These examples where disciplinary action was taken over deliberate breaches of safety rules aside, the attitude adopted by RAAF authorities was not always as disapproving as was to be expected. When low level aerobatics by a RAAF aircraft above house rooftops in the Melbourne bayside suburb of Brighton in August 1925 led to questions in the federal Parliament about whether legal safety limits had been breached, the Air Board was directed to ascertain the identity of the pilot. In compliance with this requirement the Board established that the culprit was an experienced RAAF officer, but took no action other than to issue instructions restricting aerobatics practice to certain defined areas.[65]

A further example of this ambivalent attitude is seen in the case of Air Chief Marshal Sir Frederick Scherger, who as a young flying officer was caught out after a spot of low-flying on 2 February 1927. According to the story, the CO of No.1 FTS, Wing Commander Adrian ('King') Cole, asked Scherger to drop a message to a girlfriend of his who was departing from the port of Melbourne that afternoon on board the steamship *Canberra*. Scherger did as he was asked, but finished off the mission by taking his SE5A down to pass between the ship and the wharf before pulling up into a roll and heading back to Point Cook:

> The following morning he was called into Cole's office, to be confronted by a very angry wing commander brandishing a photograph which, however blurred, clearly showed a RAAF aircraft below the level of both wharf and ship . . . Hours later, Scherger confronted Group Captain Goble in the Melbourne office [of the Air Member for Personnel, who wanted to know if he did this sort of thing often.] . . . 'Well, there's never been a photographer there before, sir.' 'Good, I'm glad to see we've still got a few in the Air Force with spirit. Go back and tell King Cole that I beat the hell out of you.'[66]

Of course, there is no mistaking the implicit message being passed to the rest of the service by such means, whether or not the word-of-mouth accounts were strictly accurate.

Clearly there was a desire among RAAF commanders not to crush every vestige of spirited flying in a service where precisely such attributes were a valuable asset in wartime. The same attitude was apparent in the case of Pilot Officer Julius Cohen (better known later as Sir Richard Kingsland). As an air cadet Cohen had been charged in November 1935 for alleged low flying, although more than 50 years after the event he still disputes the justice of the £5 fine imposed:

> As a matter of fact, I hadn't been low flying on the day they said I was. They had someone follow me in another aircraft that day, and it was on his wrong evidence I was charged. Mind you, if they had followed me the day before there couldn't have been any disputing the charge. As cadets we often went low flying. When we were told to go off and do practice landings, we'd do hedge-hopping over paddocks, and so on. I'd say a quarter of the cadets would have made good tram drivers they were such stable types, but a quarter of them were buccaneers. I

> wasn't one of the wild ones, although I did things which would be considered wild these days.[67]

Despite this brush with authority, in late 1936 or early 1937 Pilot Officer Cohen was involved in another spot of low flying which might well have ended his career. Sent on a cross-country training flight from Laverton, he arrived over the New South Wales border town of Corowa and proceeded to beat the place up before dropping a message at a nearby property which was the home of a fellow officer, Pilot Officer William Keenan. Cohen's antics at Corowa, nosediving and roaring low along the main street, half-terrified many residents and were fully reported by local newspapers.[68] Presumably only an absence of complaints meant the Air Board did not feel compelled to take any action over this display. It is a matter of record that Cohen was awarded the Distinguished Flying Cross for daring flying involved in extracting a British diplomatic mission from Rabat in French Morocco in June 1940.[69]

Understanding, then, the context in which instructions were not infrequently contravened, there can be no real surprise that sometimes the risks taken by pilots were misjudged and resulted in disaster. A clear example of this occurred on 5 January 1930 when Flying Officer Leo Ryan, a 25-year-old Duntroon graduate on secondment from the Army, took his Moth A7–10 low flying. Having allegedly spent the afternoon in the Mess at Point Cook,[70] he was flying solo back to No.1 Squadron just on dusk when he decided to do some hedge-hopping. In the course of this illegal activity, he struck a haystack about a mile from Laverton. The plane's undercarriage was torn off by the impact, and the machine plowed into the ground about 30 metres beyond the stack and was completely destroyed. Ryan was critically injured and died in the Caufield Repatriation Hospital two days later.[71]

A similar fate befell another Duntroon graduate, Pilot Officer Hiram Scoullar, who was sent to Seymour (Vic) in Wapiti A5–20 of No.1 Squadron to carry out cooperation exercises with the Army on 26 February 1931. Allegedly deciding to beat up the place,[72] he struck trees in a low dive, crashed, and caught fire. His passenger, fitter AC1 Yourn, perished with him.

Tragedy also attended the actions of Temporary Sergeant Robert Somerville, who was piloting a DH60 Moth (A7–8) on a local training flight from Richmond on 21 January 1929. Spotting a tractor ploughing on a property at nearby Cornwallis which he knew to be that of his fiancee's father, Mr A.C. Smith, Somerville decided to buzz his prospective father-in-law. Swooping low, Somerville misjudged his pass and struck Smith with the axle of his undercarriage, decapitating him.[73] Following this collision the Moth made a forced landing, crashed and caught fire; though injured, Somerville and his passenger, LAC Leslie Milgate, both escaped. Somerville was charged with manslaughter but subsequently acquitted in Sydney's Central Criminal Court on 22 March.[74] He was, however, suspended from flying and reduced to his substantive rank of LAC. A few years later he was restored to pilot duty and became an instructor at Point Cook, where he himself became the victim of an aircraft accident on 9 December 1936 which was caused by a fellow instructor ignoring the circuit instructions in operation that day.[75]

A more tragic illustration of the consequences which flowed from failure to observe strict flying discipline can be seen in a particularly shocking accident which occurred at the Tasmanian east coast town of Swansea on 6 February 1936. Three Hawker Demons of No.1 Squadron were on a training flight intended to show off the RAAF's latest acquisitions, and were taking off after a visit to the town when one machine failed to climb sufficiently and struck members of a crowd of onlookers gathered at the edge of the airfield. A 50-year-old female spectator and her 19-year-old daughter were killed, and another woman and a man were injured. The aircraft (A1–3), which had damaged a leg of its undercarriage, returned to the aerodrome and crash landed but without injury to the crew, Pilot Officer Eric Lansell and Sergeant R.J. Tonks.[76]

At the subsequent inquest the coroner was persuaded by the evidence of witnesses that all the RAAF aircraft had engaged in low flying on the morning of the accident, 'in flagrant breach of the regulations', but suggestions put to Lansell that he might have deliberately delayed his plane's lift-off from the ground to give spectators a thrill were denied. This did not prevent the coroner from finding that the deaths resulted because Lansell had 'negligently piloted' his aircraft, nor from committing him for trial on a charge of manslaughter.[77] In reality, however, Lansell was guilty of nothing more than—to use the coroner's own words—'a miscalculation of the height that he was above the ground', which was a doubtful basis for a case of civil much less criminal negligence, a fact recognised when the charge was later dropped.

A similar situation attached to an accident to a Demon of No.3 Squadron (A1–51) at Llandillo, 13 kilometres south of Richmond, on 14 June 1938. The aircraft was being flown by Sergeant Ralph Pomeroy in a practice aerial combat with another Demon, piloted by Flying Officer Harry Durant, when it

The remains of Moth A7–31 in which Sergeant R.F. Somerville suffered fatal injuries after a mid-air collision at Point Cook on 9 December 1936. The other aircraft involved was Avro Cadet A6–9, flown by a fellow instructor who disregarded the circuit instructions in effect that morning. (*ACdre N.R. Ford collection, RAAF Museum*)

struck a tree, crashed to the ground and ignited. The observer in this machine, AC1 D.C.A. Purdy, having initially escaped with very minor injuries, suffered terrible burns to his face attempting to rescue Pomeroy from the blaze. A civilian arriving on the scene pulled Purdy from the wreckage, extinguished his burning clothing and restrained him from again attempting to rescue the pilot.[78]

At the coronial inquiry into Pomeroy's death civilian witnesses stated that the two machines had been manoeuvring around each other, gradually losing height before Pomeroy's Demon struck the 12-metre high tree. In his evidence Wing Commander Joe Hewitt, the president of the court of inquiry conducted by the RAAF, contradicted the claims regarding the height at which the two Demons had been flying, but the coroner was satisfied that the aircraft had been at an altitude much lower than allowed by regulations. However, he did not consider that culpable negligence had been involved—both pilots had simply been concentrating so intently on their aerial manouevres they had failed to realise their proximity to the ground.[79] What this and other such occurrences actually highlighted was that there was often precious little separating an act of indiscipline from an ordinary error of judgement.

Ground accidents

Of course, aircraft crashes were not the only form of accident suffered by RAAF members. At times personnel on the ground faced risks which were potentially as lethal as those confronting aircrews. One of the most dangerous moments for mechanics and aircraft hands working on the tarmac was that of starting up an aircraft's engine. This involved swinging the propeller by hand, a procedure which always carried the risk of being struck by the blade when the engine fired. One such mishap at Point Cook in 1923, caused by the member 'not keeping clear of the propeller when "sucking-in"', prompted the publication of a warning in routine orders that all ranks must keep 'as clear as possible',[80] but in practice this was not easy to do and accidents continued to occur.

During the afternoon of 4 February 1926 25-year-old AC1 Leonard Devine was struck while starting up a machine at Point Cook, receiving compound fractures of the left arm and leg. He was taken to Caufield Repatriation Hospital that evening and underwent surgery which involved amputation of his leg. He died the next morning from shock and blood loss following the operation.[81] Charles Gates, who enlisted as an airman in 1922, recalls other incidents of this nature:

> A young Irishman, George Finnie, was struck on the arm swinging a propeller and sustained an injury which the medical authorities considered made him unfit for further Air Force service, and he was discharged in 1927. At the court of inquiry another man said that in his view it was Finnie's own fault that he was hit. Finnie went to New Zealand and then returned when his arm had healed, went straight to the Medical Branch at Air Force Headquarters and demanded his job back. The medical board considered him fit for service and enlisted him as a medical clerk in 1933; during the Second World War he became a Flight Lieutenant administrative officer in the medical branch. The man who had

> criticised Finnie in relation to his accident himself lost a leg while spinning a propeller.[82]

To reduce the risks to personnel, there was one mechanical aid available in the form of the 'Hucks Starter', invented by a RFC officer during the First World War. This was a T Model Ford which had been modified to harness the drive from the vehicle's gear-box to an overhead shaft, which could then be engaged into a 'dog' on the propeller boss, and the drive from the Ford's gear box used to turn the propeller until the aircraft's engine fired.[83] The Hucks Starter was an ingenious and efficient device that remained in use until around 1937, by which time aircraft types had electric starters. As well as its primary role, the Hucks was also used to tow machines and transport personnel out to their aircraft. On the latter occasions the Hucks' driver, LAC M. Selke, had a favourite prank which he played on new students. When driving cadets out to their aircraft he would loosen the nut on the steering column and, when least expected, drop the steering wheel at the feet of his passengers:

> The driver jumped off, and as the truck passed on he yelled out 'Jump'. But Selke had the nut in his pocket and the Hucks was slowly pulling up about 100 yards away (Selke had put the brake on before he jumped himself). It was a joke that went on for some time until an officer objected and that was that.[84]

Even with machines safely in the air and on their way, there were occasionally perils for those below. There were incidents of misdirected bombing, such as during preparations for the air pageant at Laverton in November 1934 which the Duke of Gloucester was to attend. At a rehearsal the day before the actual event a practice bomb was accidentally dropped from a diving Bulldog and sent crashing through the iron roof of a hangar. Narrowly missing a group of airmen, the bomb buried itself into the concrete floor, sending up a spray of fragments and releasing a dense cloud of stannic chloride.[85]

Another peril occasionally resulted from the forgetfulness of crews who failed to wind in their radio antenna before landing. With the still primitive radio equipment used in the pre-war period, long cables with lead weights at one end were trailed beneath aircraft to achieve radio communications in the air. These could only be wound out once airborne and had to be wound in again before making a descent. One pilot of this period recalls:

> One day the whole of Newcastle came to a grinding halt because some bloke forgot and dragged the aerial with all the lead weights across the tram lines on his way in to land. There was a blinding flash, explosions and everything, and Newcastle stopped for a couple of hours while they sorted that one out. There were plenty of times when the aerials were forgotten and snapped off on landing.[86]

According to other accounts which cannot be verified, there was an incident where an aircraft was caused to crash after its antenna became snagged in a tree, and another during Army cooperation exercises at Seymour, Victoria, when several soldiers on the ground required medical treatment after a trailing cable was flown through their company parade.[87]

Starting an aircraft's engine by swinging the propeller was a dangerous procedure which caused many injuries and a few fatalities among ground crew. In this case a Bristol Bulldog is being started up at Point Cook. The risk was reduced by using a mechanical device called the 'Hucks Starter'. (*Mrs Y.C. Bonwick*)

Introduction of parachutes

Apart from the mid-air break-up of Flying Officer Gordon Rae's Bulldog in May 1930 (previously described), there were only two other instances before the Second World War where RAAF members used parachutes for emergency purposes. One was the episode involving Flight Lieutenant Balmer recounted in chapter 7, wherein it was alleged he jumped from (but more likely fell out of) an aeroplane in which he was instructing a trainee. The second involved Cadet Leslie Clisby, who later achieved fame during the Second World War by winning the DFC and becoming an air ace before his death in France in 1940. Undergoing training at Point Cook before his transfer to the RAF in 1937, he parachuted to safety when his Moth (A7–45) went out of control and crashed near Werribee on 24 April 1936.

As previously noted, however, there had been at least one other occasion—the collision over Melbourne of two DH9s in April 1927—where the availability of parachutes might have had a significant impact in saving lives. Parachutes had certainly been introduced into the RAAF by this time; it was simply that there was a reluctance among air crews to actually wear them.

The first moves to equip RAAF personnel with parachutes came in 1924, during a visit by CAS to Canada and the United States. While in the latter country Williams visited the factory of the Irving Air Chute Company, makers of the parachute designed by Leslie Irvin: 'I was impressed with the possibilities of these parachutes and sent a cable to the Air Board recommending the purchase of a sufficient number to equip the aircraft we had on our establishments—at that time about eighty'.[88] This recommendation was not immediately acted upon because the Minister for Defence (Howse) was disturbed by the idea of the RAAF buying American rather than British equipment. This reservation was subsequently dispelled when it learned that the

RAF was also buying 500 Irving chutes. Recounted Williams: 'I was then given approval for our order but the Irving factory was quite a small one and we had then to wait two years until the RAF order had been completed before we could get ours'.[89]

Actually, Williams was mistaken about the timeframe associated with delivery of the RAAF's parachutes, as Irving chutes were certainly being tested at Point Cook in January 1926. A press report late that month states that a DH9 flown by Flying Officers Vernon Burgess and Alan Charlesworth had taken up two parachutes to altitudes of 500 and 1500 feet and test-dropped them using 150-pound dummies. 'Eighteen of these parachutes have now been tested at Point Cook and each test has been successful'.[90]

While delivery of the RAAF order for parachutes was awaited, action was taken to obtain a trained instructor to introduce the equipment into the Australian service. In 1925 the RAAF already had Flying Officer Ellis Wackett—the younger brother of Squadron Leader Lawrence Wackett and a 1918 graduate of the Royal Australian Naval College—in England completing post-graduate studies in aeronautical engineering. As Wackett later recalled: 'My passage home was cancelled and off I went to a parachute training school to learn to pack and use the new equipment'.[91] In the course of this training, he also became the first member of the RAAF to make a free descent by parachute.

On his return to Australia, Wackett was responsible for training other Air Force members. He began by conducting a course for nine volunteers at Richmond. It was here, on 26 May 1926, that he made the first freefall from an aeroplane in Australia (at least from a service aircraft). This was from a DH9A (A1–10), which had been specially modified by the addition of a small ladder on the side of the fuselage to enable the jumper to climb out of the rear cockpit before letting go.[92] Over the next week the nine members of the course also completed jumps. The following month a second course was conducted at Point Cook for another fourteen officers and airmen.[93]

There were no mishaps to any of Wackett's volunteers, and only one near miss. This involved Flying Officer Harry McGilvery, who chose the very moment that he let go the ladder to thumb his nose at the pilot. His arm action brought on a spin which, when he pulled the ripcord, twisted his shroud lines together. Instead of opening his reserve chute he tried to unravel the lines as he plummeted earthwards, and appeared certain to be the RAAF's first parachute fatality when the canopy suddenly opened with a bang not more than 20 feet from the ground. Although he landed heavily, McGilvery suffered nothing more than bruising.[94]

Among those volunteering to make jumps at this time was CAS who considered that 'it would be a good example if, before issuing an order for the compulsory wearing of parachutes, I showed my own confidence in them . . .'[95] Williams made his jump over Point Cook on 5 August 1926, and according to his account of the event his first experience of parachuting was only marred:

> when it appeared to me that I was going to drop into the station water tank. I thought it would be a poor ending to drown there, or even to be pulled out dripping wet. I missed the tank but landed much too close to be comfortable to a 30 000 volt electric transmission line running into the station.[96]

The CAS having made his point, the order for the compulsory wearing of parachutes (except in planes where this was unsuitable) was issued on 5 August.[97]

Little more than a fortnight later, on 21 August 1926, the first public display of parachuting by the RAAF was given at Essendon. On this occasion Flying Officer Scherger made the jump (only his second), again from A1–10 which was piloted by Wackett.[98] Pilot Officer Stuart Campbell is believed to have made an unofficial descent from a Moth aircraft at the Coode Island aerodrome opposite Fishermens Bend, Melbourne, during one Saturday morning of the following month.[99] Apart from this there was very little done over the next four years to maintain the parachuting skills thus initially introduced.

Not until October 1930 was another parachute course conducted, though few details apart from the names of three members of No.1 Squadron who attended it are known.[100] It was August 1932 before another parachute course was held—this time a short course in parachute maintenance run at Point Cook by Flight Lieutenant Don Carroll, who had recently completed a course in this field while in England studying at the RAF School of Engineering. Over two weeks the six or seven students heard lectures and received practical training in packing, repairing and inspecting chutes. Their practical skills were tested by dummy drops, during which a 113-kilogram jarrah torso weighted with lead was taken up to 1500 feet and released from the wing bomb-rack of a Wapiti. These drops produced at least one heart-stopping moment when, as sometimes happened, the parachute failed to open. On this occasion the wooden dummy crashed through the roof of the Officers Mess billiards room, shattering the table. Air Commodore 'Paddy' Heffernan, a member of this course, recalls:

> The mess secretary of the day, Flying Officer 'Jimmy' Rennick . . . , with commendable foresight threw into the wreckage all the broken cues and any unserviceable chairs that he could find. In the subsequent claim for damages the auditors, so I heard, could not understand how so much damage could have been caused by one small dummy.[101]

The course was completed with live jumps being made by the students. For these, the RAF practice was followed of using a Southampton flying boat to make 'pull-off' jumps.[102] This involved the jumper standing on a small platform on the outer wing strut of the plane, then at a given signal allowing the chute to open and drag him off his small perch. The first two jumpers using this method were Heffernan and Carroll (who had not actually performed a jump during his course in England). Heffernan recalled the take-off on 24 August 1932:

> We worked our way along the wing, through the bracing wires to the outer strut. Here we stood with our backs to the way we were going and leant against the strut until we reached jumping height. I think that this time was the worst part of the trip, because during the take-off you could feel every movement of the wings and you felt sure that they were going to fall off . . . As we got over the southern boundary we got the signal to 'Go' . . . so we pulled our respective ripcords and

Sergeant Jack Marsden (*right*) watches as a parachute is fitted to a wood dummy in preparation for a test drop at Point Cook. Such tests sometimes went wrong when the parachute failed to open, on one occasion sending the dummy crashing through the roof of the Officers Mess! (*Mrs Y.C. Bonwick*)

the next thing we knew was that we were yanked off the platform and were floating in space.[103]

Carroll landed 'in copybook style' on the aerodrome, but Heffernan was carried by a windgust towards a large pine tree near the station sick quarters and found himself suspended by its branches about 3 metres from the ground.

Having completed the 1932 course it fell to Heffernan to conduct the next one for six students in August 1935.[104] Although another such course is believed to have been conducted the following year, this appears to have been the last pre-war training conducted. This was not the end of the parachute story in the RAAF however, because in 1937 production of parachutes in Australia was commenced by the Light Aircraft Company run by George Mills, a former member of Williams' AFC flight in Palestine in 1916–17. By agreement with the Irvin Air Chute Company of Great Britain, Mill's company began producing parachutes for the RAAF under the trade name 'Dominion'. Tests of Dominion chutes were carried out by the RAAF at Camden Park, the property at Menangle belonging to the Macarthur-Onslow family, who were partners in Mills' company. Among those involved in these flights in 1938 was

Sergeant E.W.F. Sunderland clings to the wing strut of a Southampton prior to making a pull-off parachute jump. According to one who made a descent by this method, the climb was the worst part of the experience; the jumper could feel every movement in the wings during take-off and was convinced they were going to fall off! (*SqnLdr E.W.F. Sunderland*)

Pilot Officer Charles Read, who was sent from Richmond in a Demon and recalls that a problem was experienced on the first test drop which resulted in the wooden dummy separating from the parachute:

> Looking down to the middle of the airfield I could see people running in all directions. What had happened was that the dummy had slipped out of the harness and landed about 50 feet from all those people. It could have been an absolute tragedy. One of our Air Force engineers had a look at it and found an absolutely basic design fault with it. All production had to be stopped until it was redesigned.[105]

By this time the RAAF had a two-man section responsible for acceptance testing of every parachute before it was released for use in the service. Initially located at Point Cook, the section moved to Laverton about 1937 and all major overhauls of parachuting equipment were done here from this time. One member of this section was 'Ted' Sunderland, who was present during the tests at Menangle of the Light Aircraft parachutes and confirms the initial difficulty experienced. 'Still', he says, 'I don't know of any fatalities before the war involving parachuting'.[106]

Moments of gallantry

While in no way lessening the sense of tragedy involved in accidents involving loss of life, some incidents gave rise to displays of courage and sacrifice which at least strengthened faith in the human spirit at such times. The first instance of a RAAF accident which resulted in a recognised act of bravery occurred at

Flight Lieutenant P.G. Heffernan (*at right*) ran the RAAF parachute course conducted in August 1935. Here he adjusts the equipment of one of the trainees, LAC Coghill, while another student, Pilot Officer D. McLean, looks on. (*Mrs Y.C. Bonwick*)

Canberra in February 1926, when two DH9s of No.3 Squadron were sent from Richmond to undertake an aerial photographic survey of the Murrumbidgee River for the Federal Capital Commission. The first plane under Flight Lieutenant Allan Hepburn arrived safely on 9 February, but when the second machine (A6–28) followed two days later disaster struck. Arriving over the Ainslie airfield (now covered by the Canberra suburb of Dickson) on a clear morning, the pilot, Flying Officer Philip ('Peter') Pitt, shut off the engine and turned in preparation for a landing. At this point the machine stalled and spun into the ground from about 100 feet, bursting into flames 30 to 40 seconds later.[107]

Hepburn, waiting for the aircraft to land, immediately drove to the scene of the accident. He found that the DH9 had crashed about 60 metres from where a local farmer, Walter Johnson, was ploughing. Johnson had entered the burning wreckage and already managed to pull free AC1 William Callander (the observer-photographer), but the pilot, who was hanging out of the cockpit and appeared to be dead anyway, was engulfed in the flames and incinerated. Sadly, Callander also later died from injuries, but Johnson's efforts to save him were recognised by the award of the Royal Humane Society's Bronze Medal. The RAAF court of inquiry into the accident considered that, although old, the aircraft involved had been sound and airworthy, and that the crash had been 'due to an error of judgement on the part of the pilot of the machine'.

Another accident in 1937 was the occasion for an outstanding act of bravery, subsequently recognised by the sole gallantry award to a member of the RAAF

The burnt-out remains of DH9 A6–28 after it crashed at Canberra on 11 February 1926. The pilot, Flying Officer P.M. Pitt, was killed outright and the observer, AC1 W.E. Callander, died later of his injuries. The rescue attempts of local farmer, W.E. Johnson, were recognised by the Royal Humane Society. (*Canberra & District Historical Society*)

in the inter-war years. On 31 August three Demons from No.1 Squadron made an official formation flight to Hamilton, in the Western District of Victoria, where they landed at the local aerodrome. The flight coincided with the Hamilton Show, and for half an hour the machines provided a static display for the large crowd in attendance. Then, shortly after 3.15 pm, the aircraft lined up in preparation for departure.[108] The first Demon got safely away, but as the second machine (flown by Flying Officer Ken McKenzie with Sergeant Norm Torrens-Witherow in the observer's seat) took off minutes later, the pilot attempted a climbing turn. With the loss of air speed entailed by this manouevre, the plane stalled and immediately dived to earth again from a height of 200 feet, in full view of the horrified crowd.[109]

Flying Officer Joe Godsell, the pilot of the Demon still on the ground, was preoccupied with his preparations for take-off and did not notice what had happened. His passenger, however, 27-year-old AC1 Bill McAloney, a member of the Air Force for barely a year, looked up in time to see the previous machine make its fatal dive and recalled: 'As it hit the ground, the main tank burst and the aircraft became enveloped in flame and smoke. The pilot of my aircraft had his back to this [pilot and observer sat back-to-back in the Demon] when I called out for him to stop.'[110] Godsell had no sooner braked to a standstill than McAloney, throwing off his parachute harness, was out of the aircraft and racing to the other side of the landing ground where the crashed Demon was burning furiously amid exploding Very cartridges.[111]

Although hampered by the bulky clothing he was wearing for protection against the cold encountered in flight—a pullover, a pair of overalls and a flying suit (fortunately treated against fire)—McAloney was still first to reach the wreck. He later described the events that followed:

> I couldn't see the aircraft of course, which was just alight from nose to tail, as I went in to try to rescue the two occupants. I knew I would have to be quick to

> even get one; I didn't know whether they were still alive or not. When I got to the side of the aircraft, the rear occupant—the observer—was getting out, and I looked back [to see] if anyone else was coming. This chap's clothing had only just started to burn and he was not in as great a danger as the pilot, who was lying over the top of the fire from the main tank. Hoping someone else would get to the other chap, I ran around to the other side of the aircraft and went in to get hold of the pilot. His clothes were all alight as I pulled him down onto the mainplane. One of his legs went through the fabric and his foot got caught up in this. I kept trying to pull him away, but his smouldering clothes were burning my hands. As I heaved and tugged at him, one of the wing tanks exploded and I was hit with the blast. Having run for over a quarter of a mile just previously to reach the men and breathing in all the fumes, plus the heat becoming almost unbearable, I got the blast of it down my lungs. Of course I was gasping for air and this finished me off.[112]

McAloney was pulled away unconscious from the flaming wreckage by spectators who arrived on the scene. In making his rescue bid he had sustained burns which caused his hospitalisation for over a month. It was while in hospital that he learnt of the futility of his effort to save the two crewmen. The observer, although seen by McAloney endeavouring to free himself, had been hopelessly pinned by the undercarriage, which had been driven backwards by the force of the crash and jammed his ankles. In addition, both men had suffered massive head injuries on impact and could not have survived even if their rescuer had managed to get them clear of the flames.[113]

Though McAloney had failed in his bid to save the crewmen's lives, this in no way diminished the valour of his act. At the inquest which followed the crash, the coroner commented publicly on his bravery and suggested the facts be brought to the notice of the Royal Humane Society. Air Force authorities, however, were already considering the idea of recommending him for an award, although (such was the scarcity of a precedent) that there was confusion over which award was appropriate to the circumstances. Squadron Leader Charles Eaton, president of the RAAF court of inquiry into the accident, suggested to McAloney's CO (Squadron Leader Alan Charlesworth) on 3 September that he be recommended for the Air Force Medal, but instead Charlesworth proposed three days later that he should receive the British Empire Medal.

Though the BEM recommendation was supported by the base commander at Laverton (Group Captain Henry Wrigley), when the matter was considered by the Air Board on 1 October the Air Member for Personnel, AVM Nicholl (an RAF officer on exchange), argued that the Albert Medal for Gallantry in Saving Life on Land was more appropriate.[114] It was this recommendation which went forward in a submission from the Minister for Defence to the prime minister, with the argument that 'his gallantry should be recognised by an award of an official nature rather than through the Royal Humane Society, more particularly as he was engaged on Air Force duty at the time . . . and similar awards [of the Albert Medal] have been made in the Royal Air Force in like circumstance'.[115]

In London, Buckingham Palace officials considering the recommendation pondered the RAF precedents. It was noted that in McAloney's case he had not been himself involved in the crash; nonetheless, it was concluded that 'on the precedents it would seem difficult to turn the case down' and it was accordingly recommended to the King.[116] The award of the Albert Medal was announced from the palace on 4 February 1938, McAloney receiving the news while on leave at his sister's home in Adelaide. He was presented with the Medal by Lord Huntingfield, the Administrator of the Commonwealth, at an investiture held at Parliament House, Melbourne, on 30 May 1938. McAloney remained with the RAAF until he retired in 1967, and in 1972, when the Albert Medal was made obsolete, his award was converted to the George Cross.[117] His Albert Medal, the last such award made to an Australian who survived to receive it, is now on display at the RAAF Museum, Point Cook, and remains unique in the annals of the RAAF.

The Fallon affair

On 29 November 1937 nine Demons from No.1 Squadron set off from Laverton on a northerly training flight to Queensland and back. In mid-afternoon of 5 December, on the last stage of the 3200 kilometre flight, the engine of one of the eight remaining aircraft began overheating shortly after take-off from Cootamundra (NSW). The pilot, 20-year-old Pilot Officer Jack Fallon, attempted a forced landing in a paddock alongside the Ironbong Road about 16 kilometres south-west of the town. After making a normal approach, the pilot failed to round out his glide, and the aircraft was seen to strike the ground with its airscrew and wheels, skid along the ground for some 30 metres, then flip onto its back and instantly burst into flames. The airman in the rear cockpit, LAC Henry Fitzgerald, managed to kick out a side panel and escaped with non-fatal burns, but Fallon was pinned unconscious beneath the wreckage and burned to death.[118]

Authorities immediately moved to institute the usual procedures following any serious accident: an RAAF court of inquiry was assembled at Cootamundra on 6 December and a separate inquiry by the Air Accidents Investigation Committee was also initiated. The work of both these bodies was in addition to the coronial inquiry which would normally be held. But from the first it was apparent that these steps might not be adequate to deal with the blaze of publicity which followed. Newspaper reports were quick to point out that the crash which resulted in Fallon's death was one of seven mishaps which had plagued the training flight by No.1 Squadron. In addition there had been an accident involving a Demon at Canberra a week before the squadron flight began, and—making matters worse—on 7 December another Demon was extensively damaged during a forced landing near Gosford, New South Wales. In total there had been nine incidents within the space of three weeks.

Questions were asked in Parliament about the airworthiness of Demon machines and the desirability of suspending their operations, and whether RAAF pilots possessed the experience needed to fly them safely. When the Minister for Defence, Victor Thorby, responded that yet another inquiry was being launched by a special technical committee of three senior air force officers to investigate the problems experienced by No.1 Squadron, it was argued that this body should include members from outside the RAAF to allay the fears of parents of pilots as well as the public. One member of Parliament went so far as to suggest that the special committee might 'very well consider whether pilots are receiving sufficient training, and may even consider the activities of the heads of the Air Force themselves'.[119]

Although most of the press reports of the air accidents issue were blatantly sensationalist—so much so that one member of Parliament suggested to Thorby that he should appeal to the press to keep their treatment 'in something approaching proportion to that given to accidents associated with other forms of transport'[120]—the pressure on the government continued to build. Fuelled by calls from the dead pilot's father, C.G. ('Clarrie') Fallon, secretary of the 50 000-strong Queensland branch of the Australian Worker's Union, there were insistent demands for a public inquiry. The question was debated in Parliament as a matter of urgent public importance on 8 December, but defeated along party lines.[121]

This was not the end of the matter, however, as Fallon senior was adamant that his son had been required to operate a machine fitted with tyres worn smooth, maintaining that it was the bursting of one of these on landing that had resulted in his son's death.[122] As CAS at the time (Williams) later pointed out, this was a misguided accusation for the reason that tyres without tread were invariably used on RAAF aircraft and came that way from the manufacturer.[123] Nonetheless, in January 1938 Fallon senior achieved the unanimous passage of a motion at the annual delegate meeting of his union, expressing alarm at the 'hopeless inefficiency and inadequacy' of the RAAF and calling for an open inquiry into the whole organisation of that service, not just its safety record.[124]

Investigators examine the lonely paddock near Cootamundra (NSW) where Demon A1–33 crashed on 5 December 1937, with the resultant death of Pilot Officer J.F. Fallon. This accident triggered a furore regarding RAAF safety which produced the Ellington Report and the eventual removal from office of the long-serving CAS, Air Vice-Marshal R. Williams. (*AA, CRS A705, item 32/10/1918*)

By February the results of the three inquiries were all in. Neither the Air Accidents Investigation Committee nor the special RAAF technical committee had been able to detect any common underlying causes for the 'unprecedented' series of mishaps the previous year. The problems themselves had not been all of a kind, and there was no reason to suspect anything was fundamentally wrong with the RAAF's Demon machines, which were all relatively new (having entered service in Australia only from 1935) and of a type in successful operation with the RAF and elsewhere. It was also pointed out that all the No.1 Squadron pilots involved had been trained to a sufficient standard to undertake a long distance flight of the kind attempted; even Pilot Officer Fallon, although having finished his flying course less than six months before his death, had flying experience totalling 260 hours—half of which were on Demons. When No.3 Squadron had made a similar northerly flight over much the same route a few weeks earlier, using Demon aircraft and pilots at the same standard of training, there had not been a single incident of any kind.[125]

In specific regard to the Cootamundra crash, the RAAF court of inquiry had found no direct evidence of why the engine coolant had boiled, causing the overheating which prompted Pilot Officer Fallon to make a forced landing, but gave as its opinion that it may have been due to incorrect use of the radiator by the pilot. The Demon was fitted with a retractable radiator which had to be manually lowered into the air flow to achieve cooling, and it was considered that Fallon had forgotten to do this.[126] Similar evidence emerged at the coronial inquiry held at Cootamundra on 15 February. Here it was heard that as Fallon's Demon neared the ground, the pilot had told his passenger through the communication tube connecting the two cockpits that he was 'just about out to it', indicating that he was physically disabled at this critical stage of the landing sequence, probably as a result of being overcome by heat. In his findings, however, the coroner declined to point to this as the cause of the crash; on the evidence available it was impossible to say that the pilot had been to blame, but there was also no evidence to show that air force authorities had in any way contributed to the accident. According to the coroner, the crash had been just 'one of those unaccountable accidents that occur at times'.[127]

Despite the fact that exhaustive enquiry had been carried out into both the Cootamundra crash and the wider phenomenon of so many accidents occurring around the same time, the government still found itself under harassing fire in Parliament from the opposition Labor Party. Questions continued to be asked about the RAAF's accident rate, which undoubtedly showed a disturbing upwards trend. As revealed by figures tabled by Defence Minister Thorby in response to one questioner, whereas Air Force deaths for the five years since 1932 had been between one and three annually (except in 1934 when there were none), there had been seven fatalities in 1937. And another five deaths were recorded in the first five months of 1938 alone![128]

As recounted in earlier chapters, it was in response to the pressure over this issue that the government finally decided to invite Marshal of the RAF Sir Edward Ellington to Australia. When Ellington arrived in Melbourne on 16 June 1938, he was handed a letter from Prime Minister Lyons outlining the government's purpose in seeking his advice. This asked him to report on the existing organisation and the lines of the RAAF's proposed expansion program, and also sought comment on the efficiency of the existing force. The letter particularly drew attention to the air accidents issue, noting the parliamentary and press criticism attached to this aspect; the government, he was told, 'would appreciate your candid opinion on a matter which is of a highly technical nature'.[129]

The report Ellington produced demonstrates that it is unfair to imagine that he saw his function as constrained by the emphasis given to this one aspect of a much wider brief, but the expectation of his report held by ministers seems unmistakable. Ellington submitted his report on 16 July, shortly before he departed for New Zealand. This contained findings on the cause of the series of accidents—the segment described by one historian as being 'for all directly concerned the crucial part'—which outraged the Air Board. Using statistics, Ellington concluded that the RAAF's accident rate 'is definitely worse than in the United Kingdom', even allowing for fluctuations which had to be expected in a small Air Force. He considered that between a quarter and a third of crashes had been due to disobedience or bad flying discipline, and pointed to a need for more extensive training of pilots after they completed their initial flying course. More than this, however, he had concluded that a significant factor in the problem had been a failure of those in authority (at what level was not specified) to ensure 'a strict enforcement of the regulations'.[130] When the report was publicly released on 31 August, the country's principal newspapers uniformly condemned the RAAF's top leadership and suggested the need for a change.

In a response dated 21 September, the Air Board challenged the methodology employed by Ellington to arrive at his conclusions, particularly his use of statistics to effect a comparison when the situations of the RAF and the RAAF were not parallel. The answer to the Australian accident rate, the Board claimed, lay in the greatly increased number of hours being flown with expansion, and the enormous proportion of RAAF pilots who, being products of that expansion, had less than two years flying experience—the group acknowledged by Ellington to be most prone to accidents.[131]

As has been fairly pointed out in one published account of these events, the tone of the Air Board's remarks was indicative that the Ellington report had become 'a vehicle more for the conflict of opinion between the Government and the board than between the board and Ellington'.[132] More than this, however, the report also brought out into the open the longstanding antagonism between the senior air members of the Board, Williams and Goble. At this crucial juncture in affairs Goble took steps to distance himself from any responsibility for the Board's perceived failings. In a letter to the Minister (Street) he pointed out that, although training and discipline matters ordinarily came under him as AMP, he had actually been in England with the RAF for the two years prior to January 1938. Moreover, CAS had made himself responsible for operational training in 1934. Finally, he recounted that, on his return to Australia, he had objected to the RAAF's extensive commitment to public air displays on the grounds that 'practically the whole of the energies of the Service were being devoted to practising for them, at the expense of what I considered to be more important operational training'. His representations to Williams on these matters had been, however, 'without success'.[133]

The consequences of Goble's action became evident in December, when Street placed before Cabinet a memorandum on the whole affair which formed the basis of the prime minister's press statement the next month announcing that Williams was to be exchanged to the RAF for two years (see chapters 3 and 4). Street's memorandum showed that Goble's letter provided him with all the evidence he needed for concluding that 'the Air Board has not been functioning in a fully satisfactory and co-operative manner' and that 'a better standard of higher direction would reflect itself in the Air Force generally'. In view of the transfer of duties, and AMP's absence overseas for much of the crucial period, the Minister declared that 'it would appear that the main responsibility should be placed upon the Chief of the Air Staff'.[134]

In the context of the safety issue, what mattered was that Ellington had—however clumsily—put his finger on a valid deficiency which appeared to have become buried in the public brawl between the Air Board and the government. In its counter-report the Board emphasised the shortness of Ellington's visit as evidence of the superficiality of many of the marshal's findings. The Board had agreed with the criticism that training in service squadrons could be improved, but also pointed to the difficulty in securing equipment ordered from England. Doubtless there was merit to the Board's argument on both aspects, but it was far from providing the whole truth.

The fact was that the procedures used to convert RAAF pilots to the new Hawker Demons when these began entering squadron service left a lot to be desired. One pilot states that his conversion training consisted of 35 minutes as a passenger in the rear cockpit while his officer commanding took him on a flight—and there were no dual controls available—after which he 'just got in and flew off'.[135] Probably many people would have seen such practices as inadequate had they become public knowledge at the time. Given such introductions to new aircraft, Pilot Officer Fallon's alleged oversight with his radiator becomes a rather more understandable occurrence. It may have been that RAF procedures were no better in this regard—the point was never addressed in the ensuing debate—but the point made by Ellington regarding the need for more extensive training of pilots after they completed their initial flying course appears to have been perfectly reasonable.

Adequacy of training

In his autobiography Williams complained at his removal from the CAS post in February 1939 as a direct consequence of the ructions triggered by Ellington, alleging that he had simply been made a scapegoat for events which were never within his power to remedy. 'There was no change whatever in either the organisation, administration, distribution or training of the Service as a result of the Ellington report', he complained[136]—all that had happened was that he was pushed out of the CAS post to go overseas. Williams was not entirely justified in summarising fall-out from the Fallon affair in this way, just because there were no sweeping changes. In fact, the government now clearly suspected that a major cause behind the number of accidents being experienced in the RAAF was the standard of training young pilots received. Accordingly, the situation with regard to RAAF training began to be closely monitored, particularly when a disturbing spate of accidents now began with Avro Anson aircraft.

The first serious mishap to an Anson occurred at Green Hills, near Liverpool (NSW), on 29 May 1938 when A4–27 from No.22 Squadron crashed after completing bombing practice and its three-man crew were killed. A year later the AAIC was still unable to assign a specific cause to this accident.[137] By this time a second serious mishap now needed investigation. On 28 April 1939, A4–32 on a navigation training flight from Richmond crashed at Riverstone, just a few minutes short of its return, killing all four men on board. Weather at the time was fine, but the evidence of witnesses indicated that some problem appeared to have been experienced with the engines prior to the plane entering a dive. Although the engines revived and the machine seemed to recover, it had then suddenly gone into a steeper dive and crashed.

A RAAF court of inquiry was appointed to investigate the causes of this accident, comprising Wing Commander Joe Hewitt (president), Squadron Leader Harold Seekamp, and a flight lieutenant who was an experienced Anson pilot. As Hewitt records, however, this committee found itself under considerable pressure:

> There was a public outcry in the Press and Parliament ... We were urged by RAAF Headquarters to expedite proceedings and our findings. Such a Court of Inquiry requires time and concentration on the problem, but the urgings of Cabinet on the Air Board to determine the cause was passed down the line to me ultimately by telephone calls by the Air Member for Personnel.

The court completed its report on 2 May, concluding that during the Anson's landing approach some member of the crew had operated the handle trimming the elevators, which required 'a light and cautious hand', having mistaken it for the handle that lowered the wheels and which needed much vigorous winding.[138]

On 2 May 1939 (the same day that Hewitt's panel handed in its findings on the Riverstone crash), another Anson (A4–11) went down in Port Phillip Bay with the loss of a further three lives. This aircraft was operating in company with A4–45 on a local training flight to give No.21 Squadron personnel practice in radio communications and lamp signalling. When a thick fog suddenly descended, the planes broke formation. The pilot of A4–45 safely made his way back to Point Cook and landed, but the crew of A4–11 apparently became disoriented; they were attempting to take bearings when they hit the water about two miles south-west of Point Ormond at Elwood, mid-way between Gellibrand and Brighton. Killed were Pilot Officer Ronald Davis, Corporal Gordon Peake, and AC1 John Quinn, a star rugby footballer from Newcastle.[139]

As was the case after the Fallon crash, RAAF accidents became a hot political issue. Adding to press interest in such incidents was the fact that a journalist attempting to photograph the Riverstone crashsite had been manhandled by RAAF guards at the scene.[140] The Minister for Defence (Street) now ordered an inquiry into the efficiency of Anson aircraft and also the training of personnel, but rejected suggestions that there might be 'any weakness or lack of airworthiness in the type'. On 4 May, however, 30 Ansons not fitted with blind flying instruments were grounded until these were installed. The next day Prime Minister Robert Menzies indicated that the government was considering establishing a permanent court of inquiry to investigate the causes of such mishaps.[141]

On 19 May the government announced that the circumstances of the accident to A4–11 would be investigated by Justice Harold Piper of the Commonwealth Arbitration Court, assisted by Group Captain Henry Wrigley and Squadron Leader Allen Walters as expert assessors. Judge Piper's inquiry opened at Melbourne on 25 May. Included in his terms of reference was a requirement to consider 'whether the syllabus and training of pilots for the RAAF are of such a standard as to ensure that the trainees are reasonably capable of carrying out the duties and responsibilities which they are required to perform'.[142] Such was the sensitivity of the issues associated with Air Force accidents that the previous RAAF findings into the Riverstone crash were now also referred to Justice Piper and his two assessors for reinvestigation, with the same terms of reference as before.

In a finding released on 14 July, the judge agreed with Hewitt's panel that the cause of the accident to A4–32 was probably human error, involving the incorrect setting of elevator tabs.[143] He considered, however, that this had probably occurred after the plane's engines suddenly cut out through lack of fuel, due to a failure to switch petrol tanks earlier in the flight. As the aircraft entered a dive, the crew had identified the problem and restored petrol flow, causing a surge in the engines which would have normally required an adjustment to the setting of the elevator tabs. In the stress of the moment, the Judge considered that a crewman other than the pilot may have over-reacted and turned the elevator handle too far.

So far as the question of adequate training might have been a cause, Justice Piper noted in a full report dated 2 October that the RAAF adhered closely to an RAF training syllabus: 'It appears to be very comprehensive and I have no reason to doubt that it is of a high standard'. He could find no evidence that training in the particular aspect thought to have been behind the crash was in any way deficient, but did consider that pilots needed more practical experience rather than oral instruction in how to cope with an unexpected cut-out of engines due to a failure to switch fuel tanks.[144]

Although the Anson accidents of 1939 and the subsequent exertions of Judge Piper clearly showed that the rapidly expanding RAAF continued to have its problems in training pilots to full proficiency, it was interesting that the revelations of shortcomings had none of the administrative consequences which had attended the Ellington report a year earlier. The new CAS, Goble, was not held accountable in the way that Williams was, nor was there the level of recrimination which brought about the latter's downfall. The fact was that by mid-1939 it was realised that the world was not far away from war.

12

Problems of administration

Effective administration plays an essential role in the smooth running of any organisation, and for this reason the subject of administration deserves attention along with other, more exciting, aspects of the Air Force story up to 1939. The RAAF was notably weak in some areas of its administration, but—as is shown here—this was a matter frequently beyond the control of the uniformed officers with authority over the service. To a large extent this situation reflected the circumstances in which the RAAF was born and which saddled it with arrangements less than conducive to an orderly conduct of affairs. A factor of considerable importance here was the determination of the other two established services to maintain a degree of control over the third member of the defence triad and ensure that it remained, at best, a junior partner (see chapter 3).

Lack of legislative provision

A deficiency in the administrative arrangements made for the RAAF, important because of its effect on the Air Force's status as an co-equal third service, was the absence of proper parliamentary provision of the sort that covered the naval and military forces. As appropriate legislation to create an air force had yet to come before Parliament when the decision was made to bring the new service into being from 31 March 1921, the Australian Air Force was initially constituted by a proclamation issued under the *Defence Act 1903–1918*. Under this arrangement the Air Force was technically part of the military forces.[1]

Even before the new service was brought into existence by this means, it had become apparent that such an arrangement was unsatisfactory. Advice received from the attorney-general in March made it clear to the Air Board that although the Air Force could be legally formed as part of the naval or military forces, and other ranks enlisted, 'Section 148 of the Defence Act, however, limits the granting of permanent commissions, with certain exceptions, to those who are already permanent officers or graduates from the Royal Military College'. As Williams realised, this meant that:

> officers cannot legally be appointed even temporarily to the Permanent Forces ... The only legal way of obtaining the services of the required officers is to appoint them to the Citizen Forces and call them up for duty pending the passing of ... [an] Air Defence Act. This can only be done owing to a state of war which may cease before the passing of the Air Defence Act if that Act is delayed for any reason.[2]

An Air Defence Bill was presented to Parliament almost immediately but failed to gain passage. Introduced in the Senate on 8 April, this Bill followed the pattern of the Naval Act because of the intrinsic mobility of the force it covered and, in addition to the general provisions of the Defence Act, sought to apply the Imperial Air Force Act for disciplinary purposes both in peace and war, subject to any modifications enacted in Australia. Debate on the Bill, however, came straight after debate in the Senate in April on an amending Bill to the Defence Act regarding discipline in the Australian forces. This prior debate had revealed the existence of a strong belief that the discipline required for Australian servicemen was very different to that imposed on Britain's own forces, and a feeling that imperial Acts were too harsh and severe in the penalties they prescribed. Accordingly, the Air Defence Bill was modified to provide for the Imperial Air Force Act to apply only in time of war. In this form the Bill was passed by the Senate but it had received only its first reading in the House of Representatives when Parliament was prorogued in May, and the Bill thus lapsed along with the rest of the government's unfinished legislation. For the next two years, therefore, the RAAF remained constituted only under the general terms of the Defence Act.[3]

On 29 June 1923 the Bruce government presented the Air Defence Bill to Parliament again, this time to even stronger objections regarding the application of imperial legislation. Speakers from the Labor Party benches, including James Scullin, opposed tying the Australian definition of 'active service' to that which might be embodied from time to time in the relevant British Act, and decried the creation of a situation where a Commonwealth Act could be automatically changed by legislation enacted in London.[4] In the face of such criticism, the government withdrew the Air Defence Bill and on 24 August presented in its stead an Air Force Bill, which passed all stages and received the Governor-General's assent on 1 September.[5]

Because the *Air Force Act 1923* was only intended to be a stop-gap measure until a more comprehensive Bill could be introduced, expected to have been during the 1924 session of Parliament,[6] it was a very brief piece of legislation. The new Act did nothing more than constitute the RAAF as an autonomous arm of the defence forces and provide for the Air Force to be governed by the Defence Act and regulations made under the Air Force Act itself. An amendment provided that the Army Act of the British parliament did not apply to the RAAF.[7]

Another version of the Air Defence Bill was, however, never submitted, and the Air Force Act was therefore the sole legislation specifically dealing with the RAAF until after the start of the Second World War. As Air Vice-Marshal Geoff Hartnell observed in his post-retirement study of problems of command in the Australian defence structure:

> The greater part of RAAF law was never covered by legislation but only by regulation until after the outbreak of war in 1939. It was ironical that the only guidance available to the makers of the regulations was the Imperial Air Force Act which, therefore, provided the basis of the regulations despite Parliamentary objection to it.[8]

The Air Board was increasingly unhappy with this situation, and in 1938 again began wrestling with the problem that all other air forces of the British Commonwealth used the legal code laid down by the Imperial Air Force Act.[9] Recalls Wing Commander Frederick Knight, a Melbourne barrister who served with the CAF as the RAAF's legal officer from 1926:

> For thirteen years I had been complaining that the law under which we were working was inadequate and unsatisfactory, and that it would break down in time of war. I had urged that all members of the Air Force should be bound to serve anywhere and that the Imperial Air Force Act should be applied. The Air Board was sympathetic, and so I believe was the Attorney-General's Department. But the question was a political 'hot potato' which no government would touch in time of peace.[10]

With the start of the Second World War in 1939 Parliament finally passed a special Air Force Act which *did* apply the Imperial Air Force Act to the RAAF, subject to certain Australian amendments.[11]

Underscoring the inadequacy of legislative cover for the Air Force was the suggestion made in 1927 that the Australian Constitution needed amendment to recognise the Commonwealth's power to legislate on air defence and aviation matters generally. A Royal Commission had been appointed to review the Constitution and report on changes considered necessary. In response to a general request for suggestions, the Air Board pointed out on 7 September that there was no mention at all of air forces or air defence in the *Commonwealth of Australia Constitution Act 1900*, for the simple reason that the development of aircraft had not been anticipated by the document's framers. While legal opinion appeared to confirm that the Federal Parliament already had power to make laws providing for air defence, in the same way as it was specifically empowered in the military and naval spheres, the Board urged that the present review represented a suitable occasion for the Defence Department to propose changes to sections 51 and 114 that removed any possibility of challenge.[12] In view of the inherent difficulty of altering the Constitution by the required means of referendum, the suggestion was not taken up—with the result that this fundamental Act still fails to acknowledge the existence of the Air Force to this day.

An underlying problem which flowed from the lack of legislative provision was essentially one of authority. AVM Hartnell points out that Chief of the Air Staff was an appointment 'bearing precisely the same title as existed in the Air Ministry in London from which it was obviously derived'. However, whereas the British CAS was a member of a statutory authority called the Air Council which, as a corporate body, was charged with the command and administration of the Royal Air Force, the CAS in Australia was—after the formal abolition of the local Air Council in 1928—a member of a statutory body (the Air Board) which was not charged with any authority under the Air Force Act: 'In essence it was an advisory body to the Minister'.[13]

Because the Air Board lacked any formal command authority, control of the Air Force effectively rested with the officer who was the most senior in rank. From the moment of the RAAF's creation, this was invariably Richard Williams. Hartnell accordingly holds that Williams as CAS was 'the only person who could lawfully execute control decisions taken at the highest levels', and was in effect Air Officer Commanding the RAAF. But Williams was usually not consulted during the decision making process and was regarded as subordinate in matters of control. As well as the Air Board, RAAF Headquarters was also left with an undefined status. As Hartnell notes: 'The best that could be said of it was that it was the headquarters of whatever officer happened to be holding the appointment of CAS'.[14]

Status of the Air Board and CAS

Quite apart from Hartnell's observations regarding the Air Board's standing, there were other factors which reflected on its status and authority to run the Air Force efficiently. As a rule, these factors had little bearing on the Board's capacity to rule its own house. But when it came to representing Air Force interests in dealings with the other services, the Board always found itself in a position of inferior strength.

As outlined in chapter 1, the first Air Board constituted in the Defence Department actually had no administrative function to perform at all. Formed in January 1920, it was tasked to consider an air policy for Australia in conjunction with requests put to Army and Navy to prepare separate proposals for the defence of the country. When the second Air Board came into being later that same year, however, it was specifically intended to be the counterpart of the naval and military boards which ran the two established services.[15] There was, however, a significant difference in the position of the Air Board and the other boards, for the Air Board had a superior body placed over it in the form of an Air Council.

The Air Council, formed at the same time as the Air Board, consisted of the Minister for Defence, the CNS, the CGS, two members of the Air Board, and the CCA, and was responsible for deciding all matters of air policy.[16] A 1920 document entitled 'Appreciation of the Position Occupied by the Air Board, in the General State of Administrative Arrangements' made clear the relationship meant to apply between the two bodies, as the Air Board was 'charged with the control and administration of the Air Force upon the policy laid down by the Air Council, and is responsible to and subject to the control of the Air Council'.[17]

The reason for such an arrangement was to be seen most directly in the junior ranks and young ages of the men who were the new service's most senior officers. At the time the Air Board was constituted, Williams was a 30-year-old major (honorary lieutenant-colonel) while Goble was a squadron leader (honorary wing commander) aged a year younger. The other uniformed member of the Air Board, Captain Percival McBain, was eight months younger than Williams. When warning against forming an independent air force in October 1918, the then CGS, Major-General Legge, had pointed to the fact that the wartime AFC was composed solely of young men with practically no experience in peace-time administration and command, and had stressed that for some time such men would need direction from those who knew what was meant by expenditure under a responsible government.[18] Subordinating the Air Board to the authority of the Air Council was doubtless expected to provide this direction, but it was a device which also ensured that the Army and Navy had a strong say in Air Force matters—not necessarily to the RAAF's benefit.

Williams' youthfulness as the head of his service was most evident in forums such as the Air Council, and the advisory Council of Defence where he found himself among generals and admirals aged in their 50s and 60s. Not that he allowed himself to be overawed at such gatherings. Williams recounted his being approached by the federal Treasurer, Mr Edward Theodore, after one Council of Defence meeting in 1929 at which he, Williams, had successfully deflected an attack on the RAAF's existence. Having answered all the arguments put by Generals Sir John Monash and Sir Harry Chauvel, Lieutenant-General Sir Brudenell White, and Rear-Admiral Munro Kerr, Williams was told by Theodore that he could 'pick a fight in that company at any time'.[19]

Williams' very capacity to stand up for himself told against him, however, and there is clear evidence that he was regarded as something of a brash upstart. That General White had little time for him is evident from a letter Williams wrote to the then-CGS while attending the British Army staff college at Camberley in 1923. Reporting that on arrival he had been given a briefing by the adjutant which included cautionary advice not to try 'teaching the instructors', Williams wondered what was behind this and said he imagined that either an Australian or an Air Force officer had at some time endeavoured to push his own views. White noted laconically in the margin opposite: 'Rather do I think that they have heard of Williams'![20]

To a large extent, these were problems which time would steadily overcome. In fact, the Air Council became increasingly redundant after its first two years. After meeting on 19 days in 1921 and eight in 1922, it met only twice in 1923, not at all in 1924, and twice in 1925; there were no meetings after 2 April 1925.[21] Williams made it a prime objective to see the corpse buried. As he told Sir Hugh Trenchard in a letter dated 11 August 1926, he considered that the Council 'was of use when the Air Force was first formed but, in my opinion, has now served its purpose'. More than this, though, he disliked the fact that 'as constituted it had a good deal of administrative responsibility which was detrimental to efficiency' and he wanted to get the Air Board on a similar footing to its naval and military counterparts.[22]

Attainment of this object was by no means smooth. The CNS, Rear-Admiral Hall-Thompson, in particular was determined to remind the Air Board of its subservience to the Air Council and insist that this situation should continue. In January 1925 he had raised the point that, under the original regulations constituting it, the Council was to include two members of the Air Board—one nominated by CNS and the other by CGS. In March 1926 Hall-Thompson

This photograph of the staff of RAAF Headquarters at Victoria Barracks, Melbourne, shows all that it took to administer Australia's air force in 1930. One wonders what a photograph of the RAAF's central staff today would look like, or whether a single photograph would even be practicable. (*Miss G.E. Joyce*)

wrote to the Minister drawing attention to his previous correspondence on this point, maintaining that despite the existing regulations making 'the Air Board subsidiary to the Air Council'—purposely, as he understood matters, to prevent the RAAF from becoming independent of Army and Navy control—it now appeared to him 'that a drift in the direction of complete separation has started'. The Air Board had not been submitting all matters of principle for the Council's decision for a very long time, he complained, and he wanted to know whether it was planned to alter the present regulations accordingly. He considered they should stand 'and that it is desirable to carry them out rather than to allow the matter to go by default, and the present move of the Air Force towards independence (actual even if not nominal) to continue'. As it was, he argued, both CGS and himself were 'saddled by regulation with responsibilities over which we have actually no control'. Sent a copy of this letter by Hall-Thompson for comment, Chauvel had replied as CGS 'My views coincide entirely with yours'.[23]

Williams had countered this challenge by pointing to the outcome of this same debate in England (see chapter 3). He also argued the extent to which circumstances had changed since the Council was formed, principally that the three services were now all under one Minister, and the Air Force was represented on the Defence Council. He proposed forming a committee comprising the three chiefs of staff, an arrangement already operating informally, and doing away with the Air Council entirely. As he went on to recount, the regulations dealing with the appointment of the Air Council were originally brought about:

> . . . purely by reason of the fact that both the Navy and Army wanted a certain flying officer appointed to the Council when it was first formed. The absurdity of the regulation now is shown by the reasoning . . . of the First Naval Member's memo., which in effect would mean that the Air Board shall be appointed by the Chiefs of Staff of the Navy and Army.[24]

Some progress on this front was made when a Defence Standing Committee was established on 15 May 1926. Williams continued to press for abolition of the Air Council, making formal recommendations to this effect in May 1927. The following October the Defence Standing Committee met at the direction of the Minister specifically to consider the question of control of the Air Force. At this meeting CGS and CNS pointed out that, under the existing Statutory Rules, they shared responsibility for the administration of the Air Force but, as the Air Council never met, they had no voice in respect of such questions. After discussion, they finally agreed that there was no objection to abolishing the Air Council, provided its functions were transferred to the Defence Committee, and that body given formal recognition of its existence and role.[25]

On 18 January 1928 Cabinet agreed to the Air Council's abolition. In its place there would now be a Defence Committee, consisting of the three chiefs of staff and a finance member. Whereas the previous Defence Standing Committee had no statutory authority, the new body was created by statutory rule (on 8 March 1929) and charged with the responsibility for advising the Minister on matters of policy generally, and on all matters concerning more than one service. The RAAF was, at last, placed on equal footing with the other services.

Williams' pleasure at seeing this development brought about was matched by other successes during 1928. After finally defeating a Navy bid to split the RAAF by taking from it the naval cooperation role (see chapter 3), he and Goble were also promoted—Goble to group captain rank in April and himself to be air commodore from 1 July. As he told Trenchard, the relatively junior ranks of Air Board members had long placed the Air Force at a considerable disadvantage:

> One would need to be inhuman and without ambition to ignore the personal view when promoted in one's service, but in our position at present, I think that the greatest value is obtained eventually from the result on the Force as a whole of raising the status of its senior officers as compared with the other services.
>
> I cannot but feel some satisfaction in that during the past twelve months, we have succeeded in doing away with the Air Council, and the Fleet Air Arm and in raising the status of the senior officers of the Air Force here.[26]

Although these promotions now gave the Air Board a status approaching that enjoyed by the other boards, it was another six-and-a-half years, until January 1935, before Williams was promoted air vice-marshal and the RAAF finally had the equivalent of an admiral or general in its ranks. At this time Williams also found himself no longer the junior chief of staff, being actually senior to the CGS though on a rate of pay only half that of the CNS and three-quarters that of the CGS.[27] Nonetheless it could be said that by then, fourteen years after the RAAF's formation, the Air Force had shaken off much of its former status as a perpetual Cinderella among the services.

In one sense it was arguable that the initial youth of the RAAF's hierarchy was far from being a wholly negative factor. Williams later came to see the length of time he was thus able to spend in the CAS office as useful to the Air Force. He argued that this not only ensured a continuity of experience at this level, in contrast to the succession of chiefs in the other services, but—as he told serving officers in the last years of his life—enabled the RAAF to match the long tenures enjoyed by senior civilian officers of the Defence Department. This was an advantage, he claimed, not found in the post-war RAAF because his successors mostly had only three or four years in the post before retirement.[28] The reverse side to this argument was, of course, that any shortcomings of its senior administrators remained with the Air Force for a long time. Williams certainly had his limitations, as noted elsewhere in this study, and so, too, did the two next most senior officers who sat longest with him on the Air Board.

Goble had shone in his pre-1914 career with the Victorian railways, which included a three year period at head office engaged on administrative duties,[29] and by all accounts he was amiable and capable in carrying out his Air Force duties as Chief of the Administrative Staff and later AMP. But he strongly disliked the environment in which he had to operate at RAAF Headquarters. Family members recall him stating his belief that he had been wrong to return

The Air Board pictured in November 1930. *At rear, from left*: Major P.E. Coleman (Secretary), Mr A.C. Joyce (Finance Member); *front row:* GpCapt S.J. Goble (Air Member for Personnel), ACdre R. Williams (Chief of the Air Staff), and WgCdr W.H. Anderson (Air Member for Supply). Because of their relative youth, shortcomings in the administrative ability of the RAAF's three most senior officers remained with the Air Force for a long time. (*Miss G.E. Joyce*)

to Australia in 1918 instead of continuing with the RAF 'because from the time of his arrival, as a Wing Commander, he was involved in interaction with civil servants and politicians'.[30] His distaste for this aspect of his role, a crucial one for an officer in his position if he was to function successfully in his environment, remained with him even when appointed to replace Williams as CAS in 1939. One young officer at that time who, with some others, offered Goble congratulations on his appointment at an informal occasion, still recalls the shock of the reply received: 'he rather shook us by saying that he . . . found the politicians far too strong for him, and that he was going to resign' (see chapter 15).[31]

William Anderson, the RAAF's third most senior officer, several times acted in Goble's stead on the Air Board in the early 1920s and in 1929 joined the Board as AMS, a post he held (with some interruptions) until 1940 when he was appointed acting CAS on Goble's resignation. He had a creditable record from the First World War as a gallant and skilful pilot,[32] and seems to have been generally liked despite a personal manner which often made him an object of fun. He was a bachelor who reportedly shunned women other than his sister, a spinster with whom he resided. This aversion to female company did not go unnoticed, and one officer recalls that at social functions some Air Force wives used to try to bail him up, obliging him to retreat to a safe haven.[33] Another recalls:

> He was walking up the Domain Road [in Melbourne] on one occasion when coming the other way were three Air Force wives who'd just been to a meeting about starting a fund. They spotted him and thought to ask him about it, but he saw them and shot into the nearest shop—which happened to be a baby linen place. They thought that was terribly funny and more or less picketed the shop till he came out.[34]

A more important issue, of course, was Anderson's grasp of the supply function for which he held responsibility in the Air Force. One of his subordinates who had the task from 1934 of developing a 'vocabulary' of stores and spare parts, along with tables of establishment, recalls:

> He was not quite on the same wave length as others and he didn't really realise what was going on in my work. He used to send for me every Monday morning and ask how the vocabulary was getting on. Well, actually they're still working on it to this present day. He didn't understand that it was a never-ending thing.[35]

No less important than any individual inadequacies of Air Board members was the working relationship that existed between them in carrying out the Board's functions. Here, of course, the rivalry noted in previous chapters between the senior members, Williams and Goble, exercised a powerful and detrimental influence on Air Force affairs. The image of the RAAF's administration was not helped by widespread knowledge that these two officers 'could not work in the same headquarters', necessitating the alleged practice of keeping one or other of them out of the country at any given time.[36] Unfortunately for the RAAF, their inability to work effectively did not lessen with time, notwithstanding the fact that Williams enjoyed steady success in outstripping Goble in this contest and could have afforded to let up.

Matters actually worsened after the RAAF's inauguration, following changes to the Board's composition in October 1922 which removed the Director of Equipment from membership.[37] The Parliamentary Joint Committee of Public Accounts noted in June 1923 that this was a decision—made for reasons of economy—in which 'more has been lost than gained by the change', but it was thinking only of equipment issues.[38] The fact was that the new titles which accompanied the restructuring of the Board, with Williams becoming Chief of the Air Staff (CAS) and Goble made Chief of the Administrative Staff, still gave the Air Force the appearance of having two chiefs. What was more, with membership reduced to three, only the Finance Member (a public servant) separated them from deadlock at sessions of the Board. Only when Goble ceased to be known as Chief of the Administrative Staff from December 1927, and an Air Member for Supply (AMS) was added to the Board's membership two years later, could it be said that some of the heat had dissipated from this feud. By then the perception had become entrenched that Goble was, as one member of the period says, 'always a shadow of Dicky, never a strength'.[39]

Personnel problems

The shortage of capable and experienced administrators was a deficiency initially felt throughout the Air Force, not just at the Air Board level. The problem which had arisen in recruiting officers for the RAAF was precisely that foreseen by General Legge in 1918: although there was a wealth of talent available among former members of the wartime air services, their experience was almost exclusively on the flying side. It had soon become apparent that brave and skilful pilots did not necessarily make good administrators. Many of these characters joined the post-war Air Force to pursue their love of flying and showed scant interest in and little patience for 'paperwork'.

A classic illustration of the problem was the almost legendary figure of 'Jerry' Pentland, who by most accounts was 'a good fighter pilot' but 'an auditor's nightmare'.[40] It is alleged that in 1924 he found one of his flight's four DH9As was missing from the hangar during a stocktake of items on Articles-in-Use (AIU) ledgers. The solution seemed simple to him. He just penned the words 'Written off' across the appropriate ledger page and signed it, without any regard for the procedures normally entailed in striking off a major item of government equipment like an aircraft, such as a board of survey, or even troubling to obtain the approval of his commanding officer. As it happened, the aircraft had only been sent to the workshops by Pentland's flight sergeant who had earlier discovered a fault needing attention![41] The thrust of this story is confirmed by one who saw the actual ledger concerned some years after Pentland had quit the Air Force and was engaged in civil flying in New Guinea.[42]

The shortcomings of the service under such management were only too apparent to Williams as CAS, and he was determined to tackle the problem. He did so by two means. First, increasingly tight controls were introduced intended to improve administrative efficiency and curb some of the high-spirited showmanship which distinguished much RAAF flying in the 1920s. This effort had the effect of causing the departure from the service in 1923–24 of some of the pilots comprising the Air Force's original officers. According to Pentland's biographer, it was the fact that Williams was 'beginning to run a tight ship', combined with cut-backs in areas such as flying hours allowed to pilots, which got 'under the skin' of Pentland, and others, and prompted them to leave.[43]

The loss of many of these officers can only have been one borne willingly by Williams, given the defiant mood in which some of them took their farewell. A case in point was the dazzling display of aerobatics given by Vigers on the day of his discharge in 1923, later described by Flying Officers Ulex Ewart and Francis Bladin.[44] Ewart recalls that the whole demonstration was performed by Vigers with a pipe clenched firmly between his teeth, although smoking in aircraft was strictly forbidden: 'And was he charged with disobeying regulations? Certainly not. Of what use would that be on this his last day in the RAAF, and of what merit after such a spectacular display.' Vigers went to QANTAS, where he became, in the words of Hudson Fysh, 'perhaps the most eccentric and remote chap we ever had'.[45]

The second means used by Williams to deal with the slack approach towards administration adopted by many of his officers was to obtain the services of professionally trained junior officers from the other two services (see chapter 7), beginning with five officers from the Army and one from the Navy in 1923. These were joined by a further three Army officers the following year, and no fewer than five Army and three Navy officers the next.

The impact of these first imports was evidently all that Williams could have hoped it would be. On completion of flying training, Ulex Ewart found himself made assistant adjutant at Point Cook for a time before being posted to Headquarters as Director of Personnel Services. He recalls that the officer he replaced 'hadn't a clue about handling personnel matters. I was astonished when I went into his office to find him making a most beautiful rug'.[46] Others found similar opportunities for making their presence felt. Flight Lieutenant Eric Woodward made a considerable name for himself as adjutant at Point Cook, reportedly earning the nickname of 'Hackenschmidt' (after a wrestler). An ex-RMC officer who followed Woodward into the Air Force recalls that 'he was really tough, no holds barred. Boy, he really straightened the place up'.[47]

Predictably, there was a reaction against Williams' policy from serving officers who saw their own prospects being effected. One airman of the period frankly declares that the newcomers were 'universally labelled the Duntroon Bastards'.[48] Several of the former Army men who made careers in the Air Force have spoken of the feelings of hostility, of varying intensity, which they encountered, attributing this reaction essentially to jealousy of their greater professional background.[49] Only one indicates there might have been any justification for the reaction they aroused, suggesting that there was a suspicion that 'many of them [Duntrooners] regarded themselves as so valuable as staff officers that they appeared to be reluctant to accept postings which they believed were not appropriate to their training and status'.[50]

Regardless of the unrest caused, Williams' resolve to continue was unshaken. In all, some 46 Army and fifteen naval officers were trained as pilots during the years 1923–32. Those from the Army were by far the most valuable to the Air Force, for the simple reason that a far higher proportion of them accepted commissions with the RAAF at the end of their initial secondment. Only four of the attached Navy officers chose to so transfer, as against 27 from the Army. Of the remaining nineteen Army personnel trained, four had been killed in accidents before completing the secondment period, seven had not completed flying training before being returned to the Army or discharged, and eight elected to return to the Army once their secondment was over.

Some of those returned to the Army went on to achieve prominence in their original service, the most notable being Sir Eric Woodward who became a lieutenant-general and Governor of New South Wales. Others, however, were soon lost to the Army as well as the RAAF. Colin Simpson, a member of the first flying course, having decided that promotion in the Air Force was too stagnant to offer any future, decided that the Army also promised little: 'I gave it away altogether—especially as I wanted to get married and the situation with married quarters was absolutely hopeless'.[51] It was Scherger, the most successful of the seconded Army officers, who observed that many of his contemporaries lacked his passion for flying:

> That was one of the problems with the fellows seconded from the ranks of the Duntroon graduates. Some of them were not keen. They were good planners, good at man-management, but they didn't have that keen edge of keenness—if I could call it that—that [you] wouldn't get them out of an aeroplane.[52]

Whatever the deficiencies of the Duntroon graduates as a group, it was nonetheless notable that the Air Force received a lasting legacy of the secondment arrangement in the form of a succession of ex-RMC officers—McCauley, Scherger, Hancock and Murdoch—who held the RAAF's top post of CAS in the years 1954–69.

Though the import policy undoubtedly proved of lasting value to the Air Force, the numbers of RMC and RANC graduates involved meant that they were, as one of them observed, no more than 'a leaven of more fully trained officers to complement the war-time ones'.[53] The latter remained, for better or worse, the mainstay of the RAAF well into the 1930s when the effects of major expansion began to be felt. Accordingly, the service had to endure poor standards of administration for quite some time, as illustrated by the recollections of one officer of the period (perhaps significantly an RMC graduate):

> I remember CAS's inspection of Richmond at Christmas 1930. The CO, Harry Cobby, was supposed to sign our logbooks every month, but things were such a shambles that Cobby was running around with a pen signing logbooks against the walls of the Mess at the last minute, saying 'Get your logbooks, quickly, quickly!' 'Dickie' lined the whole station up and said it was the worst he'd ever seen—he said this in front of Cobby and the whole base. He said he'd be back in a week, and unless the place was cleaned up there would be no Christmas leave. I spent the next week driving a grader around, cleaning out the drains. That will give you an idea of the horrible standards that applied.[54]

Accounting and audits

The situation which existed in 1921—with the Air Force in the process of forming at the same time as great quantities of gift aircraft and equipment were arriving from Britain—contained the seeds of an administrative nightmare which was to take years to sort out properly. The only accommodation available to receive these valuable items was in hangars at Point Cook, which naturally were not equipped with bins or other facilities necessary for proper stores management. Other facilities around Melbourne were also utilised for storage purposes (see chapter 5), along with a number of wartime canvas hangars—themselves part of the gift equipment—which were erected at Point Cook to provide shelter.

These arrangements left a lot to be desired. In some instances the accommodation afforded was not even weatherproof, which meant that the equipment stored could not be unpacked but had to remain in the cases in which it was shipped to Australia. According to a brief history of RAAF stores administration written in 1929, the imperial gift equipment was brought on to ledger charge to meet ordinary accounting requirements in the best manner possible. This was

> ... in spite of the difficulties due to the lack of any arrangements for unpacking

> and binning of materials, and the great shortage of personnel who had any acquaintance with the nature of the material and any experience of accounting for stores. The men employed were of course almost entirely returned soldiers who had no better jobs to go to.[55]

Steady improvement was made to this situation over the next few years. The bulk of the gift equipment was transferred from the initial temporary accommodation and sorted into some form of order. As items were progressively put into use more space became available, enabling yet other items of gift equipment to be unpacked from the cases in which they had been deteriorating since their arrival and placed into better storage in temporary bins. Personnel were also trained in provisioning of stores from overseas and local sources, storekeeping methods and in account keeping, so that a stores accounting system similar to that used by the RAF could be introduced. Very little was attempted, though, in the way of auditing of stores accounts, with that little which was done being carried out mainly by officers of the Auditor-General's office.

In 1923 a stocktake was undertaken for the first time. This produced a result which was anything but satisfactory, due in the main to the poor accommodation with which the stores organisation still made do and the inaccurate manner in which accounts had originally been opened in 1920–21. Large discrepancies were revealed which, according to the 1929 history previously cited, 'were of course apparent only'. Not all the losses were on paper, however, and it was discovered that there had been pilfering of the gift equipment cases while these lay in makeshift storage. Among the things found to be missing were the timepieces from the SE5A aircraft, described as 'an indispensable item that was extremely reliable, virtually indestructible, and above all highly prized by souvenir hunters'.[56] Official investigations resulted in only a few of the missing watches being returned.

A great step forward was taken in 1926 with the completion of permanent buildings intended to house No.1 AD at Laverton. The great bulk of materiel not immediately required for use by the training and operational units in Victoria and New South Wales were transferred there, sorted out and properly identified, before being binned and otherwise arranged in an orderly fashion for the first time. At this point, too, accounting was greatly facilitated by the introduction of mechanical ledger-keeping and the gradual centralisation of accounts under a specialist staff. Accounts were reportedly still unsatisfactory, however, because there was scant checking of transactions into and out of store, and between units, and the amount of audit which could be done by inspectors of the Auditor-General's office remained limited.

With the problem of housing and handling of stores about to be remedied, the Air Board prepared to focus attention on improving the aspect of accounting. About 1925, steps were made to appoint an internal checking staff. Francis Hakanson, then an airman in the station accounting section at Point Cook, was selected as a member of a three-man RAAF audit section to be based at Headquarters. He recalls that he and the others were sent into Victoria Barracks for training on the ledger machines which had been bought commercially for the section's use:

> But the public servants got wind of all this, and said it wasn't on because audit was their job. So after three months of waiting to start operation, it was decided not to go ahead with forming the section and the three of us were reposted. Actually it did go ahead, but the civilians took it over.[57]

When formed some 18 months later, the Internal Checking Staff consisted of five public servants. The creation of this staff as an additional element to the system cannot have been universally welcomed, as its questioning of various transactions was a complication which personnel in units must have frequently felt they could do without. The explanations given in response to enquiries by the internal auditors sometimes stretched the bounds of credibility. One of the classics of its kind came from the CO of the Experimental Section at Randwick, Wing Commander Lawrence Wackett. Asked to explain a discrepancy in the quantities of casein cement purchased, used, and still in store, Wackett allegedly replied along the following lines:

> Usually his establishment used imported casein cement, but recently when there was none available and his need was urgent he had placed an order for a hundredweight for local manufacture to his own specifications (which incidentally proved inordinately expensive). The product was not satisfactory. Then . . . an infestation of rats, undoubtedly from Randwick racecourse, had invaded his establishment. To combat the infestation he was finally compelled to import some cats. However, by the time the cats had killed the rats they had devoured a considerable quantity of casein cement and . . . turned what was left into a form of 'phosphatic rock', which his men used as missiles to drive out the cats, which, by this time, had increased in such numbers as to become a nuisance. By the time the cats had been driven out only eight pounds of casein cement was left.[58]

This ingenious explanation was rewarded by being accepted.

Another classic case reportedly arose from the involvement of RAAF aircraft in the survey of the Great Barrier Reef (see chapter 14). As no thought had been given to providing moorings for the Seagull planes before these arrived at Bowen, an improvised arrangement was adopted using 40-gallon petrol drums in which aircraft fuel was received from oil companies. Once empty of petrol, these were resealed so they would float and secured by a length of chain to an anchor on the sea-floor. The problem was, however, that the drums were accountable items. The appropriate vouchers were raised to cover the initial conversion of the drums to use as moorings, and, once the moorings inevitably rusted and sank to the bottom of Bowen harbour, vouchers were used to 'write off' the drums—of which only two were involved, costing a comparatively minor amount. When these transactions came to the attention of the auditors, however, questions were asked as to who had authorised them, and why approval had not been sought from RAAF Headquarters. There followed a lengthy and futile correspondence on all these matters, of which one officer recalls:

> By the time I saw the Bowen files they had grown to two folders, each about an inch-and-a-half thick, with about 10 minute sheets attached to the front covers. The final minute on top was signed by the Chief of the Air Staff, Air Commodore Williams, and read 'This correspondence will now cease'. It put an end to a saga of accounting which to my knowledge will never be equalled.[59]

The appointment of internal audit and stocktaking staff also did not bring to an end the antics of the unscrupulous. Observing that 'pre-war accounting had its moments', Air Commodore 'Paddy' Heffernan recalls that it was the practice at Point Cook to dispose of material which was struck off but could not be burnt or broken up by dumping it into the sea:

> It was placed in the motor boat, taken some 100 yards out into the bay and cast into the waters. Once the splash subsided, the . . . item was officially written off. But . . . if it was obvious to certain airmen that what was dropped into the bay could possibly be of some future use, a fishing line was attached and, as the item descended . . . the line unwound and a cork floated to the surface. A certain airman of Seaplane Squadron, who had been a Navy diver and could hold his breath for FOUR minutes, would after-hours collect a couple of pals, row out to where the cork was, dive down—it was only 30 feet deep—attach a rope to the item and it was hauled up. I personally witnessed the recovery of a brand new Clerget rotary engine in this manner. Why they were ordered to be dumped I'll never know, but I do know that a scrap yard would pay 50 pound for them, thereby improving the said airman's finances.[60]

Apart from the appointment of a checking staff, steps were also taken to improve the vocabulary (that is, the list) of the thousands of items of Air Force stores in use, and the system itself was simplified. In view of the previous inaccurate condition of the accounts, and the impossibility of checking and correcting them back to the time of the previous stocktake owing to the delay in appointing and training the Internal Checking Staff, 'it was decided, with the reluctant approval of the Auditor-General, to commence afresh, in respect of some portion at any rate of the stores',[61] on the basis of a stocktake to be carried out in 1926. When this was duly held, however, its outcome was not satisfactory, principally because it was hurried and not made by an independent board.[62] It was accordingly decided to have another stocktaking in 1928, with a special staff appointed for the purpose.

This new stocktake was actually underway while Sir John Salmond made his inspection of the RAAF in 1928, and he reported that he was satisfied this time it was 'accurate and genuine'. He still found plenty to comment upon which did not please him regarding the working of the directorate of Supply. He did not consider the stores and accounting staff within the directorate yet had 'a firm footing on which to work'; the keeping of ledgers could not even be said to have been satisfactory until September 1927, when the system was instituted of maintaining central ledgers at RAAF Headquarters under a Stores and Accounting Section. Although the work appeared to be efficiently staffed and run, the section was 'working under such difficulties—through various causes—that the well-being of the Service in general is being effected'. He strongly recommended that the whole process be simplified by disposing of the surviving stores from the imperial gift. As this equipment was now almost entirely obsolete or worn out, it was, he said, 'an encumberance, and should be got rid of'. Finally, to ensure that the supply branch had a direct voice in the counsels and decisions of the Air Board, he suggested that the Director of Supply should be formally added to the Board's membership.[63] Salmond's recommendations were an important spur to further development of the supply organisation, and most of what he proposed to set this side of the Air Force's house in order was acted upon. In July 1929 auctions were held at Laverton to dispose of the remnants of the imperial gift equipment.[64] The following October Wing Commander Anderson joined the Air Board in the new position as Air Member for Supply.

Although Salmond had found considerable fault with the Air Force situation which he saw, interestingly this was not the case with a committee of audit officers which made an inquiry at the same time as himself. Appointed by the auditor-general in July 1928, this body had been formed to examine systems of stores accounts generally 'with a view to submitting recommendations for the adoption of a system suitable to the requirements of Commonwealth Departments'. Its terms of reference also called for it to fully explore 'the position of these accounts at the Air Force and Defence Military Units'. The report of the committee, completed in October, indicated that the RAAF had overcome the worst of the problems initially involved in setting up its administrative structure. Stores were found to be in good order and properly organised and run, stocktaking was being carried out carefully and efficiently, and accounts work was performed well. It concluded:

> In summing up our views on the Stores Accounts Systems of the Air Force, it may be said that we consider the system is now established on satisfactory lines (subject to completion of stocktaking at the Depot [at Laverton]) and that there is evidence of efficient supervision and administration.[65]

Despite this apparent acceptance that the state of Air Force stores administration was now satisfactory, this did not stop the auditor-general delivering a stinging report on the RAAF in 1930. The basis for his criticism was focused essentially on a single incident—the loss of the Wackett 'Widgeon II' amphibian in an accident in January 1930 (see chapter 9)—and his suggestion that this matter reflected on administrative standards observed in the service were hotly rejected by the Air Board. A second element of the auditor-general's findings, however, touched on the numbers of personnel employed in the Air Force when measured against the number of aircraft being operated.

The Board was clearly apprehensive about what the latter observation portended, no doubt bearing in mind that government shaving of the defence estimates in 1929 had required the Air Force to find direct cuts in its manpower strength in the order of 70 airmen[66]—in a force only ten times that number. Bill Linnett, then the chief clerk in RAAF Headquarters, recalls that this had been a painful process achieved by firstly letting go those who wished to leave the service, and then by discharging the 'ne'er-do-wells'.[67] In the light of this, the Board was clearly sensitive to any suggestion that the RAAF was extravagant in personnel. One newspaper reported in March that 'senior pilots' regarded the auditor-general's opinion that the Air Force was not economically administered as 'a specious plea for retrenchment [under which] a determined effort will be made to bring control of the force under the Army and Navy':

> When the Air Board receives the report it will answer the charges seriatim in a special statement to be made to the Minister. To the animadversion that the Air Force machines are used for other than service purposes the board will probably reply that in this respect Government policy is being followed, and for purposes of

> economic value . . . [where the value of] services rendered greatly outweighs the cost. The Auditor-General states that there are 15 men on the ground for every machine flying. It will be the board's aim to show that the figures for the Royal Air Force and the American Air Force are 20 to 1 and 24 to 1 respectively. A general protest against the sweeping nature of the report will be made.[68]

In fact, when the Board's comments went forward on 23 May, it suggested the only reasonable comparison that could be made was with the RAF contingent deployed in Iraq. This force operated an almost identical number of aircraft (63) yet utilised personnel numbers which gave a ratio of men to machines more than double the RAAF's. The Board went on to bluntly imply that only when the auditor-general had acquired detailed and technical knowledge of the service, and all aspects of its operations, would he be in a position to 'authoritatively state that there is room for improvement and economy in the administration and organization of any of the fighting services'.[69]

Public servants and servicemen

An important factor affecting the administration of the RAAF was the state of relations between the civilian and service elements of the Defence Department. This was a factor with its greatest impact at the senior level—that is, between the Air Board and the civilian chiefs of the Defence Department—and at this level it was an aspect which boiled down to Williams' relations with the three men who filled the post of secretary between 1921 and 1939.

The first of the permanent heads with whom Williams had to deal as Air Force chief was Thomas Trumble, brother of the famous Victorian cricketer Hugh Trumble. From the tone of Williams' comments in his autobiography it is plain that the relationship that existed was a relatively harmonious one. He understandably valued Trumble's cooperative attitude, finding him to be 'the greatest help' during what was obviously a testing time for a comparatively junior service chief suddenly entrusted with enormous responsibilities; at least he took the trouble to keep CAS informed of what was happening.[70]

For his part, Trumble reciprocated the regard in which he was held at RAAF Headquarters. Leaving the department in July 1927 to become Official Secretary at Australia House in London, he wrote to Williams from the P&O liner *Naldera* to express his thanks for the farewell compliment paid to him by the Air Force:

> It touched me greatly, and I felt honoured indeed to receive it on severing my connection with Defence. May the Air Force go on and prosper. I am sorry that I was not able to help the service much more than I did but through difficulties in the way one had at times to be content with little successes.

Williams replied on 30 August that he was only sorry that 'we could not do more on the occasion of your departure to show our appreciation for what you have done for us since we were formed into a separate service. I take this opportunity, however, of formally thanking you for it'. Throughout the time Trumble spent in London before retiring in 1932, Williams maintained a friendly but still formal correspondence—his letters always began 'Dear Mr. Trumble'—regarding developments within the Department.[71]

In contrast to this experience, Williams' relations with the new departmental head were less open and happy—a change which was very quickly apparent. It was only in November 1927 that Williams ruefully wrote to Trumble: 'There are quite a few times when I wish you were over in your old chair so that I could have a talk to you'.[72] Trumble's successor, Malcolm Shepherd, was a former head of the Prime Minister's Department and was Trumble's predecessor in London. He brought a far more autocratic style to the Defence Department, adopting the practice of keeping service chiefs and others away from the Minister and making the latter dependent on himself for advice. This was a very successful tactic which gave Shepherd enormous power. Williams complains that in the case of A.E. Green, first Minister for Defence in the Scullin Labor government, the Minister refused to sign anything, no matter how routine, which had not first been approved by the secretary.[73]

More important than Shepherd's manipulation of ministers was the fact that he showed himself to be quite unsympathetic to the Air Force and its interests. Williams claims that he was met by the new secretary with the words: 'Look here, we don't want any of these bloody air marshals around here as they have in London. It's bad enough having to deal with admirals and generals without having air marshals as well.'[74] Such an attitude was predictably not well-received and there was probably much satisfaction felt in uniformed circles when, in late December 1928, Shepherd was suspended by the Minister (Glasgow) over allegations by a disappointed contractor, Herbert ('Jimmy') Larkin of Larkin Aircraft Supply Company (LASCo.), that he had sought bribes.[75] A public service inquiry held in February 1929 found Larkin's charges 'not proved'.[76]

Though cleared and restored to duty, remaining as secretary for a further eight years, Shepherd reportedly 'took his suspension hard'.[77] If, as was possible, he believed that Williams or other senior RAAF figures had a hand in this incident, this would go a long way towards explaining the attitude he displayed later in 1929, when he took a leading role in attempting to have the Air Force disbanded as a separate service and absorbed within Navy and Army. Air Vice-Marshal Denis Creal recalls that he was standing alongside Captain Henry Feakes, commanding officer of HMAS *Albatross*, when Shepherd came on board for a visit in 1930. He heard Shepherd state that he was pushing a submission to Cabinet calling for the RAAF's amalgamation with the other services,[78] a claim which confirms Williams' own belief about where the pressure for this move originated (see chapter 3) and is supported by other evidence. On 25 February 1931 Shepherd wrote to Frederick Shedden, a senior officer in his department, about discussions he had with the Minister at which the latter agreed to put proposals to Cabinet for the RAN to be handed over to Britain and the Air Force to be amalgamated with Army: 'Will you now prepare papers on these [points] . . . The Minister was much impressed with the presentation of the case and I told him you were the author. If this Minister stays in Defence for a while we will get somewhere and get our problems settled.'[79]

Notwithstanding a certain coolness towards Williams, Shepherd's attitude was not simply one of antagonism towards all Air Force officers or servicemen generally; the youngest son of Goble stayed with the Shepherd family for a time, which suggests that 'relationships were probably not bad' in that quarter.[80]

Shepherd retired in 1937 and was succeeded by Shedden, who was to retain this post for the next 19 years. He had come up within the finance branch of the Defence Department, having been assistant to the finance secretary, and Accountant from 1924, before being appointed secretary to the Defence Committee in 1934. According to Williams, he was a figure very much in his predecessor's mould, having been well-schooled in the art of maintaining influence and control over ministers.[81] Whether this was a fair and accurate assessment matters less than the paramount fact which is clear, that Williams' relations with the new secretary left a lot to be desired. Certainly Williams was later in no doubt that Shedden had played a crucial behind-scenes role in the events during 1938–39 which resulted in his own removal as CAS.[82]

The public servants with responsiblities most directly relating to the administration of the Air Force were only two in number. These were the secretary of the Air Board, who was also secretary to the Air Council, and the Finance Member on the Board. The first holder of the secretary's post was 32-year-old Charles Auty, who was appointed in 1920 only in a temporary capacity. When the position was advertised late that year for appointment on a permanent basis, 101 applications were received. Auty was among those who applied, furnishing references from two former government ministers, Hon. William Kelly and Hon. King O'Malley; the secretary, Works and Railways; and George Swinburne. Other applicants were Captain William Hodgson (secretary of the Aeroplane Construction Committee and Advisory Committee on Aeroplane Construction, carrying the recommendation of the former CGS, Major-General Legge), G. Wentworth (a former RFC captain who had been secretary to the Minister for Munitions in Britain, private secretary to the CAS (RAF), and was now private secretary to the Governor of Tasmania), no fewer than 16 other former members of the AFC or RAF—including air ace 'Harry' Cobby, Raymond Brownell, and A.J. Mangan (who had references from Williams and another noted AFC commander, Lieutenant-Colonel Oswald Watt)—along with several lieutenant-colonels with AIF experience.[83]

From this very impressive field, the man selected was 28-year-old Patrick ('Johnny') Coleman.[84] A public servant from 1910, he had spent six years in the AIF during which he rose from the rank of staff sergeant to major. Serving on the headquarters of 1st Division in 1914–15, then AIF Headquarters 1916–18, he had been mentioned in despatches twice and awarded the OBE in 1919. With this background, Coleman functioned smoothly in the environment at RAAF Headquarters although having only a minimum of staff—in 1932 only one clerk and one typist. In time, he became thoroughly identified with the administration of the Air Force and retained the post of secretary of the Board until 1939, when he became an assistant secretary in the Department of Air upon that department's formation in 1940.

According to Williams, sometime early in 1920 he had been advised by Trumble that it was proposed to appoint 47-year-old Thomas Jolliffe, who had achieved the rank of lieutenant-colonel in the pay corps of the AIF and was now the second most senior officer of the Defence Department's finance staff, as the first Finance Member of the Air Board when that body was established later in the year. He soon found himself locked into a bitter and futile argument with Jolliffe over the preparation of finance regulations for the Air Force. The latter, acting on instructions from the Finance Secretary, Colonel T.J. Thomas, refused to consider anything else but a pay scale based on Army rates with an additional £50 flying pay and no allowances.[85]

This was a potentially disastrous start to a vital relationship on the Board, and it was highly fortunate that it was averted. As Williams records:

> This frustration [with Jolliffe] became so bad that I had to take up the matter with Mr Trumble who took me to the Minister. It was decided that a change should be made and that the Accountant of the Department, Mr A.C. Joyce, a much younger man, should be appointed Finance Member of the Air Board when formed. He was a great help and we made progress.[86]

Aged 34, Albert Joyce had been in the public service from 1903 when he joined the Auditor-General's office. He had gained experience since then on the central staff of the Postmaster-General's Department from 1911, before transferring to Defence in 1919. Though undoubtedly easier to work with than Jolliffe, he was by no means simply a pliable tool in Williams' hands—as the latter was soon to discover.

The question of RAAF rates of pay was certainly not resolved as quickly as might have been assumed from Williams' remarks. In fact, when this matter was being debated by the Board in January–February 1921 Joyce proved no more sympathetic to the approach Williams wanted adopted than had Jolliffe. In consequence he was soon pitted against the Board's uniformed members, who pressed for a pay structure based on RAF rates and which included allowances as well as deferred pay. Joyce opposed using the RAF as any sort of yardstick, pointing out that the British army's higher rates were not matched in the local military forces. He also considered the proposed RAAF rates gave an increase over the Army ones which was excessive, being 50 per cent more for officers on the general list (the General Duties Branch) and 25 per cent more in the case of the quartermasters list (later the Stores and Accounting Branch)—even though conditions of service for the latter, and the nature of their employment, were practically the same as their Army counterparts.[87]

In defence of their proposals, the rest of the Board argued that RAF rates had been used only as a guide as to what was fair and reasonable for the work required of Air Force personnel. In any event, the result had not been the acceptance of RAF rates in their entirety as the proposed RAAF allowances were not as good. And although RAAF personnel would be better off than their Army colleagues, the same was not true in comparison with the RAN because senior Air Force ranks would still be less highly paid. In particular they urged that rates for ranks other than officers had to be matched against those applying in technical trades in civil life, arguing that 'an efficient Air Force cannot be maintained with inferior mechanics ... [so] this application of the paring knife to the rates for other ranks is considered most dangerous.' They also considered Joyce's objection to allowances unacceptable, because it worked against members not provided with service quarters, and challenged his opposition to a system of deferred pay for officers meant to take the place of a pension or retiring allowance. Pointing out that the RAN received deferred pay and Army members a retiring allowance, they disputed Joyce's assumption that the government had a superannuation scheme under consideration 'which may (or may not) apply to the Air Force and that some provision will probably (or probably not) be made for a retiring allowance for the Air Force'.[88]

The matter was ultimately settled outside the Board, as questions of this nature in any event required the approval of the Air Council. When the latter body considered the question, it accepted the rates supported by the majority of the Air Board. This included allowances and deferred pay, although the latter was 'subject to revision if and when the Government introduces a Superannuation Scheme which would apply to the Air Force'.[89] After consultation with the Treasurer and Cabinet, and consideration of the views of Joyce and Goble, the Minister for Defence (Pearce) decided on 16 February to approve rates of pay as recommended by the Air Council, with some alteration of arrangements regarding deferred pay.[90]

The finance branch at RAAF Headquarters was not a large empire. In April 1932, the year Joyce left to go to Canberra as an assistant secretary in the Treasury, the branch numbered just 10 people apart from the Finance Member—seven clerks, one assistant and two typists—even though it provided all finance and accounting services for CAB as well as the RAAF.[91] When Joyce left the Air Board, his departure was a clear loss to RAAF Headquarters. Although his view of his financial responsibilities had sometimes produced differences with service officers with whom he had to work, his underlying approach had been sympathetic rather than obstructionist. The impression he gave was that he considered himself a member of the team acting in the best interests of the Air Force. Perhaps one evidence of this attitude was the fact that all four of his sons served in the RAAF during the Second World War.[92]

The difference was quickly noticed when Joyce was replaced in July 1932 by Colonel Thomas. In his autobiography, Williams reveals the existence of a long-standing antipathy between himself and Thomas,[93] so that the previous cooperative spirit that prevailed evidently came under strain. This situation was not relieved when Thomas was succeeded as Finance Member of the Board, in July 1936, by Melville Langslow. Another public servant with pay corps experience in the AIF, Langslow had reached the rank of major and punctiliously still used that rank as a form of address. By most accounts he was a baleful influence in the Air Force's administration, and as such was thoroughly despised by service officers who came into contact with him.[94] Not all officers saw him in this light, however, and at least one credited him with being 'a very able and dedicated person'.[95] Despite his 'poor press', and serious criticism of his negative and niggardly attitude in financial matters connected with the *Kyeema* disaster of 1938 (see chapter 10), when the Department of Air was created in the early months of the Second World War, it was Langslow who became first permanent head of the new department.

Undoubtedly there were problems in the administrative process which flowed from the sometimes thorny relationship between servicemen and public servants within the Department of Defence. As later recalled by Air Vice-Marshal Henry Wrigley, an officer with long experience in RAAF Headquarters before 1939:

> [S]ome of the civilian side of the show had an unhappy knack of interfering in what were purely service matters, matters which they did not understand. That unfortunately caused a lot of friction from time to time . . . because it slowed up very, very many matters, and where decisions on things should have been given and promptly implemented we found that they had been side-tracked. It delayed the smooth process of administration. Things like gazettals and notifications in orders—which should have taken only a matter of a day at the most—were sometimes held up for weeks before being dealt with. That not only interfered with the smooth flow of administrative action, but also, when it came down to questions concerning personnel, sometimes caused a lot of ill-feeling among some of our people who didn't know the background and how the delays occurred. I think some of the actions of these civilian administrators were the result of inexperience to some extent, but apart from that the Secretaries of the Department always felt that they were the king-pin, the lords of their particular service.[96]

After an interval of more than 50 years, it is practically impossible to form confident judgements upon the respective merit of complaints regarding the behaviour of senior public servants and Air Force figures. What may be concluded, however, is that then—as indeed today—relations between civilian and service personnel within the Defence Department were governed by personality. Shepherd, Shedden and Langslow were, no doubt, difficult men to get on with, but then much the same charge was often levelled at Williams. Clearly, too, the situation that existed defied the simple explanation that public servants and servicemen invariably disliked each other, as the relative warmth attaching to Trumble, Coleman and Joyce attests.

Though the public service mode of operation was often a frustration to Air Force officers used to functioning within the command-style which characterises the services, it did not follow that one was right and the other wrong. Equally, there can be no clear-cut apportionment of blame for failings that occurred, as the comments of at least one Air Force officer of the period demonstrate:

> The whole development of the Air Force was more or less retarded by a Public Service which really wielded a great deal of power . . . But I think the Air Force people themselves may have felt perhaps not confident in the administrative stuff, and were inclined not to argue on points of policy.[97]

For the purposes of this study, it need only be recognised that the (now entrenched) habit of mutual criticism of the civilian and service elements in the Department of Defence has a long history, and was a factor—an affliction even—in the story of the pre-1939 RAAF.

13

Polishing the public image

The peacetime RAAF was not solely concerned with preparing against the possible eventuality of war. Many activities were undertaken because they fulfilled wider, often non-defence purposes which were nonetheless valuable—to the wider community if not directly the Air Force. The rationale to such activities was essentially that enunciated by Squadron Leader William Bostock, CO of No.3 Squadron at Richmond, during a talk he gave in 1933 to the aeronautical engineer branch of the Institution of Engineers in Sydney. On this occasion he said:

> Primarily the service aims at efficiency in training. In peace time a portion of this training can be merged with operations which are not strictly of a service nature, but which, while providing useful exercises in many branches of Air Force work, at the same time have some commercial or other civil purpose. It is, therefore, possible to give, in some material measure, a concrete or productive aspect to public money expended on Air Force activity . . . To this end, public organisations, semi public bodies and even private individuals may, if the proposed work is of national importance, secure the services of the Air Force for flying operations, provided civil enterprise is unable, or is unwilling, to undertake the work, or, if the cost of the proposed work would be prohibitive if performed by a civil organisation, but is of sufficient public importance.[1]

The point which Bostock's remarks did not acknowledge was the obvious one that, quite apart from the actual benefits of the tasks undertaken and the incidental training derived by RAAF personnel in the course of carrying them out, there was an important public relations dividend to be gained from performing such work—not least in winning acceptance of the Air Force and helping to counter opponents of its existence as a separate service. That the RAAF's senior commanders—starting with Williams as CAS—were alert to this fact seems obvious from the frequency with which aircraft and crews were engaged in tasks of this nature. Indeed, for an Air Force obviously lacking in significant operational capability, or fulfilling any role much beyond training, activities of this sort could sometimes appear to have become a substitute for the defence role which formed the service's basic reason for existence.

Aiding civil agencies

The pattern of the RAAF's involvement in tasks of a civil nature was established early. A list of assistance given to various government departments and agencies during a single 12-months period (the 1923–24 financial year) makes quite interesting reading, including: a search for a fishing boat missing on Port Phillip Bay (July 1923); aerial photography of the Quarantine Station at Portsea and the Serum Laboratory at North Melbourne required by the Director-General of Health (Aug. 1923); carriage of the superintendent of police and a doctor to Broken Hill and back, for the investigation into a shooting 'affray' (Sept. 1923); photographic mosaics of Victoria Street, Melbourne, for the Town Planning Commission (Oct. 1923); vertical photography of the Melbourne University campus for the university authorities (Nov. 1923); photographs of the City of Melbourne, for display at the British Empire Exhibition (Nov. 1923); aerial photography of Torrumbarry Weir, Burrinjuck Reservoir and Albury Weir (three separate flights), for the Inland Water Commission (Nov. 1923); aerial photography of Canberra, for the Federal Capital Director of Works (Dec. 1923); and a mosaic of Port Melbourne (an area of 40 square kilometres), for the Town Planning Commission (1924).[2]

The extent and range of requests made to the RAAF which is evident in the above listing was by no means unusual, and in subsequent years a heavy program of further commitments fell to the service. In November 1925, for example, No.1 Squadron began a photographic survey of the Kinglake area, near Healesville, north-east of Melbourne, and this work continued throughout the whole of 1926. In the same period the unit also carried out aerial photography tasks over Geelong and the Melbourne metropolitan area for the Town Planning Commission, and areas at Seymour and Ringwood (Vic) for the Army survey section. In the last months of 1928, a photographic survey of Renmark (SA) was carried out for the Waite Institute of Adelaide, and one at Canberra for the Federal Capital Commission.[3]

Some of these undertakings seem—after more than half a century—mainly of incidental interest, appearing to be worthy but rather prosaic. But there could usually be no doubting their importance for a young, rapidly developing country such as Australia in the post-war era. Also, while the majority of these tasks were focused on Victoria, and specifically Melbourne, this was only to be expected in view of the RAAF's confinement at this stage to a single base at Point Cook. In later years, with the opening of Richmond, the work of this nature which the Air Force performed greatly expanded in geographical range.

At times it was government policy which actively pushed the Air Force in the direction of taking on more civil tasks. For example, in March 1924 Goble (as acting CAS) was suddenly confronted in the Council of Defence with Prime Minister Bruce's desire that the RAAF should take over certain duties such as carrying the mails. When he attempted to resist such a proposal, it was only to find the PM adamant and supported by the Minister for Defence (Bowden), who 'added sweetly that the RAAF was not likely to get more money unless it indicated its willingness to carry out civil duties . . . Bowden's threat must have impressed the entire Council'.[4] Nothing came of Bruce's wish regarding the mails at this time, but the fact that the government favoured such a role can only have operated as a powerful incentive for the RAAF to make itself useful in similar ways.

When the Labor ministry of Prime Minister James Scullin took office in late 1929, the question of the RAAF being employed on civil work was again raised in earnest. The notion previously favoured by Bruce for the Air Force to take over the carriage of mail was considered once more but again rejected (see chapter 3). Nonetheless, the government's desire to see movement in this direction was not without effect, and a number of fresh activities of a civil nature now came the RAAF's way.

For example, it was arranged for a Seagull III to be held in constant readiness at Point Cook during the summer months of 1929–30 to go to the rescue of fishing boats or other small craft in distress on Port Phillip Bay, and to search for boats reported missing. The aircraft was called out twice that summer, and the arrangement was repeated in 1930–31.[5] In fact, the Air Force was quite frequently called out on maritime search operations which took it well beyond Port Phillip Bay. In the space of just two months in 1936, for instance, RAAF Seagull Vs were called out to assist in trying to locate vessels which had gone missing during gales in Bass Strait, and en route from Lord Howe Island. In both cases the search effort was conducted in continuing bad weather, and in one instance required a Gannet to patrol 225 kilometres out to sea.[6] Other searches over land for missing aircraft, lost prospectors, etc., were an equally frequent occurrence—some notable cases being described in chapters 10 and 14.

Bushfire patrols

As a result of the government's desire that the Air Force should be seen to be more useful to the civil community, beginning in February 1930 the RAAF was also called upon to conduct air patrols over state forests in Victoria during the summer bushfire season. For the initial trial of this arrangement, which continued into March, a single Wapiti from No.1 Squadron was used to fly over areas to the north and east of the Dandenong Ranges, looking for first signs of smoke which would enable an outbreak to be dealt with promptly.[7] After the experiment concluded, a letter was received from the Forestry Commission stating:

> Now that the fire season has finished and a review of same made, the excellent work of your officers on fire patrol by plane becomes increasingly evident. The air work carried out with the aid of the new and improved plans forwarded, shows an accuracy of detail of which your officers may be justifiably proud ... [T]he psychological effect of these patrols has been a decided deterrent to those persons who have in the past endangered the safety of the forests owing to their fire activities.[8]

As a consequence of the success of the previous season's effort, bushfire patrols were resumed in November 1930 in the same area as before and continued until February 1931. The patrols were continued in later years also, although the area was not always the same and the time of the Laverton-based No.21

Squadron was also taken up with this work. In December 1932, for example, the focus shifted to East Gippsland. In 1937 an improved system of identifying localities in the area covered by the patrols was devised and instituted. Recalls Group Captain John Graham, then a CAF pilot officer who was engaged in this work:

> In those days in the forests of Gippsland they had a lot of logging and saw mills. These had beside them cleared areas with a large number displayed, mostly in whitewash, which could be seen clearly from the air. Using these numbers it was possible to instantly give the location of a fire anywhere over eastern Victoria.[9]

As an example of the patrols' effectiveness, just five flights by No.21 Squadron during January 1938 resulted in a total of 90 fires being detected and reported.[10]

Aerial dusting

On 15 October 1930 the Defence Minister (Green) announced that the RAAF would take part in what was apparently the first crop-dusting activities conducted in Australia.[11] The Victorian Forestry Commission had sought help in combating the infestation of a 120 hectare pine plantation at Yarrowee, near Ballarat East, by case moth caterpillars.[12] Adapting the technique pioneered in the United States ten years earlier, the front cockpit of a DH60G Moth (A7–30) was sealed after removing the seat, so as to carry 200 pounds of lead arsenate. With the aid of ventilators forcing a current of air through the powder and a controllable vent at the bottom of the aircraft, this powder could be distributed beneath the plane as a fine film.[13]

After trials were carried out at No.1 Squadron beginning on 31 October under the direction of Squadron Leader Henry Wrigley, Director of Operations and Intelligence at RAAF Headquarters, the required dusting operation was conducted on 18 November.[14] With the Moth flying about 10 feet above tree-level, it was reported to have taken only three minutes to cover the whole area. The operation was repeated in December 1930 and again during December 1931.[15] The results were most successful, it being reported that after two aerial applications of the chemical fully 99 per cent of the grubs died, thereby saving the reafforestation scheme valued at £90 000.[16]

The extent to which the RAAF continued to be involved in this work is not clear from surviving records. It is clear, however, that the capability to perform such tasks was retained, as at both the Laverton and Richmond air displays in November 1934 a demonstration of crop-dusting was given in A7–30 by Flight Lieutenant Colin Henry.[17]

Increasing demand for surveys

Even before the pressure began at the end of 1929 for the RAAF to step up its assistance to civil agencies, a number of quite notable survey undertakings had been carried out or were in the planning stage. Among those already accomplished was a major survey of the remote central region of the Australian continent undertaken for purely scientific research purposes. The first flight of this kind had actually taken place barely a year after the RAAF's inauguration, when Gerald Halligan, a noted civil engineer,[18] obtained approval early in 1922 for an aerial reconnaissance of Lake Eyre in South Australia as a prelude to a land expedition to examine the hydrography of the lake. Two DH9As piloted by Squadron Leader I.N.C. Clarke and Flying Officer Loris Balderson, with Flight Lieutenant Ern Mustard and Halligan as observers, proceeded to Marree (80 kilometres from the southern edge of Lake Eyre) and from there flew a circular 640 kilometre route around the lake on 1 March.[19]

Although Halligan reportedly succeeded in mapping the lake's contours and fixing its physiographical details, according to one account the reconnaissance was not particularly successful. The RAAF pilots refused to fly lower than 8000 feet, considering this to be the minimum altitude to provide a gliding range that would safely carry a machine away from the lake in the event of engine failure, and Halligan was disappointed at being unable to inspect the surface as closely as he wished. Halligan described other problems encountered:

> As this was my first trip in a flying-machine, the work of taking photographs in a 120-mile an hour wind, making notes and sketches, and taking compass bearings and angular measurements to ascertain distances was very difficult, while the excessive vibration, the overpowering roar of the exhaust, to say nothing of the continual dropping into and being jerked out of air-pockets, was certainly not conducive to mental calm.[20]

On the return flight to Adelaide to next day a similar reconnaissance was made of Lake Torrens.[21]

Now, seven years later, Cecil Madigan—a lecturer in geology at the University of Adelaide—sought to cover the same area of central Australia in a much broader aerial reconnaissance using RAAF aircraft and crews. He had already undertaken a geological research expedition on the ground in 1927, but wanted to survey about 78 000 square kilometres of largely unknown country. With the support of the state branch of the Royal Geographical Society of Australasia, he gained the Air Board's cooperation for what was the first attempt at systematic aerial strip-photography in the country.[22]

On 8 August two Wapitis of No.1 Squadron departed Laverton. On board machine A5–7 were Flying Officer Andrew Gerrand (pilot), AC1 L.V. Thomas (photographer) and Madigan, while Flying Officer Frederick Wight (pilot), Flight Sergeant T.H. Stewart (fitter) and AC1 H. Bain (radio operator) crewed A5–8.[23] After photographing the mines at Broken Hill (NSW), the aircraft flew across Lake Frome to Marree, then over other dry salt lakes to the north: Callabonna, Blanche, Gregory, Torrens, and Eyre.

Shifting the focus of their operation to the area between the Queensland border and the railway running south from Alice Springs, the party flew to Birdsville in Queensland, taking in the margin of the Simpson Desert. The aircraft next travelled to Alice Springs, with Gerrand making a forced landing at Undoolya (north-east of Alice Springs) on 15 August, although his aircraft was soon fixed and operational again.[24] From Alice Springs two flights were made to investigate geological features of the MacDonnell Ranges.[25] To complete the survey, the aircraft then flew across the expanse of Lake Eyre several times while attempts were made to analyse the lake's surface, before returning to Laverton on 26 August. Madigan followed up this air survey by undertaking extensive camel treks in 1930 and 1931, which made his name as an exploration geologist in Australia's centre.[26] Clearly, however, the cooperation of the RAAF had been important to the success of this program to further knowledge of this previously little-explored region.

Even while this attempt to learn more about Australia's interior was underway, steps had already been initiated to obtain the assistance of the RAAF in meeting the development needs of the southern island State. This primarily meant undertaking a survey of Tasmanian forests, the first such survey attempted in Australia. Calls in this direction had originally been heard in the latter half of 1927, when the Commonweath Development and Migration Commission initiated a preliminary survey into the cost and methodology of making a complete topographical survey of Tasmania to assist in development planning.[27] A committee appointed by the commission to make this initial investigation considered a proposal that the Air Force be asked to provide mosaics to assist the topographers.[28] A subsequent conference of senior officials at Hobart on 15 February 1928 decided on a definite program which included air photography to be carried out over two seasons, after beacons had been laid out.[29]

Despite acceptance of the idea of using the Air Force, it was some time before things began to move. In November 1929 Squadron Leader Ray Brownell and Flight Lieutenant F.A. Briggs flew to Hobart in Southampton A11–1, and from there made an aerial inspection of Port Davey on the south-west coast, to determine the suitability of that place as a base for seaplanes or amphibians.[30] This examination did not appear to provide a readymade answer to the question of finding a suitable location from which to operate, but nonetheless actual survey operations got underway on 19 April the next year using landplanes.

On 19 April 1930 two Wapitis from No.1 Squadron flew to Stanley, on Tasmania's north-west coast, accompanied by a Southampton (A11–1) from Point Cook, piloted by Squadron Leader Arthur Hempel.[31] While the Wapitis commenced to survey a 2300 square kilometre area between Rocky Cape and Arthur River, the Southampton went on to the west coast town of Strachan, from where it began flights over forests surrounding Macquarie Harbour and in the Gordon River valley, before ranging south to take in coastal areas between Macquarie Harbour and Port Davey (which was examined further as a site for future seaplane operations). The area north was then covered, working back towards Stanley, from which the big flying boat left to return to the mainland on 25 April.[32]

The Southampton carried members of the State forestry department on all survey flights, and their verdict was entirely favourable:

> The Forestry Officials state that they were very satisfied with the information collected on these flights, and think an immense amount of time would be saved to the Forestry Department as a result. They found that they could easily identify the various types of timber and also make a fairly accurate estimate as to whether it was of commercial use.

For the Air Force there was a direct operational benefit in the discovery by the visiting party that this part of the country was virtually unrecognisable from available charts and maps. 'Many more rivers, headlands and bays exist than appear on these charts', they reported, 'and those that appear are incorrectly shaped'.[33]

Attempts to follow up this beginning were made in subsequent years, but with only mixed success. In February 1931 three Wapitis and a Moth from No.1 Squadron were despatched to Launceston and Hobart to perform a variety of tasks, just one of which was forest survey.[34] After providing a display at an air pageant at Western Junction (Launceston), and cooperating with the Army during the annual militia training camp at Mona Vale, attention finally turned to the forest survey, this time in the Huon River valley. In the event, the photographic mission was an almost total wash-out. On only one day out of the fortnight spent in Tasmania was the target area clear of cloud below 14 000 feet, resulting in only about 400 square kilometres being photographed.[35]

Much the same thing happened the following year, when three Wapitis of No.1 Squadron and a Bulldog from No.1 FTS embarked on a similar visit to the State under the command of Flight Lieuteant D.E.L. Wilson. Leaving Laverton on 12 February, they were under instructions to participate in aero club pageants at Western Junction, Brighton (Hobart) and Wynward, photographing existing aerodromes and an area selected by the Forestry Department for compiling a mosaic. Again, only limited photographic work could be carried out due to the unsuitable weather conditions encountered on most days.[36] Despite these failures, further efforts were made in subsequent years to progress this work. Air Commodore Bill Garing recalls being involved in another attempt to photograph the timber resources at the headwaters of the Huon River during 1932–33 using a Southampton.[37]

The efforts being made in Tasmania were only a small portion of the photographic survey work being done from the end of 1929, as there was now an increasing demand for work of this nature to be performed. As one RAAF publication stated matters in 1931:

> In the good old days before Sir Otto [Niemeyer—the British banker who advised the Commonwealth on how to deal with its economic problems], it must be confessed that life was sedate and not unpleasant. But the great depression has given us a revised training scheme and has produced a volume of activity not unlike a league football final . . . [38]

Among the tasks undertaken in 1930–31 were the production of a mosaic of Sydney and suburbs for the Metropolitan Water Board, to assist in planning the water and sewage services of Australia's largest city (undertaken by No.3 Squadron between January and April 1930, this mosaic ended up comprising 23 sections); a further mosaic of the port of Melbourne for the Harbour Trust; coverage of about 1500 square kilometres in the Murray valley, near Albury; and a soil survey at Barmera, near Renmark (SA).

Commitments of this nature were in addition to defence survey work which was still being carried out at the same time. For example, in 1933 Wapitis from Richmond were sent on a photographic survey of the Liverpool Range and Helidon area of Queensland for the Army survey section.[39] In May 1934 a

Moth (A7–62) from Seaplane Squadron at Point Cook was flown to Darwin, where it was converted to a seaplane to enable it to cooperate with HMAS *Moresby* on a marine survey of the Van Diemen Gulf and Melville Island.[40] The Moth would appear to have been not particularly successful, as in August the following year a Seagull V was sent from No.5 squadron.[41] While all these separate survey tasks were being carried out, the RAAF was still maintaining a major commitment of men and personnel to supporting the North Australian Aerial Geological and Geophysical Survey (see next chapter).

Meteorological flights

During 1930 the Air Board agreed to a proposal for the RAAF to undertake meteorological flights for weather forecasting purposes. These flights were done initially in a Bristol Bulldog from Point Cook at 9 am each morning (weekends included), and required an aircraft with wet and dry bulb thermometers strapped to the outer wing struts to fly to an altitude of 16 000 feet, while the pilot noted the temperature and humidity at 1000-foot intervals. This information, noted on a xylonite knee-pad during both ascent and descent, along with details of cloud formations observed and the existence of rain or snow, was then telephoned through to the Weather Bureau on landing and formed the basis of the daily report.

According to a claim made in 1931, these flights were 'probably the most comprehensive yet undertaken in the Southern Hemisphere'.[42] Whether pilots flying them were impressed by their personal contribution to that signal claim may be doubted, however, as the flights—undertaken in rain or shine—were a chore of frequently the most unpleasant kind. The Bulldog was, of course, an open-cockpit machine and the pilot was fully exposed to the worst the skies had to offer for the hour that the mission usually lasted. In winter months the unfortunate pilot could find himself enduring 30 degrees of frost—or worse.[43] The record noted in only the first two years of operation was −14.2°F (−25.6°C)![44]

Quite apart from the aspect of exposure during winter months, flying conditions were often extremely difficult and offered a considerable safety risk. One pilot who flew these missions recalls that a flight was only cancelled:

> when the weather was too adverse to take off—and by 'adverse' I really mean adverse. The standard used was, if the windsock on the workshop hangar could be seen, then the pilot took off. In other words, a visibility of 150 yards was regarded as the normal minimum.[45]

The consequence of this rule of thumb was that pilots might usually be able to get airborne, but could find themselves floundering about in overcast conditions. Instruments in RAAF aircraft to handle murky conditions initially consisted solely of a spirit level (to help judge if the aircraft's wings were horizontal to the ground) and a compass. It was another two years before the first bank and turn instruments became available, and these were so often unserviceable that most pilots disregarded them. Not until 1935 were RAAF aircraft equipped with a more accurate and reliable blind-flying instrument.

In light of this situation it was hardly surprising that pilots occasionally became disoriented in cloud and lost control of their machines. One story has it that a pilot went into a spin at 11 000 feet from which he was able to recover after losing some 3–4000 feet in altitude. Quizzed about the apparent irregularity of his temperature readings when he rang the Weather Bureau, his reported reply was, 'Sorry, but that was the best I could do as I spun past those levels'.[46]

The real risk of flying in these conditions arose, however, when attempting to get back down safely. Varying wind currents at different altitudes made it difficult for pilots to gauge in which direction they were being blown. Provided they were pushed south, out over Port Phillip Bay, there was little risk of hitting anything before reaching sea-level, but if blown north or east beyond Melbourne there were mountain ranges up to 4000 feet to contend with and to the west, also, there were hills up to 2000 feet. In the light of the hazards associated with the meteorological flights, it is extraordinary that accidents were not frequent. Indeed there was only one really serious mishap in the ten years before the flights ceased to be necessary.

On 14 December 1936, 21-year-old Pilot Officer Eric Read was the rostered pilot for the weather flight. The Bulldogs had recently transferred to Laverton and take-off time for these flights had been moved forward from the previous 9 am start. Before leaving the officers quarters for the hangars to prepare for take-off, the last room that he passed was that of a colleague, Julius ('Dick') Cohen, and he paused to ask, 'Have you got a watch?' Cohen lent him his, adding 'but don't crash—it wouldn't survive one of your crashes'.[47] Shortly after 8 am Read was airborne in A12–7, on his way to the usual ceiling of 16 000 feet. Conditions were poor—scattered rain and a cloud base of 1000 feet—but not unflyable. At 2500 feet he came out of the cloud cover and this break continued until he reached 4000 feet when the overcast began again. Unknown to Read, however, conditions had closed in on the ground shortly after he took off, with drizzling rain becoming almost a fog in density, while the wind had shifted from the west and was now blowing from the south.

Having completed the upwards reading, Read began coming back down. As he emerged from the overcast into the break before the ground cover, he executed a number of figure-of-eight turns. This was standard practice to warm up the engine again after using low power at the higher altitudes, and was intended to ensure quick response in case engine power had to be suddenly applied. Then he descended. Fifty years later he recalled the events which followed:

> The next thing I knew I was in a valley, with trees either side of me, so I applied power to get up out of it. Coming out, I hit a tree and my wing was ripped open. I got the aircraft back up to about 2000 feet but then it went into a spin. As it came back out of the cloud cover it hit the trees with one wing and sort of splurged into the ground. [Read was knocked unconscious, waking at 9.10 which he estimates was about 15 minutes after the crash.] I spat out a mouthful of teeth and took stock of myself. I went around and found all the broken bones. My left foot was moving but just sitting there, and I couldn't move. So I spent all that day in the wreck. It was a filthy day—low cloud and rain—and no aircraft could come out looking for me.[48]

In fact, nine aircraft had been sent up to begin searching after he failed to return. These continued for three hours until visibility became too bad to continue. Appeals for information were broadcast over commercial radio stations in an attempt to narrow the area in which he was thought to have come down. By the next day these had produced numerous reports of an aircraft having been heard in cloud over Bacchus Marsh, to the west of Laverton, where it transpired that Read had carried out his figure-of-eight turns. Early the next morning Demons were despatched to comb the area, but initially without success.

After a miserable night still trapped in the wreckage of his plane, Read woke and, in a rough log scratched on the Bulldog's fuselage noted (evidently to his own surprise) that he was 'still alive'. The weather had cleared, revealing that he was on the crest of a hill: 'The only trouble was that the hill I was on still had a cloud sitting on top of it. I could see the Demons come up the valley and turn off when they came to the cloud.'

Eventually, about 2.30 pm, a Demon piloted by Flying Officer Joshua McDonald spotted the wreckage from fairly high up and radioed the Bulldog's location atop Mount Wallace in the Brisbane Ranges some 26 kilometres south-west of Bacchus Marsh. However, this plane reported, there was no sign of life. Shortly afterwards Pilot Officer Ralph Wiley in another Demon arrived to confirm the sighting, and this time—circling tightly overhead—saw Read waving from the cockpit. Racing back to Laverton, Wiley collected Flight Lieutenant Robert Dalton and, switching to a lighter Moth aircraft, flew back and landed at the base of the mountain. He then took off again, and guided Dalton, who was carrying bolt cutters and hacksaws, towards the wreck from the air.

Meanwhile, other ground searchers had nearly reached the site. Four local farmers arrived on the scene around 4.30 pm, followed not long afterwards by two Bacchus Marsh women (one of whom had dreamt Read was on Mount Wallace) who had made their own way through the rough country carrying milk, sandwiches, bandages and whisky, and Flight Lieutenant Dalton. These found Read conscious and, while cutting him free, attended to his thirst and injuries, which included compound fractures of both legs between ankle and knee, a fractured ankle, a crushed hand, a broken nose, a fractured jaw and the loss of all his front teeth. He had also received savage attention from the local mosquito population.

Making an improvised stretcher from their coats, the rescuers set out on an excruciating two-and-a-half hour journey to bring Read—who remained conscious throughout—down off the mountain. At the foot of the mountain they were met by a RAAF medical team who gave further attention to Read's injuries. Around this base camp were other RAAF personnel who had been called out to join in the search, among them Cohen. 'Hey, Dick!', called Read, jaunty in spite of his 31-hour ordeal, 'Here's your watch'. He was then rushed to the Caufield Military Hospital in Melbourne, where he would spend the next five months recovering. Four months after his release from hospital he was back on full flying duties.[49]

Despite the circumstances of yet another Air Force crash, Read's mishap did not have the negative impact which it might well have. Though there was press criticism of the Air Board policy preventing journalists from obtaining information from RAAF personnel on the spot, there was general admiration for Pilot Officer Read's remarkable spirit and fortitude. On balance, the episode probably did much to enhance the public image of the Air Force.

Certainly there was no pressure to terminate the meteorological flights, notwithstanding their sometimes risky nature. In fact, they were actually extended to Richmond from April 1937, and Pearce from November 1938.[50] One of the Sydney flights also produced some drama, when Demon A1–10 became lost and crash landed at Avoca Beach, near Gosford (NSW), on 7 December 1937, resulting in extensive damage to the aircraft but minimal injury to the crew.[51] There were, however, no other serious accidents.

The scene at Laverton in December 1936 as CAS, Air Vice-Marshal Williams (*in civilian clothes, foreground*), talks to one of the pilots engaged in the search for Pilot Officer Eric Read, posted missing while carrying out the daily met flight. Standing beside CAS is Group Captain H.N. Wrigley, the station commander. (*Mr E.V. Read*)

Pilot Officer Read is brought down on an improvised stretcher from the crash-site to a RAAF base camp at the foot of Mount Wallace. (*Mr E.V. Read*)

Epic flights

Another activity which advanced the public profile of the RAAF was the undertaking of major long-distance flights which broke new ground and established new records of aviation achievement. The first of these significant flights was a round-Australia attempt undertaken in 1924 by Wing Commander Goble and Flying Officer Ivor McIntyre in a Fairey IIID seaplane (A10–3). Starting out from Point Cook on 6 April, the aircraft was back in Melbourne 44 days later, having followed the Australian coastline in an anti-clockwise direction and covered 13 700 kilometres in 90 hours flying time.[52]

The purposes of this flight (as listed in Goble's report) were given as being: to reconnoitre the east coast with a view to organising a seagoing aircraft defence route to Thursday Island; examine the coastline and harbours for possible permanent bases for seaplanes; ascertain the suitability of the Fairey IIID to cooperate with the Navy in the Great Barrier Reef survey; determine the effect of tropical conditions on the performance and durability of the seaplane; report of the practicability of using seaplanes for the carriage of airmails; and determine at what height the south-east trade wind in tropical areas holds its surface direction.[53]

Worthy as these objectives may have been, it was debatable whether a survey flight of this nature was of such importance as to require the personal participation of the chief of staff of the RAAF—for that was the post Goble held in the absence of Williams overseas during 1922–25. On this flight he acted as observer and navigator to McIntyre, who was an ex-RNAS seaplane pilot recruited from the RAF just the year before. Given the difficult nature of his

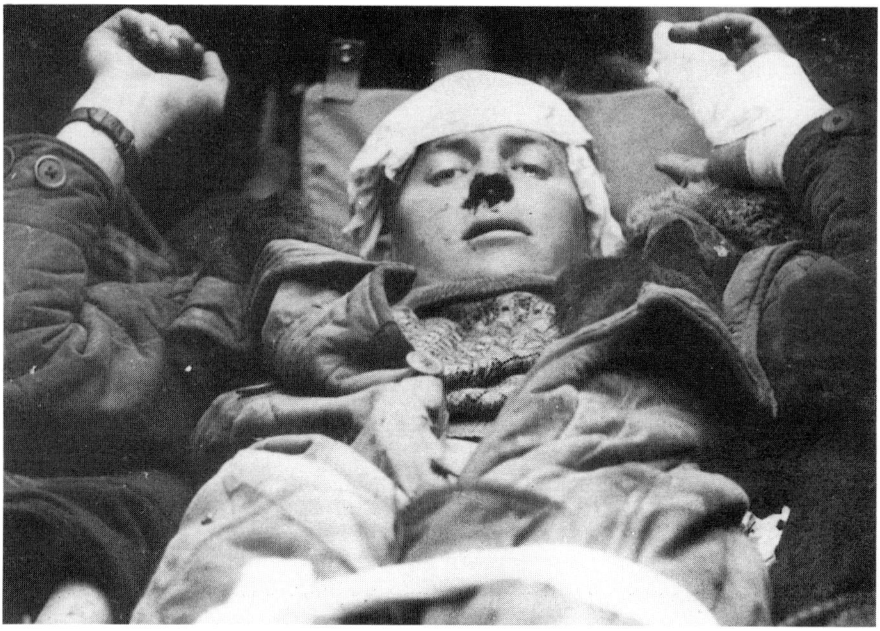

Despite injuries which included a broken nose, broken jaw and broken teeth, broken legs and a crushed hand, Read remained conscious throughout his 31-hour ordeal. (*Mr E.V. Read*)

relations with Williams, it can only be wondered whether the real purpose behind this trip had more to do with establishing superiority of credentials over a rival and incidentally promoting the naval cooperation element of the RAAF, which was practically idle since the unsuccessful attempts to devise means of adapting the RAN's cruisers to embark the large and heavy Fairey seaplanes (see chapter 7).

Such a pioneering flight needed extensive preparation, including the prepositioning of fuel supplies at 37 locations around the coast, the despatch of spare engines and parts to Thursday Island and Perth, and modifications to A10–3 itself which included an additional fuel tank, an extra radiator and extra air vents on the front cowling. Supporting the airmen was a four-man ground party, with mechanics positioned on Thursday Island, Broome and two at Perth.

Notwithstanding all this preparation, the flight once underway was soon confronted with problems. Bad weather, an exploding pilot's compass, a damaged float, and more bad weather—all before reaching Sydney. Difficulties of this sort would continue throughout the flight, to which would be added biting insects, a septic finger suffered by McIntyre, torrential downpours which at one point actually forced A10–3 onto the sea, and the unceasing task of attending to the aircraft. Commented Goble at the end of the flight:

> [W]e were always tired by the time we got into the air, owing to the strain of

Goble and McIntyre pause for a few minutes for this photograph on the sand at Elcho Island (NT), during a stop-over there on 23–24 April 1924. The two aviators were constantly fatigued from the unremitting tasks that needed to be done on the ground, quite apart from time spent in the air. Note the RAAF/RAF ensign flying from the cockpit (now on display in the RAAF Museum). *(Cdre J.D. Goble)*

> watching the machine, the manual labour involved in doing running repairs and humping drums of petrol. Each day after landing we put in about seven hours work baling floats, cleaning plugs, adjusting magnetos and valve springs, straining petrol through chamois leather, oiling up, filling the radiator etc., in addition to the long and irksome watching to ensure the safety of the machine on the water, and also to avoid being left high and dry. Several times we had to dig her out when the tide played dirty tricks on us.[54]

At Darwin (reached on 24 April) the machine was lifted from the water onto the jetty for servicing which took two days. Once more on their way on 27 April, the flight almost suffered a set-back at Mission Bay when the seaplane was stranded by the tide and a float was damaged in getting off. With one float half-full of water, the aircraft barely managed to become airborne and make it to Broome where Corporal George Gottschalk awaited them. When they left Broome on 29 April the leaking float was still taking water, and the machine now had Gottschalk on board as an additional member of the crew.

By the time they reached Carnarvon the plane had major problems with its engine, which was constantly losing power due to sticking valves.

After several days work failed to rectify the problem, the decision was made to send for the spare engine available at Perth, 1100 kilometres away. This duly arrived four days later, accompanied by the two mechanics, and work immediately commenced to fit the new engine. By 11 May, A10–3 was airborne again and reached Geraldton. Here the party had to beat off the unwanted attentions of swarms of unruly children, who very nearly brought about the undoing of the whole air party. Shortly after setting off again bound for Perth, a problem with the auxiliary fuel tank brought about a forced landing in the open sea which was running a high swell. It was discovered that inquisitive hands had interfered with the drain-cock on the fuel tank, causing a gradual loss of fuel during flight. Taking off again now proved a hair-raising experience, with the seaplane's lower wings awash in the sea. McIntyre recalled this moment as 'the narrowest squeak we had during the whole of the flight, and it might have cost us the machine'[55]—not to mention their lives.

Once Perth was reached, the machine's crew reverted to two with the departure of Gottschalk. The next day Goble and McIntyre began the easterly transit of the continent, heading for their starting place. Around 2 pm on 19 May, A10–3 arrived over Point Cook and was joined by twelve RAAF aircraft which escorted it to St. Kilda. Awaiting the triumphant airmen was a crowd of 10 000, who chaired them along the pier. Their achievement was universally recognised as a major aviation achievement, and the two were given extensive recognition. Both were made Companions in the Order of the British Empire (CBE)—a highly unusual award in the case of an officer of McIntyre's rank (he was also now promoted flight lieutenant, backdated to 31 March)—and both received monetary grants from the government.

Although of signal importance in the context of Australia's own aviation progress, the flight was widely acclaimed overseas as well, being hailed in Britain as 'the finest flight in the history of aviation'. Early in 1925 the Royal Aero Club announced the award of the Britannia Challenge Trophy of 1924 to Goble and McIntyre for 'the most meritorious performance in the air during the year'. Goble had the satisfaction of being in London by this time, having proceeded to England on Williams' return to Australia, and received the trophy at a special function given in his honour. On hand for this occasion was C.J. Fairey, chairman of the aircraft company which manufactured A10–3. He declared that Goble's flight 'had re-established in the eyes of the world the prestige of British Aviation which now outshone that of foreigners after their initial post-war supremacy'.[56] These were remarks which suggest the real context in which the achievement of Goble and McIntyre derived its significance. As Fairey himself had told Goble in a letter dated 27 June 1924:

> Your achievement was not less welcome for the reason that British Aviation had suffered a temporary eclipse by the performance of [Captain] Pelletier D'Oisy [of France] on the Breguet coinciding unfortunately with the failure of, or at least delay in, the Round the World trip of [RAF Squadron Leader A.C.S.] MacLaren. The effect of your achievement on British Aviation is highly appreciated here and recognised in the Press.[57]

While the achievement of the round-Australia flight was in no way belittled by its promotion in the terms suggested by Fairey, it becomes clear that Goble and McIntyre were lionised as loudly as they were at least partly because an achievement in this part of the empire helped recover British prestige lost elsewhere.

Not to be outdone by his arch-rival, barely two years later Williams also hit on the idea of an epic flight. The route he proposed to follow was to take him to Papua, New Guinea, Solomon Islands, New Hebrides, New Caledonia, Fiji, Tonga and Samoa, then return—a distance of some 27 300 kilometres. The justification for this undertaking was flimsy compared with that advanced by Goble, as, apart from the obvious case based on gaining knowledge of geography and flying conditions in an area of strategic interest to Australia (did this really warrant the personal attention of CAS?), the principal reason advanced was press reports that 'certain Foreign Powers' were planning just such a flight for the end of 1926 and 'it was desired that British Service aircraft should, if possible, be the first to visit by air the British possessions in the Pacific'.[58]

Unfortunately for Williams' plans, his timing came unstuck when the approval of the Minister for Defence was not obtained until August 1926. The DH50A (A8–1) which he intended to use on the flight, newly arrived in Australia, had not then been erected or tested in the seaplane role. As he admitted in his report, the delays consequently suffered 'prevented the originally proposed crew getting the experience in handling the machine necessary before starting out on a long flight'. By the time full-load tests were possible it

Wherever the Fairey Seagull touched down on the round-Australia flight, it attracted the interest of curious residents. Here it is Southport (Qld), as Goble and McIntyre prepare to leave on 10 April 1924. Interference with the aircraft by unruly children at Geraldton (WA) nearly had tragic consequences, when the aircraft was forced to land in the open sea. Again at Albany (WA) the attentions of curious children became so troublesome that police protection had to be sought. (*Cdre J.D. Goble*)

Ten thousand Melbournians were on hand to welcome Goble and McIntyre as they beached A10–3 at St. Kilda on 19 May 1924, at the end of the 44-day flight which took them right around the Australian coast. (*Cdre J.D. Goble*)

was already September, and chances of beating the start of the north-west monsoon season were slipping. For this reason Williams surrendered his hope of flying this trip himself and added an experienced seaplane pilot to the crew in the person of Flight Lieutenant McIntyre, who had flown Goble two years earlier. This was a change which relegated Williams to the role of passenger riding in the interior cabin, which he accepted reluctantly and, according to some accounts, with such bad grace that there were subsequently bitter feelings between McIntyre and himself following the flight.

Setting off from Point Cook on 25 September, accompanied by a mechanic, Flight Sergeant Les Trist, the flight was soon experiencing a share of adventure to match that of the earlier round-Australia effort. The engine had to be replaced at Sydney, and the new engine also developed a problem when the aircraft was 10 kilometres short of Southport (Qld) on 29 September, suddenly losing power and forcing the heavily laden seaplane to land on the open sea. The problem was quickly located in the throttle control and remedied, but it was then impossible to fly off again because of the ten to twelve-foot swell running. Instead it was decided to taxi to Southport, although there was not a lot of petrol remaining and darkness was approaching. Recorded Williams later:

> We were 1 to 2 miles out to sea moving parallel to the coast with huge ocean rollers breaking between us and the shore. We endeavoured to make the entrance to Southport but did not arrive in the vicinity of that entrance until nearly 6 o'clock when the breakers and the spray from them together with the failing light prevented us from picking it out. The position was by no means envious [sic]. The DH50A had no wing tip floats and a wave over the wing would certainly have capsized the machine. Flight Lieutenant McIntyre by watching the waves and

> skilfully operating engine and controls avoided this, but it can be imagined that the machine was at some peculiar angles at times.

It was decided to attempt to take the aircraft through the surf and onto the beach. This was accomplished, though not before McIntyre was twice thrown out of the aircraft and regained the cockpit only with difficulty. Fortunately they had landed only 300 to 400 metres south of the Southport entrance and the firing of a Very light soon brought helpers to the scene.

Although safely ashore the party now discovered that the aircraft needed a new propeller. None was immediately available, although there was a half-finished one at the Experimental Section back in Sydney. This was sent for, but had to be completed before despatch and was not finally received until 4 October. While at Southport Williams now received a telegram from Melbourne advising that the Acting Minister for Defence (the Minister—Howse—was in England for an Imperial Conference) was concerned that the various delays might have 'jeopardised the possibility of carrying the flight to a successful issue', and wanted a report from Thursday Island on the prospects for success if the flight was permitted to proceed, and—if approved to go on—once more when Rabaul was reached.

In the event Williams cabled the Department from Rabaul on 18 October, stating that he considered the flight had time to get to the Solomon Islands before the wet season commenced. A reply was received approving this course but requiring a further report once that place had been reached. A complete engine overhaul was carried out while at Rabaul and it was not until 29 October that the DH50A finally embarked on the next leg across New Ireland to Nissan Island. Arriving at this place an inspection of the engine revealed an internal leak which necessitated the replacement of the cylinder block. But it was realised that if the party returned to Rabaul to fit a new block 'further time would be lost and it seemed probably that our permission to do the Solomons might be cancelled'. Accordingly they decided to press on, reaching Tulagi on 3 November.

By this time the troublesome engine failed completely and the party was obliged to spend nineteen days awaiting the arrival of a replacement from Australia and installing it. At this stage, too, the decision was taken in Melbourne that—because of the lateness in the year—they should proceed no further but instead return to Point Cook, which they reached on 7 December. On their arrival they were welcomed by a formation of twelve Air Force planes, and a 300-strong guard of honour was drawn up as CAS and his crew came ashore.[59]

The 126 hours spent in the air during the flight had covered a distance totalling 16 000 kilometres, and 23 places in Papua, New Guinea and the Solomons had been visited. Most of these localities had never seen an aircraft before, but a few had been called on by one operating from the US fleet during the previous year's visit to the region. The RAAF flight was, in fact, the first out of mainland Australia by a service aircraft, and the DH50A was the first aircraft to bear the RAAF legend on its fuselage to provide national identification beyond Australia's boundaries.

In terms of Williams' personal aims, the flight—although far short of achieving all that had been planned—was also quite successful. It was surely no

The scene on the beach at Southport (Qld) the morning after the DH50A taking Williams, McIntyre and Trist to the northern islands struggled ashore through the surf on the night of 29 September 1926. In the background is the Grand Hotel, with the town beyond that. The entrance from the sea is around the point of land visible over the tail of the plane. (*RAAF Historical*)

coincidence, given the clear 'show the flag' element to the whole undertaking, that CAS had been presenting an Australian presence out in the Pacific Islands at the very time that Dominion heads of government were meeting in London. This allowed a wonderful moment of theatre from the British CAS (Trenchard) who interrupted his presentation to the Imperial Conference on empire air defence to refer to 'the present flight of Group Captain Williams, concerning which, if Mr. Bruce will allow me to show you I have a map here which will show you where, as far as we know he is this morning, and where he is going—to Samoa and back, and that shows the value there'.[60]

Prime Minister Bruce was presumably only too glad to have Australia placed on the map in this way, and understandably cabled from London to congratulate the three airmen on their 'splendid achievement'.[61] Also satisfying Williams' objectives, he was made a CBE—just as Goble had been two years previously; McIntyre was awarded a bar to his Air Force Cross, and Trist received the Air Force Medal.[62]

Even after this success, Williams was not personally finished with epic flights. A year after his Pacific Islands venture, he used the DH50 and two DH9s which had been formed into a Northern Survey Flight to carry out a long reconnaissance around and through Central Australia. The object of this undertaking was to gain 'first-hand knowledge of aerodromes and landing grounds prepared throughout the Commonwealth, the nature of the country along prepared routes and of other areas passed over, flying conditions in the Northern and Central Australia, and of existing organisations for the handling and operating of aircraft as well as their value for defence'. Again, it is surprising that a CAS saw the need to involve himself so directly in an operation of this kind—important though it undoubtedly was—but at least this time Williams was able to pilot his own aeroplane. With him he took Squadron Leader Allan Hepburn, Director of Works and Buildings, and a technical officer, Flight Lieutenant Arthur Murphy, two NCO pilots and two mechanics.[63]

Leaving from No.1 AD on 21 July, the three aircraft went first to Adelaide, and from there to Richmond. After delays occasioned by damage to all three aircraft, on 5 August the restored Northern Survey Flight set off for Brisbane, from where they flew to Darwin, and from there began the long stretch down the coast of Western Australia to Perth. The eastward passage was notable chiefly for the bad weather experienced, but by the time they reached Adelaide the aircraft engines were in need of overhaul. The final phase of the survey began on 2 September, and took the three planes north again, back into Central Australia. Passing through Alice Springs, the Flight picked up its outward route at Anthony Lagoon (NT) and turned east into Queensland. Travelling through central New South Wales, by 10 September they had returned to Laverton. In total the flight had traversed over 19 000 kilometres.

Just three years had passed since Goble and McIntyre first circled the continent, and it was only a year since Williams had travelled a lesser distance but beyond Australia's borders. The 1927 survey was, therefore, still a notable achievement, yet practically nothing is usually heard about it—in contrast to both the other flights mentioned. Such was the rapidity with which Australia was growing used to the strides made in aviation. For his part, Williams felt he

The seaplane undergoing a complete engine overhaul in a Japanese-owned shipyard on the northern shore of Rabaul Harbour, New Guinea, in October 1926. By the time Rabaul had been reached it was clear that the flight's original destination of Samoa was unattainable before the onset of the wet season. (*RAAF Historical*)

appreciated for the first time 'the immensity of this country of ours . . . [and] now had a reasonable picture of the country for whose defence the Royal Australian Air Force should be developed'.[64]

Perhaps the final saga of its kind is the flight undertaken to Singapore and back in February 1938. Again this involved Williams, who this time found himself delegated as the Australian representative at the official opening on 14 February of the recently completed floating dock at the naval base. He later expressed surprise at receiving this news, observing that:

> The Qantas service to Singapore was infrequent—once a week if I remember rightly—and the time of the journey by air from Melbourne was such that I could not be in Singapore by 14 February if I travelled by civil airline. I found myself suspecting that the Government had failed to make a decision as to who was to represent Australia on this occasion until it was too late for anyone to reach Singapore in time for the ceremony except possibly in a Service aircraft, and that was the reason for my selection. I cannot otherwise explain the selection of an Air Force officer as the Government representative on a naval occasion.[65]

He nonetheless succeeded in turning what might have been a non-event into a notable one, by choosing to make the trip in a Gannet aircraft (A14–3). Piloted by Squadron Leader Allen Walters, and with Sergeant K.P. Connolly and Corporal W.T. Taylor on board as fitter and radio operator respectively, this machine successfully completed the task between 6 and 23 February.[66] The trip's significance derived less from the event which prompted it, or occurrences along the way, than from the fact, noted by Williams, that this was 'the first time an aircraft designed and built in Australia had completed such a flight overseas'.[67] Actually, A14–3 was a brand new machine taken on delivery from CAC only two days before departure, and just two days after Walters made his first solo flight on what was for him an aircraft type which he had not previously flown.[68]

Special occasions

Quite apart from the one-off history-making flights just described, there were public occasions which also drew an Air Force involvement. In part, such calls on the service arose because the RAAF was actually the principal element of the permanent defence forces with a presence in proximity to Melbourne. Accordingly, even for naval occasions—such as the visit of the British special service squadron in 1924 (see chapter 2), or the US Pacific fleet the next year—it was the Air Force which provided the guard of honour, complete with band, for the dockside welcoming ceremonies.

The visit by the US Pacific Fleet in July 1925 afforded a special opportunity for the RAAF to establish itself in the public eye in other ways as well. It was reportedly CAS's idea that a Fairey IIID should be flown by Flight Lieutenant McIntyre (Goble's companion of the previous year) and Flying Officer William Walne from Point Cook to intercept a squadron of the fleet off Gabo Island as it proceeded to Melbourne for a scheduled port-call. The weather was bad—so bad that the American admiral had cancelled flying for the seaplanes carried aboard the warships. One historian records that: 'Much to the delight of the

The RAAF Band and guard of honour parade on Prince's Pier for the official landing of Admiral Coontz, commanding the visiting squadron of the US Pacific Fleet in July 1925. The ship in the background is USS *Seattle*. The RAAF guard was under Flight Lieutenant A.H. Cobby and Flying Officer H.A. Austin (*at left, marked B*). (*Mr R. Austin*)

Fairey crew—and surprise and chagrin of the Americans—the US Fleet was discovered, shadowed and contacted by wireless. McIntyre then emerged from the cloud, and flew cheekily around the fleet before returning to Point Cook.'[69]

There was, admittedly, an enormous element of luck, not to mention risk, involved with this successful surprising of a fortunately friendly naval force. McIntyre's machine was, in fact, only one of two Faireys sent on the mission, the other being crewed by Flight Lieutenant Hempel and Sergeant Charles Gates. Gates recalls that:

> On Sunday we took off from Eden before dawn. We had no navigation lights or anything like that, just followed the flames of McIntyre's exhaust. We had the most appalling flying conditions imaginable. When daylight came the sea looked like the top of a mug of beer. I don't think anyone ever flew in that sort of aeroplane in that sort of weather. It was shocking, though I wasn't frightened—I had a job to do. But the two aircraft were changing position, 4000 feet to sea level in a matter of seconds. After about an hour and a half Hempel called that we had lost the other machine, and were returning. McIntyre went on . . . and took a photograph which showed so much foam you could hardly see the ships.[70]

The visit of the Duke and Duchess of York in 1927 was another occasion calling forth an enormous RAAF commitment. To provide a sufficient spectacle for the royal party's arrival in Sydney on 26 March it was judged to be necessary for the aircraft numbers of the local squadron to be supplemented with machines sent from Point Cook.[71] Two DH9s and two DH9As were flown to Richmond to take part in the aerial escort, returning in time to fulfil the same role when the visitors reached Melbourne on 21 April. Unfortunately, this occasion was marred by a horrendous mid-air crash (see chapter 11) which did nothing for the RAAF's image. An informal visit paid to Point Cook by the duke, wearing the uniform of an RAF group captain, helped restore the service's public profile—especially when it was announced that he had also accepted appointment as an honorary group captain in the RAAF.[72]

Early in May the focus of attention shifted from Melbourne to Canberra, where the duke was to open the new seat of Federal Parliament and later take the salute at a mass military review marking the occasion. The air escort for this event was planned to involve five aircraft of each of the RAAF's three main combat types, with a further two of each type on hand as spares if required. This was, in fact, close to the whole of the RAAF's operational strength. To put this number of planes in the air, and still have enough troops to take part in the march past, meant a considerable strain on the resources of so small a service—especially when other personnel had also to be found for a guard at Government House, Yarralumla. The problem was solved by holding No.3 Squadron's annual CAF camp at the same time, so that the Richmond unit was at full strength for the period that the various service contingents were encamped at Canberra. This expedient ensured that the Air Force would have a presence of 410 officers and airmen.[73]

No.3 Squadron arrived first at the national capital on 30 April, occupying the Air Force camp which had been established by an advanced party under Squadron Leader Allan Hepburn on the flats below the military college at Duntroon (now the site of Canberra's airport). The twelve aircraft (four DH9s, four DH9As, four SE5As) arrived minus one of the small SE fighters which suffered a forced landing near Goulburn. The contingent from Point Cook arrived two days later, bringing four DH9s, three DH9As, and three SE5As. On 6 May three Seagull IIIs also arrived from Point Cook, bringing the number of aircraft present to 24.[74]

May was not a good time of year to be sleeping under canvas in Canberra, and apart from frosts there were other hardships to be borne. There was little to the national capital in 1927, and not much for members of the large service contingents to do with their spare time once the day's rehearsals were over. Recalls Bert Gardiner, an airman in the RAAF contingent: 'Canberra was a "dry" town in those days. Each night we were allowed to visit Queanbeyan for a bath and drinks. All three services were exemplary in conduct and [there were] no inter-service brawls.'[75] As the local Canberra newspaper reported, however, some of the visiting troops were passing the time by riding the bus between the railway station and the centre of the town, and back—as there was nothing else to do.[76]

For most of the week preceding the ceremonies on 9 May the practising proceeded in the air and on the ground. The Canberra conditions were by no means pleasant, judging from a report of the rehearsal on 5 May, during which:

The Air Board and other senior RAAF officers pictured before departing from RAAF Headquarters to attend ceremonies connected with the arrival in Melbourne of the Duke and Duchess of York, 21 April 1927; *from left*: Mr A.C. Joyce (Finance Member), Squadron Leader W. Palstra, Group Captain R. Williams (CAS), Wing Commander R.S. Brown (acting Chief of the Administrative Staff), Squadron Leader A.T. Cole, Mr P.E. Coleman (secretary, Air Board). Hopes that the royal visit would be a fillip for the Air Force's public image were largely dashed by a string of aircraft crashes. (*Miss G.E. Joyce*)

> flights of aeroplanes in formation circled overhead, and provided thrills by flying low down over the troops. At times they surged, with throttles wide open, into the teeth of the 30 mile an hour gale, making heavy weather. Then, turning, swept by again on the wings of the gale. Some daring pilots took their machines high up towards the racing clouds, and carried out private rehearsals of 'stunts'.[77]

On the day of the actual ceremonies, the RAAF planes took off shortly before 10 am to provide an air escort for the royal party as it made its way from the rural isolation of Yarralumla. When the royal carriage arrived at Parliament House, the RAAF Band played the salute and the duke inspected the assembled guard of honour from each of the services before proceeding to the building's portico. Records one account of the subsequent events: 'As Dame Nellie Melba stepped forward to sing the National Anthem, an RAAF squadron flew low overhead, drowning out most of the diva's performance. The Prime Minister made himself heard, but when the planes passed over a second time, the Duke's reply was all but lost.'[78]

During the afternoon an estimated 35 000 people watched the military review, with one newspaper report remarking that the marching of the RAAF contingent 'equalled that of a crack infantry regiment'.[79] Unfortunately the event was again marred by the fatal crash of one the aircraft in the RAAF formations overflying the parade ground (see chapter 11). When all was said and done, the crashes associated with the 1927 royal visit would be remembered long after the rest of Air Force's contribution had been forgotten. And, sadly, the trail of wreckage had not yet reached its end.

On 10 May Sergeant Orm Denny was flying back to Point Cook in SE5A A2–11, carrying with him official photographs of the previous day's ceremonies at Canberra which were to be included in an album for presentation to the Duke of York to commemorate his Australian visit. Unfortunately, after leaving Cootamundra Denny encountered a storm, which he attempted to avoid. His detour took him over rugged country around Mount Buffalo and at this most inopportune time he noted his oil pressure begin to drop. With no clear area visible in which to attempt a landing, the seizure of the engine a short time later presented him with no option but to hope for a heavily foliaged tree to cushion his impact.[80]

Denny awoke some 30 minutes after coming to earth, the wreckage of his aircraft strewn about him but with no more injury to himself than a cut forehead. When the storm he had been attempting to avoid swept the area a short while later, he was forced to seek the shelter of a hollow log for the night. The next morning he attempted to find his way out of the bush, soon abandoning his heavy flying suit (with the photographs safely wrapped inside) which he placed in a hollow log. After walking all day without food or water, he stumbled across a trail and was following it when he came across a man tending grazing stock. Given shelter in the man's nearby farmhouse, the following day Denny and his host rode into Mansfield (Vic), some 30 kilometres to the west, to advise RAAF Headquarters of his safe whereabouts.[81]

Subsequent efforts to locate the crash-site were not initially successful, and it was some time later that a bushman involved in the first attempt to find the wrecked SE5A reported that he had come across it. Denny then led another party into the site and retrieved all salvageable parts, but without finding his flying suit containing the photographs. These remained one more casualty of the less-than-glowing saga surrounding the royal visit.

Other notable events in which the RAAF took part included the celebrations to mark the opening of the Sydney Harbour Bridge in March 1932. All the aircraft available at Richmond (nine Wapitis, three Moths, three Seagulls) were involved, providing formations which would dive in salute at the moment that the premier cut a ribbon across the roadway.[82] The signal for the orbiting aircraft to play their part would be a large ribbon which would fall automatically from beneath the bridge deck to the water below, and when that signal came the salute was duly given. Ulex Ewart, commanding the RAAF aircraft on that day, recalls that they had no way of knowing at the time of what had happened on the bridge, with a New Guard horseman armed with sword beating the premier to the ribbon![83]

There were, of course, very many other special occasions which achieved the object of showing Australian taxpayers that they had an Air Force operating for national defence purposes. It has not been possible to do more than pick a few of the more notable, and to indicate that when things went other than as was expected there was always the risk of the resulting effect being one of embarrassment rather than kudos.

A RAAF formation swoops low over the mass military review held at Canberra before the Duke of York on the afternoon of 9 May 1927; a following formation can be seen as blurs against the hill in the background. The review took place to the east of Capitol Hill on which the new Parliament House now stands. The machines pictured appear to be Avro 504Ks, which records do not indicate were actually present during the Canberra ceremonies—raising the question of whether this photograph is genuine or a composite produced later. (*NLA*)

Air displays and races

Another means—closely allied to the ceremonial occasions described above—of letting the Australian public know about the existence of their Air Force, was through participation in highly popular aerial pageants. Apart from taking part in the annual display of the Victorian section of the Australian Aero Club at Essendon, the RAAF went to events further afield, including State capitals outside Victoria and New South Wales, as well as smaller country centres. Annual training flights undertaken to Western Australia and Tasmania from 1932 afforded other opportunities to present displays of Air Force machines and skills to the general public in places where there was usually a scant defence presence, thereby demonstrating a point about the flexibility of air power which the service was only too pleased to make.

Again, the frequency with which these events were staged prevents more than cursory treatment of this aspect, however, this very frequency was itself a comment on the Air Board's willingness to accede to requests from organisers of such events. Clearly it was felt that—quite apart from the publicity generated—participation in these displays provided pilots with good practice of relevant air skills, in a setting somewhat more exciting than that obtained by simply practising over their home base, while attendance at a more distant locality meant that they were afforded additional experience in cross-country flying and greater familiarity with a diversity of flying conditions. Certainly this was the view of Squadron Leader John Summers who led three Bulldogs of Fighter Squadron across to the west in October 1932. Flights of this nature were, he said:

> of the utmost value for it teaches the personnel their limitations, gives them confidence and they gain most useful experience in the handling and maintenance of their aircraft under adverse conditions. Experience gained under these conditions is never forgotten.[84]

Certainly if mechanical problems or forced landings were experienced in remote areas, personnel were likely to learn more about adverse conditions than they would care to remember. When Flight Lieutenant P.G. Heffernan had a forced landing in a Bulldog (A12–7) 16 kilometres from Fisher, on the Nullabor Plain, during the 1935 training flight, a major headache was presented in recovering the machine to somewhere affording facilities for repair.[85] As Summers (again in command of this flight) remarked, this experience 'definitely shows the dire results that might occur with machines of small

This view is from an authorised RAAF flight between the spans of the uncompleted Sydney Harbour Bridge. Air Force machines from No.3 Squadron were also on hand to provide an aerial salute during the official opening ceremonies in March 1932. (*Mr J.C. Miles*)

Flight Lieutenant F.R.W. Scherger refuels his Bulldog (A12–1) at Nhill (Vic) on 15 October 1931, on his way to South Australia for the Aero Club pageant. A gale was blowing when he landed at Nhill, causing him to stand the machine on its nose—though with no more damage than slight bending of the right lower wingtip and engine bearer plate. Scherger completed the flight to Adelaide, won the aerial derby (against another Bulldog) with a speed of 160.98 mph, and returned to Point Cook without any further trouble. (*Mrs J. Gray*)

endurance and absurdly low cruising speed. It is considered that the maximum safe range under all conditions for Bulldog aircraft is 250 miles'.[86] Considering the risks entailed to pilots on such training flights, it is reasonable to wonder whether enthusiasm had not outstripped common sense for the sake of being able to mount effective air displays in Perth.

Apart from participating in other people's airshows, the Air Force occasionally organised some of its own. As mentioned in chapter 2, the first air display staged by the RAAF was held as early as 1924. Ten years later it was responsible for another memorable event organised in conjunction with centenary celebrations of the State of Victoria and the City of Melbourne. These celebrations had actually involved the RAAF in a variety of ways, beginning with the AMS on the Air Board, Wing Commander Adrian Cole, who became one of the organisers of the Melbourne Centenary Air Race from London. Recalled Cole in his unpublished memoirs, when he agreed to the suggestion of the Lord Mayor in March 1932 he little realised that 'I would get myself twenty months' hard work, without pay, [and] with loads of scurrilous and other criticism'.[87] Among the charges he would have to defend was a complaint that he was not cooperating with the Air Board but rather showing preference to the Centenary Council.[88]

One facility of enormous benefit to the Air Race committee was the RAAF's shortwave radio link with London, which Cole says handled the transmission of over 600 long cables regarding race arrangements. The RAAF's radio technicians also proved invaluable in providing communications for the race contestants as they transited Australia in October–November 1934. A mobile

The recovery of Flight Lieutenant P.G. Heffernan's Bulldog (A12–7) after it made a forced landing on the Nullabor Plain, ten miles east of Fisher (SA), in September 1935. Although Heffernan landed only 100 metres north of the railway line, it was found that the rail flat-top trucks were too narrow for the aircraft's undercarriage and had to be widened by lashing planks across the truck. The plane's length also meant that the front wheels and tail wheel rested on separate trucks! The overhang of the wings presented additional difficulties in getting past railway signals. (*RAAF Historical*)

direction finding station was despatched to Charleville (Qld) to provide bearings for those race aircraft fitted with radio and able to obtain this information.[89]

Once contestants arrived in Melbourne their aircraft were housed in hangars at the RAAF base at Laverton, which on 10 November hosted an air display involving some 40 aircraft. An additional drawcard to the event was the Duke of Gloucester, who was on hand to present the Air Race prizes. The crowd attending the airshow was vast, resulting in what one RAAF member attending has described as the 'greatest public relations fiasco of all time':

> The crowd was estimated at being a quarter of a million, which had not been expected. Very little had been arranged in the way of extra toilet facilities. Traffic was so congested that to open the show the Duke had to be taken up Ballarat Road to Bacchus Marsh, across country to Werribee, and in by a back road. At nine o'clock that night cars were still banked up as far as North Melbourne station trying to get to Laverton for an evening barbeque that was put on. Then it rained. I was told that next day local farmers were making a fortune using tractors to tow out cars at seven pounds a time. For a long time afterwards 'barbeque' was a forbidden word at Air Force headquarters.[90]

Possibly to mollify Sydney-siders following press complaints the previous year that the RAAF was 'snubbing' that city when it came to providing air shows there,[91] the display given at Laverton was repeated at Richmond on 26 November. Learning from the experience of the eager Melbourne crowd, a large attendance was planned for and special train and bus services were provided to enable people to get to the airbase. In the event an estimated 15–17 000 people turned up to watch the show. When heavy rain intervened mid-way through the program of events, there was a stampede by spectators for the shelter of the open hangars, but by most accounts the problems experienced were nowhere near on the scale of Laverton.

Four years later, the Air Force was moved to mount another air display—only this time the Melbourne venue was the Flemington racecourse! Over 80 aircraft took part in this event, staged on 9 April 1938, which was—probably correctly—hailed in press reports as 'the largest and most spectacular aerial demonstration yet witnessed in Australia'. An impressive program of events lasting four hours was executed without mishap in front of an audience estimated at 170 000—probably only half of whom paid to enter the racecourse, in preference to watching from vantage points outside. This display, too, was repeated at Richmond two weeks later, on 23 April. Unfortunately, during rehearsal the day before this replay, there was a fatal crash in which Pilot Officer Lance Sutherland, the adjutant of No.2 Squadron, was killed practising a slow roll close to the ground in a Bulldog.[92] On the actual day of the display, this time it was Richmond's turn to experience crowd difficulties. Recalls one officer of the unit:

> It was an absolute shambles because it was next to impossible to get there. I know my mother tried to come out to see the display, and she never even got off the train. The traffic jam was so great, all along the Windsor Road, that people were turning around and going back. Apart from that, the displays were enormously successful. It was amazing how many people would come to see them. Of course, prangs were pretty regular in those days and I reckon a lot of people used to go in the hope of seeing one.[93]

In November 1934 the RAAF held an air display at Laverton in conjunction with celebrations marking the centenary of the State of Victoria and the City of Melbourne. Involving some 40 aircraft, shown here lined up on the grassy verge of the tarmac, the display drew a vast crowd which was entirely unanticipated by Air Force organisers and scenes of chaos ensued. (*Mr P.C. Eaton*)

RAAF flypast over Sydney for the departure of the Duke of Gloucester on board HMS *Sussex*, 1935. (*SqnLdr W.J. Linnett*)

RAAF participation in air shows was not just confined to demonstrations of formation flying, aerial combat, aerobatics and the like. Air Force pilots and machines were often competitors in the speed event, or 'aerial derby', which was a usual feature of aerial pageants of the period. The 1928 New South Wales Aerial Derby at Mascot, for example, was won by Flying Officer Norman Mulroney in a DH9A (A1–28) which had been specially rigged as a single-seat machine. An officer, Flight Lieutenant George Wells, also took part in the Victorian Aero Club's aerial derby at Essendon in May 1928 using a modified DH9A. He enjoyed less success, however, sustaining severe head injuries when his machine struck high tension wires and crashed. While it was not unreasonable for Wells and Mulroney to compete in events like this, the practice of modifying an Air Force aircraft for the purpose seems questionable. As one ex-RAAF member of the period observes:

> It must have been sheer vandalism for someone in authority to have authorised the mutilation of a piece of military equipment—one of the best aeroplanes we had—to participate in a Sunday afternoon show lasting a couple of hours. Wells' accident may not have been attributed to the alterations to the aeroplane, but these may have contributed in some way by altering the handling characteristics.[94]

Despite Wells' misfortune, others continued to enter races of this kind, such as Flight Lieutenant Charles Eaton, who flew a Gypsy Moth in the air derby conducted between Sydney and Perth in conjunction with Western Australia's centenary celebrations in late 1929.[95] In March 1930 a RAAF Wapiti flown by Flying Officer Andrew Gerrand had a 'brilliant win' in the aerial derby at Parafield (Adelaide).[96] Squadron Leader John Summers won the 1931 Victorian Aerial Derby in a Bulldog (A12–3), achieving an average speed of 163 mph,[97] a success emulated in the same competition the next year when Flying Officer Robert Dalton also won in a Bulldog (A12–6) with an average speed of 156.5 mph.

There was, however, sometimes a bad impression created by RAAF participation, even when successful, as Flying Officer Wilfred Compagnoni discovered when he competed in the 1935 New South Wales Aerial Derby in a Hawker Demon from No.3 Squadron. Although he won the event it was with the comparatively slow average speed of 143.6 mph.[98] He recalls that after coming in to land he was asked by some of the aero club members 'What was wrong with you . . . Why were you so slow?' In reply he could only point out the constraints which prevented his Demon from giving maximum performance, such as the 1000 feet altitude imposed for the race and the short length of the course to be flown, all of which ensured that the machine was 'lucky to be doing 160' let alone the 200 mph top speed claimed for this type.[99] Regardless of his explanation, his 'win' was liable to be viewed in some circles as a poor reflection on the aircraft operated by the Air Force in the nation's defence.

In fact, participation in events like this often produced situations which cast the Air Force in less than favourable light. As commander of the three Bulldogs from Fighter Squadron involved in the 1933 annual training flight to Tasmania, Flight Lieutenant Fred Scherger was later spurred to complain about aspects of their program which required them to participate in pageants at Smithton, Wynyard, Western Junction (Launceston) and Brighton (Hobart). Following his return, Scherger reported that organisers of the Launceston and Hobart displays had advertised that the RAAF Bulldogs would perform aerobatics in formation. This was, however, beyond the standard of training reached by pilots of his squadron, 'and the subsequent non-performance of formation aerobatics must have led to some comment'. It should, he suggested, be a requirement for aero clubs to clear their advertised programs with RAAF Headquarters beforehand, because such instances as this harmed the service.[100]

Scherger's advice was evidently not acted upon, because almost an identical thing happened during the next year's training flight to the same State. As before, in addition to cross-country flying training, it was planned that the RAAF party would participate in pageants at Smithton, Launceston, Wynyard, Latrobe and Hobart. Following his return, the commander of the three Bulldogs on the flight, Flight Lieutenant Frank Wright, commented on precisely the same problem as experienced the previous year. Pageant organisers had programmed events which he had not been prepared to undertake. At Latrobe, it was discovered that the Bulldogs were expected to engage in balloon bursting, which he refused to allow because he considered it wrong 'to belittle the training of air force pilots by taking part in useless displays of this kind'. Wright also noted that, although the aero clubs were always very anxious to have RAAF aircraft included in the main race of the day, 'they invariably handicap them out of the race'. He regarded this as an entirely undesirable practice, both for its effect on young enthusiastic pilots, who pushed their aircraft in the attempt to win and thereby did damage to their engines, and also for the false impression of the speed of Air Force machines that was conveyed to the general public which generally did not stop to consider the impossible handicap.[101]

By 1937 there was a quite strong feeling against the extent of the RAAF's involvement in certain of the civil activities being undertaken. As Williams discovered at a CAS Conference at Headquarters in April, COs were concerned that, while civil tasks had definite training value, they were also very disruptive of progressive training being carried out of units as a whole. The CO at Richmond (Group Captain Cole) was even prompted to suggest the formation of a special flight to handle these tasks. Williams was apparently not disposed to take these objections very seriously. The minutes record that he responded that 'these tasks were of value from a political and propaganda point of view, and he thought their training value was very considerable'; there was, he said, no financial prospect of forming a special flight.[102] Within a year of making these remarks, Williams was to find himself forced to bear the consequences of the inadequate unit training of young RAAF pilots. Interestingly, his arch-rival Goble would point to the amount of time spent on practising for air displays as taking away from more important training in operational skills (see chapter 11).

14

Furthering national goals

In addition to the non-defence activities described in the previous chapter, from time to time the RAAF was called upon to carry out tasks which, by virtue of their scope or impact, or the nature of the interests involved, meant that the Air Force was effectively employed as an instrument of national policy, either by 'showing the flag' or by assisting the national government to advance development goals and objectives. Some activities and events already mentioned can be said to have contained elements of this purpose—such as the early plan to have the RAAF pioneer civil airmail services, the historic flights by Goble and Williams, and the geological surveys of Central Australia. This chapter describes the major instances in which the RAAF became involved in such activities.

Great Barrier Reef survey

Soon after formation, the RAAF found itself involved in a program to chart the Great Barrier Reef. In 1921 the federal government decided to commit defence resources to surveying these treacherous waters, which in most places had not been examined since Matthew Flinders' passage in 1802. The RAN survey sloop HMAS *Geranium* was put to the task, starting in waters south of Cairns, but it was quickly appreciated how massive was the task involved. Accordingly, the assistance of the Air Force was sought with the provision of a seaplane to enable large areas to be covered more rapidly.

Planning for the RAAF's involvement began in 1923, with a Fairey IIID being sent to Sydney to take part in photographic experiments conducted between April and June.[1] Not until the following year, however, was a start actually made. Necessary alterations to the *Geranium* were undertaken in February to provide space for an embarked aircraft and the group of two officers and eight airmen who would operate it. Trials using an aircraft followed in June, which showed that, while the Fairey could be safely borne in sheltered waters inside the reef, carriage in open seas had a marked destabilising effect on the *Geranium*. Consequently, when the vessel sailed for northern waters late in June, it did so without an aircraft on board. After the *Geranium* had arrived at Townsville early in July, the machine assigned to work with the sloop was flown up from Point Cook later that month and handed over to Flight Lieutenant Ern Mustard.[2]

Despite the interruption of two months of bad weather, the Fairey performed valuable work before the *Geranium* returned to Sydney in December and the aircraft to Point Cook. Flown by Mustard and Flying Officer Tom Swinbourne, some 2000 square kilometres of uncharted water was surveyed and mosaic photographs from an altitude of 10 000 feet were also taken of 19 reefs. Although the undertaking appeared to have been quite successful, the results accomplished were not all that had been hoped for. The original intention had been that the production of photo mosaics would permit the preparation of charts. It was found, however, that air photographs gave inadequate definition beyond a depth of five fathoms, and the aircraft had to be employed as an adjunct to more conventional survey methods. The commanding officer of the *Geranium* also considered the air party accompanying the Fairey was more than he cared to have on board, and he also complained that the RAAF tender was too cumbersome and slow.[3]

Mustard and Swinbourne were again on board the *Geranium* when it sailed for Townsville in June 1925. This time two Faireys were shipped by cargo steamer and required assembly before one of the aircraft was flown out from Townsville to join the survey sloop on the Reef. The RAAF party spent the next five months with the Navy, making two or three photographic flights each week.[4]

In July 1926 the RAAF formed No.101 (Fleet Cooperation) Flight with newly acquired Seagull IIIs to work with the Navy's survey ship HMAS *Moresby* on the Barrier Reef in the region of Flinders Passage. A ground party of 36 men under Flying Officer Herbert Warland moved first, arriving at Bowen by rail on 26 August 1926.[5] The three aircraft of the flight (A9–1, A9–2, and A9–3), piloted by Flight Lieutenant A.E. ('George') Hempel, and Flying Officers David Ross and Joe Hewitt (both RAN officers on secondment to the RAAF), flew up from Point Cook on 23 August along an inland course.

Prior to departure, Ross and Hewitt had practised landings and alightings in confined spaces, the latter writing that they capped off this training with a display of their abilities in the heart of Melbourne, by:

> flying over Church Street Bridge, South Yarra at about 200 feet and 50 feet over Anderson Street Bridge, touching down on the Yarra with sufficient distance ahead to clear Princes Bridge below the level of the top storey of Flinders Street Station. (The Swan Street Bridge was not there in 1926.) . . . the media gave our flights some indulgent criticism [and] I was called to account as leader . . . Group Captain Williams, who asked me why we had touched down in the Yarra, was reservedly pleased about the publicity, and Hempel too.[6]

Only two of the Seagulls made the flight north without mishap, arriving at Bowen around 5.30 pm on 27 August; the third machine damaged a wheel and one wing on landing at Styx River to refuel, and did not complete the trip for

From August 1926, the top floor of the Port Denison Sailing Club's clubhouse on the beach-front at Bowen provided living quarters for airmen of the RAAF detachment engaged on the Great Barrier Reef survey. (*Warland collection, RAAF Museum*)

another four days. The appearance of the first two planes over Bowen was an event witnessed by a large part of the town's population who crowded the foreshore. The local paper reported that:

> The planes came along the south-west side of the bay and then across the middle of the harbour and alighted in shallow water on the North-East side of the jetty. The second plane to alight lost the use of its engine temporarily and was blown against the jetty, damaging its nose and both wings. It was towed out of its precarious position by the pilot launch and later on came in over the flat, the tide being out, in great style, as did the other one before it.[7]

The first priority was to establish a shore base for the flight. The Port Denison Sailing Club's two-storey clubhouse on the town beach was leased and, supplemented by some tents, provided airmen's quarters; the sailing club retained the ground floor of the building as a boatshed throughout the RAAF's occupancy. The officers (Hempel and Hewitt subsequently joined by their wives) found hotel or guest house accommodation around the town. The local drill hall nearby was taken over to house the headquarters and photographic sections, while a large timber and iron hangar was erected on the foreshore for the aircraft with a path of railway sleepers laid from the high water mark to the hangar entrance.[8]

While these arrangements were generally satisfactory, most other facilities for the storage of petrol, spares, and tools, as well as aids for maintaining the Seagulls, all had to be improvised.[9] Other difficulties had to be surmounted arising from the fact that Bowen was a highly unionised town, with perpetual strikes particularly on the waterfront when the tomato harvest was ready for shipment. A member of the flight recalls one occasion when the *Moresby* made its weekly call into the port on Sunday, and the two airmen sent to deliver the week's mosaics and the ship's mail were refused admittance to the wharf by union pickets. Hempel promptly formed a guard carrying bayonets and belts, and headed for the wharf, where he walked straight through the picket line without any obstruction.[10]

The principal work of the flight at this time was in fixing the positions of the multitude of reefs in the area from Hook Island, off Bowen, to the outer edge of the Barrier Reef north-east of Townsville, and photographing the islands in the Whitsunday Passage and the coast from Cape Palmerston as far north as Townsville. One map of Flinders Passage which survives from this period bears notes indicating the method employed in the work of fixing reef locations:

> The Reefs were sketched on Field Boards whilst flying above them at altitudes of 5000 to 8000 ft., the sketches being later transferred to a Collector Tracing. The Relative Positions of Reefs were found by Dead Reckoning and Compass Bearings during flight. Positions marked were fixed by Astronomical Observations after alighting there.[11]

The purpose of the photographic missions was to produce mosaics for the survey staff on the *Moresby*, although the mosaics remained only an aid. The *Moresby* still carried out its own soundings, using the mosaics to check its information to identify discrepancies or suspicious objects requiring special examination. Photographic flights were generally carried out at 8000 feet because this was the greatest height attainable by the Seagulls while carrying the necessary load, and it was found that at lower altitudes the mountainous terrain near the coastline produced distortions which hampered interpretation of the prints.

In December the *Moresby* sailed back to Sydney, leaving the Air Force personnel to catch up on maintenance and other tasks, but by February the ship was back on the reef to begin the task of laying out control points at roughly 16-kilometre intervals along the coast. These points were needed for triangulation purposes and obtaining scales, and for determining the exact overlap present in photo runs produced by the RAAF's continuing program of sorties. Additional to this work was the effort to discover any navigable passages through the Barrier Reef between Sandy Cape, off Bundaberg, and Holbourne Island, particularly in the area north and south of Hook Island. As Hewitt had observed of this search for passages, 'It was something of a hide and seek game from the air, much more so for the surveying team in the *Moresby*'.[12]

The work of the survey continued throughout 1927 and 1928, though by the latter year the pace was beginning to show signs of tapering off. At times the RAAF detachment's strength had been as high as nine officers and 46 airmen, at others as low as three officers and 17 airmen. For the 1928 season there were now only two Seagulls in use, and initially only two pilots to fly them: the CO (Hempel) and Flying Officer Frank Crowther, another RAN officer on secondment. Crowther recalls that, as Hempel did not do many photographic missions, the bulk of this work fell on him alone.[13] Later additions to the officer strength during the 1928 season included Flying Officer Denis Creal (who acted as administrative officer for the detachment), Flight Lieutenant Gerald Packer (who supervised the work of the photographic section), and Flying Officer Stuart Campbell, who was, fortunately, another seaplane pilot.

In this period the principal task was coastline photography for the *Moresby* survey staff of the area between Cape Conway, off Proserpine, and Slade Point, just north of Mackay. Mosaics were also made of St.Bees, Keswick, Carlisle, Brampton and Cockermouth islands.[14] Writes Crowther:

> [I]t was my job to take vertical pictures from 8000ft, from which our photographic section made a mosaic to the scale of 1/24 000 and subsequently produced a photographic outline of the coastline. It was in fact a very boring task and took several months to complete, as quite often the area was covered with cloud which frequently remained for days.[15]

In September 1928 the flight formed an advanced base of two officers and seven airmen on St.Bees Island, some 30 kilometres north-east of Mackay. This was done to facilitate a reconnaissance survey of the outer edge of the Barrier Reef, approximately 160 kilometres due east of Hook and St. Bees islands, searching for navigable passages and to enable preparation of a sketch plan of the main reefs in the area. To quote Crowther again:

> All the personnel lived under canvas and the Brooke boat was brought from Bowen for air rescue work if required. A small fuel dump was made at Penrith Island for use in emergency. *Moresby* used to bring out our aviation spirit in 65 gallon drums, which were rolled over the ship's side and towed to the beach. In all about ten sorties were made but no through passages were found.[16]

Operations were ended in November 1928 and all RAAF personnel were withdrawn to Point Cook in preparation for embarkation in the Navy's new seaplane tender, HMAS *Albatross*. According to the *Bowen Independent* on 1 December 1928, the local community was sorry to see them go:

> The members of the Royal Australian Air Force 101 Flight are leaving for Melbourne. A number went by the steamer Kanowna and the remainder will go by plane in a few days time. The Air Force will be greatly missed here, from a social and also from a business standpoint . . . Included in the Kanowna's passenger list were airmen E.J. ('Blue') Littlejohn and Bob Kennedy and their wives, sisters Jenny and Myrtle (nee Miller), also airman Madden and his wife Molly (nee Blake), three of the Bowen girls to marry members of the Flight.

The work of the flight in making the waters off North Queensland's eastern coast a safer area for navigators has gone largely unheralded over the years. Probably the most visible legacy of RAAF involvement was that created by the survey staff of the *Moresby*, who gave the many previously unrecorded reefs the names of the ship's officers and Air Force personnel. Accordingly, features such as the Hewitt, Ross, Warland, Creal and Packer Reefs, still appear on maps of the Great Barrier Reef Marine Park, along with reefs presumably named after the Southampton flying-boats operated by the RAAF and the Bugatti motor-car privately raced by Hempel.

New Guinea oil survey

Even while the Barrier Reef survey was underway, the RAAF became involved in another such exercise beyond the Australian mainland. Since 1920, geological exploration had been going on in the Australian territory of Papua and the mandated territory of New Guinea in an attempt to uncover deposits of oil. This search was carried out after 1921 under a series of temporary agreements between the Australian government and the Anglo-Persian Oil Company. Under an agreement made in 1927, another survey team was to be sent in, and to assist this expedition the Department of Home Affairs and Territories asked the Air Force to conduct a preliminary aerial reconnaissance to identify areas worth close inspection by field parties and to produce maps of these areas by aerial photography.

For this essentially experimental undertaking, a Papuan Survey Flight was formed at No.1 AD on 2 August 1927.[17] It comprised six personnel—Flight Lieutenants Ellis ('Punch') Wackett (OC) and Ross (pilots); Flying Officer Packer (photographer); Corporal G. Simpson and LAC Ball (both fitters aero), and AC1 Bray (carpenter rigger and boat builder)—with two Seagull IIIs. The flight was ordered to reconnoitre the coastal districts of Papua and New Guinea as far north as Aitape during the interval between the dry and wet monsoon seasons. If time permitted, it was also to cover Bougainville Island for the administration of the British Solomon Islands Protectorate.

On 27 September the two aircraft (A9–5 and A9–6) set off from Laverton and headed north. Though stripped of all fittings not vital to flight—including radio equipment—the machines were still heavily loaded, as the flight was required to be self-contained. Petrol supplies had been placed at points around the coast from which they would operate, and stores and spare parts were prepositioned at Samarai. But even so, with four hours' fuel on board, in addition to the three-man crews, engine kit, rations, photographic accessories, medical requirements and personal baggage (limited to 30 pounds per man), the machines still weighed just 250 pounds below the Seagull's maximum load.

Arriving at Thursday Island on 15 October, the actual work of the survey was commenced the next day as the aircraft passed up the Papuan coast west of Daru. A faulty magneto forced A9–6 to remain at Daru on 17 October while the other machine flew to Port *Moresby*, then on to Samarai, to collect a replacement from the stock placed there in advance. When A9–5 reached *Moresby* with the new part on 19 October, it took on board Mr J. Nason-Jones, a geologist employed by Anglo-Persian Oil who had proceeded to Papua ahead of the main expedition.[18] Having previous flying experience, Nason-Jones was to act as observer during the survey and would help direct the work to be performed; to make room for him, Aircraftsman Bray was left at Samarai.

The procedure followed during the survey was for Nason-Jones to decide on the area to be examined and the specific objects of each day's flight before take-off. This was necessitated by the fact that the photographer and Nason-Jones could not be carried in the same aircraft, as in addition to the pilot each machine needed to carry a mechanic in case of a forced landing. Consequently while Packer operated the camera equipment from A9–5, flown by Wackett, Nason-Jones accompanied Ross in A9–6.

Completion of each day's flying was far from being the end of the working day. In his report of the survey, Packer recorded that:

> After arrival at the destination, usually a small government station, several hours would elapse before the machines were refuelled and baggage ferried ashore in

The advanced base at St. Bees Island, North Queensland, established to facilitate reconnaissance of the outer edge of the Great Barrier Reef in September 1928, shortly before the RAAF involvement in the survey was terminated in November. (*Warland collection, RAAF Museum*)

> native canoes. After dusk, the days photographs were developed—a matter of from two to three hours' work—and hung up to dry. At the same time the position of each exposure was plotted on the chart. As a rule, flight was resumed about nine o'clock the following day, in order to take advantage of the best light for photography.[19]

By 20 November the Seagulls had reconnoitred as far as the north-west town of Aitape, 160 kilometres from the Dutch New Guinea border, but the early onset of the northeast monsoon and the absence of petrol supplies beyond that place prevented continuing operations towards Vanimo, on the western border of the territory. The aircraft therefore returned to Madang three days later and there received orders to return to Australia. In taking off, A9–6 developed a problem with its engine which necessitated the lengthy process of obtaining a replacement from Sydney. In the meantime, A9–5 left Madang on 2 December and retraced its route southwards, arriving in Melbourne on the day after Christmas; the second machine did not finally return until 19 January 1928. The aircraft and crews had covered some 17 700 kilometres in the course of the survey, having examined nearly 130 000 square kilometres of country and produced 350 air photos.

According to Packer, the results of the reconnaissance had not been up to expectations. The jungle-covered terrain, particularly along coastal areas where the vegetation was dense, continuous and uniform, usually meant it was impossible to deduce anything about the nature of the underlying soil and strata, and low cloud and rain had prevented photography of many areas. Other disappointments were experienced on account of the limited range of the aircraft. This frequently meant that interesting formations which were observed could not be pursued sufficiently to derive any systematic conclusions. In any event no previously unexplored areas were actually inspected during the reconnaissance; what had been obtained was essentially additional information on tertiary deposits already known, along with up-to-date information on the progress of drilling operations.[20]

The limitations of the Seagulls had, Packer reported, been felt in another significant regard. Although mechanically quite reliable, these aircraft had proven themselves unsuited for service in the tropics. By the time New Guinean waters were reached, their performance operating under full load conditions had steadily deteriorated. After a month it 'became difficult to stay at 400 ft. in still air with full load at normal full throttle . . . On bumpy days it was not possible to climb higher than three or four hundred feet for the first couple of hours of flight'.[21] On the flight's last stages, it was found that 'it took nearly 30 minutes' flying to rise above the level of the tree tops in the Sepik Valley'.[22]

Despite these negative aspects the survey had nonetheless been an important and pioneering venture, given that it took place just a year after Williams' flight through the same region. And although there had been limitations to the results achieved, these had still been sufficiently encouraging for geologists to seek repetition. Packer recorded that during the flight the RAAF party had encountered at the Vailala River the Commonwealth geological adviser, Dr Walter Woolnough, who told them that a single air photograph 'gave him more insight into the nature of the local structure than several weeks on the ground'.[23] After the main Anglo-Persian Oil expedition reached *Moresby* at the end of January 1928 to commence field work, informal conferences were held with the RAAF in Melbourne regarding a survey to provide accurate topographical maps. However, estimates of the costs involved showed that this further work was probably prohibitive.[24] In October 1929 the Australian government decided to cease further geological investigations.

Antarctic expeditions

Another field of endeavour furthering important national objectives took RAAF personnel and aircraft into the frozen reaches of Antarctica. In February 1929 the Australian Prime Minister (S.M. Bruce) announced that, resulting from a decision of the 1926 Imperial Conference, it was planned to despatch a scientific expedition to the southern continent. Sir Douglas Mawson, professor of geology and mineralogy at the University of Adelaide, was invited to organise and lead this joint British, Australian and New Zealand Antarctic Research Expedition (BANZARE), having had experience of two previous expeditions. The stated purpose of this new undertaking was study of the 'Australian Sector' (between the Ross Sea and Enderby Land) in extension of work done during 1911–14, but the real purpose not admitted by Bruce was to claim formal possession of this area.[25]

As part of his preparations Mawson decided to equip his team with two aircraft for reconnaissance purposes, but in the event space limitations aboard his vessel, the Royal Research Society's ship *Discovery* provided by Britain, restricted this to just one. This was not the first time he had attempted to

The two Seagull IIIs involved in the oil survey of Papua New Guinea—A9–6 in foreground, A9–5 at rear—on the beach at Karema on 26 October 1927. This was only a year after the first RAAF flight to the northern Islands. Karema was examined for its suitability as a future seaplane base, with a hangar proposed for where the machines are shown standing. (*RAAF Historical*)

utilise the technology of flight in the Antarctic. For his 1911 expedition he had obtained a monoplane from the Vickers company of Britain; unfortunately this crashed during a trial flight at Adelaide and was rendered unflyable before the expedition even got underway, although it was shipped with the expedition anyway and employed as a tractor-sledge—minus its wings which Mawson offered to the CFS at Point Cook in 1914.[26]

Mawson was possibly not seriously expecting his frail aircraft of 1911 to withstand the sort of flying conditions associated with the Antarctic's harsh climate, judging from his later remark that it had been planned for the machine's capabilities to be 'chiefly exercised for the purpose of drawing public attention to the expedition in Australia'.[27] By 1929, however, he was certainly convinced that aircraft promised to realise new techniques in polar exploration, providing improved methods of negotiating pack-ice and greatly increasing the range of observation.[28]

Perhaps the most difficult aspect flowing from the decision to take an aircraft concerned the choice of type that was best suited to the expedition's needs. Visiting England in the first half of 1929, Mawson was deluged with approaches from practically every manufacturer of small aircraft, with the result that when he departed London to return to Australia he left this question in the hands of R.G. Casey, the Commonwealth government's liaison officer to the British Foreign Office. Casey sought the personal advice of the CAS, Sir Hugh Trenchard. On the latter's recommendation a DH60 Gipsy Moth seaplane was embarked on the *Discovery* when the ship sailed from Cape Town in October 1929.[29]

Although the BANZARE's Moth carried the civil registration markings VH-ULD, its aircrew came from the RAAF. In her biography of her husband, Lady Mawson wrote that Sir Douglas interviewed many applicants for the position of pilot.[30] It seems, however, that Mawson—aware of the success of his Adelaide colleague Cecil Madigan in obtaining Air Force support for the Central Australia survey earlier that same year—intended from the outset to seek his flier from the RAAF.[31] There was reportedly only one applicant for this position, 26-year-old Flying Officer Stuart Campbell, but, as it was considered advisable to have a second pilot on board, Sergeant Eric Douglas was also selected on Campbell's recommendation.[32] Both men sailed from Melbourne on board the *Nestor* to join the rest of the expedition in Cape Town.

When the time came to uncase the Moth and assemble it for flying operations off the Antarctic coastline, problems were encountered with the aircraft's magnetoes which had become damp on the long sea voyage from London. Not until the last day of 1929 did the Moth seaplane become airborne, but now it produced an encouraging success when Campbell and Douglas reported the probable appearance of land on their first test flight. Continuously circling the open pool in which the *Discovery* lay—the ship's master, Captain J.K. Davis, having vetoed their flying over the surrounding pack-ice for fear of being unable to reach them should they suffer a crash—they reached an altitude of 5000 feet. From this height they could see roughly 50 kilometres towards the south, and in this direction observed 'darker shapes standing out boldly against the deadly whiteness'.[33] This sighting was reported on alighting, but the lack of a passage through the ice forced the *Discovery* to retreat to the north-west in pursuit of an easier approach.

On 5 January 1930 the Moth was again sent aloft to reconnoitre, this time with Campbell and Mawson on board. In an article written not long afterwards for fellow members of the RAAF, Campbell recorded that: 'From a height of 2000 feet a long line of bare, black mountain peaks could be seen protruding through the ice-covered slopes of what was undoubtedly the Antarctic Continent'.[34] Mawson was convinced that what they had seen was a whole new section of coastline and accordingly named the area MacRobertson Land, in honour of one of the main patrons of the expedition, Macpherson Robertson, who was Australia's largest confectioner.

On returning to the ship, arrangements were made for another flight early the next morning to locate a suitable landing place for raising the Union Jack and proclaiming British sovereignty over this new find. But before this could occur a raging blizzard set in. Campbell noted that:

> for three days anxious eyes were focussed on the frail seaplane which, with its air speed indicator standing steadily at 65 mph, threatened every moment to be torn from the deck and disappear in the driving snow . . . At length the wind abated and all on board breathed a sigh of relief. But the forces of nature were not yet baffled. During the blizzard the wind-driven snow had collected all over the masts and rigging and coated them with a shell of ice an inch or more in thickness and with the first appearance of the sun this ice thawed and came crashing down from aloft on to the unprotected surfaces of the wings and tail unit and wrought havoc with the fabric and ribs.[35]

Fortunately the Moth was repairable so that later, along the coast of Enderby Land, a running survey was commenced, supported by flights by the Moth operating from sheltered water in the lee of icebergs. On 25 January a number of flights were undertaken in the vicinity of Proclamation Island, several to enable the expedition's photographer, Captain Frank Hurley, to obtain historic still-shots as well as movie footage. During the final flight for the day, Mawson ceremoniously dropped a British flag onto the land ice below, after reading a proclamation extending the expedition's claimed discoveries as far west as the 45th meridian. The next day the *Discovery* turned north and left Antarctica, heading for Kerguelen Island to replenish the ship's bunkers. The Moth was reassembled here and further flying was undertaken. On leaving Kerguelen the aircraft was dismantled for the voyage to Australia. On 31 March the ship reached Adelaide, where the expedition was disbanded. Six days later the *Discovery* arrived in Port Phillip, to be greeted by an escort of three RAAF aircraft from Point Cook.[36]

A second BANZARE led by Mawson departed from Hobart on 22 November 1930 and proceeded south, accompanied again by the Moth seaplane and Campbell (now a Flight Lieutenant) and Douglas (now Pilot Officer) as pilots. The aircraft was assembled four days before Christmas but the aircraft did not actually commence flying operations until after the *Discovery* reached Cape Denison in King George V Land on 4 January 1931. From the first flight—an ice reconnaissance undertaken by Campbell and Douglas on 7 January—a large program of exploration and coastal survey was carried out, despite constant bad sailing and flying weather.

This effort was not carried out without incident or some danger, and on 27 January Douglas and Mawson suffered a mishap after landing and attempting to hook onto the ship's lifting tackle while the ship steamed slowly forward into a swell. After several unsuccessful tries, Mawson managed to get the sling over the hook. But the Moth was suddenly upset by an unexpected roll of the *Discovery* and was left bumping into the ship's side at the end of the lifting cradle with the occupants dangling from it. Mawson and Douglas were quickly rescued and escaped a cold ducking. Although the aircraft was a sad sight when brought on board, things were not as bad as they first appeared and the two aviators soon had it mended.[37] The machine was aloft again on 6 February, displaying no ill effects apart from a slight list to the right. By this stage the expedition was nearing the end of its season, and the final flight of the Moth took place on 11 February.

A week later Campbell and Douglas celebrated their part in the second BANZARE voyage by hoisting the RAAF ensign during a proclamation ceremony at Cape Bruce; Campbell later passed this ensign to RAAF personnel accompanying Australia's 1947–48 Antarctic expedition.[38] The aviators' participation was also commemorated by the naming of parts of the Antarctic landscape in their honour: Campbell had a distinct rocky head on the coast of MacRobertson Land named for him, Douglas a 5000-foot peak lying south-west of Mount Codrington.[39] Both pilots also became the first RAAF personnel to receive the Polar Medal.[40]

Little more than a month later the *Discovery* was back in Hobart and the

Flight Lieutenant S.A.C. Campbell accompanied both the BANZARE expeditions of 1929–30 and 1930–31 as senior pilot. In the opinion of Sir Douglas Mawson the aircraft operations, conducted with determination and skill by Campbell and fellow RAAF aviator, Pilot Officer Eric Douglas, were a most important factor in the success of the geographical exploration program. Sightings made from the air later formed a vital element of Australia's territorial claims. (*RAAF Museum*)

expedition was over. In all, a total of eighteen flights had been undertaken by the RAAF pilots, the success of these having thoroughly proved the value of the Moth seaplane as an observation platform. Mawson himself commented:

> The aeroplane proved a most important factor in the success of the geographical operations. That so much use was made of the machine, operating under difficult conditions, is owing to the determination and skill of the aviators, Campbell and Douglas, whose capacity and ability is of the highest order. They deserve very great credit.[41]

In July 1931, the Moth and its spare engine was sold to the West Australian section of the Australian Aero Club for £650, to help Mawson defray the costs of the two expeditions; the same machine was impressed into the wartime RAAF in July 1940, becoming A7–94.[42] As pointed out in one later account of the 1929–30 and 1930–31 BANZARE voyages, sightings from the Moth had formed an essential basis for the claims made by the expedition. Several of these were crucial elements in a later dispute with Norway over the extent of the BANZARE discoveries and nomenclature given to parts of the continent following Australia's formal annexation in 1933 of the areas it had explored.[43]

Late in 1935 the Air Force again became involved in activities in Antarctica, but this time it had nothing to do with exploration or research. Instead, the RAAF was called on to undertake an aerial search for American explorer Lincoln Ellsworth, who had gone missing in his Northrop Gamma all-metal low-wing monoplane, piloted by Englishman Herbert Hollick-Kenyon, while making his third attempt to fly across the Antarctic continent. Ellsworth and Kenyon had taken off in their aircraft *Polar Star* from Dundee Island, 800 kilometres south of Cape Horn, on 23 November, intending to reach the Bay of Whales, in the Ross Sea 3540 kilometres away. They planned to wait here at the Little America base, where Commander Richard Byrd of the US Navy had wintered in 1929 while attempting his flight to the South Pole, while their support ship *Wyatt Earp* sailed back around Antarctica to pick them up. Instead of the 14 hours planned, the flight ended up lasting 22 days and involved five landings on the ice, the last of which, brought about when the plane ran out of fuel, was just 26 kilometres from their destination. The two men left the aircraft and walked until they stumbled, by luck almost, onto the underground and unattended base at Little America.[44]

The fact that the aerial party was safe was unknown to Sir Hubert Wilkins, an Australian heading the expedition's support group embarked in *Wyatt Earp*, as the plane's radio had ceased operating some twelve hours after take-off. Wilkins became increasingly anxious, and when nothing more was heard from *Polar Star* he initiated the pre-arranged rescue plan. Issuing a general radio appeal for aid on 30 November, he turned the *Wyatt Earp* towards Chile to collect a relief aircraft with which to establish food caches.

On 2 December the Australian prime minister (J.A. Lyons) announced that his government had suggested to Britain that a relief expedition be organised using the Royal Research Society's vessel, *Discovery II*, then engaged on its fourth scientific commission, a circum-polar cruise.[45] The ship was already on its way to Fremantle when agreement to participate in the Australian expedition was given by Britain and New Zealand, and she was diverted to Melbourne.

The first intimations of what this meant for the RAAF came late on 4 December, when Flight Lieutenant Eric Douglas at No.1 FTS was ordered to report to RAAF Headquarters for a conference in the office of the secretary of the Defence Department at 9.15 the following morning, bringing with him data on the types of seaplane currently in service with the Air Force and the period required to make these ready. After the meeting a warning order was given to No.1 Squadron at Laverton to prepare their 'best Wapiti . . . fitted with Jupiter VIIIF engine'. At this moment the squadron was actually on deployment to Cootamundra (NSW), so that this instruction meant a machine with long distance fuel tanks (A5–37) was specially ordered back for the purpose on 6 December. At the same time Point Cook was told that the aircraft being returned was to be completely overhauled, and the special Wapiti float undercarriage held in storage at No.1 AD was to be despatched to Point Cook immediately for fitting and testing.[46]

At this juncture it was decided that, in addition to the Wapiti, a DH60X Gipsy Moth floatplane (A7–55) of the Seaplane Squadron, along with a second DH60, would also be going with the relief expedition. It was envisaged that while the Moths were to be used purely for local reconnaissance, the Wapiti would be called on to make three search flights over the ice up to 300 kilometres from the ship. The Wapiti certainly, and possibly one of the Moths, might—depending on the ice conditions—need to operate from land, rigged as a ski-plane. Drawings were accordingly hastily prepared for the manufacture of two sets of wooden laminated skis to adapt the Wapiti and a Moth for take-off and landing on ice; these had to be made and fitted in readiness for embarkation in about eight days. Squadron Leader Arthur Murphy, in charge of the Laverton workshops, was directed by the Air Board to take charge of all work to modify the two aircraft and provide the special equipment required. All facilities of both the FTS and AD workshops were to be placed at his disposal, and whatever hours needed were to be worked to have everything ready in time.

Work on the skis was a major undertaking involving a great many men, both at Laverton and Point Cook. Air Commodore Rex Taylor recalls:

> I can claim some little association with this venture as it was my job, back at No.1 AD, to help in the design and supervise the manufacture of the skis, particularly the splicing of the heavy steel cables and fabrication of the shock absorbers which were an essential part of the design.[47]

A.J. Skinner, then in the FTS workshops, also remembers the long hours associated with constructing the 'two pairs of extremely large and long skis plus a modified undercarriage to maintain the extra weight on landing': 'We knew it was urgent, so we worked three 8-hour shifts and all turned out very well'.[48] After all this, it was discovered that there was insufficient space on the *Discovery* for three planes, so that finally just the Moth floatplane, the Wapiti, and one set of skis were loaded aboard for the voyage.[49]

Initially, the RAAF party accompanying the two aircraft was to comprise just six men: Eric Douglas was to have charge of flying operations, with Flying Officer Alister Murdoch as his second-in-command; the others were Sergeants S.F. Spooner (airman pilot & fitter), E. Easterbrook (rigger), J.W. Reddrop (signals), and Corporal (acting Sergeant) N.E. Cottee (rigger). Subsequently, a decision was taken to add a second fitter to the list, provided room was available on board the ship. Naturally many of those involved in the extensive preparations hoped to actually accompany the expedition. A member of Seaplane Squadron at the time, George Grendon recalls: 'From very limited knowledge I had made the undercart to put the skis on the Wapiti, rigged all the strutting and cross bracing. It looked as if I would land the trip, and I was very pleased about that . . .'[50] In the event AC1 C.W. Gibbs—erroneously reputed to be the nephew of Squadron Leader George Jones, the Director of Training at Headquarters—was selected to go.

The *Discovery* cast off from Williamstown on 23 December, with the Wapiti stowed on deck and the Moth housed on the flat roof of the ship's hospital.[51] During the passage south to the Ross Sea, the expedition took on some curious overtones, for there was a strong suggestion that it was neither wanted nor particularly welcomed by either Wilkins or the US government. When he heard of the relief operation, Wilkins had sent a radio message on 9 December thanking the Australian government but declaring there was no urgency as his arrangements were adequate.[52] Cabled advice received from the American authorities in mid-December also recommended that *Discovery* go only as close to the Bay of Whales as was necessary to find out if Ellsworth's radio was in action; if it was, and Ellsworth's situation was in accordance with the plan, nothing need be done to disturb the arrangement for the two men to be collected by *Wyatt Earp* in late January, unless Wilkins was himself unable to perform this task.

In announcing the Australian initiative, Prime Minister Lyons had stated that it was a gesture in return for the American search for Charles Ulm and his crew in December 1934,[53] but, as one account suggests, it seems more probable that the Australian authorities decided to press on with the relief operation 'to emphasize Australian and New Zealand interest in the Antarctic'.[54] The voyage south accordingly developed into something of a race, especially once it was learnt that Wilkins was speeding up his return to the Bay of Whales. Recorded Douglas in a rough log of the trip: 'but I think we will arrive first'.[55]

Two days after entering pack ice on 7 January 1936, about 1600 kilometres south of Dunedin, the captain of *Discovery* began looking to the embarked aircraft to help him find a passage to avoid sustaining damage to the ship's controls. The Moth—its fuselage now painted bright yellow to make it more

The Moth floatplane taken to Antarctica on board *Discovery II* to assist in the search for missing American aviator, Lincoln Ellsworth, in December 1935, shown with a mantle of snow after a storm. A Wapiti floatplane was also carried on the ship's deck (*visible in background*) but was not used. (*SqnLdr A.E. Moore*)

visually conspicuous (in the process obliterating its RAAF serial number)—was in readiness, but there was no open water from which a flight could be attempted. Not until 12 January was it possible to undertake the first reconnaissance flight, the observations obtained from this and another flight the next day proving to be 'of the utmost value' to the ship's captain.[56]

The 10-metre-high cliffs of the Ross Ice Shelf (usually called the Barrier) were sighted ahead at noon on 15 January. The West Cape of the Bay of Whales was rounded a little after 7.30 pm, and shortly afterwards a tent was spotted on the barrier ice at the eastern side of the bay, with two orange signal streamers fluttering close beside it. Steaming as close as possible, the ship's siren was blown and rockets fired but these drew no visible sign of life or movement ashore. At the captain's request, Douglas now prepared to fly over the tent and then on over the charted position of the Little America base 10 kilometres south of the marker flags, to ascertain whether the missing men were sheltering there.

Just before 9 pm on 15 January the Moth, crewed by Douglas and Murdoch, was hoisted out over the ship's side, nose-first and with the engine running, and lowered to the water. With quick take-off a necessity to avoid the sea-spray freezing to the floats and undersurface of the wings, the aircraft became airborne after a run of half a mile and slowly climbed to 1000 feet. Approaching the Barrier, the glare from the ice merged with the glare from the clouds in a 'white-out' effect which obliterated the horizon. It was through this wall of yellowish light that the airmen sighted another flag about two miles in from the Barrier's edge, and shortly after this they saw what appeared to be a patch of crevassed ice. As the aircraft approached nearer at an altitude of about 500 feet, this area was seen to suddenly change shape: the black cracks appeared to 'stand up' and were discovered to be wireless masts and poles rising out of the snow. In his written report of operations Douglas later recounted:

> We then realised that we were over Little America. I carefully circled the area and we then noticed orange coloured strips near the poles. Suddenly we saw the figure of a man appear as if out of a hole and he started to move his arms. This caused great excitement between us as we realised it must be either Ellsworth or Kenyon.[57]

Murdoch dropped a bag of food and cigarettes over the side by parachute and the man on the snow was seen to pick it up. In the bag was also a letter from *Discovery*'s captain congratulating the two men on their magnificent flight and suggesting that, if fit enough, they should move to the edge of the Barrier ice where they would be met by a party from the ship. Turning for *Discovery*, the airmen sighted the wing of an aircraft which they took to be Ellsworth's machine[58] (it was, in fact, one of Byrd's Fokkers which had been left behind in 1929). By 9.35 pm the Moth was back alongside the ship with the good news.

In the next few hours ground contact was established with the missing aviators. According to one Australian account of events:

> When the overland party from *Discovery II* reached Little America, Ellsworth remarked that it was decent of them to look him up. He and his pilot looked so well that the leader of the *Discovery II* party did not have the heart to say that they had come to rescue them. Ellsworth always denied that he had been rescued

> or aided or retrieved. The British and Australians settled for 'located' . . . [59]

Three days later *Wyatt Earp* arrived, and Douglas noted in his log that he was looking forward to meeting Wilkins and learning his plans. 'Personally', he said, 'unless we can give a hand in the salvaging of the Northrop Gamma I am anxious to begin our voyage home because to me, now most of the interesting job has gone and apart from a proposed visit to Little America by foot this region is very uninteresting'.[60]

Meeting with Wilkins evidently did little to put Douglas in a happier frame of mind; he recorded that when he sat next to him at tea on board *Discovery* Wilkins had made clear his annoyance with the Australian relief mission:

> his reasoning appeared to be quite OK until I found that he had wired Washington asking the British Government to help with the *Discovery II* by this ship steaming south and listening for W/T signals. Well, he could hardly blame the Discovery committee and the Australian Government for wanting to carry it out properly or not at all.

As these undercurrents seemed to dictate, the air party's role was effectively ended. The team on *Wyatt Earp* needed no assistance in recovering *Polar Star* and embarking it. (The *Polar Star* is today on display in the National Air and Space Museum of the Smithsonian Institution, Washington.) The RAAF contingent consequently busied itself on board *Discovery*, unshipping the rudder from the Moth to reduce the strain on it and the aircraft fuselage, before visiting the American base on 20 January.

Only two more flights were undertaken during the return passage, both being hour-long ice reconnaissance missions in the Moth on 28 January,[61] and for the rest of the time the RAAF contingent carried out essential aircraft maintenance and general shipboard duties. Still on board, however, was Ellsworth, who was returning with the ship as a guest of the Australian government. Douglas in particular struck up a friendship with the American during the trip, so warmly that on his return he changed his four-month-old son's names from Ian Enderby to Ian Ellsworth.[62] The RAAF men presented the explorer with a model of their Moth aircraft which Douglas and Easterbrook had made. They also presented him with the parachute used on 15 January to drop stores, which was inscribed with the compliments of 'the personnel of the Royal Australian Air Force aboard the RRS *Discovery II*' and signed by detachment members.[63]

On berthing at Melbourne on 16 February, Ellsworth received an official welcome from the Minister for Defence, Archdale Parkhill, the following day. For the RAAF party there was a busy time in getting off the Moth and the unused Wapiti, which was still unassembled, and in clearing and accounting for all Air Force stores from the ship before they could proceed on leave. Douglas summed up his feelings about the expedition by noting that he would not by choice travel in a scientific ship. The atmosphere on board *Discovery* did not compare favourably with his trips with the BANZARE expeditions. He, for one, felt it was 'good to be back'.[64]

Lincoln Ellsworth (*centre of front row*) pictured with the RAAF party on board *Discovery II* after the success of the attempt to locate him and his pilot, Hollick-Kenyon, at the Bay of Whales in January 1936. Ellsworth was insistent that he was in no need of rescue, and in fact his own support ship arrived three days after the Australian relief expedition. He nonetheless returned to Melbourne with *Discovery* as a guest of the Australian government. (*SqnLdr A.E. Moore*)

Search for potential oilfields

Despite what appeared to have been a slightly disappointing result with the 1927 Papuan survey flight, there remained strong interest in the possibilities of air photography for furthering national development—and in the role which the Air Force might play in such activity. By this time the RAAF had become involved in assisting the Army's survey section in a military mapping program and was proposing to establish specially equipped flights within both its squadrons to undertake the necessary work.

As early as March 1927 CAS was seeking approval for a new RAAF unit specifically for survey purposes, arguing that even a single flight of four aircraft would allow a considerable amount in this line to be done, 'which would provide maps of great value and also prove of much assistance to numerous Federal and State Government Departments dealing with mapping, water conservation, railway construction, forestry, etc'.[65] The government appeared slow to take up this suggestion, and there was no approval of the proposed survey flight. By the following year, however, it seemed that Williams' advocacy was about to bring results. In November 1928 he told a farewell banquet for the retiring Lord Mayor of Melbourne, Sir Stephen Morell, that: 'A Plan is being prepared now to map from the air many sections of Australia, which will redeem it from being, as it is at present, the worst mapped country in the world'.[66]

Fulfilment of such an undertaking still necessitated the RAAF developing its air survey capability. Following recommendations from a conference of surveyors-general in 1929, the CAS directed the preparation of new plans for raising a dedicated photographic unit with a capacity to provide coverage of 50 000–130 000 square kilometres per year. In December that year, Flight Lieutenant Gerald Packer—still the RAAF's foremost specialist in the aerial photography field—prepared a submission for forming an Air Survey Flight.

Probably drawing on his experience with the Barrier Reef survey the previous year, Packer envisaged that the new unit would need to be manned by three officers, two airmen pilots, and 20 other ranks. He considered it should be capable of operating from landing grounds remote from towns or railways, and have a range of about 300 kilometres, and therefore suggested the flight be equipped with a minimum of four Wapitis, with two in immediate reserve to permit periodic overhauls and to meet contingencies. To transport personnel and stores from point to point, he proposed providing two six-wheel Morris lorries (one for flight stores, the other with a trailer for use as a mobile photographic workroom), with an additional heavy lorry if tentage and camp equipment needed to be carried.[67]

Although this scheme apparently foundered on the rock of economic cutbacks which accompanied the onset of the Depression in 1930, this did not prevent some work being done. In 1932 three significant flights were flown for the purpose of demonstrating the value of photographic surveys as an auxiliary in the search for oil. The first of these, undertaken early in the year, was to southern Queensland (taking in the area around the towns of Roma, Charleville, Tambo, Blackall, Barcaldine and Longreach) and involved two Wapitis from No.1 Squadron piloted by Flight Lieutenant Alan Charlesworth and Flying Officer John Miles. Following Packer's departure from the service in February of that year with eye problems, Charlesworth had become the Air Force's principle air photo specialist, having undergone training in England which included service in 1929–30 with an RAF survey flight working on the border between British and Italian Somaliland.[68]

The Longreach oil survey ended on a less than auspicious note when, on the return trip, the flight put down at Bourke (NSW) after receiving a report that bad weather was on the way. The Wapitis were tethered at the aerodrome and the crews departed into town when gale-force winds struck. Recalled Miles: 'Things were flying all around the street. We thought we'd better go and look at the aircraft, and arrived in time to see Charlesworth's machine taken up about 100 feet by the wind then dropped backwards on its tail. My aircraft was wrecked too'.[69]

The second survey which followed later in the year took the flight right around Australia. Again two No.1 Squadron Wapitis were used (A5–35 and A5–36), piloted by Charlesworth and Miles; the other members of the flight were Sergeant F. Wilson (photographer, from No.3 Squadron), LAC G.S. Brinsley (photographer), and LAC T.L.J. Jackson and AC1 C.T.P. Jones (both

A photographic survey to assist in the search for oil was undertaken in Southern Queensland, around Longreach, early in 1932. Two Wapitis from No.1 Squadron had completed the task and were on the return flight south when they were wrecked on the ground during a fierce storm at Bourke (NSW). (*Mr J.C. Miles*)

fitters aero). Accompanying them was Dr Woolnough, the Commonwealth geological adviser, who was to select the areas to be photographed. To make room for him in the aircraft, which could carry a maximum of three, it was necessary to assign one of the photographers to travel from point to point by rail or steamer with the stores and equipment. Sergeant Wilson was accordingly sent on ahead to Darwin from Sydney on 7 July, on board SS *Marella*.

The two aircraft left Laverton on 12 July and headed north for Richmond, where Dr Woolnough joined. Departing there the next day, the members of the flight were away for two months, arriving back at Laverton on 13 September, by which time they had spent over 250 hours in the air. The initial work of the survey covered much the same territory as that photographed on the flight earlier in the year, but the route then passed on from the Roma–Longreach area to Camooweal and Newcastle Waters and eventually Darwin. Several days were spent at the northern capital, with reconnaissances flown to Melville and Bathurst islands and photographic sorties flown over the town itself, before moving on to Wave Hill, Fitzroy Crossing and Wyndham. From here, the route took in Derby, Broome, Port Hedland, Onslow, Carnarvon, Geraldton, and Perth, before heading east for home.[70]

Numerous diversions were made along the way, usually to allow examination of areas thought worthwhile in terms of their geological interest although occasionally other factors intervened. In his official report, Charlesworth remarked that after leaving Wave Hill and passing into Western Australia they travelled to Fitzroy Crossing via Ord River and Halls Creek: 'A good many photographs were taken over this area, one or two points of which aroused Dr Woolnough to a state of extreme interest and excitement'. On reaching the Crossing, a telegram was received from the Air Board requiring a diversion to Wyndham to inspect a proposed aerodrome site. Between 16–23 August various sorties were flown over the Byro Plain, situated inland from Carnarvon and between the Wooramel and Murchison rivers. Further photography had been planned during the eastern leg of the flight, but bad weather prevented coverage of the target areas at Mount Gambier (SA) and Portland (Vic) and brought the survey to an early end.

In December a third flight in the program to identify potential oilfields was undertaken to Tasmania. Again, two No.1 Squadron Wapitis flown by Charlesworth and Miles were used, with Sergeant J.A. Campbell and AC1 G.P. Hellwig (both fitters) completing the crews. Sergeant Wilson was again photographer, and proceeded to Tasmania with the stores and equipment by sea. Before the main survey party followed by air on 16 December, a one-day flight to South Australia was made on 30 November, to pick up the areas of interest around Mount Gambier which had been previously missed.[71] By 29 December the aircraft were back at Laverton, having completed what was undoubtedly an extremely arduous mission. The following year Charlesworth was awarded the Air Force Cross for his role in leading the air party 'under conditions . . . providing no facilities for the care of aircraft and photographic material, nor for the development of photographs or the comfort of personnel'.[72]

The oil survey was also a highly significant undertaking in national terms. As Charlesworth noted in his report at the conclusion of the main flight in September: 'The results obtained throughout the flight are thought to be good and in Dr. Woolnough's opinion they have thoroughly justified the expense of the operation.' Although the operations this year led to no immediate discoveries of oil, they were an important milestone in the long saga of exploration towards this end.[73]

During the 1932 round-Australia oil survey: the Commonwealth geological adviser, Dr Walter Woolnough (*in centre, with goggles*), with ground survey personnel on Byro Plain (WA) in August. The pilot sitting in the cockpit of the Wapiti was Flying Officer John Miles. (*Mr J.C. Miles*)

Mapping the north

With such survey operations now a proven success, in August 1933 an interdepartmental committee was formed to examine the question of photographic air survey in Australia. The committee, headed by Wing Commander Harry Cobby of the RAAF, included representatives of the Military Board, the Civil Aviation Branch, the Department of the Interior, and the Development Branch of the Prime Minister's Department. Its terms of reference called for consideration of the possibilities of air survey and air photography for furthering national development, the extent to which this could be done under existing arrangements by the RAAF responding to requests subject to service requirements, or whether civil contractors—with or without government assistance—offered a preferred means of progressing such work.[74]

In conducting its enquiry the committee took oral evidence as well as written submissions from individuals in the aircraft industry and others with an interest in the subject. Despite views put to it in favour of private firms doing the work that might be required by departments, the committee considered that a pattern of duplication and extravagance was already evident, and that a means had to be found to avoid haphazardness and uncoordinated effort. It concluded that the wider national interest was better served by having a definite scheme of aerial mapping carried out by a government agency such as the RAAF, and that building on the existing arrangement by the establishment of an Air Survey Unit would help meet defence requirements in wartime, allow for the extension of peacetime military survey activities, and as well enable the undertaking of tasks for federal and state authorities.

The findings of the IDC's report dated 18 April 1934 were not, however, unanimous. Dissenting from them in a minority report, the CAB representative, Arthur McComb, claimed that the proposal for forming a special RAAF unit flew in the face of the evidence presented to the committee favouring the use of private contractors. He cited views from an RAF authority on the subject to discredit notions that service resources were suitable for work of the nature contemplated.

Disagreement on this fundamental aspect served to block government approval for the proposed Air Survey Unit to be raised, but it did not prevent the initiation of new major undertakings. During 1934 the governments of Queensland and Western Australia entered into an arrangement with the federal government to share the costs of a three-year program to make a scientific assessment of mineral bearing regions of Northern Australia, as a means of advancing the development of this part of the Commonwealth through its mining industry. Available funds would permit about 80 000 square kilometres to be surveyed under this program, involving a preliminary air reconnaissance followed by inspection by ground geological survey parties. Given the national significance of this effort, the Minister for Defence, Sir Archdale Parkhill, agreed to make available the resources of the RAAF, including its aerial photography branch.[75]

An administrative committee of ministers met for the first time in Melbourne in May 1934, and decided to form an executive committee to direct and control the work of the survey. This body met for the first time the following month, with the Commonwealth's consultant on development, Sir Herbert Gepp, in the chair as director of the North Australian Aerial Geological and Geophysical Survey. At this meeting, attended by Flight Lieutenant Charlesworth (Staff Officer Photography at Headquarters) as the RAAF representative, the disagreements which marked the earlier IDC re-emerged: 'Different views were expressed by the Committee regarding the method of carrying out the photographic work. Some members inclined towards the use of the Royal Australian Air Force, while others favoured the employment of private companies'.[76]

In November, tenders were called for the carrying out of the photographic work of the survey. Three bids were received, but the size of these quickly settled the question of who should carry out the task. When the executive committee met again in February 1935, it was decided to inform ministers that commercial companies would charge about £26 000 to carry out the first year's planned operations, whereas consultations with Williams had disclosed that the RAAF could perform this for only £4000. The committee of ministers had no difficulty in approving a recommendation that Air Force resources be utilised.[77]

As a result of this decision, a North Australian Survey Flight was formed from aircraft and personnel of No.1 Squadron. The machines were Wapitis A5–37: Flight Lieutenant Robert Simms (pilot and OC), Corporal Kelly, LAC Langford, and A5–34: Flight Lieutenant Tony Carr (pilot), AC1 C.T.P. Jones, AC1 A. Bain. Leaving Laverton on 29 April, the flight went to Cloncurry (Qld) which served as survey headquarters. Here a printing section was also established under Sergeant Ryan to handle the work of mosaicing, but its original four members were soon found to be insufficient to handle the volume of work and another two RAAF personnel had to be provided.

The flight carried out photographic operations around Cloncurry—initially in conjunction with a third aircraft (A5–23) flown by Charlesworth and AC1 G.P. Hellwig—then moved on to Alice Springs. A setback was experienced on 13 May when A5–34 suffered an engine failure which forced it down 30 kilometres south of Tennant Creek and necessitated a new cylinder being flown up from Laverton. Though repaired and able to continue with work at Port Hedland (WA) and Alice Springs, the continuing unsatisfactory state of the engine of A5–34 forced the replacement of this aircraft by A5–32 before the flight moved to Darwin, where a reconnaissance of the Bynoe Harbour–Cape Hotham area and Bathurst Island was flown with members of HMAS *Moresby* as passengers. Simms reported that during this flight on 20 July, 11 Japanese luggers were found anchored in Apsley Strait, and rather cryptically recorded that: 'Natives associating with crews on Melville Island threw spears at aircraft when dived upon—no hits'. After returning to Cloncurry the flight headed east to Brisbane, then north to reconnoitre areas between Cairns and Cooktown, before returning to Laverton on 16 September on completion of the year's photographic objectives.[78]

The flight had flown about 80 000 kilometres and photographed a total of 8990 square kilometres (3290 in Qld, 2850 in NT, and 2850 in WA). The survey report presented to Parliament pointed out that 'the photography for the Survey was the largest aerial work ever undertaken by the Royal Australian

Pictured against the vastness of the Australian landscape, a Wapiti engaged in aerial photography in 1935 captures the enormity of the task which the North Australian Aerial Geological and Geophysical Survey program entailed. The absence of safe landing spots led to a recommendation after the first season's flying that only twin-engine aircraft capable of single-engine performance should be used. (*SqnLdr W.J. Linnett*)

Air Force, and that it was the largest operation carried out by the Air Force away from its base'.[79] The report also commented on the fact that the survey involved flying over difficult country in single-engine machines, observing that the successful accomplishment of this work without casualties reflected great credit on the Air Force.

Not that flight members relished this element of risk. Simms included in his report a strong recommendation that two-engined aircraft capable of flying on one engine should be used for the survey in future, remarking that the fact no damage had been sustained so far should be considered lucky in view of the type of country flown over—especially in Queensland, where dense jungle offered no safe emergency landing spots.[80] This recommendation was of particular significance given the decision of the survey authorities to increase the area to be covered the following year in order to speed up the pace of work, and to ask the RAAF to undertake the task again in 1936.

The day after the flight touched down back at Laverton, another aircraft took off on a further flight connected with the survey. This was A3–1, the DH89 Dragon Rapide which was the first machine of this type to enter service, having been delivered just the previous month. It was flown by Squadron Leader George Jones, now on the staff at Headquarters, who had previous flying hours on the Rapide gained while in England. With LAC G. Harford (fitter) as his other crew member, Jones was assigned to transport a geological survey party comprising Sir Herbert Gepp (dubbed 'Sherbert'), James Stopford (the Queensland Minister for Mines), P.B. Nye (chief executive officer associated with the survey) and L.C. Ball (a senior Queensland geologist) on a tour of inspection of the survey area.[81]

In A3–1 the inspecting party flew north to Queensland, then west through the Northern Territory and northern Western Australia. At various landing places, Gepp and his mining experts visited the survey camps or other points of special interest in the area before moving on. During a stop-over at Wave Hill, Jones was intrigued to discover the remains of a Liberty engine belonging to one of the RAAF DH9As which had taken part in the 1929 search for *Kookaburra*.[82] The party inevitably experienced some adventurous times, such as the rough ride endured while battling fierce head-winds on the leg to Broome which saw them reach their destination with just ten minutes fuel remaining, or the anxious moments caused by a jammed throttle at Tennant Creek. Unaware that dust was clogging the magnetoes of the Rapide, Jones and his mechanic prepared for a test flight after making adjustments in the rigging. As Jones later recounted the moment:

> When I took off . . . I could only only get the throttles half open. At that time the old landing ground near the telegraph station allowed a run of about six hundred yards. I shut off, applied the brakes hard, and finished up amongst the ant hills on the edge of the field . . . Fortunately, no real damage was done.[83]

Potentially more disastrous were the events of 13 October, while A3–1 was broadly retracing its route on return to Laverton after a month's absence. Being keen to reach Brisbane as quickly as possible, Jones decided to depart from Cloncurry at 4.30 am. It was still dark and raining heavily when the crew and passengers reached the airfield at the designated time. Stopford almost walked into one of the revolving propellers. Recalls Jones:

> I had to get the mechanic out onto the ground in front to guide me along onto the runway with a pocket torch—I couldn't see a darned thing. Once we got off I had to fly wholly by instruments—it was just like being shut up in a light-proof box. I had arranged for the beam lights to be turned on at Moscow Downs and McKinley, but even so it was hazardous. We got to McKinley just on daybreak. That was the end of the hazardous part and we got back to Melbourne without mishap.[84]

For the season beginning in April 1936, it was hoped to triple the area photographed the previous year to around 23 300 square kilometres.[85] Accordingly the RAAF formed two special North Australia Survey Flights—one from each of No.1 and No.3 Squadrons. The No.1 Squadron flight, under Flight Lieutenant Carr, left Laverton on 14 April for its base at Cloncurry to begin photography of selected parts of Queensland, particularly the area of the Croydon goldfields some 160 kilometres south-east of Normanton. The No.3 Squadron flight was under Flight Lieutenant Bill Hely, and departed Richmond on 11 April; this flight would be based at Port Hedland and was to focus on areas in Western Australia. Both flights were equipped this time with twin-engine aircraft but, while Hely was flying a Rapide, Carr had a Gannet which—because it lacked the ability to fly on one engine—was only a marginal improvement over the Wapiti.

In addition to a flying staff of three (photographer, mechanic and radio operator), each flight had an attached printing and mosaic section of five NCOs and airmen. In Port Hedland the section was set up in the former residence of the customs officers. When the flights had finished in the States to which they were initially assigned, both were to be engaged on work in the Northern territory. As well as these aircraft and crews, it was planned that Seagull Vs would be sent north in September to provide photography of an area inland from Portland Roads, on Cape York Peninsula, which was considered to show 'great possibilities'.

Unfortunately for the high hopes of making more rapid progress that year, things got off to a hopeless start. The Rapide (A3–2) disappeared in the desert between Newcastle Waters and Wave Hill, after Hely became lost and had to make a forced landing when the plane ran out of fuel. This triggered an intensive search by two Demons from Richmond, Carr's Gannet, and two civil aircraft engaged from Qantas and MacRobertson Miller Co. The Demons, piloted by Flight Lieutenant Simms and Flying Officer Bernard Cresswell, found the Rapide about 145 kilometres north-west of Newcastle Waters on 22 April, after it had been missing ten days. The crew (Hely, LAC H. Walkington and LAC R.P. Sherwood) were uninjured, but the aircraft was badly damaged and had to be dismantled for shipment back to Richmond.[86] This mishap threw out the program planned for Queensland, because—quite apart from the time lost—the Gannet sustained damage to one engine during the search and was unavailable to commence photography until 8 May. The delay inflicted on the program in Western Australia was much worse. Two Wapitis were sent to replace the Rapide, flying to Port Hedland via Perth. Engine trouble and sickness further plagued this flight however, and the program in the West was not finally begun until June.

The impact of these events showed up in the results achieved for the year. By the time the last RAAF personnel withdrew from Port Hedland in November, only 15 500 square kilometres had been photographed instead of the 23 300 anticipated and this included a much smaller area in the Northern Territory than had been hoped for. And while two Seagulls from No.5 Squadron at Richmond had been sent in August to Lloyd Bay to obtain the required coverage of Portland Roads, their effort had been completely abortive. Though these aircraft (A2–7 crewed by Squadron Leader Leon Lachal (pilot and OC), Corporal Richardson (metal rigger), AC1 Barnes (W/T operator mechanic) and AC1 Wilcox (photographer), and A2–8 crewed by Flying Officer Bill Gibson (pilot), Corporal Tomkins (W/T operator mechanic), AC1 McLean (fitter), and Corporal Cole (photographer)) had stayed in the area until mid-October, they had been unable to carry out their task. Lachal reported that: 'During the whole of the period the detachment was at Lloyd Bay, there was not one day on which the sky was not covered with great masses of cumulus cloud . . . On twelve occasions fruitless efforts were made to commence the survey.'[87]

As a consequence of this outcome, Gepp was obliged to defer plans to repeat the aerial tour made with Jones the year before, to make contact with the different ground parties engaged in economic, structural and other surveys. The fact was that the RAAF had no suitable machines available for this purpose, due to 'delays in delivery of aircraft from Great Britain'.[88] Despite this there were, as the survey report to Parliament pointed out, positive benefits to the RAAF's continuing participation in the survey program:

> In previous reports, we have emphasised the value, from a defence point of view, of the experience which the Air Force is obtaining by participating in the Survey. In addition, considerable knowledge is being acquired for defence purposes of portions of Australia about which little, if anything, was known.[89]

For 1937, the last year of the original program, the use of two detachments was repeated but these were not drawn from Nos. 1 and 3 Squadrons as in 1936. In February the Director of Operations and Intelligence at RAAF Headquarters, Squadron Leader Tom Swinbourne, proposed forming a standing flight for survey photography using the Gannet and Rapide aircraft which had entered service in the previous 12 months. These were presently organised as a practice and communications flight of No.1 AD and Swinbourne suggested renaming it the 'Communication and Survey Flight'. This proposal was supported by CAS, who instructed Swinbourne to prepare a draft establishment for the new unit. Approved by the Air Board on 13 April, the Communications and Survey Flight came into formal existence as part of Recruit Training Squadron of No.1 AD on 3 May 1937.[90]

The new Flight comprised Western and Eastern Air Detachments: the former was under Hely and comprised LAC Bill Taylor (W/T operator), LAC H.W. Johnson (photographer) and AC1 Alan Moore (fitter aero); the latter was under Flying Officer Don McLean and included Corporal Thompson, LAC Everingham, and AC1 Ellerton. Again, a photographic section (Corporal A. Bain, LAC R.M. Brown, AC1 C.J. Robertson and AC1 K.L. Porteous) was established at Cloncurry, and another (Temporary Sergeant J.B. Stirling, AC1 B.E. Hughes, AC1 J. Barker) at Port Hedland. But, whereas in 1936 the photographic work of the Queensland detachment was handled at Cloncurry, in 1937 a separate photographic section from Richmond was sent to Cairns.[91] Both detachments were to operate Gannets this time, as the Rapide was taking Gepp on his planned inspection tour, but the pilots were not happy about this arrangement. After flying A14–2, Hely later complained that, as a type the Gannet proved structurally 'very solid and reliable' in operations from rough surfaces, but from a flying point of view was 'unsatisfactory—being unstable in all controls—particularly directional and fore and aft controls'.[92]

The Eastern Air Detachment left Laverton for Cairns in A14–1 towards the end of April. Its tasks included obtaining photographic coverage of the Portland Roads district which had previously proved so difficult to achieve. To improve the chances this year, a landing ground had been constructed from survey funds at the Lockhart River Mission on Lloyd Bay. In the event this airstrip was found to be unsafe for use by the Gannet and photography of this district was reluctantly abandoned. Cloudy weather interfered with operations in other districts too, but by the end of June some 900 square kilometres of the Herberton area had been covered.

Setting off from Laverton, too, on 19 May, the Western Detachment found

Engine maintenance still had to be performed during the lengthy periods that aircraft were away from their normal bases while engaged on survey work. Here AC1 A.E. Moore (*left*) and LAC W.T. Taylor carry out a top overhaul on the Western Air Detachment's Gannet A14–2 at Port Hedland (WA) during the 1937 season. (*SqnLdr A.E. Moore*)

itself immediately diverted on a search mission for the Rapide A3–2 carrying Gepp and his chief executive, which had gone missing in the Northern Territory. Crewed by Flight Lieutenant Carr, Sergeant W.C. Blakeley and LAC M. Hanrahan, the Rapide had left Tennant Creek on the morning of 21 May bound for Tanami, 480 kilometres west. At 1.30 pm that day, a radio message from the aircraft was picked up at Wave Hill, stating that they had become lost and due to shortage of fuel were preparing to land on the edge of a dry lake which was either Landers Lake or Lake Mackay.[93] A telegram advising of this situation reached Laverton later that afternoon, and action was swiftly initiated to rescue the downed party.

The occupants of A3–2 were in no immediate danger. A radio watch mounted at Laverton received transmissions from the aircraft stating that everyone on board was alright and the machine itself was intact, but assistance was required in the form of petrol, and men and tools to clear a runway for take-off. It was at this point that Hely's party in A14–2 were diverted from Oodnadatta. Two Demons of No.21 Squadron (A1–12 crewed by Squadron Leader Charles Eaton and Corporal A.M.J. Clark, and A1–13 with Pilot Officer Ralph Wiley and AC1 G.P. Hellwig) were also despatched from Laverton on 22 May, under orders to wait at Alice Springs to support the search operations of Hely.

DH89 Dragon Rapide A3–2 is recovered from the site of its forced landing on the edge of Lake Mackay (NT) in May 1937, caused when the crew became lost and the machine ran out of fuel. On board was the survey director, Sir Herbert Gepp, and his chief executive, who were making an inspectional tour of the various ground parties engaged in the survey. After a ground team arrived with fuel, the aircraft was able to fly off from an improvised runway and resume the tour apparently none the worse for its stranding. (*RAAF Historical*)

In the event the search proved to a very brief affair. Arriving at The Granites on 22 May, Hely found the Rapide at 2.30 pm the next day and dropped blankets, food and water nearby. The missing aircraft was on the eastern shore of Lake Mackay, which straddles the Territory border with Western Australia, more than 300 kilometres south of its original destination. A ground party was immediately despatched in a six-wheeled truck and a utility, and the task of the Gannet became that of guiding the ground rescuers over the intervening country and continuing drops of supplies to the marooned group with A3–2. On 30 May the ground party reached the Rapide, and the drama was over. The next day the aircraft, evidently none the worse for its forced landing, took off from an improvised runway and flew to The Granites before resuming its inspectional tour and returning safely to Melbourne on 8 June. The Demons, which had been held at Alice Springs in case required, were ordered back home that same day. Hely remained in the area to make several reconnaissances before proceeding on his way to Port Hedland on 10 June.[94]

The search for the missing Rapide proved to be just the beginning of a very eventful season. Hely's Gannet had returned from Western Australia and was operating in the Alice Springs area when a radio message was received from Laverton on the evening of 9 August, warning of the advisability of inspecting the Gannet's tail plane for cracks where the ribs were welded to a cross member. AC Alan Moore wrote to his wife:

> So on Tuesday morning we cut the fabric and inspected these parts, hoping to fly in the afternoon, but to our dismay we found that there were cracks and Mr. Hely

During the unscheduled stop-over of A3–2 at Lake Mackay, the plane's occupants overcame a shortage of drinking water by improvising a still from aircraft parts to purify brackish water from a clay pan dubbed Lake Rapide, about 400 metres from the aircraft. Here Sir Herbert Gepp (*centre*, with pipe and topee with fly net) and other members of the air party watch the still in operation. (*SqnLdr A.E. Moore*)

> had to send a wire to the Air Board to this effect. This stopped all flying for the time and we had to wait for a reply.[95]

To get things moving again, Rapide A3–2—the same machine as Hely had helped rescue less than three months earlier—was sent as a replacement for the Gannet, arriving on 22 August. This was very much to the liking of the air crew, who considered the DH89 'very much better than the Gannet to fly in'.[96]

This was not the end of the problems, however. On the morning of 12 September, LAC Bill Taylor and AC Moore were swinging the propellers of the Rapide preparatory to take-off from Tennant Creek when one of the blades kicked back, striking Taylor on the wrist. Suffering a bad cut, Taylor was taken to a doctor; but it was also found that the knock had broken the propeller, and a delay of a week could be anticipated while a new blade was sent. The replacement duly arrived on 20 September, and the Rapide was no sooner mobile again than a new drama arose—this time involving a search for the flying doctor, Dr Clyde Fenton, who went missing the very next day in his Gipsy Moth while answering a medical summons to O.T. Downs, over 200 kilometres north-east of Newcastle Waters.[97]

The Rapide was called into the search on 25 September, and spent several days of fruitless effort. In response to Hely's recommendation for more aircraft, two Ansons with four-man crews were sent from Richmond at dawn on 27 September piloted by Flight Lieutenant Gordon Rae and Sergeant Bill Whetters.[98] These additional aircraft arrived at Newcastle Waters just on dusk, but it was Hely himself who found the doctor at 10 am the next morning about

30 kilometres north-east of Tunnumbrini Station. Fenton was uninjured and had spent his six days of loneliness preparing a landing ground near his machine, flagging the runway and putting out a ground sign.

Hely used this improvised airstrip to land and collect Fenton, who was suffering from exposure. The risks of this undertaking became evident a few days later to Alan Moore, who was on board when the Rapide returned Fenton to refuel his aircraft and fly it off. He recorded: 'We were a bit lucky when we landed as the ground was very rough and the clearing not over long and so, after doing a lot of bumping about when first we landed, we stopped with a tree a couple of yards away from each wing tip and another some 20 yards in front of us'.[99] Fenton himself paid a public tribute to his rescuers: 'I have only the highest praise for the efficient manner in which Hely conducted a difficult search and the skilful way in which he located me and picked me up'.[100] Hely's skill and determination was recognised the following year when he was awarded the Air Force Cross for his part in the rescue of Gepp and Fenton, as well as the unsuccessful search for a prospector named Russell who went missing in Central Australia in November.[101]

These various events ensured that the 1937 program again achieved rather less than anticipated—only some 6820 square kilometres (2480 in Qld, 3157 in NT and 1183 in WA), instead of the 12 170 planned. Nonetheless, progress was sufficient to prompt an extension of the program by a further year. Reduced funding provided by the cooperating governments meant, however, some curtailment of the work undertaken so that only 3880 square kilometres were to be photographed in 1938.

Only one aircraft, a Gannet of the Communications and Survey Flight flown by Flight Lieutenant McLean, formed the North Australia Survey Flight this

In September 1937 the flying doctor, Dr Clyde Fenton, went missing north-east of Newcastle Waters (NT) while responding to a medical call. Dragon Rapide A3–2 was diverted from survey work to mount a search, and additional aircraft were sent from Richmond. Fenton was found after a difficult search, his Gipsy Moth undamaged and the doctor suffering only exposure and a broken nose, the latter injury caused by the recoil of a flare pistol he attempted to aim at some ducks he was hunting for food. (*SqnLdr A.E. Moore*)

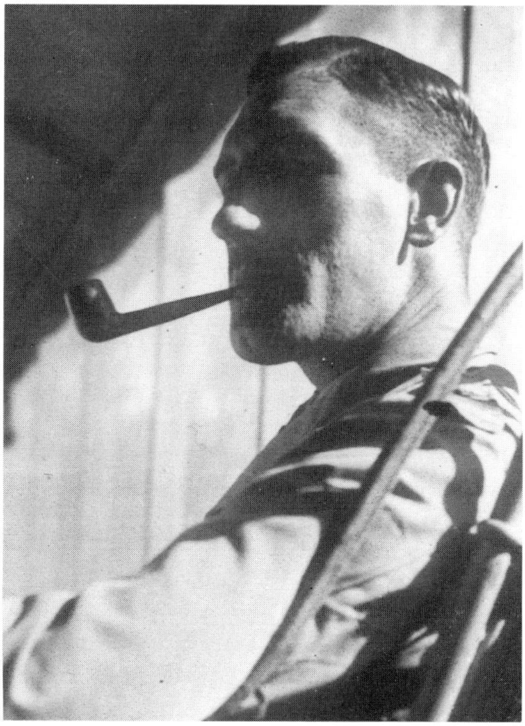

Flight Lieutenant W.L. Hely pictured during a relaxed moment at Alice Springs in 1937. For his role in the Lake Mackay rescue, and in finding Dr Fenton, Hely was awarded the Air Force Cross. (*SqdLdr A.E. Moore*)

year. A second detachment comprising a Seagull from Richmond, flown by Flight Lieutenant Gibson, was to operate from Thursday Island and Cairns later in the season, to provide coverage of areas in eastern Queensland, and photographic sections were still established at both Cloncurry and Cairns.[102] Inevitably, there were problems during the 1938 season—for example, cloudy weather yet again frustrated the detachment working on the tip of Cape York Peninsula and they returned to Richmond in November without success—but by the time the main photographic operations were concluded early in July some 1800 square miles had been covered nontheless.[103] In recognition of his meritorious service performed as a photographer on the survey at this time, Sergeant Alexander Bain was awarded the British Empire Medal in 1939.[104]

Plans were made to continue the work of the survey for another two years, to the end of 1940, though Western Australia withdrew from participation so that operations were to continue only in Queensland and the Northern Territory and cover around 3100 square kilometres in 1939. International events intervened before the end of the projected two-year program. Nonetheless RAAF resources were again committed to the survey during April–May 1939 in the form of a Gannet from Laverton under Flight Lieutenant Arthur Charlton, and a photographic section at Cloncurry under Sergeant G.A. Everingham. The Air Force was requested to once more attempt to photograph those portions of

Cape York Peninsula which had defied earlier efforts, but this time replied that service requirements would prevent this being done.[105]

By the time the war began in September, the RAAF could look back with considerable satisfaction at the part they had played for five years in the major enterprise to map the nation's northern region. The North Australia Survey program became doubly significant, of course, providing in fact the most up-to-date maps for the conduct of military operations once this part of Australia came under enemy attack from 1942.

Pelagic fish survey

In the mid-1930s calls were heard for the development of a fishing industry utilising the nation's allegedly vast fisheries resources. To support the contention that Australia had fish stocks capable of commercial exploitation, the Council for Scientific and Industrial Research (the forerunner of CSIRO) set out to obtain reliable data on the geographical movement and quantities of pelagic (open-sea) species occurring in waters off the south-east coast. The Commonwealth commenced construction of a fisheries research vessel to undertake this work, but in the meantime the RAAF was approached in June 1936 about assisting with the task by aerial survey. Under arrangements with CAS, an aircraft was provided (with personnel and fuel, at no cost to CSIR) to enable a test to be conducted.[106]

On 19 October 1936 a Seagull V (A2–11) of No.5 Squadron, piloted by Flying Officer Athol Xavier Richards (known as 'Axle') and with Sergeant A. Kerr (fitter) and AC1 J. Barnes (radio operator) as its other crew members, left Richmond for Point Cook. Here it took on board Mr Stanley Fowler, a fisheries officer who had been singlehandedly carrying out aerial survey and other investigations before transferring to CSIR the previous year. The flight was also joined at Mallacoota by Mr Reg Pearse, a government cinematographer, before embarking on reconnaissance and photographic sorties which lasted until 1 December and took in the coastal waters of New South Wales south of Sydney, south-eastern Victoria, Bass Strait, and down the whole east coast of Tasmania as far south as Port Davey.[107]

As a result of this series of observations, a second program was undertaken between 25 February and 28 March 1937. This time Seagull A2–7 was used, piloted by Flight Lieutenant James Alexander until 15 March and thereafter by Pilot Officer John Bell, again with Kerr and Barnes making up the crew. Fowler was also on board for this series of flights, which were sometimes accompanied by fisheries inspectors of the Tasmanian and NSW governments too. A third survey was carried out between 5 July and 8 August 1937, on this occasion employing A2–4 crewed by Pilot Officer Bell, Corporal Bill Bradley and AC1 A.F. Thorley. This series of flights covered much the same area as before but extended into Queensland waters as far north as Cairns.

In his report of the survey's results, Fowler commented on the value of aircraft for spotting shoals otherwise undetectable by fishing boats and remarked on the enormous size of some of shoals observed. One sighted east of Bicheno (E. Tas) comprised tuna 3 to 4 feet long, and was clearly visible from the air at a distance of 8 nautical miles: 'The sight presented by the many thousands of great fish when viewed from above would move the most phlegmatic and so, despite the roar of the engine, we could not restrain a cheer'.[108] Another picked up in Wagonga Inlet (Narooma, NSW) was at first taken to be a bed of sea grass or weed, but was then discovered to be a vast shoal of Australian salmon conservatively estimated at 1000 tons. Fifty years later, James Alexander still recalled the stirring experience of overflying at 500 feet a shoal of fish a mile across, with a school of dolphin cutting through them and surrounded by thousands of wheeling sea-birds.[109]

Fowler also reported the use made during the survey of 20-pound Cooper bombs. Four of these were dropped on a shoal in the hope of yielding dead or stunned fish on the surface which, by landing to scoop them up, would provide positive identification of the species encountered. Unfortunately, he recorded, these experiments were not particularly successful, because Cooper bombs were direct contact weapons and burst on the surface, whereas what was required was a type which penetrated 10 to 20 feet before exploding.

Another problem encountered had been the weather:

> Consideration of the limited time available for the tests and the keen desire on the part of all to carry the job through were, I am afraid, largely responsible for our attempting flights under conditions which, though not hazardous in a physical sense, were nevertheless not quite suitable to the ends in view.

This was especially true in Tasmania, where the weather experienced was bad to the point where it seriously interfered with the work program. Over New South Wales the weather was much better as a rule, but here poor visibility was often experienced due to a heavy pall of smoke from bush fires.

The result of all these efforts has been summarised by the historian of CSIR in the following terms:

> After allowing for the limited scale of the project and the magnitude of the task, the attempt to assess the commercial potential of the tuna and salmon schools of the south-east was unsuccessful. Many sightings were made of blue fin tuna and Australian salmon . . . But the schools appeared and disappeared unpredictably. In one year October seemed to be prime month, but the next year at the same time few sightings were made.[110]

Quantitative assessments of the commercial potential of Australia's fisheries had still not been produced before the CSIR's research vessel was requisitioned in 1942 for wartime naval purposes.

15

Transition to war

When Britain declared war on Germany on 3 September 1939, it was in many respects the culmination of a process which many, if not most, people felt had been fully foreseen. Australia, like Britain, had been rearming for nearly five years, in recognition of the increasing uncertainties of the international situation. The adequacy of those arrangements may—with the benefit of hindsight—seem questionable, but certainly not their rationale.

At the end of 1934 the Australian government had announced increased spending to expand the nation's defence forces after the protracted period of neglect during the Depression. For the Air Force, this meant embarking on a three-year program—belatedly implementing much of the 1928 Salmond report—which would add one CAF and three general purpose squadrons to the RAAF order-of-battle. By June 1936 the Minister for Defence was already talking of the next defence program to follow after 1937, indicating that it was the government's intention to complete the first part of the Salmond plan.[1]

In March 1938 Prime Minister Lyons announced a revised plan allocating a sharply increased amount of money—£12.5 million—towards expanding the RAAF far beyond the modest objectives previously envisaged. Under this new scheme an additional nine squadrons would be brought into being by June 1941, along with other headquarters and support units. First-line strength was expected to increase to seventeen squadrons and 198 aircraft, with the number of permanent personnel growing to 500 officers and 4500 airmen.[2] Even these intentions were altered before the year was out, with spending increased to £16.4 million mainly to allow the formation of another CAF squadron at Townsville (Qld), the establishment of a base at Port Moresby, and the construction of some new airfields in three States. Plans were also developed for the establishment of a second flying training school at Wagga Wagga (NSW), this being formally approved in August 1939.[3]

Proposals to alter the program yet again were made in June 1939, when the CAS, Air Vice-Marshal Goble, brought forward the 'Z Plan' calling for an expansion of the RAAF to 32 squadrons and 360 aircraft. This scheme was not approved by government, yet even so the underlying trend was clear. The

The facilities at the new RAAF station at Pearce were impressively modern, testimony to the larger allocations of money devoted to defence purposes in the five years before the outbreak of the Second World War. (*RAAF Museum*)

steady rise in Australia's defence spending throughout the period ensured that, by the time war finally came, the Air Force actually had twelve squadrons fully or partially formed towards a projected total of eighteen squadrons, with 310 officers and 3179 airmen in uniform—all from a base of just two such units and a total personnel strength of less than 1300 three years earlier.

Despite the climate in which such preparations were considered necessary, the actual declaration of hostilities found the Air Force—at least at unit level—with very confused notions as to what this event portended. The reaction within No.3 Squadron is recalled by Wing Commander Dick Cresswell, then a pilot officer who had been with the unit just two months since completing flying training:

> It was all pretty exciting—more like panic really. The second night we all slept on the hangar floor next to our aircraft. Imagine it: there was war over in Europe and here we were at action stations! The stewards brought us down a bottle of champagne from the Mess.[4]

Another who remembers the atmosphere at Richmond is Air Marshal Sir Charles Read, then a flying officer:

> The day war was declared there was a bit of a scurry on looking for real instead of practice bombs. We didn't use HE [high explosive] bombs very much for our training, they were too costly. But I recall Gordon Steege and me sitting in front of a Demon with 250-pound bombs ready to go somewhere—we didn't know where we were going! After all, we were pretty remote from Europe.[5]

Such a situation raised, of course, important questions regarding the extent to which thinking about the likely application of the air power which Australia was acquiring had kept pace with the flow of funds provided for this purpose. What, after all, was Australia expecting to do with its air force in the event of war, and what was the nature of the conflict which the RAAF had been shaped to meet?

Recognising the threat

The plain fact of the RAAF's situation at the outbreak of the Second World War was that it was a force structured for two different and entirely separate purposes. The policy pursued since the 1923 Imperial Conference, of developing the empire's air forces along common lines of organisation, training and equipment, meant that Australia's Air Force—whatever its capacity for local defence—had inherent value as an imperial air reserve. Many members of the force before the war still hold the view that it was the latter role which was central to the RAAF's existence. Says one:

> I always had the feeling that our role was to train for cooperation with the RAF in case we had to fight an Empire war and to cooperate with the other services, possibly in Australia, but more probably in an overseas theatre of war. The idea then was that there was no real threat to Australia.[6]

Certainly this understanding of the RAAF's primary role reflected the expectation of the British Air Ministry, as revealed by a briefing paper prepared by the Air Staff in London for a meeting with the Australian Defence Minister (Parkhill) during the 1937 Imperial Conference. This argued that Australia need fear nothing more than sporadic raids whilever Singapore remained available for use by a Royal Navy fleet, and that, apart from local defence, air forces in Australia were of 'peculiar value' as a means of rapidly augmenting the defences of the Singapore base. Judged from this viewpoint, the Air Staff concluded that the development program then being implemented for the RAAF included 'too great a diversity of [aircraft] types, and too many squadrons trained *primarily* for tasks ancillary to surface operations'. It was the Air Staff view that:

> a more homogenous air force, in which squadrons equipped with modern multi-engine medium bomber types predominated, would be more suitable to Australia's needs. This type of aircraft with its long range, suitability for oversea operations and reconnaissance tasks, and its effective striking power, would be most valuable both for local defence purposes and as a potential Imperial air reinforcement.[7]

Much the same thinking permeated the view adopted by the Air Staff, also in 1937, about arrangements for the higher direction of Australian forces in war. A major debate had developed within Australia's Defence Committee over Army proposals to have the functions of the CGS redefined to match those of the CNS. Supported by CAS (Williams), Army argued that the circumstances of a future war could justify having a commander-in-chief to advise the government on strategical direction and a separate operational commander for the field Army which would probably be formed. This debate resulted in a decision by the Council of Defence that the functions of both CGS and CAS should be revised to accord with those of the naval chief, but only after consultation with authorities in London. The British CAS (Ellington) explained in June 1937 to Field Marshal Sir Cyril Deverell, the Chief of the Imperial General Staff, that while the Air Staff 'favour the enlargement of the CAS's functions in his capacity as a Chief of Staff to accord with those of the CNS, we are against his appointment as a supreme Commander-in-Chief in war'.[8] Though the reasons for this conclusion were not made clear, presumably the role expected of the RAAF did not, in the British view, warrant a separate font of strategical advice on air matters becoming available to an Australian government.

Despite the slender likelihood attached by the British Air Staff to the prospect of Australia becoming engaged in a Pacific conflict, the Australian Air Staff quite naturally did not exclude such a possibility from its planning consideration. Throughout the 1930s there is evidence that the RAAF attempted to focus on the likely imperatives of air operations fought over or in proximity to Australia. As in the case of the expansion program pursued throughout the latter half of the decade, the adequacy of these attempts may be questioned in hindsight—but not the rationale which inspired them.

The principal problem affecting the RAAF's preparations against a local war was the insufficient number of intellectually able officers to form a thinking base for the service. It was not simply a matter of British doctrine of air operations dominating RAAF thinking; the problem was also that the RAAF lacked the means to adapt or formulate new doctrine to meet Australia's particular circumstances. At RAAF Headquarters, the Air Staff of which Williams was nominally 'Chief' was not only pitifully small but it was also so heavily committed with practical administration of the service's expansion that it could hardly be expected to devote great attention to contingency planning as well. When, in August 1938, the Council of Defence charged each of the service boards with preparing a comprehensive statement outlining the state of progress in preparing plans for mobilisation in the event of an emergency, the Air Board could only reply:

> Plans for the mobilisation of the Air Force are in the embryo stage only. Mobilisation Regulations are in first draft form but require considerable work on them before being suitable for issue. On Air Force stations and in Air Force units, however, mobilisation schemes have been drawn up to deal with local matters as they arise. Progress with mobilisation plans can only be achieved as experienced staff becomes available. This will not be in the near future as it is more important at present that the maximum of experienced staff should be employed in dealing with the Development Programme.[9]

As this response indicated, the Air Force's inability to produce what was required was governed both by priority and also availability of experienced (for which one can read 'capable') staff. The plain fact was, however, that in the pre-war officer corps of the RAAF there were few individuals who demonstrated a coherent conception of air power adapted to Australian circumstances, indeed few who appeared to have any interest at all in theoretical issues affecting the employment of their service in time of war. Among the few may be counted Air Vice-Marshal Frank Bladin, who, as a flight lieutenant at the RAF Staff College in 1931, produced an article for the *Royal Air Force Quarterly*;[10] however, this went little further than arguing familiar platitudes regarding the need for 'greater effort [to] be made to co-ordinate, or rationalize, the efforts of the several parts of the Empire to provide for the Empire's defence'. Another was Air Vice-Marshal Henry Wrigley, who published a history of No.3 Squadron of the AFC during the First World War (*The Battle Below*, 1935), and left upon his death in 1987 volumes of personal notebooks resulting from his private study of the history of air conflict.

For all his other claims to have nurtured the development of the RAAF, Williams, too, had failed to provide a theoretical foundation for his service—as one historian has noted: 'One looks in vain [to Williams] . . . for any original contribution towards providing a solution to the problems of Australian defence'.[11] Indeed, others consider him to have been almost indifferent when attempts were made to highlight the difficult issues associated with local defence. Recalls Air Marshal Sir George Jones:

> In about 1934, when I was Director of Training, I arranged a strategic exercise for the senior officers conference which the Air Board used to hold every year. This was based on a scenario that the Japanese wanted to send migrants to Queensland, the Australian government refused to accept them so Japan appealed to the League of Nations, which agreed that they should be allowed to do so. The Japanese then organised an expedition consisting of ships of migrants defended by aircraft and naval vessels. Our job was to prevent them taking Port Moresby, or getting onto the Queensland coast around Bundaberg, and we very soon discovered we didn't have enough aircraft—or good aircraft at any event—to do the job, so they got right down into the Coral Sea. Well, at that time the exercise finished because we had no means of stopping them.
>
> I gave an address on the exercise in which I pointed out the lack of aircraft, and said that to do a strategic exercise without adequate aircraft was just about as good as having a naval exercise without ships. I remember Williams heard this, and eventually he appointed me and George Mackinolty, who was head of the Equipment Branch, and Harold Seekamp, an engineer officer, to do an investigation into aircraft production in Australia. We went into it very thoroughly, taking about two or three weeks, then put our report in to Bill Anderson, who was Air Member for Personnel at the time. He had a little stub pencil and went all over it putting a few notes in the margin, and he said 'Damned good, damned good exercise' before sending it on to Williams. And that was the last I ever heard of that—I doubt if Williams ever read it. That was all in 1934, running into 1935.[12]

Although lacking a recognisable 'brains trust' of the kind which was behind the original General Staff concept—itself the precedent for a so-called Air Staff—

the RAAF was well-served by the appointment to headquarters of a number of officers with particular aptitude for staff work. Among the men who did most to fill the void during the late 1930s should be counted Air Marshal Sir John McCauley, at that time a squadron leader. Later assessments of this officer have too often focused on his reserved personality, and a shaky reputation as a pilot which won him the nickname 'Crasher',[13] rather than the dedication and intellectual strengths which earned him appointment as CAS in 1954. One of the few serving Air Force officers of the pre-war era who took the trouble to further his education by private means, gaining a Bachelor of Commerce degree from Melbourne university in 1936,[14] McCauley is described by one of his subordinates in the Directorate of Training during 1936–38 as 'a pretty intelligent officer, a great leader, with a great deal of force'; in the opinion of this officer, McCauley did not deserve the sneers of his colleagues who valued little beyond flying ability.[15]

It was McCauley who, in September 1935, was behind the sending out of instructions from CAS's Branch to all RAAF stations calling for staff at each base to draw up local defence schemes[16]—the same schemes which three years later appear to have been the sole measure of mobilisation planning actually carried out. At the same time he placed a requirement on units to prepare papers for submission to RAAF Headquarters by 16 October on specified subject areas: for No.101 Flight, the subject was 'Naval Co-operation'; for Seaplane Squadron at No.1 FTS, 'Coast Reconnaissance'; for Fighter Squadron at No.1 FTS, 'Fighting'; No.1 Squadron, 'Striking'; and No.3 Squadron, 'Army Co-operation'. The point of this order was that the papers prepared were to serve as a basis for the formulation of policy to be embodied in Air Staff Memoranda and subsequently tested in staff exercises. As a separate requirement, all officers below the rank of flight lieutenant in service units were instructed to write a 10 000 word essay on 'The Manner in which Air Forces could co-operate with the Fixed Defences to Protect the Vital Areas of Australia from Enemy Raids'; the three best essays from each unit were also to be sent to RAAF Headquarters by 1 February 1936. The culmination of this process, which was a crude but nonetheless first-ever attempt to formulate operational-level doctrine, was an Air Staff Memorandum issued on 15 April 1936 dealing with the tactical employment of air forces in the local defence of Australia.[17]

In September 1938, an attempt was made to strengthen senior staff arrangements at RAAF Headquarters to facilitate attention to contingency planning. Within the CAS's Branch, a new appointment of assistant to the CAS was created to assume responsibility, 'under the Chief of the Air Staff, for all Defence Council and Defence Committee matters, organisation, and plans for the employment of the Air Force in war, etc.'. The officer appointed to fill this post was Wing Commander Thomas Swinbourne, whose former post within the Branch as Director of Operations and Intelligence was filled by Wing Commander Bill Bostock—both officers considered by their contemporaries as among the Air Force's best brains.[18]

Despite these improvements there were major deficiencies in the command and control arrangements at RAAF Headquarters for coping with operations entailed by any local emergency situation. For example, in April 1939, when the RAAF became involved with Navy in the major defence exercise held off the Victorian coast 'to test the ability of the existing sea and air forces to keep open Australia's trade shipping routes' (see chapter 7),[19] it was found that the volume of exercise signals soon swamped the capacity of the Signals Directorate which, in the absence of an operations room or staff, was expected to handle the decoding and distribution of this traffic.[20]

If the RAAF's leadership did not seem especially imbued with a sense of urgency at Australia's strategic position, this was not wholly its own fault. The fact was that Australia was heavily, though by no means entirely, dependent on Britain for strategic intelligence—including assessments of the capabilities and intentions of the most likely Pacific aggressor, Japan. Thus when the RAAF considered which aircraft type would equip the fighter squadron planned under the development program, the Assistant CAS (Swinbourne) instructed that information be sought from London regarding the main Japanese aircraft types which the fighter might conceivably meet in combat, 'the more important being those normally carried in cruisers'.

It was advice received from the Air Ministry in September 1938 that Australian fighter aircraft would only face reconnaissance floatplanes having a maximum speed of 170 miles per hour which encouraged a belief by the Air Board that a Wirraway powered by a double-row Wasp C engine instead of the normal Wasp H would satisfy the Australian fighter requirement. Such aircraft would have an expected speed of 280 miles per hour, 'which would be sufficient to deal with enemy aircraft operating from raiders at sea'. Though it was acknowledged that better types were in the development stage, the Board considered that the Wirraway fighter was acceptable based on the assumption that the threat to Australia was as postulated by London, 'ie. seaplanes of no great performance'. As Williams advised the Minister for Defence in February 1939, the object was to provide a fighter to 'meet the needs of our local problem' while reducing the number of aircraft types in service, achieving self-containment in production, and cost-saving.[21] As events were to show, the Wirraway fighter was actually an untenable option in the light of the aircraft types already operating with the Japanese navy, and this must have been realised had accurate intelligence been forthcoming from London.

As it happened, however, the RAAF had no excuse for being as much in the dark as it was about the nature of the threat it might face in the air in a future conflict with Japan. Since invading China in 1937, the Japanese had been refining combat techniques and testing the capabilities of their equipment in operations against the Chinese Nationalist forces of Chiang Kai-shek, and clearly there was much that could have been learnt from close and careful observation of this conflict. Unfortunately, there was little disposition—in Australia or elsewhere—to view this theatre of war very seriously, and those conclusions which observers did draw were often the wrong ones. Group Captain Wilfred Compagnoni, then an RAAF flight lieutenant, recalls:

> I remember we were all herded together one day to listen to an American correspondent who had been in China and had observed Japanese air attacks on the Chinese mainland. He told us that their aircraft were slow and did not appear to have any great performance. As far as he could see they could never be a real

> threat to a British-designed aircraft. I heard him say that Japanese had poor eyesight and . . . their pilots all wore glasses. He gave us to believe that there was not much to worry about; they were a second-rate air force and not a threat. Where he got his information from I don't know, but this was the sort of stuff we were given.[22]

The RAAF had access to better information than such misleading nonsense, however, because it was in the singular position of having a member of its own Air Force Reserve, Wing Commander Garnet Malley, serving as aviation adviser to Generalissimo Chiang Kai-shek (see chapter 8). In a letter to Williams as CAS, addressed from Hankow on 19 May 1938, Malley provided some idea of the nature of the air war being fought by and against the Japanese. He especially deplored the fact that the Australian government had not seen fit to send Williams to China following on from CAS's visit to Singapore the previous February:

> . . . I cannot understand the short-sighted policy of the Minister [in] not asking you to continue your journey to China, when you were so far East, considering that there is a first-class war on here, which must eventually affect Australia and its Defence schemes, and obviously you were the man to investigate the situation, and advise the Australian Government . . . From my observations of the present war conditions, I have observed many points of *vital importance* in future aerial warfare, and would very much like to have discussed these matters with you.[23]

Australia's inability to obtain prompt delivery of aircraft orders from Britain led to the adoption of stop-gap measures to enable the RAAF to continue with its program of expansion—such as utilising the Wirraway advanced trainer in a fighter role. The decision can be criticised in retrospect, but seemed a reasonable course to pursue at the time when it was expected that the most serious threat the aircraft would need to combat was that posed by Japanese 'seaplanes of no great performance'. (*Hertz collection, NLA*)

Malley claimed that he was 'the only British officer connected directly with the Chinese Government', and given the unique opportunity afforded the RAAF to gain first-hand insights on a subject which within a few years would become of crucial importance to Australia's security it can only be deplored that better use was not made of him. In the sole exchange of letters known to have occurred with Williams, he provided some useful intelligence of Japanese tactics. The ceaseless harassment of Chinese airfields by Japanese bombing, he argued, meant 'that the days of well-established aerodromes for military purposes are numbered', and he outlined a solution—which he claimed as 'purely my own original idea'—using specially constructed straight stretches of road as improvised runways. The idea was then only being implemented in China, so Malley said he could not give a report on their efficiency, but he went on to suggest that: 'In Australia the road engineers could be preparing these runaways [sic] while they are on road construction work, which could be used as emergency landing grounds until such time as they were required for service use'.

But despite the valuable foretaste of things to come which such comments provided, the general tenor of Malley's comments as observer seems rather pedestrian and at worst his reports tended to confirm the low opinion of Japanese capabilities contained in reports from British diplomatic sources. At one point in his letter, he remarked:

> I might add that the Air Force in China is a Service to be reckoned with these days. It has developed enormously . . . and the Japanese have by no means wiped it out the many times they would have the world believe. As a matter of fact, they had a very bitter lesson right over our heads last week, when they lost over 30 machines in one of the most spectacular battles I have ever seen. There were over 130 machines involved, and I quite felt I was back again in 1918, but as a spectator this time!

Other information was, of course, available regarding China, but it appears that these sources were disregarded when they conflicted with Malley's accounts. Compagnoni recalls that when he became Director of Signals at RAAF Headquarters in 1938:

> . . . one of the things I had to do was interview somebody who had just come back from China and said he had information which could be of vital interest to Australia. I remember going down to a Sydney beach and talking to this man, who did not want me to take any notes. I've forgotten his name now, but he told me about . . . Japanese designs on Australia and expansion in the Pacific, and he discounted all this nonsense about them not having a good air force. He said that they had some extremely fine aeroplanes and that a lot of the propaganda we were hearing was obviously of their own making, to lull us into a false sense of security. It was rather a startling story. I reported to the CAS, 'Dicky' Williams, and he got in his assistant CAS, Tom Swinbourne. They thought it was a lot of nonsense. Finally, 'Dicky' turned to Swinbourne and said 'Get in touch with Malley' . . . No-one had much of an opinion of Malley, and I thought that getting in touch with him to check the story wasn't going to get us very far.[24]

On the basis of this evidence it might be concluded that Malley's presence in China did not provide the benefits in preparing for a Pacific war which might have been expected. Compagnoni's mention of the low regard in which Malley was held personally must be counted as an additional factor behind Williams' apparent failure to exploit to full advantage this conduit for crucial intelligence about the Japanese. On the evidence of Malley's letter, the man was conspicuously full of his own importance:

> Some time ago I had a very interesting trip by air all over China. I went on an inspection tour, and can just imagine how much you would have enjoyed it—landing on aerodromes that were all polished up for inspection. Being met by provincial dignitaries and lavishly entertained ... When landing on most large Airports, it was most impressive seeing hundreds of Officers and personnel lined up in parade formation, and smartly turned out in white. Bands were playing, and then one was whisked away by motor-car to some sumptious [sic] banquet.

Williams' reply to Malley, dated 27 February 1939, was perfectly friendly in tone but leaves an impression of nothing more than polite interest in what he had to say. As Williams noted, however, the eight months since receiving Malley's letter had been taken up with the extraordinary business of the Ellington Report. It was a period in which, CAS admitted, 'I found myself deferring everything that had not to be done at once', and he was now writing the day before he left for England. Given these circumstances it was not remarkable that Williams did not ply Malley with eager and interested questions, but it remains odd that he did not press him to keep up a more regular correspondence with others at RAAF Headquarters about the situation in China. As it was, Williams merely said he saw possibilities in the 'road-cum-landing runway' idea and would be glad to know how it worked in practice, and ended by remarking casually: 'We find the information reaching us from China most interesting, but still most confusing, and we really don't know what's happening there ... There are lots of questions I could ask you verbally, but not here ... Who is really winning your war?'[25]

Notwithstanding the failure of the RAAF leadership—almost unbelievable in retrospect—to capitalise on an intelligence windfall of immense significance, serious planning for air operations in Australia's own region nonetheless did proceed in the years before the war. This much is evidenced by the steps taken to identify locations for future airbases outside the mainland. According to one account, the islands of Timor and Ambon in the Netherlands East Indies were selected as suitable sites as early as 1936 or 1937 by a secret RAAF mission comprising Squadron Leader Charles Eaton and another officer in civilian clothes. At Koepang, in Dutch Timor, the two were detained by local authorities, who placed them under house arrest for three days.[26]

In mid-1938 the RAAF began conducting a survey of landing grounds and seaplane anchorages in the Pacific Islands close to Australia, namely Norfolk Island, the New Hebrides and the Solomon Islands. The northern cruise by Australian warships with embarked seaplanes was expressly expected to provide the opportunity for inspections to be made of the latter two areas, and in February 1939 an officer from Richmond (Flight Lieutenant Thomas Curnow) was sent by commercial ship to make a similar examination of Norfolk

Island.[27] Planning underway in March 1939 envisaged that this work was to be extended to New Guinea during the RAN's northern cruise during August, following the announcement of proposals to establish an air base at Port Moresby.

The importance of this task of identifying potential operating bases for future use was recognised outside purely Air Force circles. In fact, in March 1939 RAAF Headquarters received representations from three academics at Sydney university, L.A. Cotton (Professor of Geology), T.D.J. Leech (lecturer in Civil Engineering) and K.R.M. Hart (lecturer in Mechanical Engineering) for an examination to be made into the suitability of Lake Kutubu in Papua, some 480 kilometres north-west of Port Moresby, as an advanced seaplane base for the defence of Australia. Information from Mr S.W. Carey, a geologist employed by the Australian Petroleum Company who had recently returned to Sydney on a visit after four years spent in the territory, prompted the three to urge that the lake offered unique defence advantages. Not only would a base here be very difficult to attack without thorough local knowledge, but from it Australian seaplanes of the Sunderland type would be able to attack Japanese territory in the Caroline Islands and Palau.[28] There is, however, no indication that any notice was taken of this suggestion.

Australia was not alone in realising the need for preliminary planning. In May 1938 the New Zealand government had suggested the holding of a defence conference on Pacific matters, and, after agreement on attendance was obtained from Britain and Australia, a meeting was scheduled in Wellington for March 1939. The agenda proposed by the host country foreshadowed that the conference should cover the whole range of issues likely to arise from simultaneous conflicts in Europe and the Far East, including the possibility of enemy action against empire interests in the South-West Pacific before the arrival of promised naval reinforcements; measures that might be needed to defend the Pacific Islands (specifically Fiji, Tonga, the New Hebrides and the Solomon Islands), shipping trade, and Australian and New Zealand territory; effective means for cooperating with Britain, including schemes for mutual reinforcement; and coordination of defence policies in peace and war. Britain was plainly uncomfortable at the idea of discussing such issues at anything other than a set-piece Imperial Conference where it was totally in charge, and argued for narrowing the scope of the conference. This was eventually agreed and the conference went ahead in April, Australia being represented by the CNS (Vice-Admiral Sir Ragnar Colvin), Colonel V.A.H. Sturdee of the Army, Wing Commander George Jones, and Captain E.C. Johnston of the Civil Aviation Department.[29]

Among the recommendations on air matters adopted by the conference was that Australia should accept responsibility for aerial reconnaissance from the New Hebrides to New Guinea, providing flying boat facilities (and, if possible, a landing ground too) in the former island group. Meanwhile New Zealand would be responsible for the arc from Tonga to the New Hebrides, and was to undertake the expansion of flying training facilities of the RNZAF in order to train 280 pilots a year—220 of these for short service commissions in the RAF. In line with the conference proposals, in June 1939 the Air Board recommended that the Commonwealth accept responsibility for air reconnaissance between New Guinea and New Hebrides, and begin work to provide flying boat facilities and landing grounds at Tulagi (in the Solomons) and Vila (New Hebrides).[30] By May, the Air Board recognised that there was likely to be difficulty in locating suitable landing grounds at the two designated places, and the Naval Board was asked to have embarked RAAF personnel on warships make the necessary examinations to see if a suitable landing ground could be found in the northern portion of Guadalcanal Island, in the Solomons.[31]

The rush to re-equip

Notwithstanding the recognition which certainly existed within RAAF Headquarters, to greater or lesser extent, regarding the likelihood of war occurring in Europe or the Pacific—or both—a major impediment to the making of effective preparations for such an eventuality was the inability to obtain the supplies of equipment, especially aircraft, needed to sustain a major development program. Delays experienced in having orders for aircraft filled by British manufacturers was the spark which finally ignited Australian resolve to embark on local production (see chapter 9). Even this step did not offer an immediate solution to the problem, however, as the inevitable lead-time meant it was July 1939 before the first machines from this source reached the RAAF. In the meantime, attempts were made to speed up deliveries from Britain.

During his time in London to attend the 1937 Imperial Conference, the Australian Minister for Defence (Parkhill) held discussions with the Air Ministry regarding the problems the RAAF was experiencing over its aircraft orders. The British Air Staff was not particularly sympathetic to representations on this score, as the briefing paper prepared for the Air Ministry delegation made perfectly clear. Dominions already received priority over foreign customers, it maintained, and delays were often caused by Australian requirements for modifications. Moreover, aircraft sales overseas were of such importance to Britain's trade that the Air Ministry reserved the right to supply limited numbers of planes to maintain the interest of foreign customers until their main requirements could be satisfied:

> On the whole the Air Ministry has met the Dominions very well, and if the authorities in Australia had themselves had experience of production difficulties, they might not be so ready to complain about deliveries. When their own factory is established they will be better informed of the difficulties which arise in the production of Service aircraft.[32]

The line taken with Parkhill during the meeting which took place on 18 June was not quite so brutal, though the Minister was still left in no doubt as to the RAAF's low priority in the Air Ministry's eyes. When he complained that Australian orders were not filled until after RAF reserves had been provided, and even foreign orders, he was bluntly told that this was usually caused by Australian procrastination, with RAAF orders following long after initial inquiries had been made, during which time firm foreign orders had been received. The CAS (Ellington) even argued that it was often not in the interests of Dominions to receive early delivery of new types, as these frequently required modifications; he nonetheless agreed in principle that Australia 'would always get delivery of their aircraft before aircraft were taken for stored and war reserves in this country'.[33] All of which was of little comfort to a service desperate for aircraft, not just to replace obsolete types but for expansion purposes.

By mid-1938 the RAAF's position was parlous. New squadrons raised during the previous year had only been equipped by taking the reserve machines of existing units. This had been seen as purely an expedient pending delivery of twin-engine Bristol Bolingbroke aircraft on order from Britain, but subsequently the Air Ministry advised the cancellation of the development program for this type in preference to an even newer type, the Bristol Beaufort. Production of this replacement was scheduled to begin in 1939, but the Air Board had been advised that an Australian order for 50 could not be met until 1940. Special high-level representations had succeeded in bringing forward the expected date of delivery to the latter half of 1939. Until that time the RAAF was obliged to accept further numbers of Avro Ansons (ten on purchase and 40 on a charter basis from the British government) even though this type was recognised to be of very inferior performance to what was required.[34]

These were extraordinary circumstances, and it was in this context that the Air Board looked long and hard at its alternatives in the event that either the Beaufort development proved a failure or its export was prevented by the outbreak of war. The only options open to Australia were to consider other aircraft types which might be available in the United States, or measures which could be taken locally. In May 1938 the Board reported to the secretary of the Defence Department that CAC had proposed the development of a twin-engined prototype able to be manufactured within the existing Fishermans Bend plant:

> The cost of such an aircraft is quoted at approximately £60 000, which is little more than the cost of two Beauforts, and on this basis seems reasonable. It is promised within twelve months. Expenditure on a proposal of this sort would only be justified as a partial insurance (partial to the extent that this type similarly may not be completely successful) against [anticipated difficulties]. The possibility of any of these difficulties arising cannot be known for some months yet, and those months would be lost in the absence of some insurance against it. On the other hand, such expenditure could only be justified provided no design suitable for local manufacture and for service needs is already available anywhere else.[35]

Although the Board considered the development of a type to Australian requirements 'would be wise', it preferred to await the outcome of a British mission already in the United States to investigate the availability of service types from that source for the RAF. Despite this, some unofficial planning evidently went ahead within CAC, with Lawrence Wackett embarking on the preparation of a design of a suitable machine. This was a low-wing monoplane powered by twin Rolls-Royce Kestrel engines, the undated plans of which still survive. With a single tail fin, a dorsal machine-gun and another single gun placed to port in the nose, and able to carry four 120-pound bombs under the centre-fuselage, the machine is said to have had 'an appearance (other than the tail assembly) of a BF.110', a fighter-bomber produced by Messerschmitt for the German Luftwaffe.[36]

In the event, nothing further was heard of the CAC type because a suitable aircraft was discovered to be already in production in the United States. This was the Lockheed Hudson, a five-seat twin-engine medium bomber. In the wake of the war crisis narrowly averted by talks at Munich in September 1938, the Minister for Defence approved of the Air Board considering the purchase of eighteen of these machines (enough to equip a single squadron) as insurance against non-delivery of Beauforts from Britain,[37] and on 2 November, Prime Minister Lyons announced the actual placement of an order for 50 of these machines.[38] In the meantime the RAAF would have to make do with more Wirraways, with the Minister for Defence increasing the order for these from 40 to 100.[39] It was in this context, too, that the Air Board suggested to the Minister that a fighter version of the Wirraway might meet the immediate need for aircraft with which to equip the fighter squadron, which was planned to be raised under the development program. This course was also approved, with two prototypes being ordered at a cost of £30 000 in February 1939.[40]

Notwithstanding the factors which brought about the placing of orders for Hudson bombers, this was a decision by the Australian government which swiftly produced a reaction from London. The original selection of the NA–33/Wirraway still rankled in British aviation circles, with Sir Edward Ellington petulantly refusing even to watch a flying demonstration put on for his benefit during his 1938 visit of inspection of the RAAF.[41] In his report to the Australian government Ellington had placed himself in the nonsensical position of urging that Australia defer making a decision about the aircraft type to be built by CAC until Britain had a suitable type to offer, a course which would assuredly have only compounded all the difficulties already experienced over British non-delivery of aircraft orders.[42]

Now, in a last-ditch attempt by London to counter a process which it feared would see Australia totally switching from British to American aircraft, an Air Ministry delegation was despatched to Australia in January 1939 to examine the prospects for local production. Members of the delegation were Sir Hardman Lever, Colonel Sir Donald Banks and Air Marshal Sir Arthur Longmore (who was actually an Australian by birth). The report submitted by this party in March 1939 recommended the setting up of a new organisation, entirely separate to CAC, to undertake manufacture of Beaufort bombers in Australia to meet the needs of both the RAAF and the RAF east of Suez.[43] Both the British and Australian governments accepted the scheme, with the result that the RAAF could look forward to having two local sources supplying it with operational aircraft where there had been none at all just a few years previously.

Thus, by mid-1939 it appeared that Australia had gone a long way towards addressing the serious deficiencies which had hindered its rearmament program for the RAAF. The reality of this situation was, however, that substantial lead-times were involved before Australia saw either the first Hudson (January

1940) or the first locally built Beaufort (August 1941). When war actually came, the RAAF's frontline strength still consisted of obsolescent Ansons and Demons, and just a handful of Wirraways.

The Air Force was not significantly better off than it had been during the Munich crisis of September 1938. Then, the RAAF had been confronted with the discomfiting reality of going to war with what the service already had on hand. According to Wing Commander Charles Gates:

> CAS called a conference in his room. I recall I was present, probably because my Director of Signals was away. Williams said we'll do this and that, and send a squadron of Ansons to Darwin to do anti-submarine patrols. Goble asked, 'What will you use for bombs?' Dicky replied, 'We'll get the ordnance factory to make some.' 'Caught with our pants down, aren't we,' said Goble.[44]

In the event the Minister for Defence decided that no action need be taken to form an emergency squadron for despatch to Darwin.[45] This was probably just as well, as it appears that just a single flight of four Demons from No.1 Squadron was all that was immediately available to be sent. These had been 'completely fitted with war load' and were placed under armed guard in the squadron hangar, while the crews detailed to fly them to Darwin remained at 'a minutes notice in the event of hostilities'.[46]

Action to rectify the deficiency in bombs noted at this time was actually already in hand, but—as with aircraft—entailed considerable lead-time to overcome fully. The shortage of munitions had been clearly recognised in the Council of Defence in February 1938, when the RAAF claimed that remedial action through local production had not been possible before because the necessary designs had still not been received from Britain.[47] As a result of the Munich scare, an order for an additional 1000 250-pound bombs was approved. It was, however, the end of November 1938 before the RAAF received its first supplies of 500-pound general purpose and 20-pound anti-personnel bombs.[48]

Despite all these last-minute efforts, the general equipment position of the RAAF in the early months of the war is neatly encapsulated in the recollections of one junior pilot based at Richmond:

> We did anti-submarine patrols at the time, which were in response to reported or alleged sightings. Imagine doing this in a Demon six miles off the coast—there was an awful lot of water below. We used to increase our altitude to make sure we had enough height to glide to land in the event of engine failure, at least, I know I did. We sometimes, but not always, carried bombs—one 112-pound bomb under each wing. These were old First World War stock and were actually leaking slightly. They were so dangerous we ditched them rather than land with them again. I have an idea there was an in-house order 'Don't bring the bastards back'.[49]

Such was the state of Australia's air defence in late 1939!

The old order passes

When Australia followed Britain's lead in declaring war on Germany on 3 September 1939, the government of Prime Minister Robert Menzies was initially uncertain about the best form for its contribution to the Allied war effort to take. When advice was sought from London, the response of the British government on 8 September was—predictably—to suggest that, provided Japan remained friendly, Australia might consider preparing an expeditionary force for service in Europe, but that in any event formations should be readied to reinforce Singapore or other Allied territories in the western Pacific.[50]

This advice was viewed somewhat sceptically by Air Vice-Marshal Goble, who had taken over from Williams in February as 'acting' CAS. He pointed out that:

> the Dominions Office cable makes the assumption that our immediate object is how to assist Great Britain in the European theatre and the inference appears to be that we should reorganize our development programme so as to support a large effort overseas . . . but I assume that at the moment, the assumption is not correct, local defence being the immediate objective.[51]

Notwithstanding this significant reservation, and presumably at specific government behest, CAS initiated planning at RAAF Headquarters for raising an air expeditionary force to take part in operations in Europe or another theatre of importance. The scheme developed by Goble and presented to War Cabinet by the Minister for Defence, Geoffrey Street, on 20 September proposed offering a complete air contingent to Britain. Underlying this plan was a concern that if only individual pilots and other aircrew members were sent, these would be distributed throughout the RAF and the Australian identity of this contribution thus submerged.[52] The offer was duly conveyed to the British government and planning for the despatch of the air expeditionary force progressed on this basis accordingly.

Under Goble's scheme, the force proceeding overseas was to comprise two bomber wings each of two squadrons and a fighter wing of two squadrons, with a headquarters and ancilliary units.[53] The 96 aircraft needed to equip the units could not be provided from Australia and were expected to be supplied by Britain on the force's arrival at its destination. Even without the equipment worry, this was an ambitious plan. A force of six squadrons was half the effective unit strength of the RAAF as it then existed, and the personnel required to man it—3200 officers and airmen—was little short of the number comprising the whole of the pre-war PAF. Notwithstanding the calling-up of reservists and CAF personnel, and the plentiful supply of new recruits offering for enlistment, it was clear that a sizeable proportion of the RAAF's experienced personnel would need to be taken for the expeditionary force, thereby making it harder to meet the government's stated requirement for the operational efficiency of the RAAF to be maintained while provision was also made for expansion.

Other aspects of the air contingent scheme have been subjected to later criticism. Williams, for example, could not resist attacking in his autobiography his former rival's choice of Group Captain H.F. De La Rue to command what was to be the largest RAAF formation ever organised. Pointing out that this officer was seventh on the Air Force seniority list, he complains that De La Rue was also 'our only experienced seaplane pilot whose seniors were more experienced in landplane operations in war and were, in addition, graduates of the Staff College or of the Imperial Defence College or both'. On this criteria, he clearly considered the command ought to have gone to any of Group Captains Cole, McNamara or Wrigley. Williams claims that the choice had been justified on the basis that the more senior officers could not be spared—'surely an odd one for deciding on the commander of a force preparing for active service'—and practically accuses Goble of patronage in promoting the interests of an ex-RNAS colleague.[54]

Cabinet itself possibly harboured some doubts about the wisdom of sending away such a large air contingent, because even while approving the Goble plan it stipulated that ground maintenance staff were not to be offered for service until a survey had been made of the possible effect of the loss of such personnel on munitions production.[55] Compounding these uncertainties were further developments which soon changed the whole planning basis. On 26 September the British government cabled proposals inviting the Dominions to participate in a pooled training scheme aimed at providing aircrew for squadrons to be formed in England. The idea for this scheme was said to have been that of the Australian high commissioner in London (Bruce),[56] but—as one recent historian has pointed out—such a claim was almost certainly a fallacy maintained for presentational purposes. The Air Ministry had kept such a principle in mind for 20 years, and had been actively seeking Dominion concurrence to it at least since 1936, when a conference was convened which Goble (then on exchange with the RAF) had attended as Australia's representative.[57] Regardless of its origins, when considered by the Australian War Cabinet on 5 October the proposal was accepted without reference to the Air Board, though it was stated that this agreement did not alter the government's desire to continue with raising the air expeditionary force.[58] Nonetheless, it quickly became apparent that the new arrangement created requirements for trained personnel which competed with the Goble plan.

On 20 October the War Cabinet considered a cable from Bruce in London suggesting that the detachment from the RAAF's No.10 Squadron (already in England to take delivery of the new Sunderland flying boats and fly them to Australia) should be retained to serve alongside the RAF. Bruce also advised that the air expeditionary force should be abandoned altogether, to enable Australia to concentrate its effort on what would become known as the Empire Air Training Scheme (EATS). By this stage the government's doubts about whether personnel could be spared for the air contingent scheme had become more concrete and it was therefore inclined to accept Bruce's advice, although the high commissioner was told that it would prefer to receive a formal call to this effect directly from British authorities. When a letter was duly received from the British high commissioner in Canberra conveying the advice as being from the Air Ministry, the government announced the news on 31 October that the air expeditionary force would not proceed 'for the present' but that additional air and ground crews were to be sent to England to upgrade the RAAF detachment taking delivery of the Sunderlands to full squadron status as soon as possible. Instead of the six-squadron national force taking its place alongside the RAF, No.10 Squadron would now be the sole RAAF presence in the European theatre.

Group Captain H.F. De La Rue was chosen to command the air expeditionary force offered by Australia for overseas service on the outbreak of war in 1939. The capacity of the RAAF to raise the six squadrons promised came into question, however, and Britain made clear that it preferred Australia to join in the Empire Air Training Scheme, with the result that the proposed air contingent was abandoned. (*Mrs Y.C. Bonwick*)

Other hopes of a distinctive RAAF contribution to the war effort were raised at this time, only to be immediately dashed too. On 6 November Group Captain Bostock—who had accompanied the Treasurer, Casey, to London for defence discussions—cabled Melbourne to suggest that, as Army was preparing to despatch an AIF division (the 6th) overseas and raise another immediately to form a corps, the RAAF should consider the requirements involved in sending Army cooperation squadrons with these formations. Such units, he said, might be formed in Australia using Wirraways for service training until the Air Ministry could arrange for their re-equipment with a standard type of RAF aircraft on their arrival overseas. On the basis of this advice, branches within RAAF Headquarters were notified on 27 November that a recommendation had been put to the government that No.3 Squadron should go with the 6th Division to the Middle East and a new squadron formed to accompany the next division being raised. Within a few days, however, it was realised that there were serious problems confronting the despatch of even such a modest force as had now been proposed.[59]

In July 1939 a detachment of No.10 Squadron was sent to England to take delivery of the first three Sunderland flying boats for the RAAF and fly them to Australia. Following the outbreak of war the Australian government agreed to the machines staying in England, and the detachment, shown here in late 1939 at Pembroke Dock, was built up to full unit strength to serve with the RAF. The officers pictured in the front row are (*from left*): I.S. Podger, W.N. Gibson, C.W. Pearce, L.V. Lachal, W.H. Garing, J.A. Cohen, H.M. Birch, K. Campbell. (*ACdre I.S. Podger*)

As staff of the AMP Branch pointed out on 30 November to their chief, Air Commodore John Russell (an RAF officer who had been sent to Australia in exchange for Williams), the RAAF already had 120 aero fitters and 100 metal riggers less than it needed for its current units at war strength, and formation of new units for overseas service would only accentuate the problem of aircraft maintenance across the whole force. A deficiency of radio mechanics and operators in No.3 Squadron would also have to be met by taking personnel from other units which were short in these areas too. The conclusion was that it was impracticable to form even one Army cooperation squadron on a war basis for overseas operations before April 1940.[60] Russell immediately advised CAS of the problem, noting the impact which the deficiencies identified would have on the function and formation of training units. Whether to send No.3 Squadron immediately 'or wait until such time as it is practicable to do so' hinged, he observed, on a 'question of the degree of importance' attached to having a unit abroad. On 2 December the Air Liaison Officer in London was advised that the government had decided against sending No.3 Squadron, and he was instructed to suspend negotiations with the Air Ministry on arming and equipping the squadron overseas.[61]

These decisions cancelling out the hopes of proceeding on active service were, inevitably, an immense disappointment to the whole of the RAAF. Nowhere would that sense of let-down been more keenly felt than in the office of the CAS, for, although Goble had not been imbued with the desire to rush to England's assistance with total disregard of local defence requirements, he was firmly wedded to the idea that any contribution to the war effort sent from Australia should be organised as a self-contained national entity and not simply form a training resource for the RAF. According to at least one historian, it was disgust at the government's negation of that principle which on 19 December prompted Goble to request that he be relieved of both his RAAF commission and the appointment as CAS.[62] This became a political bombshell when, due to departmental mishandling of his letter, the news of his extraordinary action appeared in newspapers before Prime Minister Menzies had been advised of it.[63]

As a reflection of the frustration felt at many levels throughout the RAAF, Goble's decision was supported by many other officers of the service. Family members recall Air Force visitors to the house applauding Goble's stand and some even indicating a preparedness to also resign in sympathy.[64] There is, however, now some misunderstanding about the issues which the resignation was meant to protest. The reasons involved more than simple objection to the government's preference for EATS over the planned air contingent—as the two-month interval between the War Cabinet's announcement and the resignation attests.

The fact is that Goble had found himself in an impossible position arising from a personal conflict with Air Commodore Russell, the AMP and one of the three service members of the Air Board. The problems in CAS's relations with this officer had been developing for some months, but came steadily to the fore from October 1939. When he was spoken to by the Minister for Defence (Street) on 11 October about the unsatisfactory functioning of the Air Board, Goble set down in writing the next day the fact that he attributed the problems to the attitude and conduct of Russell:

> . . . I am sorry to record that the Air Member for Personnel has been a great hinderance to progress. I regret that it is necessary for me to refer in such an unfavourable strain to a senior officer of the RAF but the fact remains that he has stubbornly refused to adjust himself to local conditions and regulations and has shown marked inconsistency and unreliability in his opinions and statements on Service matters . . . I find it impossible to carry on any longer unless there is stability in the direction of the Branch of the Air Member for Personnel. I consider that Air Commodore Russell's temperament and outlook make him unsuitable for a Board or Command appointment in this Service to which I am proud to belong but which he constantly belittles.[65]

Notwithstanding the clear warning given by Goble about his likely future action at this stage, Street evidently gave no support in handling the politically awkward situation of a fractious subordinate from the British service. On 2 November Goble sent the Minister a four page note documenting instances of Russell's erratic and unreliable behaviour, and adding 'I should be grateful if an early decision can be obtained on the last paragraph of my letter [of 12th

October]'.[66] When no decision proved forthcoming Goble took action of his own, sending a signal to London on 14 November asking his RAF counterpart, Air Chief Marshal Sir Cyril Newall, to immediately recall Russell. As he explained in a letter sent the same day, this request was one 'forced through sheer desperation'. Citing Russell's troublesome record, Goble expressed a direct fear that 'it will be practically impossible to implement the Empire Training Scheme if he remains in the appointment of AMP where stability and sound training knowledge are essential qualifications': 'I am very sorry indeed about all this and about troubling you at such a critical time, but it simply cannot go on any longer and I cannot permit Russell to wreck this Service'.[67]

The response to these representations was, plainly, less than he expected and felt he deserved. When Menzies—angered at first reading of Goble's resignation in the press—wanted to know the reasons for it, CAS cited no other issue than his impossible relations with Russell and the government's failure to act in the only way which circumstances indicated was required. Replying to Menzies' call for reasons through the secretary of the Defence Department, F.G. Shedden, Goble said:

> Considering the urgent representations I have been making in vain during the last few months ... I am surprised to read that the Prime Minister now wishes to know the reasons for my resignation, especially in view of his own directions communicated to me through the then Minister for Defence following my representations of 12th October and the 2nd November ...
>
> The Cabinet apparently sees fit to prefer to accept directly or indirectly the advice of an Officer considerably junior to me and who is in my opinion anti-Australian in his outlook, insufficiently experienced for his appointment, lacking in loyalty, and unreliable in his statements and judgement. In consequence the morale, contentment and efficiency of this Service, to which I am proud to belong but which he constantly belittles, are seriously impaired.
>
> I have done everything in my power to find a diplomatic solution to this problem and to offset its unfortunate effect on the outlook of the Service. The fact that this Officer has recently become increasingly discourteous to me is mentioned merely in passing ... He has declined to accept a Group Command to which he was appointed with the concurrence of the Air Ministry, and the Prime Minister has indicated that he must stay where he is and 'we have just *got* to make things work'.
>
> I must again stress that unless a change is made promptly a breakdown is inevitable and this will again reflect adversely on a Service already operating under serious disabilities ... I have constantly stressed these views to the then Minister for Defence and if the Cabinet is not prepared to accept my advice as the present Head of the Service (thereby implying lack of confidence in me) I must decline to accept any responsibility.
>
> I am gravely apprehensive for the wellbeing of this Service, which it has been my proud privilege to serve and lead, and as the Cabinet are determined to retain the officer concerned in his present appointment, I have no option but to ask that I be relieved of mine ... [68]

The extent to which Goble's complaints about Russell were justified is difficult to determine completely. There is, however, evidence that he had conducted himself more as an agent of a foreign government than as the third most senior officer at RAAF Headquarters. His actions had included going behind CAS's back in urging the Australian government to abandon plans to send complete RAAF units overseas. Air Commodore Ulex Ewart, then director of personnel services as a wing commander, recalls:

> Unbeknown to Goble, Russell went to Shedden and told him that the EATS was seen in London as vital to Britain's war effort and Australia could forget about any air expeditionary force. He obtained Shedden's consent to signal the Air Ministry through the secretary's office to avoid using the RAAF channel and risking his communications coming to CAS's notice. I know that this was what was arranged because Russell took me to the meeting with Shedden. I wasn't very impressed with Russell, and consider him to have been an RAF cast-off.[69]

Another who had doubts about Russell was the British high commissioner to Australia, Sir Geoffrey Whiskard, who reported back to London on 1 January 1940: 'Incidentally I met Russell for the first time at dinner at Government House, Melbourne, the other night; and both my wife, whose judgement in these matters I value, and I, took an instant dislike to him'.[70]

Somewhat belatedly, Prime Minister Menzies was also convinced that Russell was a cankerous element on the Air Board and even took the opportunity to castigate London for sending him to Australia in the first place. Whiskard recounted to London that in a discussion with the prime minister on 31 December 1939, Menzies had told him that:

> he regarded Russell as mainly to blame ... Russell not only talked a great deal too much in Clubs and other congregations of tittle-tattlers, but he was undoubtedly disloyal to his colleagues. Furthermore ... while he must assume that Russell had been a success in the command which he had held prior to coming to Australia, he was convinced that (apart from the disabilities of character noted above) he was not suited for any job in which he was required to advise on matters of policy ... [W]hile he was satisfied that the Admiralty and War Office now appreciated that they must not use Australia as a wastepaper basket for their duds he was by no means sure that the Air Ministry had yet learnt this lesson. It was most important that they should. The sending out of unsuitable men here was not only bad for the Australian services (which he admitted were, in peace time at any rate, from the United Kingdom point of view unimportant) but it lowered the prestige of the United Kingdom services in Australian eyes.[71]

A willingness to take the necessary action of sending Russell home—which the government now did—might have ended this deplorable situation, but the government had another trump card to play. As Menzies told Goble in an interview on 23 December to discuss the latter's resignation, the government had actually decided 'some time ago' to replace him as CAS with an RAF officer with greater seniority and experience.[72] In essence, the government had also been working behind Goble's back in negotiating with the Air Ministry for a replacement. According to family sources Goble had already got wind of this development, having been warned privately by the civilian secretary of the Air Board, Coleman.[73] Such a warning was probably unnecessary, however, because as soon as news of Goble's resignation appeared in the press there had been reports that an RAF replacement was planned.[74] Although Goble evidently went into his interview with Menzies prepared to withdraw his resignation with Russell's departure, he found the prime minister had no intention of allowing him to do so; he could retain his RAAF commission but not remain as CAS. As Menzies told Whiskard, 'Goble's resignation was undoubtedly very convenient, as it saved ... the unpleasantness of publicly superseding him'.[75]

In the wake of their interview, Menzies cabled London to advise the Air Ministry of Goble's desire to obtain a command with the RAF and the government's willingness to second him if his services were so required. Before anything came of this approach, however, on 4 January 1940 War Cabinet settled on different arrangements. A senior officer was required to look after the implementation of the EATS scheme in Canada, and it was decided that Goble should fill this post.[76] This same meeting also heard from James Fairbairn—appointed in November to the new portfolio as Minister for Air—who had represented Australia at the EATS conference in Quebec before proceeding to England where, on Menzies' instructions, he had interviewed prospective appointees to the CAS post.

According to accounts, the officer originally proposed to Fairbairn by Casey, already in London, was Air Chief Marshal Sir John Steel, who was then aged 62 and had been retired for two years.[77] Fairbairn travelled by car to Portsmouth to interview Steel, and was accompanied on the trip by Air Marshal Sir Charles Burnett, one of the RAF's two inspector-generals, who went along to make the necessary introductions. As Fairbairn reportedly told friends later, he got on so well with Burnett during the trip that he decided to seek his guide's services instead.[78] Aged 58, Burnett was two years off retirement; he was also not a well man, suffering from a stomach ulcer. Given this officer's age, health, and present (mainly titular) appointment, the Air Ministry was prepared to release Burnett to Australia. On 4 January War Cabinet approved the recommendation put to it by Fairbairn that Burnett, given a step in rank to air chief marshal, should be appointed for one year but with the option of extension.

Burnett arrived by air to take up his new appointment in February 1940. Accompanying him on the flight out to Australia was none other than Williams, whose early return from his exile on exchange had been arranged by Fairbairn. Agreeing to this, too, on 4 January, the War Cabinet was at pains to stress that 'he should clearly understand that ... [this does] not imply any right of reversion to the post of Chief of the Air Staff'.[79] Granted temporary rank of air marshal, he was expected to fill either the post of AMP or else a new position in charge of a Training Command to manage the EATS in Australia. In the event he rejoined the Air Board as Air Member for Organisation and Equipment.

With the implementation of these changes at its top levels, the transition of the RAAF to a wartime footing was complete in its most important elements. What had occurred was, in fact, less a matter of transition than of abandonment. The government had effectively turned its back on the pre-war Air Force structure, believing it to be inadequate or incapable of meeting the challenges brought by the war without outside assistance. Many historians are inclined to agree with that judgement,[80] though clearly no government of the inter-war period remained blameless for any of the deficiencies which may be considered to have been revealed in 1939. What was more, members of the force might with some justification consider that their service had been 'sold out' in 1939–40 by a government which, from its prime minister down, was only too eager to ingratiate itself with London.

In a very real sense, the discrediting and banishment of first Williams and then Goble represented the visible passing of the established order of the pre-war RAAF, a situation which—in view of its unsatisfactory features—might well have been of benefit to the service as a whole. However, with the appointment as CAS of Air Chief Marshal Burnett—who one historian describes as 'the first "recruit" into the Australian section of the Empire Air Training Scheme'[81]—the basis had been laid for the plundering of Australia's air defence resources for the benefit of Britain's war effort in Europe. The penalties of this process would be alarmingly exposed with Japan's drive into the South Pacific region in 1942. There was, of course, nothing new in the subordination of Australia's national interests on the altar of the presumed benefits of imperial defence, a theme which dominates the history of Australian defence policy in the first half of the twentieth century. It might be argued with reason, therefore, that the old order which passed in the first months of 1940 was simply making way for the re-establishment of an even older order.

This has been the story of the Royal Australian Air Force in the interval between two world wars. In recounting this tale, it has not been the author's desire to claim for the RAAF a significance beyond the reality. A small service operating in an environment far removed from the mainstream of world affairs must inevitably be 'small beer' in global terms, though it was—as the record shows—of considerable importance in national terms. Before laying aside this volume, however, the reader is invited to reflect on the fact that it was Australia's puny air force of 1939, for all its problems of inadequate equipment and fractured leadership, which proved capable of a fifty-fold expansion during the Second World War to reach a peak strength of 182 000 officers, airmen and airwomen. At the end of the war in August 1945 the RAAF had a personnel strength of almost 132 000 in the Pacific theatre and operated nearly 3200 operational aircraft, many of which were among the most modern types available; almost another 3000 training machines were also in use.[82] The claim is made that at this time the RAAF was the fourth largest air force in the world, smaller only than those maintained by the United States, the Soviet Union and Britain.[83] Such a massive growth in Australian air power was possible, it is suggested, only because of the basis which had existed in the pre-war RAAF.

Appendix 1
Ministers for Defence

Senator G.F. Pearce	17.9.1914 — 21.12.1921
W.M. Greene	21.12.1921 — 5.2.1923
E.K. Bowden	9.2.1923 — 16.1.1925
Major-General Sir Neville Howse	16.1.1925 — 2.4.1927
Senator Major-General Sir William Glasgow	2.4.1927 — 22.10.1929
A.E. Green	22.10.1929 — 4.2.1931
Senator J.J. Daly	4.2.1931 — 3.3.1931
J.B. Chifley	3.3.1931 — 6.1.1932
Senator Sir George Pearce	6.1.1932 — 12.10.1934
Sir Archdale Parkhill	12.10.1934 — 20.11.1937
J.A. Lyons	20.11.1937 — 29.11.1937
H.V.C. Thorby	29.11.1937 — 7.11.1938
Brigadier G.A. Street	7.11.1938 — 13.11.1939
R.G. Menzies	13.11.1939 — 7.10.1941

Appendix 2
The Air Board

9.11.1920

Brevet Major (Hon. Lieutenant-Colonel) R. Williams, DSO, OBE (Director of Intelligence & Organisation)

Squadron Leader (Hon. Wing Commander, RAF) S.J. Goble, DSO, OBE, DSC, (Director of Personnel & Training)

Captain P.A. McBain, MBE (Director of Equipment)

Mr A.C. Joyce (Finance Member)

[On the formation of the Australian Air Force on 31.3.1921, Williams and Goble became wing commanders and McBain became squadron leader; Major P.E. Coleman was appointed as the civilian secretary to the Air Board on 1.1.1921]

24.10.1921

Wing Commander R. Williams, DSO, OBE (Director of Intelligence & Organisation)

Squadron Leader W.H. Anderson, DFC, (Director of Personnel & Training) (Acting)

Squadron Leader P.A. McBain, MBE (Director of Equipment)

Mr A.C. Joyce (Finance Member)

2.10.1922

Wing Commander R. Williams, DSO, OBE (Chief of the Air Staff) (First Air Member)

Squadron Leader W.H. Anderson, DFC, (Chief of the Administrative Staff) (Second Air Member) (Acting)

Mr A.C. Joyce (Finance Member)

15.12.1922

Wing Commander S.J. Goble, DSO, OBE, DSC, (Chief of the Air Staff) (First Air Member) (Acting)

Squadron Leader W.H. Anderson, DFC, (Chief of the Administrative Staff) (Second Air Member) (Acting)

Mr A.C. Joyce (Finance Member)

10.2.1925

Wing Commander R. Williams, DSO, OBE (Chief of the Air Staff) (First Air Member)

Wing Commander S.J. Goble, CBE, DSO, DSC, (Chief of the Administrative Staff) (Second Air Member)

Mr A.C. Joyce (Finance Member)

1.7.1925

Group Captain R. Williams, DSO, OBE (Chief of the Air Staff) (First Air Member)

Squadron Leader (Temp. Wing Commander) W.H. Anderson, DFC (Chief of the Administrative Staff) (Second Air Member) (Acting)
Mr A.C. Joyce (Finance Member)

1.3.1926
Group Captain R. Williams, DSO, OBE (Chief of the Air Staff) (First Air Member)
Squadron Leader (Temp. Wing Commander) A.T. Cole, MC, DFC (Chief of the Administrative Staff) (Second Air Member) (Acting)
Mr A.C. Joyce (Finance Member)

6.7.1926
Group Captain R. Williams, DSO, OBE (Chief of the Air Staff) (First Air Member)
Wing Commander R.S. Brown, AFC (Chief of the Administrative Staff) (Second Air Member) (Acting)
Mr A.C. Joyce (Finance Member)

31.12.1927
Group Captain R. Williams, CBE, DSO (Chief of the Air Staff) (First Air Member)
Wing Commander S.J. Goble, CBE, DSO, DSC (Air Member for Personnel) (Second Air Member)
Mr A.C. Joyce (Finance Member)
[Goble was promoted group captain on 1.4.1928]

1.10.1929
Air Commodore R. Williams, CBE, DSO (Chief of the Air Staff)
Group Captain S.J. Goble, CBE, DSO, DSC (Air Member for Personnel)
Wing Commander W.H. Anderson, DFC (Air Member for Supply)
Mr A.C. Joyce (Finance Member)

11.7.1932
Air Commodore R. Williams, CBE, DSO (Chief of the Air Staff)
Group Captain S.J. Goble, CBE, DSO, DSC (Air Member for Personnel)
Wing Commander W.H. Anderson, DFC (Air Member for Supply)
Mr T.J. Thomas (Finance Member)

7.12.1932
Group Captain (Temp. Air Commodore) S.J. Goble, CBE, DSO, DSC (Chief of the Air Staff)
Group Captain W.H. Anderson, DFC (Air Member for Supply)
Mr T.J. Thomas (Finance Member)

3.1.1933
Group Captain (Temp. Air Commodore) S.J. Goble, CBE, DSO, DSC (Chief of the Air Staff)
Group Captain W.H. Anderson, DFC (Air Member for Personnel)
Wing Commander A.T. Cole, MC, DFC (Air Member for Supply)
Mr T.J. Thomas (Finance Member)

13.6.1934
Air Commodore R. Williams, CBE, DSO (Chief of the Air Staff)
Group Captain (Temp. Air Commodore) S.J. Goble, CBE, DSO, DSC (Air Member for Personnel)
Wing Commander A.T. Cole, MC, DFC (Air Member for Supply)
Mr T.J. Thomas (Finance Member)
[On 1.1.1935 Williams was promoted air vice-marshal and Cole became a group captain; Williams was appointed AMP in addition to CAS on 29.9.35]

7.12.1935
Air Vice-Marshal R. Williams, CB, CBE, DSO (Chief of the Air Staff)

Air Commodore H.R. Nicholl, CBE (RAF) (Air Member for Personnel)
Group Captain A.T. Cole, MC, DFC (Air Member for Supply)
Mr T.J. Thomas (Finance Member)

20.4.1936
Air Vice-Marshal R. Williams, CB, CBE, DSO (Chief of the Air Staff)
Air Commodore H.R. Nicholl, CBE (RAF) (Air Member for Personnel)
Group Captain W.H. Anderson, CBE, DFC (Air Member for Supply)
Mr T.J. Thomas (Finance Member)

6.7.1936
Air Vice-Marshal R. Williams, CB, CBE, DSO (Chief of the Air Staff)
Air Commodore H.R. Nicholl, CBE (RAF) (Air Member for Personnel)
Group Captain W.H. Anderson, CBE, DFC (Air Member for Supply)
Mr M.C. Langslow, MBE (Finance Member)

[Anderson was promoted air commodore on 1.1.1938]

27.1.1938
Air Vice-Marshal R. Williams, CB, CBE, DSO (Chief of the Air Staff)
Air Commodore S.J. Goble, CBE, DSO, DSC (Air Member for Personnel)
Air Commodore W.H. Anderson, CBE, DFC (Air Member for Supply)
Mr M.C. Langslow, MBE (Finance Member)

28.2.1939
Air Commodore (Temp. Air Vice-Marshal) S.J. Goble, CBE, DSO, DSC (Chief of the Air Staff)
Air Commodore J.C. Russell, DSO (RAF) (Air Member for Personnel)
Air Commodore W.H. Anderson, CBE, DFC (Air Member for Supply)
Mr M.C. Langslow, MBE (Finance Member)

9.1.1940
Air Commodore W.H. Anderson, CBE, DFC (Chief of the Air Staff) (Acting)
Air Commodore J.C. Russell, DSO (RAF) (Air Member for Personnel)
Mr M.C. Langslow, MBE (Finance Member)

11.2.1940
Air Chief Marshal Sir C.S. Burnett, KCB, CBE, DSO (RAF) (Chief of the Air Staff)
Air Commodore J.C. Russell, DSO (RAF) (Air Member for Personnel)
Air Commodore W.H. Anderson, CBE, DFC (Air Member for Supply)
Mr M.C. Langslow, MBE (Finance Member)

11.3.1940
Air Chief Marshal Sir C.S. Burnett, KCB, CBE, DSO (RAF) (Chief of the Air Staff)
Air Commodore W.H. Anderson, CBE, DFC (Air Member for Personnel)
Air Vice-Marshal (Acting Air Marshal) R. Williams, CB, CBE, DSO (Air Member for Organization and Equipment)
Mr M.C. Langslow, MBE (Finance Member)

18.3.1940
Air Chief Marshal Sir C.S. Burnett, KCB, CBE, DSO (RAF) (Chief of the Air Staff)
Air Commodore W.H. Anderson, CBE, DFC (Air Member for Personnel)
Air Vice-Marshal (Acting Air Marshal) R. Williams, CB, CBE, DSO (Air Member for Organization and Equipment)
Mr R. Lawson, OBE (Director-General of Supply and Production)
Mr M.C. Langslow, MBE (Finance Member)

Appendix 3
*Defence expenditure**

	RAN	Army	RAAF	Civil aviation	Defence Total
1924–25	2393(44.4)	1558(28.9)	398(7.4)	105(1.9)	5393
1925–26	2621(40.8)	1548(24.1)	429(6.8)	114(1.8)	6431
1926–27	2765(35.0)	1526(19.3)	572(7.2)	112(1.4)	7891
1927–28	2598(35.2)	1494(20.2)	517(7.0)	113(1.5)	7386
1928–29	2396(36.7)	1466(22.4)	549(8.4)	98(1.5)	6536
1929–30	2167(44.3)	1239(25.4)	498(10.2)	103(2.1)	4886
1930–31	1748(45.3)	1195(30.9)	392(10.2)	150(3.9)	3859
1931–32	1444(45.3)	995(31.2)	326(10.2)	132(4.1)	3185
1932–33	1496(47.3)	978(30.9)	320(10.1)	137(4.3)	3160
1933–34	1646(39.6)	1237(29.8)	409(9.8)	134(3.2)	4157
1934–35	1998(36.6)	1328(24.3)	536(9.8)	164(3.0)	5458
1935–36	2371(33.8)	1811(25.8)	783(11.2)	186(2.6)	7014
1936–37	2577(31.9)	2232(27.7)	1163(14.4)	426(5.3)	8065
1937–38	2960(30.3)	2182(22.3)	1930(19.7)	247(2.5)	9773

* All amounts are expressed in £000s and denote expenditure charged to the individual services and Civil Aviation Branch under ordinary Votes and Appropriations; figures in brackets denote percentages of the total Defence outlay.

Note: These figures are drawn from Commonwealth Year Books Nos. 22–32 (1929–1939). Earlier issues of the Year Book provide figures for years before 1924–25, but their basis of calculation differed significantly from that used later and therefore allow no useful comparisons to be made.

Appendix 4
RAAF personnel strength

	Officers	Air cadets	Airmen	Total
31.3.21	21	—	128	149
1.4.22	49	—	252	301
1.4.23	55	5	286	346
1.4.24	59	3	297	359
1.4.25	69	4	379	452
1.4.26	76	26	594	696
1.4.27	84	9	721	814
1.4.28	92	—	734	826
1.4.29	92	18	818	928
1.4.30	97	16	757	870
1.4.31	97	18	712	827
1.4.32	100	9	717	826
1.4.33	97	8	709	814
1.4.34	96	15	706	817
1.4.35	106	40	810	956
1.4.36	134	63	1065	1262
1.4.37	160	73	1722	1955
1.4.38	199	64	2032	2295
1.4.39	274	74	2824	3172
1.4.40	859	258	7901	9018
1.4.41	2253	—	37379	39632

Appendix 5
RAAF aircraft strength

	1 Jul. 1922	30 Jun. 1924	1 Jan. 1930	1 Jan. 1935	1 Jan. 1939
Fighters					
SE5a	4(31)	8(14)			
Bristol Bulldog*			(8)		
Fighter-Bombers					
Bristol Bulldog				7	3
Hawker Demon				(17)	55
Bombers					
DH9	4(24)	7			
DH9a	3(27)	8			
Avro Anson GR1					58(26)
General Purpose					
Westland Wapiti			28(6)	25	24
Seaplanes/Amphibians					
Fairey IIID	1(5)	3			
Supermarine Seagull III			9	3	
Supermarine Southampton			2	2	1
Supermarine Seagull V				(2)	22
Empire Flying Boat					(2)
Trainers					
Avro 504K	9(44)#	28			
Sopwith Pup	4(7)	6			
DH-60 Moth (Cirrus & Gypsy)			21(23)	32	19(4)
Avro 643 Cadet				(6)	20(12)
Miles Magister					1
CAC Wirraway					2(32)
CA-6 Wackett Trainer					(1)
DH-82 Tiger Moth					(20)

RAAF aircraft strength (continued)

	1 Jul. 1922	30 Jun. 1924	1 Jan. 1930	1 Jan. 1935	1 Jan. 1939
Other					
DH-89 Dragon Rapide				(2)	
Wackett (Tugan) Gannett				(1)	2(1)
DH-86					(1)
DC2					(4)
TOTAL	25	60	60	69	207

* Type re-designated fighter-bomber in early 1935.

Includes 6 locally built aircraft delivered during 1922–23.

Note: Figures in columns not in brackets denote aircraft actually in service; figures in brackets denote other machines in course of delivery or in storage.

Abbreviations

AA	Australian Archives
AAC	Australian Air Corps
AAF	Australian Air Force
AA&ECo.	Australian Aircraft and Engineering Company
AAIC	Air Accidents Investigation Committee
AAL	Australian Air League
AC1	Aircraftsman Grade 1
AC2	Aircraftsman Grade 2
ACdre	Air Commodore
AD	Aircraft Depot
ADB	*Australian Dictionary of Biography*
Adel	Adelaide
AFC	Australian Flying Corps; also Air Force Cross
AHSA	Aviation Historical Society of Australia
AIF	Australian Imperial Force
AIS	Aviation Instructional Staff
ALO	Air Liaison Officer
ALP	Australian Labor Party
AM	Air Marshal
AMP	Air Member for Personnel
AMS	Air Member for Supply
ASC	Air Service Committee
AVM	Air Vice-Marshal
AWM	Australian War Memorial
BANZARE	British, Australian and New Zealand Antarctic Research Expedition
CAB	Civil Aviation Branch (later Board)
CAC	Commonwealth Aircraft Corporation

CAF	Citizen Air Force
CAG	*Commonwealth of Australia Gazette*
Canb	Canberra
CAS	Chief of the Air Staff
CBE	Companion of the Order of the British Empire
CCA	Controller of Civil Aviation
CFS	Central Flying School
CGS	Chief of the General Staff
CNS	Chief of the Naval Staff
CO	Commanding Officer
CPD	*Commonwealth Parliamentary Debates*
CPP	*Commonwealth Parliamentary Papers*
CTD	Central Training Depot
DFC	Distinguished Flying Cross
DFJ	*Defence Force Journal*
DIO	Director of Intelligence and Operations
DofE	Director of Equipment
DPR	Defence Public Relations
DPT	Director of Personnel and Training
DROs	Daily Routine Orders
DSC	Distinguished Service Cross
DSM	Disciplinary Sergeant-Major
DSO	Distinguished Service Order
EATS	Empire Air Training Scheme
FOff	Flying Officer
FLt	Flight Lieutenant
FM	Finance Member
FTS	Flying Training School
GD	General Duties
GMH	General Motors Holden
GpCapt	Group Captain
HMS	His Majesty's Ship
Hob	Hobart
IAF	Independent Air Force
IAL	Imperial Airways Limited
IDC	Imperial Defence College; also interdepartmental committee
IRAFB	Inspector of Royal Air Force Badges
IRAAFB	Inspector of Royal Australian Air Force Badges
IWM	Imperial War Museum, London
LAC	Leading Aircraftsman
LASCo.	Larkin Aircraft Supply Company
Lond	London
LRDU	Long Range Development Unit

MBE	Member of the Order of the British Empire
MC	Military Cross
Melb	Melbourne
MHR	Member of the House of Representatives
MSB	Munitions Supply Board
MSM	Meritorious Service Medal
NCO	Non-commissioned Officer
NLA	National Library of Australia, Canberra
NSW	New South Wales
NT	Northern Territory
OBE	Officer of the Order of the British Empire
OYB	*Official Year-Book of the Commonwealth of Australia*
PAF	Permanent Air Force
PM	Prime Minister
PMG	Postmaster-General
OOff	Pilot Officer
PRO	Public Record Office, Kew
QANTAS	Queensland and Northern Territory Aerial Services
QEA	Qantas Empire Airways
Qld	Queensland
RAAF	Royal Australian Air Force
RAF	Royal Air Force
RAN	Royal Australian Navy
RFC	Royal Flying Corps
RMC	Royal Military College, Duntroon
RN	Royal Navy
RNAS	Royal Naval Air Service
RNZAF	Royal New Zealand Air Force
SA	South Australia
Sgt	Sergeant
SLA	*Sea, Land and Air*
SMH	*Sydney Morning Herald*
Sqn	Squadron
SqnLdr	Squadron Leader
Syd	Sydney
Tas	Tasmania
UK	United Kingdom
US	United States
VC	Victoria Cross
Vic	Victoria
WA	Western Australia
WgCdr	Wing Commander
WOff	Warrant Officer
W/T	wireless telephone

References

Only works cited once or twice are listed here with full publication details. Full details of works cited more than twice are given in the bibliography.

Introduction (pp. xv–xviii)

1. *CAG*, 1911, no.97, p.2399.
2. AA, CRS A289, item 1849/8/409.
3. AA, CRS A289, item 1849/8/398.
4. AA, CRS A289, items 1849/8/339 and 1849/8/410.
5. Military Orders 234, 385 and 516 of 1913, and 329 of 1914; AA, CRS A289, item 1849/8/410.
6. *ADB*, 9, pp.214–15.
7. Military Order 383 of 1914.
8. Williams, p.28.
9. Military Orders 356, 461 and 691 of 1914.
10. AA, CRS A2023, item 38/3/130.
11. S.S. Mackenzie, *The Australians at Rabaul*, vol.10 of *The Official History of Australia in the War of 1914–1918*, Syd, 1936, pp.160–1.
12. Parnell & Lynch, p.10; E. Harrison diary, entry for 10 Dec. 1914, RAAF Museum, Point Cook.
13. AA, B197, item 2021/1/64.
14. Military Order 670 of 1915.
15. Cutlack, p.31.
16. K. Isaacs, *Military Aircraft of Australia*, 1909–1918, p.158.
17. ibid., p.13.
18. Doak & Isaacs, p.76, say that three courses were completed at the NSW Aviation School, however, pictorial evidence at the RAAF Museum leaves no doubt there were six.
19. K. Isaacs, *Military Aircraft of Australia*, 1909–1918, pp.13–14, 37.
20. Department of Air, *The Golden Years*, p.xx.
21. Cutlack, p.421.
22. S.F. Wise, *Canadian Airmen and the First World War*, vol.1 of *The Official History*

of the Royal Canadian Air Force, Toronto, 1980, p.x, and W.A.B. Douglas, *The Creation of a National Air Force*, vol.2 of ibid. Toronto, 1986, p.193.
23. K. Isaacs, *Military Aircraft of Australia* 1909–1918, p.158.
24. ibid., pp.152–7.
25. Williams, p.96.
26. Odgers, p.23.
27. C. Schaedel, *Men & Machines of the Australian Flying Corps*, 1914–19, Dandenong, Vic, 1972, p.14.
28. Wackett, pp.72–4.
29. Williams papers: RAAF Museum.

Chapter 1

1. K. Isaacs, *Military Aircraft of Australia*, 1909–1918, p.133.
2. Gillison, pp.1–2.
3. R.M. Jones, p.8.
4. File on highlights of Air Force history: RAAF Historical Section (hereafter referred to as RAAF Highlights file).
5. Gillett, *Wings Across the Sea*, p.19.
6. RAAF Highlights file.
7. Gillison, p.3.
8. ibid., p.4.
9. McCarthy, *Australia and Imperial Defence*, p.23.
10. R.M. Jones, p.9.
11. Gillison, p.5.
12. RAAF Highlights file.
13. ibid.
14. Gillison, p.6.
15. RAAF Highlights file.
16. AA, CRS A6006, item 19/4/1.
17. Gillison, p.6.
18. RAAF Highlights file.
19. ibid.
20. Gillison, pp.8–9.
21. RAAF Highlights file.
22. AA, CRS A2180, item D1821/1/107.
23. RAAF Highlights file.
24. R.M. Jones, pp.11–13.
25. Gillison, p.9.
26. Hughes papers, NLA, MS 1538/19/124–5.
27. RAAF Highlights file.
28. ibid.
29. E.H. Sugden and F.W. Eggleston, *George Swinburne: A Biography*, Syd, 1931, pp.390–1.
30. RAAF Highlights file.
31. AA, CRS A705, item 4/3/6.
32. R.M. Jones, p.11.

33. ibid., p.13.
34. AA, CRS A705, item 4/3/6.
35. R.M. Jones, p.13.
36. AA, CRS A705, item 4/3/6.
37. ibid.
38. R.M. Jones, p.13.
39. Report on the Air Defence of Australia by the Air Board, 1920, pp.8–9: RAAF Historical Section.
40. ibid., p.9.
41. Williams, p.128.
42. R.M. Jones, p.14.
43. ibid.
44. ibid.
45. ibid., p.15.
46. ibid., pp.15–16; AA(Vic), MP 1049/1, item 1920/0204.
47. R.M. Jones, p.15.
48. AA, CRS A705, item 4/3/6.
49. ibid.
50. Military Order 437 of 14 Sep. 1918: AWM.
51. ibid.
52. *SLA*, 1 Nov. 1918, p.460; E.G. Roberts, *Box Kites and Beyond*, Melb, 1976, p.38; Argus, 16 Sep. 1918.
53. *SLA*, 1 Nov. 1918, p.460.
54. Military Orders 437 of 14 Sep. 1918, and 508 of 19 Oct. 1918.
55. *SLA*, 1 Nov. 1918, p.460.
56. *Argus* (Melb), 12 Oct. 1921.
57. Roberts, op.cit., pp.38–9.
58. Highlights notes, pp.10–11.
59. ibid., p.6.
60. Cutlack, p.65.
61. Letter, Lieutenant-Colonel R.F. Oakes to author, 7 Feb. 1982.
62. Letter, Oakes to AM Sir Richard Williams, 24 Nov. 1964, Williams papers, NLA, MS 6525, box 6.
63. Oakes letter to author; Miller, p.69; SLA, Jan. 1919, p.603.
64. *Argus*, 6 Apr. 1919; SLA, May 1919, pp.70–5, 77–82.
65. AA, CRS A571, item 20/49110.
66. *Argus*, 13, 14 Jun. 1919.
67. *Argus*, 17 Jun. 1919.
68. Oakes letter to author.
69. ibid.
70. H.N. Wrigley, 'First Across Australia by Air', manuscript article in Wrigley's personal papers: courtesy of FLt P.E. Hall.
71. ibid.
72. H.N. Wrigley, 'My Flight Across Australia', *SLA*, Feb. 1920, p.733.
73. Report by Lieutenant W.H. Fysh dated 14 Feb. 1920: RAAF Historical Section; *SLA*, Jan. 1920, p.668.

74. File on Wrigley's 1919 flight to Darwin: RAAF Historical Section.
75. ibid.
76. *SLA*, Jan. 1920, p.690; *Herald* (Melb), 27 Dec. 1919; *SMH*, 29 Dec. 1919.
77. AAC Routine Orders No.26 dated 23 Feb. 1920: RAAF Museum; program for official reception, 24 Feb. 1920: RAAF Historical Section.
78. AHSA *Newsletter*, vol.4, no.1, 1988, p.17.
79. *CAG*, no.131, 1919, p.1786.
80. AA, CRS A1195, item 715/1/24.
81. ibid.
82. ibid.
83. AA, CRS A1952, item A559/28/679.
84. Letter: Lieutenant-Colonel R.F. Oakes to author, 7 Feb. 1982.
85. Miller, p.70.
86. *SLA*, Aug. 1919, p.328.
87. AA, CRS A1952, item A559/28/679.
88. ibid.
89. AAC Routine Orders No.1 dated 5 Jan. 1920.
90. ibid., No.24 dated 19 Feb. 1920.
91. ibid., No.32 dated 10 Mar. 1920.
92. AA, CRS A1195, item 759/1/1115.
93. R.M. Jones, p.16.
94. AA, CRS A1195, item 715/1/24.
95. ibid.
96. ibid.
97. ibid.
98. AA, CRS A1195, item 715/1/27.
99. AAC Routine Orders No.11 dated 19 Jan. 1920: RAAF Museum.
100. Log-book of AVM A.T. Cole, Cole papers; *Australasian* (Melb), 26 Jun. 1920.
101. Walker, p.173.
102. Letter: E.C. Johnston, CCA, to Secretary, Aero Club of South Australia, 14 Jul. 1932, AA, CRS A705, item 108/1/921.
103. *Argus* (Melb), 23 Jul. 1920.
104. Williams, pp.144–6.
105. *SLA*, Sep. 1920, pp.360–3.
106. AA, CRS A1195, item 715/2/30.
107. AA, CRS A571, item 20/35038.
108. AAC Routine Orders No.66 dated 6 Aug. 1920 and No.67 dated 11 Aug. 1920: RAAF Museum.
109. C.A. Owers, 'The Peace Loan Flights', unpub ms: copy provided courtesy of the author.
110. AA, CRS A571, item 20/39432.
111. C.A. Owers, 'The Peace Loan Flights'.
112. AA, CRS A1195, item 856/2/53.
113. *Argus*, 27 Sep. 1920.
114. AHSA *Newsletter*, vol.2, no.4, 1986, p.75.
115. AA, CRS A1195, item 856/2/53.

116. Parnell & Boughton, p.40.
117. Miller, pp.60–1.
118. *SLA*, Jan. 1921, p.653.
119. AA, CRS A1195, item 856/2/53.
120. *Argus*, 8, 28 Jun. 1921.
121. AA, CRS A1195, item 856/2/53.
122. Williams, pp.128–9, 137.
123. ibid., p.137.
124. Military Order 536 dated 4 Dec. 1920: AWM.
125. Goble papers, in possession of Dr A.J. Goble.
126. ibid.
127. AA (Vic), MP 1049/1, item 1920/0346.
128. AA, CRS A1195, item 863/2/51.
129. ibid., and Air Council agendum no.22.
130. Williams, p.138.

Chapter 2

1. Air Board agendum no.4: RAAF Historical Section.
2. ibid., agendum No.3.
3. ibid., agendum No.6.
4. Air Council agendum no.4.
5. Gillison, p.15.
6. *CPD*, vol.94, pp.7310–11.
7. ibid., p.7679.
8. File: 'Australian Air Force, Establishment & Development': in RAAF Historical Section.
9. *Argus* (Melb), 23 Jun. 1921.
10. *CAG*, no.65, 1921, p.1207.
11. AA, CRS A1195, item 751/1/46.
12. *The Golden Years*, p.2; Odgers, p.48.
13. *CAG*, no.58, 1921, p.1083; AA, CRS A1195, item 736/2/4.
14. AA, CRS A1195, item 715/2/224.
15. Air Council agendum no.62.
16. AA, CRS A705, item 6/1/1.
17. *Aircraft*, 20 Sep. 1921, p.51.
18. ibid., 20 Oct, 1921, pp.69–71.
19. ibid., 30 Nov. 1921, p.104.
20. Brownell, pp.186–7.
21. Interview: AM Sir George Jones, 21 Jan. 1988.
22. ibid.
23. C. Schaedel, *Australian Air Ace*, pp.72–3.
24. Williams, pp.141–2.
25. *Aircraft*, 20 Oct. 1921, p.71; *SLA*, 1 Nov. 1921, p.596.
26. AA, CRS A1195, item 715/1/42.
27. ibid.

28. AA, CRS A1195, item 715/2/224.
29. Nominal roll held by ACdre C.R. Taylor; DROs for No.1 FTS, 19 Sep. 1921: RAAF Museum; 1 AD unit diary: RAAF Historical Section.
30. See AA, CRS A1195, item 751/1/111.
31. Letter: Wackett to W.H. Nicholson, in possession of Mr A.W. Nicholson.
32. Nominal roll held by ACdre C.R. Taylor; 1 AD unit diary.
33. Interview: FLt F.C.H. Hakanson, 29 Jun. 1988.
34. G. Jones, p.30.
35. Ellison, p.59.
36. AA, CRS A1195, items 751/1/83, 88, 92, 101, 102, 103, 105, 106, 108, 111.
37. C. Schaedel, *Australian Air Ace*, opposite p.64.
38. *SLA*, 1 Jan. 1922, p.764.
39. Walker, pp.174–6.
40. Doak & Isaacs, p.36.
41. AA, CRS A1195, item 736/2/4; 'A History of the Royal Australian Air Force 1900–1951', unpub.ms. (hereafter referred to as 'RAAF History 1900–1951'): RAAF Historical Section.
42. RAAF History 1900–1951.
43. AA, CRS A1195, item 751/1/49.
44. AA, CRS A1195, item 751/1/46.
45. RAAF History 1900–1951.
46. 1 AD unit diary.
47. Air Board agendum no.98.
48. 1 AD unit diary; Air Board agendum no.182; *Aircraft*, 30 Apr. 1922, p.60.
49. Air Board agendum no.100.
50. *SLA*, 1 Feb. 1922, p.862.
51. ibid., 1 Mar. 1922, p.938.
52. *CPP*, 1923–24, 4, pp.372–3.
53. Williams, p.142.
54. AA (Vic), MP 153/20, item 4.
55. Gillison, p.20.
56. AA, CRS A1195, item 763/1/34.
57. *OYB*, no.16, 1923, p.624, no.17, 1924, p.595.
58. *CPP*, 1922, 2, p.2396.
59. Air Board agendum no.257.
60. ibid. no.260.
61. Gillison, p.21.
62. AA, CRS A1195, item 763/1/34.
63. DROs No.1 Station, dated 10 Oct. 1922: RAAF Museum.
64. Air Board agendum no.347.
65. *Argus*, 11 Jan. 1922.
66. 1 AD unit diary.
67. ibid.
68. G. Jones, p.30.
69. *Aircraft*, 20 Sept. 1921, p.51.
70. *Argus*, 11 Jan. 1922.

71. *Aircraft*, 31 Mar. 1922.
72. Letter: Mr B.J. Videon to author, 8 Mar. 1990.
73. Hakanson interview.
74. C.A. Owers, 'The RAAF's first fighter', *Flightpath*, vol.1, no.3, May 1989, pp.60–1; AHSA *Newsletter*, vol.4, no.3, 1988, p.56.
75. *Aircraft*, 30 Dec. 1922, p.255.
76. C. Schaedel, *Australian Air Ace*, pp.71–2.
77. *Sun* (Melb), 29 Mar. 1923.
78. Hakanson interview; 'Gwill' (Bostock) papers, vol.2.
79. *SLA*, 1 Jul. 1922, pp.283–5.
80. DROs No.1 Station, dated 29 Jun. 1922.
81. Taped recollections: WgCdr C.F. Gates, 1988.
82. Air Board agenda no.429 and 1640.
83. AA, CRS A664/1, item 415/401/2.
84. A.J. Hill, *Chauvel of the Light Horse*, Melb, 1978, pp.213–14.
85. AA, CRS A705, item 165/1/198; *Argus*, 12, 15, 17, 18 Mar. 1924; AHSA *Newsletter*, vol.4, no.3, 1988, p.56.
86. *SMH*, 10 Apr. 1924.
87. ibid., 14 Apr. 1924.
88. Written memoir prepared for author by ACdre U.E. Ewart, Jun. 1988.
89. Letter: Mr A.W. Nicholson to author, Dec. 1988.
90. AHSA *Newsletter*, vol.4, no.3, 1988, p.56.
91. *Argus*, 15 Dec. 1924.
92. Parnell & Boughton, p.33; G. Jones, p.33.
93. *Argus*, 11 Dec. 1924.
94. *SLA*, 1 Jun. 1921, p.215.
95. AA, CRS A1195, item 863/2/51.
96. ibid.; 'Some Anecdotes of Early Civil Aviation in Australia', AVM A.T. Cole papers.
97. A.T. Cole, 'Merry Old Souls', unpub.ms., Cole papers.
98. Parnell & Boughton, p.42.
99. AA, CRS A1195, item 863/2/50.
100. ibid.
101. *SLA*, 1 Jun. 1921, p.215.
102. Lukis' report on landing grounds on Adelaide–Perth route, Feb. 1922: RAAF Historical Section.
103. *Aircraft*, 20 Sep. 1921, p.51.
104. Clarke & Mustard's report dated March 1922: RAAF Historical Section.
105. Hepburn's report on landing grounds on Geraldton–Derby route, Jun.–Sep. 1922: RAAF Historical Section.
106. Cobby's report on landing grounds on Derby–Wyndham route, 7 Nov. 1923: RAAF Historical Section.
107. AA(Vic), CRS A705, item 7/36/36.
108. Air Board agendum no.440.
109. Goble's report: RAAF Historical Section.
110. Parnell & Boughton, p.57.
111. Hepburn's report, 18 Feb. 1924: RAAF Historical Section.

112. Hepburn's report on inspection of the route between Cootamundra and Brisbane, 1925: RAAF Historical Section.
113. Hepburn's report on route Narromine–Brisbane, 20 Apr. 1926: RAAF Historical Section.
114. Report to RAAF HQ by FLt F.W.F. Lukis (CO 3 Sqn), 2 Jul. 1926: RAAF Historical Section.

Chapter 3

1. H.M Hyde, *British Air Policy Between the Wars 1918–1939*, Lond, 1976, pp.29–35.
2. Wise, op.cit., pp.8–10.
3. T.G. Fergusson, *British Military Intelligence, 1870–1914: The Development of a Modern Intelligence Organization*, Lond, 1984, p.173.
4. M.J.D. Cooper, *The Birth of Independent Air Power*, Lond, 1986, p.73; Hyde, p.29.
5. B.D. Powers, *Strategy Without Slide-Rule: British Air Strategy 1914–1939*, Lond, 1976, p.11.
6. D. Nevin, *Architects of Air Power*, Alexandria, Virginia, 1981, p.23; Hyde, p.23.
7. Cooper, p.136; Hyde, p.46; Nevin, pp.29–30.
8. Hyde, p.44.
9. A.G. Serle, *John Monash: a biography*, Melb, 1982, pp.463–4.
10. Chauvel to Blamey, 14 Jul. 1924, cited in Hill, op.cit., p.209.
11. Hansard, House of Commons, 5th series, vol.192, p.719.
12. AA, CRS A705, item 4/1/35.
13. ibid.
14. Trenchard papers, MFC 76/1/293: RAF Museum.
15. ibid.
16. ibid.
17. ibid.
18. McCarthy, *Australia and Imperial Defence*, p.36.
19. Williams to Trenchard, 11 Aug. 1926, Trenchard papers.
20. Trenchard papers.
21. Williams to Trenchard, 5 Apr. 1928, Williams papers: RAAF Museum.
22. *Argus*, 12 Oct. 1921.
23. Air Council agendum no. 35: RAAF Historical Section.
24. *ADB*, 7, pp.645–6.
25. *CAG*, no.72, 1921, p.1329.
26. PRO, Air 2/1453.
27. AA, CRS A705, item 4/1/35.
28. ibid.
29. *Herald* (Melb), 15 Oct. 1925.
30. AA, CRS A705, item 4/1/35.
31. Williams papers.
32. P. Congdon, *Behind the Hangar Doors*, Woodhall Spa, Eng., 1985, p.21.
33. Williams papers.
34. ibid.
35. Trenchard papers.
36. Williams to Trenchard, 2 Mar. 1927, ibid.

37. AA, CRS A705, item 34/2/118.
38. ibid.
39. ibid.
40. Hill, op.cit., p.218.
41. McCarthy, *Australia and Imperial Defence*, p.38.
42. C.E.W. Bean, *Two Men I Knew*, Syd, 1957, p.196.
43. Williams, p.199.
44. ibid., p.200.
45. *CPD*, vol.122, p.8.
46. ibid., p.96.
47. Williams, p.200; it should be noted that Williams was the source for McCarthy's more categorical assertion (pp.37–38) that Shepherd initiated the suggestion.
48. ibid., p.198.
49. *ADB*, 9, p.89.
50. L. Ross, *John Curtin: A Biography*, South Melb, 1977, p.44.
51. I. Dowsing, *Curtin of Australia*, Melb, 1969, pp.25–6.
52. *CPD*, vol.122, pp.1233–4.
53. ibid., p.678.
54. Parnell & Boughton, p.90. On 3 Jan. 1930 Williams prepared notes on the AFC Association's statement, presumably for Green's use; these notes have been sighted (in the Williams papers, RAAF Museum) but the original AFC letter has not.
55. *CPD*, vol.124, p.2668.
56. Hill, op.cit., p.217.
57. Williams to Trenchard, 20 Aug. 1930, Williams papers.
58. S. Brimson, *The History of Australia's Airlines*, Frenchs Forest, NSW, 1988, p.56.
59. AA, CRS A705, item 4/4/61.
60. PRO, Air 9/56.
61. PRO, Air 2/648.
62. PRO, Air 9/56.
63. PRO, Air 2/648.
64. McCarthy, *Australia and Imperial Defence*, p.39.
65. PRO, Air 9/56.
66. Hill, op.cit., p.216.
67. McCarthy, *Australia and Imperial Defence*, p.42.
68. ibid., p.40.
69. PRO, Air 8/221.
70. *CPP*, 1937–38–39–40, 2, p.403.
71. Correspondence in Williams papers: RAAF Museum.
72. ibid.
73. ibid.
74. ibid.

Chapter 4

1. Air Council agendum no.1.
2. *CAG*, no.72, 1921, p.1333.
3. See introduction to C. Darby, *RNZAF: The First Decade 1937–1946*, Melb, 1978.

4. Air Board agendum no.4.
5. ibid., no.106.
6. Air Council agendum no.68.
7. PRO, Air 2/507.
8. PRO, Air 2/517.
9. Williams, p.136.
10. Congdon, op.cit., pp.29–30.
11. Williams, p.136.
12. M. Wedd, *Australian Military Uniforms 1800–1982*, Kenthurst, NSW, 1982, p.118.
13. Air Council agendum no.112.
14. Written memoir prepared for author by ACdre U.E. Ewart, Jun. 1988.
15. Letter: Mr B.J. Videon to author, 8 Mar. 1990.
16. Williams, p.136.
17. R.J. Cluley, *Unit Badges of the Royal Australian Air Force*, Adel, 1990, p.viii.
18. PRO, Air 2/432.
19. ibid.
20. PRO, Air 2/507.
21. PRO, Air 2/517.
22. Air Force Office, Canberra, file Air 454/7/2.
23. Cluley, op.cit., p.viii; *RAAF News*, Sep. 1966.
24. Air Force Office, Canberra, file Air 454/7/2.
25. PRO, Air 2/1453; Gillison, p.24.
26. Air Matters raised at the Imperial Conference, 1923: CAS file.
27. ibid.
28. PRO, Air 2/1453.
29. ibid.
30. PRO, Air 2/532.
31. Interview: AVM E. Hey, 16 Jun. 1988.
32. Interview: AM Sir Charles Read, 4 Feb. 1988.
33. ibid.; Bramson, p.13.
34. Interview: AVM J. Hobler, 29 Feb. 1988.
35. PRO, Air 2/1453.
36. Trumble to Williams, 13 Jan. 1932: Williams papers, RAAF Museum.
37. AA, CRS A705, item 14/8/475.
38. Interview: WgCdr D.J. French, 5 May 1988.
39. Gillison, pp.39–40.
40. RAF Officers of Australian Origin—Lists A, B & C: RAAF Historical Section.
41. Air Board agendum no.352.
42. A.T. Cole, 'Merry Old Souls'.
43. Air Board agendum no.308.
44. Williams, pp.155–6.
45. G. Jones, p.44.
46. Interview: WgCdr C.F. Gates, 1 Mar. 1988.
47. T.I.G. Gray, *The Imperial Defence College and the Royal College of Defence Studies 1927–1977*, Edinburgh, 1977, pp.3, 78.
48. ibid., p.32–4.

49. Air Board agendum no.295.
50. Interview: ACdre W.H. Garing, 14 Dec. 1987.
51. G. Jones, p.46; Brownell, p.193; Rayner, p.34.
52. Interview: ACdre P.G. Heffernan, 20 Jan. 1989.
53. J.E. Hewitt, *The Black One*, p.147; Bostock Papers, folder 16; M.J.F. Bowyer, *2 Group RAF: A Complete History, 1936–1945*, Lond, 1974, p.43.
54. *ADB*, 9, p.34; Bowyer, op cit., pp.43–4.
55. Letter: Air Marshal Sir Edward Chilton to author, 6 Apr. 1990.
56. AVM G.G. Banting papers.
57. Interview: Sir Richard Kingsland, 25 Nov. 1987.
58. Cutlack, pp.43n, 87.
59. Williams, p.316.
60. McCarthy, *Australia and Imperial Defence*, p. 46.
61. Hamill, p.150.
62. ibid., pp.154–6, 264; McCarthy, *Australia and Imperial Defence*, pp.61–2; Horner, pp.8, 10–11.
63. Hamill, pp.263, 265.
64. McCarthy, *Australia and Imperial Defence*, p.60.
65. ibid., pp.60, 76; Hamill, pp.157, 252, 258; R.G. Neale (ed.), *Documents on Australian Foreign Policy 1937–49*, vol.1: *1937–38*, Canb, 1975, p.90.
66. Hamill, p.265; McCarthy, *Australia and Imperial Defence*, pp.60–61.
67. McCarthy, *Australia and Imperial Defence*, pp.81, 86–7.
68. Hudson & North, p.71.
69. Williams papers, RAAF Museum.
70. McCarthy, *Australia and Imperial Defence*, p.68; Williams' 1925 memorandum: RAAF Historical Section.
71. Williams, p.182.
72. ibid.
73. Hudson & North, p.248.
74. ibid., pp.269, 325, 337.
75. Salmond Report, part 1: RAAF Historical Section.
76. Salmond Report, part 2.
77. Williams papers, RAAF Museum.
78. Trenchard to Salmond, 22 Nov. 1928, Trenchard papers, RAF Museum, MFC 76/1/266 (4).
79. Drummond to Salmond, 18 Jan. 1929, Salmond papers, RAF Museum, B2630.
80. Williams to Trumble, 23 Oct. 1928, 6 Mar. 1929, Williams papers, RAAF Museum.
81. PRO, Air 9/56.
82. Trenchard Papers, RAF Museum, MFC 76/1/266 (4).
83. *Argus* (Melb), 9 Oct. 1928.
84. ibid.
85. Williams papers, RAAF Museum.
86. J.W.R. Taylor, *A History of Aerial Warfare*, Lond, 1974, p.139.
87. Unidentified press clipping in RAAF Museum.
88. *SMH*, 29 Jan. 1928.
89. Brogden, p.94.

90. Parnell & Boughton, p.69.
91. Wixted, p.29.
92. Interview: Mr L.M. Diprose, 14 Mar. 1989.
93. ibid.; *RAAF News*, Sep. 1966, p.6.
94. Diprose interview.
95. Hepburn's report dated 17 Apr. 1932: RAAF Historical Section.
96. Summer's report dated 13 Apr. 1932: RAAF Historical Section.
97. Diprose interview.
98. Wixted, p.186.
99. *SMH*, 12 Aug. 1937.
100. AA, CRS A705, item 106/8/70.
101. J.J. Halley, *Famous Maritime Squadrons of the RAF*, vol.1, Windsor, 1973, pp.59–60.
102. Brook, p.28; *Herald* (Melb), 7 Feb. 1938.
103. Halley, op.cit., p.60.
104. AA, CRS 1969/100, item 4/22/Air.
105. AA, CRS A705, item 106/8/70.
106. Brook, pp.38–9.
107. Parnell & Boughton, p.171.
108. Brook, pp.38–9.
109. M. Donne and C. 7 Fowler, *Per Ardua ad Astra: Seventy Years of the RFC & the RAF*, Lond, 1982, p.59.
110. A.J. Cobham, *Australia and Back*, Lond, 1926, pp.98–9; Parnell & Boughton, p.59.
111. J. Batten, *My Life*, Lond, 1938, pp.243–4.
112. Taped recollections: GpCapt W.G. Compagnoni, 1989.
113. Batten, op cit., p.244.
114. Williams to Secretary, Defence Department, 25 Feb.1936, Williams papers, RAAF Museum.
115. ibid., 23 Jul. 1935, Williams papers.
116. Draft report of IDC, 9 Dec. 1935, Williams papers.
117. Williams to Secretary, Defence Department, 26 Feb. 1936, Williams papers.
118. Hempel's report to Air Board: RAAF Historical Section.
119. Gunn, p.284.
120. ibid., pp.285–6.
121. Hempel to Secretary, Defence Department, 13 Jul. 1936, Williams papers, RAAF Museum.
122. Williams to Secretary, Defence Department, 13 Jul. 1936, Williams papers.
123. Williams papers.
124. Undated draft Air Staff note, Williams papers.
125. PRO, Air 8/221.
126. Williams, pp.235–6.
127. Hyde, op.cit., pp.297, 402, 494–5; M.S. Smith, *British Air Strategy between the Wars*, Oxford, 1984, pp.38–9, 103, 129, 134.
128. J. Terraine, *The Right of the Line: The Royal Air Force in the European War 1939–1945*, Lond, 1985, pp. 15–17, 44–5, 682.
129. McCarthy, *Australia and Imperial Defence*, p.84.
130. Gillison, p.48.

131. Ellington Report, paras.2, 13: RAAF Historical Section.
132. Williams, pp.236–7.
133. ibid., p.238; Williams to McNamara, 7 Oct. 1938, Williams papers, RAAF Museum.
134. Williams to McNamara, 7 Oct. 1938.
135. ibid.
136. Williams, pp.240–5.
137. Williams papers.
138. ibid.
139. *Age* (Melb), 27 Sep. 1938; *Truth* (Melb), 15 Oct. 1938; *Sunday Sun* (Melb), 16 Oct. 1938.
140. Williams papers.
141. *Herald* (Melb), 25 Oct. 1938.
142. Williams papers; *Argus*, 15 Oct. 1938.
143. *CPD*, vol.158, pp.1993–6.
144. Gillison, p.51.
145. Williams, p.238.
146. Prime Minister's statement, published with Ellington Report.
147. McCarthy, *Australia and Imperial Defence*, pp.91–2.
148. Williams papers.

Chapter 5

1. C.A. Butler, *Flying Start: The history of the first five decades of Civil Aviation in Australia*, Syd, 1971, p.19.
2. *OYB*, no.17, 1924, p.595.
3. J. Clare, Military Architecture in Australia and Great Britain: A Comparative Study, 1987, pp.5–6, 143–4: Department of Housing and Construction (Vic–Tas Region).
4. Heritage survey of Point Cook (uncompleted), 1983 (hereafter referred to as Heritage Survey 1983), p.11: Department of Housing and Construction.
5. *CPP*, 1922, 2, pp.2387–2407.
6. Heritage survey, 1983, p.13.
7. Air Board agendum no.134.
8. Written memoir prepared for author by ACdre U.E. Ewart, Jun. 1988.
9. Heritage survey, 1983, p.13.
10. Letter: Mr C.H.G. Gardiner to author, 14 April 1988.
11. AA, CRS A705, item 24/2/143; Parnell & Lynch, p.46.
12. RAAF Point Cook Heritage Survey, Buildings and Conservation Areas, Summary and Statement of Significance, 1986 (hereafter referred to as Heritage Summary 1986), para.3.1: Department of Housing and Construction.
13. DROs No.1 Station, 17 May 1923: RAAF Museum.
14. Ewart memoir.
15. Telephone interview: Mr C.H. Simpson, 20 Jan. 1988.
16. Interview: Mrs E. Candy, 3 Feb. 1988.
17. Interview: Mrs J. Crock, 7 Sep. 1988.
18. Interviews: WgCdr C.F. Gates, 1 Mar. 1988; ACdre P.G. Heffernan, 4 Nov. 1988; Mrs J. Gray, 4 Nov. 1988; SqnLdr G.L. Grendon, 26 Apr. 1989.

19. DROs No.1 Station, for example, 19 May, 7 Jun. 1922, 26 Feb., 30 May 1923.
20. ibid., 15 Sep., 1 Nov. 1922.
21. Interview: Mrs E. Candy, 3 Feb. 1988.
22. Recollections of Lady Woodward, courtesy of Hon. Sir Edward Woodward, 2 May 1988.
23. DROs No.1 Station, 19 May, 21 Jun. 1922, 2 May 1923: RAAF Museum.
24. Letters: Mr C.H.G. Gardiner to author, 14 Apr., 28 May 1988.
25. Information of Sir Edward Woodward, 2 May 1988.
26. DPR video interview with ACM Sir Frederick Scherger, 2 Feb. 1982.
27. E.J. Richards, *Australian Airmen: History of the 4th Squadron, Australian Flying Corps*, Melb, nd, pp.71–2.
28. Ellison, pp.20–21.
29. Interview: WgCdr C.F. Gates, 1 Mar. 1988.
30. Heritage Summary 1986, para.3.1.
31. Heritage Survey 1983, p.15.
32. *Aircraft*, 31 Oct. 1922, p.228; Williams, p.233.
33. *CAG*, no.76, 1921, p.1382
34. *CPP*, 1923–24, 4, pp.473–529.
35. *Argus* (Melb), 13 Nov. 1924.
36. ibid., 6 Jan. 1925.
37. DROs 1 AD, no.42 dated 1 Mar. 1926.
38. Letter: Rev. W.F. Gilmour to author, 11 Dec. 1989.
39. 1 AD unit diary: RAAF Historical Section.
40. Air Office, file Air 454/7/2.
41. AVM F.M. Bladin, unpub.ms., p.17.
42. Interview: ACdre C.R. Taylor, 21 Jan. 1988.
43. Air Board agendum no.100.
44. Parnell & Lynch, p.44.
45. Parnell & Boughton, p.26.
46. *CPP*, 1925, 2, pp.1663, 1669.
47. 3 Sqn unit diary: RAAF Historical Section.
48. *Daily Guardian* (Syd), 29 Jun. 1925.
49. Information of FLt E.P.W. Rowson, 5 May 1989: RAAF Historical Section.
50. Taped recollections: GpCapt W.G. Compagnoni, 1989.
51. Ellison, p.175.
52. Interview: WgCdr R.C. Cresswell, 14 Sep. 1989.
53. *Smith's Weekly* (Syd), 6 Apr. 1935.
54. Interview: Mrs N. Bostock and Mrs J. Stewart, 21 Apr. 1988, and WgCdr J.H. McCorriston, 15 Mar. 1989.
55. *SMH*, 18 Mar. 1972.
56. ibid., 2 Jul. 1935.
57. AA, CRS A705, item 22/1/326.
58. *Daily Telegraph* (Syd), 21 Apr. 1933; *Windsor & Richmond Gazette*, 28 Apr. 1933.
59. *Windsor & Richmond Gazette*, 10 Aug. 1933; *SMH*, 23 Sep. 1933; *Cumberland Argus*, 29 Nov. 1934.
60. AA, CRS A705, item 22/1/326.

61. ibid.
62. *SMH*, 12 Oct. 1934.
63. Interview: AVM E. Hey, 16 Jun. 1988.
64. *Sun* (Syd), 1 Jul. 1935; *SMH*, 2 Jul. 1935.
65. *SMH*, 11 Oct. 1935.
66. A.T. Cole, 'Merry Old Souls'.
67. Ellington report, para.6.
68. Compagnoni taped recollections.
69. Salmond report, Part 1, p.16.
70. Air Board agendum no.1695.
71. Air Board agendum no.1771.
72. Brownell, p.193.
73. 23 Sqn unit diary: RAAF Historical Section.
74. ibid.
75. Brownell, pp.193–4.
76. ibid., p.194; interview: ACdre A.D.J. Garrisson, 26 Apr. 1988.
77. Garrisson's letter to his mother, 17 Feb. 1938.
78. AA, CRS A705, item 171/5/109.
79. Garrisson to his mother, 13 Mar. 1938.
80. Garrisson interview.
81. AA, CRS A705, item 171/5/109.
82. Brownell, p.194.
83. 23 Sqn unit diary.
84. ibid.
85. Brownell, p.198.
86. Garrisson interview.
87. 23 Sqn unit diary.
88. ibid.
89. 14 Sqn unit diary: RAAF Historical Section.
90. AA, CRS A705, item 171/6/41.
91. 12 Sqn unit diary: RAAF Historical Section.
92. AA, CRS A705, item 171/6/41.
93. 12 Sqn unit diary.
94. ibid.
95. *Herald* (Melb), 29 Jun. 1939.
96. *Argus* (Melb), 29 Jun. 1939.
97. AA, CRS A705, item 171/6/108.
98. 12 Sqn unit diary.
99. Letter: Mr J.L. Truscott to author, 13 Jan. 1988.
100. Hewitt, *The Black One*, p.135.
101. K.H. Trevan, 'The Darwin Mobile Force', *Army Journal*, no.275, Apr. 1972, p.12.
102. Grendon interview.
103. Truscott letters, 13 Jan., 8 Aug. 1988.
104. 12 Sqn unit diary.
105. *CAG*, no.81, 1938, p.2904.

106. 23 Sqn unit diary: RAAF Historical Section.
107. *Argus* (Melb), 29 Jun. 1939.
108. 23 Sqn unit diary.
109. ibid.
110. Letter: Mr R.W. Steel to author, 29 Apr. 1988.
111. 23 Sqn unit diary.
112. Salmond report, Part I, p.15.
113. Hewitt, *The Black One*, p.140.
114. J. Newton, *RAAF Rathmines*, privately published, p.7: RAAF Museum.
115. 9 Sqn unit diary: RAAF Historical Section.
116. Newton, op.cit., p.7.
117. Hewitt, *The Black One*, p.161.
118. Parnell & Boughton, p.181.
119. Hewitt, *Adversity in Success*, p.1.
120. Newton, op.cit., p.7.
121. ibid., p.8.
122. Newton, p.8.
123. A.J. Fitzgerald (ed.), *Canberra's Engineering Heritage*, Canb, 1983, pp.179–80.
124. Salmond report, Part I, p.15.
125. Fitzgerald, p.181.
126. H.J. Gibbney, *Canberra 1913–1953*, Canb, 1988, p.185.
127. Williams papers.
128. AA, CRS A705, item 153/2/142.
129. Gibbney, op.cit., pp.185–6.
130. Fitzgerald, p.182.
131. File on air routes in NSW and Qld, 1926: RAAF Historical Section.
132. *CPD*, vol.120, p.1265.
133. AA, CRS A705, item 161/2/235.
134. *OYB*, no.24, 1931, p.416.
135. AA, CRS A705, item 7/1/65.
136. AA, CRS A705, item 208/1/1358.
137. 1 Sqn unit diary; *Argus* (Melb), 20 Feb. 1938.
138. Interview: AVM B.A. Eaton, 26 Jan. 1988.

Chapter 6

1. K. Isaacs, *Military Aircraft of Australia*, pp.160–1.
2. *SLA*, Feb. 1919, p.667; *Argus* (Melb), 26 Feb. 1919.
3. *SLA*, May 1919, pp.69–70.
4. ibid.; *Argus* (Melb), 20, 25 Mar., 1, 9, 17 Apr. 1919; Parnell & Boughton, p.31.
5. AAC Routine Orders: RAAF Museum.
6. AA, CRS A1195, item 715/2/224; K. Isaacs, *Military Aircraft of Australia*, pp.160–1.
7. Isaacs, p.34.
8. AA, CRS A1195, item 715/2/224; log-book of AVM A.T. Cole, Cole papers.
9. Telephone interview: WgCdr C.F. Gates, 14 Feb. 1989.

10. Report on CFS stores, 30 Jun. 1919: RAAF Historical Section.
11. AA (Vic), MP 367/1, item 524/10/196.
12. Information provided by Mr C.A. Owers from private research.
13. *Argus* (Melb), 17, 18 Jun. 1919.
14. Interview: Mr C.H.G. Gardiner, 6 May 1988.
15. Cutlack, p.407.
16. K. Isaacs, in *RAAF News*, Oct. 1961.
17. AA, CRS A2/1, item 20/2265.
18. Williams, p.118.
19. ibid., p.119.
20. AA, CRS A2/1, item 20/2265.
21. Air Service Committee, folder of minutes, 1919: RAAF Historical Section.
22. AA, CRS A2, item 20/2265.
23. ibid.
24. ibid.
25. Williams, p.118.
26. AA, CRS A1195, items 715/1/24, 715/1/27; *SLA*, Jan. 1920, p.672; 'Brief History of Stores Administration of the Australian Air Force', 20 Dec. 1929, in 1 AD unit diary: RAAF Historical Section; *CPP*, 1923–24, 4, pp.368, 487.
27. *Wings*, 15 Mar. 1946, p.16.
28. ibid.
29. Hall, pp.6, 8.
30. ibid., pp.10, 17.
31. *The Golden Years*, p.2; Pentland & Malone, p.14.
32. AA, CRS A1195, item 715/2/224.
33. K. Isaacs, 'The Fairey IIID in Australia', *Air Enthusiast*, no.24 (1988), pp.40–1.
34. AA, CRS A1195, item 715/1/42.
35. Interviews: ACdre U.E. Ewart, 7 Apr. 1988, and GpCapt Sir Wilfred Brookes, 13 Mar. 1989.
36. Interview: Captain L.R. Ambrose, 5 Feb. 1988; taped recollections: WgCdr C.F. Gates, 1988; Parnell & Boughton, p.62.
37. *CPD*, vol.99, p.777.
38. Allotment of aircraft to No.1 FTS, 1 Jul. 1922: in possession of ACdre C.R. Taylor.
39. AA, CRS A664/1, item 415/401/2; see also *Weekly Times* (Melb), 24 May 1924.
40. *Argus* (Melb), 4 Apr. 1925.
41. Air Board agendum no.123: RAAF Historical Section.
42. ibid., no.79.
43. Gillett, *Australia's Military Aircraft*, p.81.
44. AA, CRS A705, item 38/1/256.
45. C.A. Owers, 'The RAAF's first fighter', *Flightpath*, vol.1, no.3, May 1989, p.61; AHSA *Newsletter*, Jul.–Nov. 1985, p.13.
46. Miller, p.114.
47. Air Board agendum no.673.
48. ibid., no.216.
49. ibid., no.284.
50. Hakanson interview.

51. Gillett, *Australia's Military Aircraft*, p.40.
52. G. Jones, p.32.
53. Parnell & Boughton, p.30; AA, CRS A705, item 9/1/47; *Aircraft*, Aug. 1926, p.307.
54. AA, CRS A705, item 9/1/47.
55. C.A. Owers, 'The RAAF's first fighter', p.62.
56. Written memoir prepared for author by ACdre U.E. Ewart, Jun. 1988.
57. AA (Vic), MP1118/2, item 69/2/331.
58. ibid., and AA, CRS A705, item 9/1/47.
59. CRS A705, item 9/1/47.
60. ibid.
61. Gillett, *Australia's Military Aircraft*, p.15.
62. *Wings*, 15 Mar. 1946, p.26.
63. ibid.
64. AA, CRS A664/1, item 415/401/70.
65. ibid.
66. *Aircraft*, 31 Jul. 1929, p.451.
67. K. Isaacs, *Military Aircraft of Australia*, p.91.
68. Sutherland, pp.16, 18, 30.
69. Log-book of AVM A.T. Cole.
70. Sutherland, p.18.
71. J.M. Bruce, 'The De Havilland D.H.9', Part 1, *Flight*, 6 Apr. 1956, pp.385–88, and Part 2, 13 Apr. 1956, p.422.
72. Cutlack, p.403.
73. ibid.
74. Bruce, 'The De Havilland D.H.9', Part 2, p.424.
75. Memorandum regarding the Air Defence of Australia, 21 Apr. 1925: RAAF Historical Section.
76. C.G. Grey (ed.), *Jane's All the World's Aircraft 1932*, Lond, 1932, p.244c.
77. *Wings*, 15 Mar. 1946, p.27.
78. Parnell & Boughton, p.77.
79. ibid., p.178.
80. Gillett, *Australia's Military Aircraft*, p.177.
81. *The Times* (Lond), 26 Jan. 1926; *Argus* (Melb), 10 Feb. 1926.
82. Williams, p.164.
83. ibid., p.196.
84. ibid., p.176.
85. K. Isaacs, 'First of the Line', *DFJ*, no.3, Mar.–Apr. 1977, p.27.
86. Parnell & Boughton, p.48.
87. *SMH*, 28 Jan. 1928.
88. Parnell & Boughton, p.64.
89. *Sun* (Melb), 26 Nov. 1929; Brownell, p.191.
90. Parnell & Boughton, pp.65–6.
91. O. Thetford, *Aircraft of the Royal Air Force since 1918*, 4th edn., Lond, 1968, p.504.
92. Parnell & Boughton, p.66.
93. C.R. Taylor, 'An Engineers view of the RAAF between the Great Wars', AHSA *Journal*, vol.17, no.3, Jul.–Aug. 1976, p.12.

94. Interview: WgCdr R.C. Cresswell, 14 Sep. 1989.
95. Interview: AM Sir Charles Read, 4 Feb. 1988.
96. Salmond report, part 1, p.16.
97. Parnell & Boughton, pp.84–5.
98. J. Bushby, *Air Defence of Great Britain*, Lond, 1973, p.93; Thetford, op.cit., p.112.
99. Bushby, op.cit.,p.94.
100. C. Bowyer, *Fighter Command 1936–1968*, Lond, 1980, p.19.
101. Air Board agendum no.1398.
102. ibid. no.1534.
103. C.R. Taylor, op.cit., p.12.
104. Air Board agendum no.1591.
105. Parnell & Boughton, p.93.
106. *Argus* (Melb), 1 Feb. 1935.
107. *Wings*, 15 Mar. 1946, p.27.
108. AHSA *Newsletter*, vol.3, no.2, 1987, p.31.
109. Williams, pp.177–8.
110. ibid., pp.209–11.
111. Thetford, op.cit., p.442.
112. Interview: ACdre J. Alexander, 28 Apr. 1988.
113. Parnell & Boughton, p.141.
114. Thetford, op.cit., p.48.
115. Parnell & Boughton, pp.131–32.
116. ibid., pp.159–60.
117. ibid., p.172; *ADB*, vol.7, p.474.
118. AA, CRS A1196, item 1/501/224.
119. G. Jones, p.79.

Chapter 7

1. 1. Allotment of Aircraft to No.1 FTS, 1 Jul. 1922, held by ACdre C.R. Taylor.
2. 1 & 3 Sqns unit diaries: RAAF Historical Section.
3. 3 Sqn unit diary.
4. 1 Sqn unit diary.
5. PRO, Air 8/221.
6. Rayner, p.29.
7. Diprose interview.
8. PRO, Air 8/221.
9. 1 Sqn unit diary.
10. Letter: ACdre P.G. Heffernan to author, 30 Dec. 1988.
11. Interview: AVM B.A. Eaton, 26 Jan. 1988.
12. AVM F.M. Bladin, unpub.ms.
13. 1 Sqn unit diary; *SMH*, 25 Nov. 1935; *Sun* (Syd), 26 Nov. 1935.
14. Bladin ms.
15. Interviews: ACdre C.R. Taylor, 20 Apr. 1988, and WgCdr C.F. Gates, 1 Mar. 1988.
16. 1 Sqn unit diary.
17. Williams, pp. 144–5.

18. ibid., p.400; see also Gillison, p.17.
19. AVM F.M. Bladin, unpublished ms.; telephone interview with Mr C.H. Simpson, 20 Jan. 1988; letter: ACdre U.E. Ewart to author, 16 May 1982.
20. Rayner, p.15
21. J.E. Lee, *Duntroon: The Royal Military College of Australia 1911–1946*, Canb, 1952, p.74.
22. ibid., p.76.
23. Letter: ACdre W.H. Garing to author, 30 Oct. 1984.
24. R.E. Frost, A History of the Royal Australian Air Force College, Aug. 1947–Jan. 1986, unpub.ms, p.3: RAAF Historical Section.
25. ibid., pp.3–4.
26. ibid., p.4.
27. *CPP*, 1926–27–28, 2, p.171.
28. Frost, pp.4–5.
29. ibid., p.5.
30. ibid.
31. ibid., p.6.
32. ibid.
33. Salmond report, part 1, pp.11–12, 15; C.D. Coulthard-Clark, *Duntroon: The Royal Military College of Australia, 1911–1986*, Syd, 1986, p.111.
34. *Herald* (Melb), 2 Jun. 1962.
35. Letter: AM Sir Valston Hancock to author, 1 Sep. 1981.
36. AA, CRS A705, item 14/8/413.
37. ibid.
38. Interview: AVM C.D. Candy, 3 Aug. 1982.
39. PRO, Air 8/115.
40. File: Development Program 1938–41, Personnel Intake & Training Figures: RAAF Historical Section.
41. Course record for 1927 'A' Flying Course: RAAF Historical Section.
42. Pentland and Malone, p.23.
43. *Nhill Free Press*, 7 Mar. 1929.
44. *Argus* (Melb), 8 Mar. 1929.
45. AA (Vic), MP 187/4, item 186.
46. Williams, p.216.
47. Casey to Williams, 19 Jan. 1938: Williams papers.
48. AVM F.M. Bladin, unpub.ms.
49. Rayner, p.21.
50. Interview: AVM J. Hobler, 29 Feb. 1988.
51. Letter: Cadet G.C. Hartnell to his mother, 31 Jan. 1936, AVM G.C. Hartnell papers; interviews: AM Sir Charles Read, 4 Feb. 1988, WgCdr D.J. French, 5 May 1988, and ACdre G.H. Steege, 2 Jun. 1988.
52. Letter: ACM Sir Frederick Scherger to author, 11 Nov. 1981.
53. French interview.
54. Read interview.
55. French interview.
56. Steege interview.
57. Interview: Sir Richard Kingsland, 25 Nov. 1987.

58. Interview: Francis James, 14 Dec. 1987.
59. Text of talk by AVM A.T. Cole, 1937, Cole papers.
60. Bramson, pp.9–10.
61. Interviews: Sir Richard Kingsland, 25 Nov. 1987, and ACdre C.R. Taylor, 21 Jan. 1988.
62. Taylor interview.
63. Taped recollections: GpCapt W.G. Compagnoni, Dec. 1989.
64. ibid.; AA (Vic), MP 187/3. Box 1; Air Board agendum no.1945.
65. Interview: AM Sir John McCauley, 4 Feb. 1988.
66. *Argus* (Melb), 13 Aug. 1937.
67. ibid., 3 Jun. 1933.
68. Letter: GpCapt H.E. Daish to author, 9 Aug. 1988.
69. Bramson, pp.9–10.
70. Unidentified press clipping dated 7 Sep. 1938, in possession of Sir Richard Kingsland.
71. Interview: ACdre W.H. Garing, 14 Dec. 1987.
72. AA (Vic), MP 187/3, Box 1.
73. Interviews: AVM B.A. Eaton, 26 Jan. 1988, and ACdre I.S. Podger, 15 Jan. 1988.
74. *Courier-Mail* (Brisb), 26 Sep. 1938.
75. *Argus* (Melb), 12 Oct. 1938.
76. ibid., 22 Nov. 1938.
77. ibid., 16, 17 Nov. 1938.
78. Letter written from Cloncurry, 18(?) Nov. 1938, AVM G.C. Hartnell papers.
79. *Argus* (Melb), 1 Dec. 1938.
80. *Courier-Mail* (Brisb), 14 Dec. 1938.
81. Unidentified press clipping dated 21 Nov. 1938, in possession of Sir Richard Kingsland.
82. Garing interview.
83. Unidentified press clipping in possession of Sir Richard Kingsland.
84. ibid.
85. AA, CRS A1195, item 929/1/134.
86. *CPP*, 1926–27–28, 2, pp.174–5; 1 Sqn unit diary; the identity of the machines is from evidence in photograph albums belonging to ACM Sir Frederick Scherger, held at the RAAF Museum.
87. Hill, op.cit., pp.214–15; Serle, op.cit., p.464.
88. 1 Sqn and 3 Sqn unit diaries: RAAF Historical Section.
89. 3 Sqn unit diary.
90. *The Bomber: The Journal of No.1 Squadron, RAAF*, vol.1, no.2, Nov. 1931, pp.30–1.
91. *Telegraph* (Syd), 31 Oct. 1932; *SMH*, 1 Nov. 1932.
92. Interview: AVM E. Hey, 16 Jun. 1988.
93. Air Office, file Air 454/7/2.
94. AA, CRS A705, item 208/3/433.
95. ibid.
96. Interview: ACdre G.H. Steege, 2 Jun. 1988.
97. AA, CRS A1195, item 715/3/85, and MP 1049/1, item 1920/0204; Gillett, *Wings Across the Sea*, p.20; R.M. Jones, p.16.

98. Air Board agendum no.32; Air Council agendum no.29.
99. AA, CRS A1195, item 715/3/85.
100. ibid.; R.M. Jones, pp.22–5; Gillett, *Wings Across the Sea*, p.21.
101. AA, CRS A1195, item 715/3/85.
102. R.M. Jones, p.27.
103. Air Board agendum no.227.
104. *Wings*, 15 Nov. 1945, p.15.
105. *CPP*, 1923–24, 4, p.369.
106. K. Isaacs, 'First of the Line', *DFJ*, no.3, Mar.–Apr. 1977, p.26.
107. DROs No.1 Station, 3 May 1923: RAAF Museum.
108. Reports by Hempel and Mustard, Jun.–Aug. 1923: RAAF Historical Section; Clune, pp.89–90.
109. Report on Goble and McIntyre flight to Hobart, Feb. 1924: RAAF Historical Section.
110. Report on Hempel and McIntyre flight, Oct. 1925: RAAF Historical Section.
111. Williams, pp.176–7.
112. K. Isaacs, 'First of the Line', p.32.
113. Report on combined land and sea exercise, Oct. 1929: RAAF Historical Section.
114. Rayner, p.28.
115. Hewitt, *The Black One*, p.128.
116. Interview: AVM D.A.J. Creal, 10 Dec. 1987.
117. Hewitt, *The Black One*, p.129.
118. ibid., pp.130–1.
119. Record of interview with SqnLdr W.J. Symons, Dec. 1977–Jan. 1978: RAAF Museum.
120. A. Payne, *HMAS Canberra*, Syd, 1973, pp.7–8.
121. Creal interview.
122. AAIC report no.74: RAAF Historical Section.
123. AA (Vic), MP 124/6, item 415/201/688.
124. Gillett, *Wings Across the Sea*, p.30.
125. Interview: WgCdr W.A. Bradley, 15 Mar. 1989.
126. K. Isaacs, 'First of the Line', p.33.
127. Gillett, *Wings Across the Sea*, p.30.
128. Report of cruise by HMAS Australia, Sep.–Oct. 1932: RAAF Historical Section.
129. Payne, *HMAS Canberra*, p.11, and *HMAS Australia*, Syd, 1975, p.13.
130. Graham's report on HMAS *Canberra* detachment: RAAF Historical Section.
131. Payne, *HMAS Australia*, p.31.
132. Interview: ACdre J. Alexander, 28 Apr. 1988.
133. Gillett, *Wings Across the Sea*, p.38.
134. 3 Sqn unit diary.
135. R.M. Jones, pp.79–80.

Chapter 8

1. *SLA*, vol.1, no.8, 1 Nov. 1918, pp.462.
2. *Report on the Air Defence of Australia by the Air Board, 1920*, pp.9–10: RAAF Historical Section.

3. Brook, p.2.
4. ibid.
5. Air Board agendum no.6, and Air Council agendum no.4: RAAF Historical Section.
6. *CPD*, vol.94, p.7311.
7. Air Council agendum no.62.
8. McCarthy, *Australia and Imperial Defence*, p.73.
9. Air Board agendum no.171; Air Council agendum no.93.
10. Provisional Regulations for the CAF, in papers of Secretary, Department of Air: RAAF Historical Section.
11. AA, CRS A705, item 4/1/60.
12. Air Council agendum no.107.
13. *Conditions of Entry to the Active Citizen Air Force and the Air Force Reserve*, 1934: RAAF Museum.
14. 3 Sqn unit diary: RAAF Historical Section.
15. Air Board agendum no.670.
16. 3 Sqn unit diary.
17. 1 Sqn unit diary.
18. L.J. Robertson, unpub.ms., p.7: RAAF Historical Section.
19. ibid.
20. ibid., pp.1, 7.
21. 1 Sqn unit diary.
22. AA, CRS A705, item 172/1/358.
23. Robertson, p.9.
24. Interview: ACdre C.R. Taylor, 21 Jan. 1988.
25. Written memoir of ACdre U.E. Ewart, Jun.1988.
26. Interview: Mr T.F.C. Lawrence, 19 Jul. 1988.
27. Brookes interview.
28. AA (Vic), CRS A705, item 224/6/32.
29. Salmond report, part 1, pp.7–8.
30. McCarthy, *Australia and Imperial Defence*, p.74.
31. ibid., p.75.
32. Brook, p.9.
33. ibid., pp.9–20, 26.
34. ibid., pp.17, 20.
35. Interview: GpCapt J.E. Graham, 6 May 1988.
36. ibid.
37. Brook, p.23.
38. ibid., p.24.
39. ibid., p.25.
40. Graham interview; A.J. Fitzgerald, *Canberra in Two Centuries: A pictorial History*, Canb, 1987, p.142.
41. Brook, p.36.
42. ibid., pp.47–9.
43. *Sun News-Pictorial* (Melb), 24 Dec. 1938.
44. Interview: Mr J.C. Miles, 15 Mar. 1989.

45. Interview: ACdre W.H. Garing, 14 Dec. 1987.
46. Taped recollections: GpCapt W.G. Compagnoni, Dec. 1989.
47. Parnell & Boughton, p.43.
48. Brookes interview, 13 Mar. 1989.
49. Robertson ms, p.3.
50. ibid., p.4.
51. ibid., p.5.
52. Brookes interview, 13 Mar. 1989.
53. Robertson ms, pp.4–5.
54. ibid., p.6
55. ibid., p.5
56. ibid.
57. Brookes interview, 13 Mar. 1989.
58. Bladin manuscript.
59. *Argus* (Melb), 20 Feb. 1926; *Aircraft*, 1 Apr. 1926, p.120.
60. Robertson ms, p.6.
61. Cole papers, item 7.
62. *CAG*, no.54, 1926, p.1021
63. Robertson ms, p.7.
64. Letter: GpCapt Sir Frederick Thomas to author, 25 Sep. 1989.
65. Air accident records: RAAF Historical Section.
66. *CAG*, no.34, 1927, p.743.
67. Salmond report, part 1, pp.9, 27.
68. 1 Sqn and 3 Sqn unit diaries.
69. 1 Sqn unit diary.
70. Letter: GpCapt J.E. Graham to author, 27 Feb. 1988; interview, 6 May 1988; telephone interview, 20 Feb. 1990.
71. Unidentified press clipping in possession of Mr T.F.C. Lawrence.
72. Interview: Mr T.F.C. Lawrence, 19 Jul. 1988.
73. Interview: AVM I.D. McLachlan, 5 Feb. 1988.
74. *Sun* (Syd), 29 Mar. 1935; *SMH*, 30 Mar. 1935.
75. *Aircraft*, 1 Feb. 1938, p.19.
76. Brook, p. 31.
77. *RAAF List*, Sep. 1938.
78. ibid, Sep. 1938, and Feb. 1939.
79. D. Hains, 'No.25 (City of Perth) Squadron's Golden Jubilee', *Wings*, Sep. 1987, p.17.
80. 23 Sqn unit diary.
81. Records of 'A' flying courses Nos.1–14 (1923–1934): RAAF Historical Section.
82. AA, CRS A705, item 14/8/475.
83. *Conditions of Entry to the Active Citizen Air Force* etc., pp.25–31.
84. *Sunday Sun*, 21 May 1933.
85. Malley to Williams, 19 May 1938, Williams papers: RAAF Museum.
86. Malley papers: RAAF Museum.
87. AA (Vic), MP 341, item 1935/5882.
88. AA, CRS A705, item 153/1/645.

89. Report by FLt C. Eaton on 'Kookaburra' search, 1929: RAAF Historical Section.
90. *Mercury* (Hob), 1 May 1929.
91. AA (Vic), MP 341, item 1935/5882.
92. Report on training flight to WA, 1932: RAAF Historical Section.
93. Report on training flight to WA, 1933: RAAF Historical Section.
94. Report by FLt D.E.L. Wilson on Tasmanian training flight, 1933: RAAF Historical Section.
95. AA (Vic), MP 341, item 1935/5882.
96. ibid.
97. ibid.; *Herald* (Melb), 13 Jul. 1933.
98. Letter: WgCdr C.F. Gates to author, 29 Dec. 1989.
99. Telephone interview: Mr K.G.T. Robey, 22 Apr. 1990.
100. *SMH*, 6 Jun. 1935; Australian Air League (hereafter 'AAL') *Souvenir History*, 1984, p.3.
101. *SMH*, 17 Dec. 1935.
102. *SMH*, 12 Dec. 1934.
103. ibid., 6 Jun. 1935.
104. ibid., 17 Jun. 1935.
105. AAL *Souvenir History*, p.5.
106. *Air League Journal*, vol.2, no.6, Jun. 1942, p.1.
107. ibid., vol.3, no.1, Jan. 1943, p.1.
108. ibid., vol.2, no.8, Aug. 1942, p.15.
109. Robey interview.
110. AAL *Souvenir History*, pp.4–5.
111. *Air League Journal*, vol.3, no.5, May 1943, p.2

Chapter 9

1. *Daily Telegraph* (Syd), 9 May 1911.
2. See articles by Captain W.O. Watt in the *Commonwealth Military Journal*, May 1912, pp.426–7, and Jan. 1913, p.64.
3. K. Isaacs, *Military Aircraft of Australia*, p.20.
4. *Report upon the Department of Defence from the First of July, 1914, until the Thirtieth of June, 1917*, Part I, Melb, 1917, p.437.
5. AA, CRS A2180, item D1821/1/6; SLA, Nov. 1918, p.461.
6. AA, CRS A2180, item D1821/1/6.
7. A.T Ross, The Arming of Australia: The Politics and Administration of Australia's Self Containment Strategy for Munitions Supply 1901–1945, Ph.D. thesis, University of NSW, 1986, pp.61–2.
8. ibid., p.173.
9. Air Board agendum no.69.
10. *CPD*, vol.93, p.4393.
11. Ross, op.cit., p.173.
12. Parnell & Boughton, p.6.
13. Air Board agendum no.172; Air Council agendum no.94.
14. Parnell & Boughton, p.35; *ADB*, 10, p.153.
15. Air Board agendum no.69.

16. Papers with Air Board agendum no.69.
17. Air Council agendum no.49.
18. Parnell & Boughton, p.35.
19. N.B. Love, *The Autobiography of Nigel B. Love: Aviation in Australia, 1915–1923*, Strathfield, NSW, 1969, p.39.
20. *Argus* (Melb), 24 Jul. 1922.
21. AA, CRS A705, item 44/3/256.
22. Ellison, p.78.
23. C.A. Owers, 'The Avro 504K in the RAAF', unpub.ms.
24. K. Isaacs, in *RAAF News*, Oct. 1961.
25. AA, CRS A705, item 44/3/256.
26. *SLA*, 1 Jul. 1922, p.290.
27. Parnell & Boughton, pp.35–6.
28. *CPP*, 1923–24, 4, p.376.
29. Air Board agendum no.458.
30. LASCo. proposals for aircraft manufacture, 1925: RAAF Historical Section.
31. ibid.
32. ibid.
33. De Havilland Company's views on local manufacture, 1927: RAAF Historical Section.
34. Wackett, p.87.
35. Air Board agendum no.73.
36. Air Council agendum no.53.
37. Air Board agendum no.286.
38. AA, CRS A705, item 199/1/7.
39. ibid.
40. Ellison, p.31n; also McCarthy, p.98.
41. Wackett to W.H. Nicholson, 25 Oct. 1920; letters in possession of Mr A.W. Nicholson.
42. ibid., 31 Aug. 1921.
43. Wackett, pp.86–7.
44. ibid., pp.87–8.
45. McCarthy, *Australia and Imperial Defence*, p.98.
46. Air Board agendum no.416.
47. Air Council agendum no.126.
48. Air Board agendum no.416.
49. Wackett, pp.89–90.
50. AA, CRS A705, item 199/1/7.
51. DROs, RAAF Experimental Section, No.1 dated 28 Jan. 1924; RAAF Historical Section.
52. Wackett, p.90.
53. DROs, RAAF Experimental Section, No.8 dated (erroneously?) 17 Feb. 1924.
54. Ewart memoir.
55. K. Isaacs, 'The Most Meritorious Performance in the Air', *Aircraft*, May 1974, p.20.
56. Interview: WgCdr W.A. Bradley, 15 Mar. 1989.
57. Wackett, p.94.

58. Ellison, p.34.
59. Brownell, pp.189–90.
60. See *Argus* (Melb), 2, 4, 8 Dec. 1924; unidentified press clippings in Wackett papers, NLA, MS 4858; Miller, pp.110–11.
61. Brogden, p.88.
62. Ellison, pp.34–5.
63. L.J. Wackett, 'Report on Australian Timbers for Aircraft Construction', undated RAAF Experimental Section document: NLA.
64. *Argus* (Melb), 25 Apr. 1925.
65. Air Council agendum no.128.
66. *Herald* (Melb), 4 Dec. 1925.
67. Letter: Mr C.H.G. Gardiner to author, 14 Apr. 1988.
68. *Daily Telegraph* (Syd), 9 Jul. 1925.
69. *Herald* (Melb), 4 Dec. 1925.
70. *CPD*, vol.113, p.2375.
71. *Sun* (Syd), 15 Oct. 1926.
72. Parnell & Boughton, p.50.
73. *SMH*, 29 Jun. 1927.
74. Interview: WgCdr C.F. Gates, 1 Mar. 1988.
75. Transcript of interview with SqnLdr W.J. Symons, Dec. 1977–Jan. 1978, held by RAAF Museum, Point Cook.
76. Parnell & Boughton, p.50.
77. *CPD*, vol.112, p.818, and Wackett, p.93.
78. DROs, RAAF Experimental Section, 28 Oct. 1926.
79. *Aircraft*, Aug. 1926, p.313; G. Jones, p.32.
80. McCarthy, p.101; undated clipping from *Daily Guardian* (Syd), in Wackett papers, NLA, MS 4858.
81. *Daily Telegraph* (Syd), 16 Dec. 1927.
82. *SMH*, 22 Feb. 1928; Parnell & Boughton, p.79.
83. Parnell & Boughton, p.79.
84. Wixted, p.29.
85. Letter from Reynolds to his fiancee, 8 May 1928; letters in possession of Mrs M. Bennett.
86. Wackett, pp.97–9.
87. Wackett papers, NLA, MS 4858; Parnell & Boughton, p.79.
88. Letter from Sgt J.E. Reynolds to his fiancee, 3 Jul. 1928; letters in possession of Mrs M. Bennett.
89. Interview: WgCdr J.H. McCorriston, 15 Mar. 1989.
90. Hewitt, *The Black One*, pp.126–7.
91. Interview: SqnLdr G.L. Grendon, 26 Apr. 1989.
92. *Advertiser* and *Register News-Pictorial* (Adel), 7 Jan. 1930; Wixted, p.47.
93. CAS file: '1928–29 Report of the Auditor–General on RAAF'; RAAF Historical Section.
94. AA, CRS A705, item 60/2/318.
95. CAS file: '1928–29 Report of the Auditor–General on RAAF'.
96. Parnell & Boughton, p.58.
97. *RAAF News*, May 1962.

98. AA, CRS A705, item 60/2/318.
99. *SMH*, 10 Dec. 1928.
100. *Aircraft*, Dec. 1928–Jan. 1929.
101. Interview: ACdre P.G. Heffernan, 1988.
102. *RAAF News*, May 1962; interview: Mr J.C. Miles, 15 Mar. 1989.
103. Salmond Report, Part I, p.14.
104. Wackett, pp.102–3.
105. Parnell & Boughton, pp.49, 79.
106. McCarthy, *Australia and Imperial Defence*, p.100.
107. AA, CRS A705, item 60/2/318.
108. ibid.
109. Interview: Mr J.C. Miles, 15 Mar. 1989.
110. AA, CRS A705, item 60/2/318.
111. ibid.; Air Board agendum no.1566.
112. Interview: AM Sir John McCauley, 4 Feb. 1988.
113. Interview: ACdre W.H. Garing, 14 Dec. 1987.
114. LASCo. quotation on aircraft for RAAF, 1928: RAAF Historical Section.
115. *Argus* (Melb), 15 Mar. 1929.
116. R.G. Parker, *Cockatoo Island*, Melb, 1977, p.38.
117. Air Board agendum no.1520.
118. AHSA *Newsletter*, vol.4, no.4, 1988, pp.83–5.
119. Williams, pp.183–4.
120. H. Purvis with J.F. Priest, *Outback Airman*, Adel, 1979, p.44.
121. Wackett, p.105.
122. *Sun* (Syd), 12 Oct. 1935.
123. *Herald* (Melb), 23 Feb. 1938.
124. Parnell & Boughton, p.143; Gillsion, p.40.
125. McCarthy, *Australia and Imperial Defence*, pp.111–12.
126. Williams, p.227.
127. McCarthy, *Australia and Imperial Defence*, pp.113–14.
128. ibid., p.115.
129. PRO, Air 8/221.
130. Statement by Parkhill, 23 Jan. 1937: NLA.
131. Notes on formation of CAC by Sir Lawrence Hartnett: courtesy of Mr C.A. Owers.
132. McCarthy, *Australia and Imperial Defence*, p.118.
133. AA, CRS A705, item 16/2/3.
134. AA, CRS A705, item 16/1/7.
135. AA, CRS A705, item 9/1/310.
136. AA, CRS A705, item 16/1/7.
137. ibid..
138. AA, CRS A705, item 16/1/9.
139. Notes provided by Mr W.A. Whetters, 4 Jun. 1989.
140. AHSA *Newsletter*, vol.4, no.3, 1988, p.60.
141. 3 Sqn unit diary.
142. *CPP*, 1937–38–39–40, 3, p. 1189; C.B. Schedvin, *Shaping Science and Industry: a*

history of Australia's Council for Scientific and Industrial Research, 1926–1949, Syd, 1987, pp. 197–8.

Chapter 10

1. *SLA*, 1 Nov. 1918, p.462.
2. *CAG*, no.6, 1921, p.80.
3. Williams, p.132.
4. See, for example, *SMH*, 12 Mar. 1934. The confusion probably arose from the fact that, when Brinsmead joined AIF Headquarters in 1917, the AFC staff supervised the 1st Wing in Gloucestershire where AFC pilots were trained. The claim that he obtained his flying 'ticket' at the Royal Navy's flying school at East church in Aug. 1915 (see Parnell & Boughton, pp.18, 47, 413) is simply a case of mistaken identity; Brinsmead was never a member of the RNAS (see *ADB*, 7, pp.415–16).
5. Air Council agendum no.3.
6. *CAG*, no.38, 1921, p.773.
7. Air Council agendum no.3.
8. Air Board agendum no.39.
9. Air Council agendum no.3.
10. Air Board agendum no.45.
11. AA, CRS A1195, item 827/1/151.
12. Air Council agendum no.88.
13. Gunn, p.58.
14. Butler, op.cit., p.18.
15. *Brisbane Courier*, 18 Nov. 1921.
16. Parnell & Boughton, p.439; D.C. Eyre, *The Illustrated Encyclopedia of Aircraft in Australia and New Zealand*, 2nd edn, Frenchs Forest, NSW, 1988, p.63.
17. DROs No.1 Station, 19 Apr. 1923: RAAF Museum.
18. AA, CRS A664/1, item 415/401/2.
19. Parnell & Boughton, p.440
20. *Man and Aerial Machines*, no.11, Jan. 1990, p. 147.
21. Bostock's logbook, 'Gwill' (Bostock) papers, vol.2.
22. Gillett, *Australia's Military Aircraft*, p.154.
23. Sinclair, p.259.
24. *SLA*, Jan. 1920, p.676.
25. Air Board agendum no.24.
26. RAAF History 1900–1951, op.cit., chap.4, p.3.
27. AA, CRS A664/1, item 415/401/2.
28. 1 Sqn unit diary: RAAF Historical Section.
29. AA, CRS A705, item 183/1/5.
30. Air Council agendum no.67.
31. *CAG*, no.9, 1923, p.214.
32. Records of flying training courses: RAAF Historical Section.
33. AA, CRS A705, item 183/1/5.
34. Parnell & Boughton, p.50.
35. Records of flying training courses.
36. Parnell & Boughton, p.52.
37. ibid., p.57.

38. *CPP*, 1923–24, 4, p.491.
39. Williams to Secretary, Department of Defence, 13 Jun. 1925: Williams papers, RAAF Museum.
40. Parnell & Boughton, pp.443–4.
41. AA, CRS A705, item 28/7/1988
42. Unidentied press clipping dated 19 Mar. 1930, in 'Gwill' (Bostock) papers.
43. Salmond report, part 1, p.17.
44. Ellington report, published with the comments of the Air Board, Civil Aviation Board and Commonwealth Government, 1939, p.27: RAAF Historical Section.
45. ibid., pp.29–30.
46. ibid., p.22.
47. ibid., p.27.
48. ibid., pp.22–23.
49. ibid., p.23.
50. ibid., pp.29–30.
51. ibid., p.25.
52. McCarthy, *Australia and Imperial Defence*, p.33, citing editorial comment in *Aircraft*, 17, (Aug. 1939), pp.7, 38.
53. C.R. Taylor, op.cit., pp.12–13.
54. *ADB*, 9, p.306.
55. ibid., pp.599–600.
56. E. Rogers, *Faith in Australia: Charles Ulm and Australian Aviation*, Frenchs Forest, NSW, 1987, p.66.
57. Williams, p.202.
58. P. Davis, *Charles Kingsford Smith: The World's Greatest Aviator*, Syd, 1977, pp.67–71.
59. Interview: AVM E. Hey, 12 Oct. 1988.
60. Brogden, pp.95–6; Ellison, pp.144–5.
61. Secretary, Air Board, to Secretary, Defence Department, 23 Dec. 1927: Williams papers.
62. Brogden, p.97.
63. AA, CRS A705, item 153/1/645.
64. AA, CRS A705, item 153/1/456.
65. AA, CRS A705, item 153/1/645; Eaton's report on the 'Kookaburra' search: RAAF Historical Section.
66. D. Smith & P. Davis, *Kookaburra: the most compelling story in Australia's aviation history*, Syd, 1980, p.60.
67. ibid., pp.68–9.
68. ibid., p.72.
69. ibid., p.76.
70. Hall, pp.20–1.
71. Report by Brain on search flight for 'Kookaburra': RAAF Historical Section.
72. Smith & Davis, op.cit., p.103.
73. *Argus* (Melb), 11 Mar. 1931; citation on Eaton's personal file.
74. Smith & Davis, pp.106–110.
75. E. Coote, *Hell's Airport*, Syd, 1934, p.235; AA, CRS A1, item 1930/10793.

76. AA, A705, item 153/1/737; Eaton's report on search operation for Pittendrigh and Hamre: RAAF Historical Section.
77. AA, CRS A705, item 153/1/758.
78. Sinclair, pp.241–2.
79. ibid.
80. ibid.
81. Hall, pp.23–4; M. Higgins, 'Solving the mystery of Australia's first major civil air disaster', *Canberra Historical Journal*, new series no.16, Sep. 1985, pp.1–7.
82. Gunn, pp.202–3, 269–70.
83. Parnell & Boughton, p.146.
84. C.R. Taylor, op.cit., p.13.
85. T. Gwynn-Jones, *Pioneer Airwoman: The Story of Mrs Bonney*, p.120; H. Purvis with J.F. Priest, op.cit., p.81.
86. Cited in L. Edmonds, 'Value for money?: civil aviation and defence between the wars', AWM *Journal*, no.15, Oct. 1989, p.30.
87. AA, CRS A705, item 208/1/547.
88. Butler, op.cit., pp.21–2; ADB, 9, pp. 506–7.
89. Butler, p.28.
90. ibid., pp.30, 36.
91. *ADB*, 8, p.42.
92. Butler, p.35.
93. ibid., pp.28–9; Parnell & Boughton, p.64.
94. Gunn, pp.202–3.
95. R.C. Meyer, *Aeradio in Australia*, Canb, 1985, p.3.
96. Hall, pp.31–2.
97. Meyer, op.cit., p.10.
98. Hall, p.41.
99. ibid., p.44.
100. AAIC report on 'Kyeema' crash, p.13: RAAF Historical Section.
101. Butler, p.39.
102. AAIC report on 'Kyeema' crash, pp.14–16.
103. L. Edmonds, op.cit., p.32.
104. Taped recollections: GpCapt W.G. Compagnoni, 1989.
105. AAIC report on 'Kyeema' crash, p.12.
106. Copley, pp.125–6.
107. Extract from report of committee inquiring into the general administration of CAB (Lawson report), 30 Dec. 1938: Bostock papers, folder 7.
108. Gillison, p.711.
109. Letter: Dr A.J. Goble to author, 2 Feb. 1988.
110. Gunn, pp.334–5, 343; Williams, p.331.
111. R. Lawson to CAS, Business Member and Secretary, Air Board, 30 Sep. 1940: Bostock papers, folder 7.

Chapter 11

1. *Argus* (Melb), 7 Apr. 1921.
2. Air Council agendum no.82: RAAF Historical Section.

3. *Argus*, 26 Mar. 1925.
4. ibid., 28 Mar. 1925.
5. ibid., 4 Jun. 1926.
6. 'RAAF Flying Accidents, Analysis and Statistics, 1921–37' (herafter referred to as 'RAAF accident statistics'): RAAF Historical Section.
7. AA, CRS A705, item 1/2/40.
8. *Aircraft*, Oct. 1926, pp.400–08.
9. AA, CRS A705, item 32/10/298; Gillett, *Wings Across the Sea*, p.23; Parnell & Boughton, p.60.
10. AA, CRS A705, item 32/10/381.
11. ibid.
12. Letter: Mr C.H.G. Gardiner to author, 14 Apr. 1988.
13. Williams, p.179.
14. Interview: SqnLdr G.L. Grendon, 26 Apr. 1989.
15. *Age* (Melb), 22 Apr. 1927.
16. Interview 13 Mar. 1989.
17. AA, CRS A705, item 32/10/381.
18. *SMH, Argus* (Melb), 10 May 1927, *Canberra Times*, 13 May 1927.
19. *Argus* (Melb), 11 May 1927.
20. *Canberra Times*, 13 May 1927.
21. *SMH*, 10 May 1927.
22. *Argus* (Melb), 9 May 1927.
23. *SMH*, 9 May 1927.
24. Gates interview.
25. *Age* (Melb), 16 May 1927.
26. *Canberra Times*, 13 May 1927.
27. ibid., 17 May 1927.
28. Williams to Trenchard, 31 Dec. 1927, Williams papers.
29. AA, CRS A705, item 32/10/381.
30. ibid.
31. Williams, pp.110–11.
32. Brookes interview, 13 Mar. 1989.
33. AA, CRS A705, item 1/2/26.
34. Parnell & Boughton, p.100.
35. See, for example, *Herald* (Melb), 8 Feb. 1933, 18 Apr. 1936; *Sun-News Pictorial* (Melb), 28 Apr. 1938.
36. Interview: ACdre P.G. Heffernan, 20 Jan. 1989.
37. AHSA *Newsletter*, vol.4, no.3, 1988, p.59.
38. *Argus* (Melb), 17 Feb. 1926.
39. Sutherland, p.32–3.
40. Written memoir prepared for author by ACdre U.E. Ewart, Jun. 1988.
41. ibid.
42. AA, CRS A705, item 32/10/298.
43. Gillett, *Wings Across the Sea*, p.23.
44. Sutherland, p.30.
45. Interview: ACdre P.G. Heffernan, 20 Jan. 1989.

46. G.G. Jefferis, 'Around Australia's Capital', vol.1, p.14, unpub. ms, 1931: Canberra & District Historical Society; Letter: Major–General R.R. McNicoll to author, 14 Jun. 1988.
47. Ewart memoir.
48. Sutherland, p.30.
49. R.K. Piper, 'Bulldog Breakup', *RAAF News*, Apr. 1987.
50. AAIC report no.77: RAAF Historical Section.
51. R.K. Piper, op.cit.
52. Pentland & Malone, p.39.
53. Letter: ACdre P.G. Heffernan to author, 12 Nov. 1988.
54. Letter: WOff A.J. Skinner to author, Mar. 1989.
55. Interview: ACdre J. Alexander, 28 Apr. 1988.
56. RAAF accident statistics.
57. AAIC report no.176B, cited in AHSA *Newsletter*, vol.3, no.2, 1987, p.34.
58. RAAF accident statistics; Parnell & Boughton, p.168.
59. Clune, pp.85–6.
60. *Wings*, Jun. 1987, p.23.
61. AA (Vic), MP 187/4, item 190.
62. L.A. Strange, *Recollections of an Airman*, Lond, 1933, pp.175–8.
63. Schaedel, *Australian Air Ace*, p.72.
64. Written memoir prepared for author by ACdre U.E. Ewart, Jun. 1988.
65. Parnell & Boughton, p.57.
66. Rayner, pp.28–9.
67. Interview: Sir Richard Kingsland, 25 Nov. 1987.
68. Unidentifed press clippings in possession of Sir Richard Kingsland.
69. Gillison, p.24.
70. Interviews: ACdre P.G. Heffernan, 9 Dec. 1983, and WgCdr C.F. Gates, 1 Mar. 1988.
71. AA, CRS A705, item 32/10/960.
72. Interview: ACdre P.G. Heffernan, 9 Dec. 1983.
73. Interview: WgCdr C.F. Gates, 1 Mar. 1988.
74. 3 Sqn unit diary: RAAF Historical Section; AA, CRS A705, items 32/10/833 and 42/3/44.
75. AA (Vic), MP 187/3, Box 1, and MP 116, S/1.
76. AAIC report No.155, cited in AHSA *Newsletter*, vol.3, no.2, 1987, p.31.
77. *Mercury* (Hob), 25 Feb. 1936; *SMH*, 5 Mar. 1936.
78. AAIC report No.188, cited in AHSA *Newsletter*, vol.3, no.2, 1987, p.35.
79. Parnell & Boughton, p.167.
80. DROs No.1 Station dated 17 Jul. 1923: RAAF Museum.
81. *Argus* (Melb), 6 Feb. 1926.
82. Taped recollections: WgCdr C.F. Gates, 1988.
83. Letter: ACdre P.G. Heffernan to author, 20 Apr. 1989.
84. Letter: WOff A.J. Skinner to author, Mar. 1989.
85. *Herald* (Melb), 10 Nov. 1934.
86. Interview: AM Sir Charles Read, 4 Feb. 1988.
87. Display notes at RAAF Museum.

88. Williams, p.159.
89. ibid., p.162.
90. *Argus* (Melb), 26 Jan. 1926.
91. *RAAF News*, Jan. 1960.
92. P.G. Heffernan, 'Parachuting for Beginners', *Stand-to*, Apr.–Jun. 1967, p.10.
93. W.W. Johnson, *Ripcord Australia*, Syd, 1984, pp.49, 51.
94. ibid., pp.10–11.
95. Williams, p.179.
96. ibid., p.180.
97. *Aircraft*, Aug. 1926, p.312.
98. Johnson, op.cit., pp.49–50; Rayner, p.27.
99. Johnson, p.50.
100. 1 Sqn unit diary.
101. P.G. Heffernan, 'Parachuting for Beginners', op.cit., p.12.
102. *War in the Air*, part 16 (vol.2), Lond, 20 Feb. 1936, p.502.
103. P.G. Heffernan, 'Parachuting for Beginners', op.cit., pp.13–14.
104. ibid., p.14.
105. Interview: AM Sir Charles Read, 4 Feb. 1988.
106. Interview: SqnLdr E.W.F. Sunderland, 14 Mar. 1989.
107. *Argus* (Melb), 13 Feb. 1926.
108. *Spectator* (Hamilton), 2 Sep. 1937.
109. S. Cauveren, 'William S. McAloney, GC, OBE: The only George Cross of the RAAF', *Transit* (Qantas staff publication), 1983: copy in possession of GpCapt W.S. McAloney.
110. ibid.
111. *Contact*, vol.42, no.1, 1987, p.3.
112. Cauveren, op.cit.
113. ibid.
114. AA, CRS A705, item 55/1/219; Air Board agendum no.2101.
115. Letter: J.A. Lyons (as Minister for Defence) to J.A. Lyons (as Prime Minister), 23 Nov. 1937; copy in possession of GpCapt W.S. McAloney.
116. PRO, HO45/17691, 4516; courtesy of Mr A. Staunton, Lyons, ACT.
117. L. Wigmore, *They Dared Mightily*, 2nd edn., Canb, 1986, pp.184, 187.
118. AA, CRS A705, item 32/10/1918; AAIC report no.184, cited in AHSA *Newsletter*, vol.3, no.2, 1987, p.34.
119. *CPD*, vol.155, pp.245, 294–5.
120. ibid., p.405.
121. ibid., pp.405–08, 409–19.
122. AA, CRS A705, item 32/10/1918.
123. Williams, p.235.
124. *Courier-Mail* (Brisb), 19 Jan. 1938.
125. AA, CRS A705, items 32/10/1915 and 32/10/1918.
126. This remains the opinion of AVM B.A. Eaton, who was the pilot who followed Fallon down out of formation and witnessed the crash: interview with author, 26 Jan. 1988.
127. AA, CRS A705, items 32/10/1915 and 32/10/1918.
128. *CPD*, vol.155, pp.545–6, 569, 738, vol.156, pp.1782–3.

129. Williams, p.236.
130. Ellington Report, paras 7, 10: RAAF Historical Section.
131. Air Board's response: held on unnumbered box file, RAAF Historical Section.
132. Gillison, pp.53–4.
133. Goble to Secretary, Defence Department (for Minister for Defence, 20 Sep. 1938: papers in possession of Dr A.J. Goble.
134. McCarthy, *Australia and Imperial Defence*, pp.90–1.
135. Interview: AVM E. Hey, 16 Jun. 1988.
136. Williams, p.244.
137. Parnell & Boughton, pp.167–8.
138. Hewitt, *The Black One*, p.162.
139. Brook, p.46; inquiry into crash of A4–11: RAAF Historical Section.
140. Parnell & Boughton, p.168.
141. ibid.
142. Inquiry into crash of A4–11.
143. Parnell & Boughton, p.168.
144. Inquiry into crash of A4–32: RAAF Historical Section.

Chapter 12

1. *CAG*, no.28, 1921, p.480.
2. Air Board agendum no.43: RAAF Historical Section.
3. G.C. Hartnell, 'The Problem of Command in the Australian Defence Force Environment', Canberra Papers on Strategy and Defence No.27, Canb, 1983, p.60.
4. J.R. Robertson, *J.H. Scullin: A Political Biography*, Nedlands, WA, 1974, pp.84–5.
5. *OYB*, no.17, 1924, p.594.
6. ibid.
7. Hartnell, op.cit., p.60.
8. ibid., pp.60–1.
9. Gillison, p.47.
10. Knight, pp.321–2.
11. A.D. Garrisson, 'Command and Control in Theory and Practice', Strategic and Defence Studies Centre Reference Paper No.111, Oct. 1983 (revised Oct. 1985), p.8.
12. Secretary, Air Board, to Secretary, Defence Department, 7 Sep. 1927, Williams papers: RAAF Museum.
13. Hartnell, op.cit., pp.64–5.
14. ibid., pp.61–2. 64–5.
15. *OYB*, no.15, 1922, p.925.
16. *CAG*, no.98, 1920, p.2097.
17. AA, CRS A1195, item 704/1/2.
18. McCarthy, *Australia and Imperial Defence*, p.28.
19. Williams, p.200.
20. Williams to White, 2 Jul. 1923: White papers, NLA, MS 5172, folder 25.
21. Air Council minutes: RAAF Historical Section.
22. Williams papers: RAAF Museum.
23. AA, CRS A705, item 4/3/6.

24. ibid.
25. ibid.
26. Williams to Trenchard, 11 May 1928, Williams papers: RAAF Museum.
27. Williams, p.218.
28. Interview: AVM C.G. Cleary, 2 Feb. 1988.
29. Goble papers, in possession of Dr A.J. Goble.
30. Letter: Dr A.J. Goble to author, 2 Feb. 1988.
31. Taped recollections: GpCapt W.G. Compagnoni, Dec. 1989.
32. Cutlack, p.181.
33. Written memoir prepared for author by ACdre U.E. Ewart, Jun. 1988.
34. Interview: AVM D.A.J. Creal, 10 Dec. 1987.
35. ibid.
36. Interview: SqnLdr W.J. Linnett, 13 Mar. 1989.
37. Air Council agendum no.105; Air Board agendum no.297.
38. *CPP*, 1923–24, 4, p.365.
39. Interview: SqnLdr G.L. Grendon, 26 Apr. 1989.
40. Interview: AM Sir George Jones, 21 Jan. 1988; AVM F.M. Bladin, unpub.ms.
41. Ewart memoir.
42. Interview: ACdre P.G. Heffernan, 9 Dec. 1983.
43. Schaedel, *Australian Air Ace*, pp.72–3.
44. Ewart memoir; Bladin ms., pp.8–9.
45. Sir Hudson Fysh, *Qantas Rising*, Syd, 1965, p.150.
46. Interview: ACdre U.E. Ewart, 7 Apr. 1988.
47. Interview: ACdre P.G. Heffernan, 9 Dec. 1983.
48. Interview: WgCdr C.F. Gates, 1 Mar. 1988.
49. Letter: ACdre U.E. Ewart to author, 16 May 1982; Rayner, p.21; ACdre P.G. Heffernan, 'RMC and the RAAF Connection', Duntroon Society *Newsletter* 1/1983, Mar. 1983, p.2; F.M. Bladin, unpublished ms., p.7.; interview: AVM C.D. Candy, 3 Aug. 1982.
50. Rayner, p.21.
51. Telephone interview: Mr C.H. Simpson, 20 Jan. 1988.
52. DPR video interview with ACM Sir Frederick Scherger, 2 Feb. 1982.
53. Letter: ACdre U.E. Ewart to author, 16 May 1982.
54. Interview: ACdre P.G. Heffernan, 20 Jan. 1989.
55. 1 AD unit diary: RAAF Historical Section.
56. Schaedel, *Australian Air Ace*, pp.70–1.
57. Interview: FLt F.C.H. Hakanson, 29 Jun. 1988.
58. Ellison, pp. 177–8.
59. ACdre P.G. Heffernan, 'So you want to be a Pilot?', *Stand-To*, Apr.–Jun. 1966, pp.24–5; interview with author, 25 Apr. 1988.
60. *Wings*, Sep. 1987, p.11.
61. 1 AD unit diary: RAAF Historical Section.
62. Salmond report, part 1, p.38.
63. ibid., pp.10, 38–9.
64. *Aircraft*, 31 Jul. 1929, p.451.

65. Report of Stores Accounts and Stocktaking Committee of Investigation, 31 Oct. 1928: CAS records, Air Office.
66. White papers, NLA, MS 5172, folder 22.
67. Interview: SqnLdr W.J. Linnett, 13 Mar. 1989.
68. *Age* (Melb), 17 Mar. 1930.
69. Report of the Auditor-General on RAAF, 1928–29: CAS records, Air Office.
70. Williams, pp.129, 195.
71. Williams papers: RAAF Museum.
72. ibid.
73. Williams, p.200.
74. ibid., p.195.
75. *ADB*, 9, p.674.
76. Parnell & Boughton, p.87.
77. *ADB*, 11, p.592.
78. Interview: AVM D.A.J. Creal, 10 Dec. 1987.
79. Horner, p.9.
80. Letter: Dr A.J. Goble to author, 2 Feb. 1988.
81. Williams, p.236.
82. Williams to GpCapt F.H. McNamara (ALO, London), 7 Oct. 1938: Williams papers.
83. AA, CRS A705, item 4/3/1.
84. *CAG*, no.6, 1921, p.80; *Argus* (Melb), 14 Apr. 1950.
85. Williams, p.134.
86. ibid.
87. AA, CRS A1195, item 856/1/23.
88. ibid.
89. Air Council agendum no.14.
90. AA, CRS A1195, item 856/1/23.
91. AA, CRS A705, item 4/4/61.
92. Telephone interview: Miss G.E. Joyce, 20 Feb.1990.
93. Williams, p.110.
94. Interviews: AVM D.A.J. Creal, 10 Dec. 1987, and ACdre P.G. Heffernan, 20 Jan. 1989; letter: Dr A.J. Goble to author, 2 Feb. 1988.
95. Interview: AM Sir John McCauley, 4 Feb. 1988.
96. DPR taped interview with AVM H.N. Wrigley, 2 Feb. 1982.
97. Interview: ACdre J. Alexander, 28 Apr. 1988.

Chapter 13

1. *Australasian Engineer* (Syd), 7 Jun. 1933, p.14.
2. AA, CRS A664, item 415/401/2.
3. 1 Sqn unit diary.
4. Hill, op.cit., pp.212–13.
5. Parnell and Boughton, p.98.
6. AA, CRS AA1969/100, item 4/4/Air; report on search for *Mystery Star*, 1936: RAAF Historical Section.
7. 1 Sqn unit diary.

8. *The Bomber*, Aug. 1931, p.31.
9. Interview: GpCapt J.E. Graham, 6 May 1988.
10. Brook, p.28.
11. Parnell & Boughton, p.98.
12. *Aircraft*, 18 Dec. 1930, p.16.
13. 1 Sqn unit diary.
14. ibid.
15. ibid.
16. *The Bomber*, Aug. 1931, pp. 30–1.
17. Program for Laverton air display, 10 Nov. 1934: RAAF Museum; *Argus* (Melb), 12 Nov. 1934; *Daily Telegraph* (Syd), 27 Nov. 1934.
18. *ADB*, 9, p.171.
19. *Australasian*, 25 Feb. 1922, p.350; *Register* (Adel), 2 Mar. 1922; file on Lake Eyre reconnaissance: RAAF Historical Section.
20. Clune, p.89.
21. File on Lake Torrens reconnaissance: RAAF Historical Section.
22. *ADB*, 10, pp.374–5.
23. Report by No.1 Sqn on Air Survey in Central Australia for S.A. Geographical Society, 1929: RAAF Historical Section.
24. Parnell & Boughton, p.90.
25. *Register News-Pictorial* (Adel), 13 Aug. 1929.
26. C.T. Madigan, 'The Geology of the MacDonnell Ranges and Neighbourhood, Central Australia', ANZAAS *Report*, vol.21, Syd, 1933, p.75.
27. Parnell & Boughton, p.69.
28. Report on 1930 Tasmanian survey flight by A11–1: RAAF Historical Section.
29. ibid.
30. Report on Port Davey reconnaissance, 1929: RAAF Historical Section.
31. Parnell & Boughton, p.69.
32. Report on 1930 Tasmanian survey flight by A11–1: RAAF Historical Section.
33. ibid.
34. Report on the 1931 Tasmanian flight: RAAF Historical Section.
35. ibid.
36. Report on the 1932 Tasmanian flight: RAAF Historical Section.
37. Interview: ACdre W.H. Garing, 14 Dec. 1987.
38. *The Bomber*, Aug. 1931, p.30.
39. Report of Liverpool–Helidon survey, 1933: RAAF Historical Section.
40. Garing interview.
41. AA, CRS AA1969/100, item 4/10/Air.
42. *The Bomber*, vol.1, no.2, Nov. 1931, p.22.
43. *Sun News-Pictorial* (Melb), 22 Jun. 1932.
44. *The Bomber*, Nov. 1931, p.22.
45. P.G. Heffernan, 'The RAAF Meteorological Flight', Stand-to, Jan.–Mar. 1966, p. 19.
46. ibid., p.22.
47. Interview: Sir Richard Kingsland, 25 Nov. 1987.
48. Interview with Mr E.V. Read, 28 Apr. 1988.

49. ibid.
50. 3 Sqn unit diary; Parnell & Boughton, p.124.
51. AHSA *Newsletter*, vol.3, no.2, 1987, p.32.
52. K. Isaacs, in *Aircraft*, May 1974, pp.19–21, 50.
53. Report on round-Australia seaplane flight, 1924: RAAF Historical Section.
54. Copley, p.63.
55. Isaacs, in *Aircraft*, May 1974.
56. *Argus* (Melb), 6 Apr. 1925.
57. Letter in papers held by Cdre J.D. Goble.
58. Williams' report on survey of Mandated Territories, 1926: RAAF Historical Section.
59. Sinclair, p.18.
60. Trenchard to Williams, Dec. 1926: Williams papers, RAAF Museum.
61. K. Isaacs, 'Historic Flight', *DFJ*, no.8, Jan.–Feb. 1978, pp.24–32.
62. *Argus* (Melb), 3 Jun. 1927.
63. Report on Central Australia reconnaissance, 1927: RAAF Historical Section.
64. Williams, pp.188, 193.
65. ibid., pp.229–30.
66. Report of flight to Singapore, 1938: Williams papers.
67. Williams, p.233.
68. Logbook: AVM A.L. Walters: in possession of Mrs J. Walters.
69. K. Isaacs, 'The Fairey IIID in Australia', p.48.
70. Interview: WgCdr C.F. Gates, 1 Mar. 1988.
71. AA, CRS A705, item 34/2/53.
72. *Argus* (Melb), 10 May 1927.
73. AA, CRS A705, item 34/2/118.
74. *Canberra Times*, 3, 6 May 1927.
75. Letter: Mr C.H.G. Gardiner to author, 14 Apr. 1988.
76. *Canberra Times*, 6 May 1927.
77. ibid.
78. G. Souter, *Acts of Parliament*, Melb, 1988, p.222.
79. *Canberra Times*, 13 May 1927.
80. C.A. Owers, 'The RAAF's first fighter'.
81. *Herald* (Melb), 10 Apr. 1963.
82. AA, CRS A705, item 57/1/621.
83. Ewart memoir.
84. Summers' report on 1932 training flight, dated 14 Nov. 1932: RAAF Historical Section.
85. P.G. Heffernan 'Forced Landing', *Stand-to*, Jul.–Sep. 1967, pp.36–40.
86. Summers' report on 1935 training flight: RAAF Historical Section.
87. AVM A.T. Cole, 'Merry Old Souls'.
88. Diary for 1933, Cole papers.
89. Hall, pp.28–9.
90. Taped recollections: WgCdr C.F. Gates, 1988.
91. *Smiths Weekly* (Syd), 30 Sep. 1933.
92. *Sun* (Syd), 23 Apr. 1938; AA (Vic), MP 187/4, item 186.

93. Interview: AM Sir Charles Read, 4 Feb. 1988.
94. Taped recollections: WgCdr C.F. Gates.
95. *Register News-Pictorial* (Adel) and *Advertiser* (Adel), 2 Oct. 1929.
96. *Advertiser*, 24 Mar. 1930.
97. Pentland & Malone, p.32.
98. *Sun* (Syd), 8, 9 Sep. 1935.
99. Taped recollections: GpCapt W.G. Compagnoni, Dec.1989.
100. Report by FLt Scherger dated 16 Mar. 1933: RAAF Historical Section.
101. Report by FLt Wright dated 13 Mar. 1934: RAAF Historical Section.
102. Air 454/7/2.

Chapter 14

1. AA, CRS A664/1, item 415/401/2.
2. Gillett, *Wings Across the Sea*, p.21; *Argus* (Melb), 21 Jul. 1924.
3. K. Isaacs, 'The Fairey IIID in Australia', *Air Enthusiast*, no.24, p.48; Gillett, *Wings Across the Sea*, pp.21–2.
4. Clune, p.90.
5. *Bowen Independent*, 28 Aug. 1926.
6. Hewitt, *The Black One*, pp.111–12.
7. *Bowen Independent*, 28 Aug., 4 Sep. 1926.
8. ibid., 11, 18 Jan. 1984.
9. Hewitt, *The Black One*, pp.112–13; Gillett, *Wings Across the Sea*, p.25.
10. Interview: AVM D.A.J. Creal, 10 Dec. 1987.
11. Folder of survey maps of Barrier Reef: RAAF Historical Section.
12. Hewitt, *The Black One*, p.113.
13. Commander F.G. Crowther, cited in Gillett, *Wings Across the Sea*, p.32.
14. File on survey of Barrier Reef, 1928: RAAF Historical Section.
15. Crowther, in Gillett, *Wings Across the Sea*, p.32.
16. ibid.
17. Report on Papuan Survey Flight 1927 by FLt G. Packer: RAAF Historical Section.
18. Sinclair, p.272.
19. Papuan Survey Flight report, p.9.
20. ibid., pp. 11–12; G. Packer, 'Around New Guinea in a Flying Boat', *The Bomber: The Journal of No.1 Squadron, R.A.A.F.*, vol.1, no.2, Nov. 1931, pp.18, 20.
21. Papuan Survey Flight report, p.16.
22. Packer in *Bomber* article, p.18.
23. ibid.
24. R.K. Richardson, 'The Use of Air Photography in the Oil Exploration of Papua and New Guinea', paper dated 29 Apr. 1929: RAAF Historical Section.
25. *ADB*, 10, p.456; Parnell & Boughton, p.54.
26. K. Isaacs, *Military Aircraft of Australia*, p.14.
27. D. Mawson, *The Home of the Blizzard*, vol.1, Lond, 1915, p.24.
28. A.G. Price, *The Winning of Australian Antarctica: Mawson's BANZARE Voyages 1929–31*, Syd, 1962, p.21.
29. Correspondence between Casey and Trenchard, May–Jun. 1929: Trenchard papers, RAF Museum, MFC 76/1/151.

30. P. Mawson, *Mawson of the Antarctic: The Life of Sir Douglas Mawson, FRS, OBE*, Lond, 1964, p.159.
31. In his letter to Trenchard dated 10 May 1929, Casey stated that the pilot would be 'probably from the RAAF'.
32. Antarctic file: RAAF Museum.
33. S.A.C. Campbell, 'With "The Discovery", 1929–30', The *Bomber*, Aug. 1931, p.10.
34. ibid., p.11.
35. ibid.
36. D.J. Wilson, 'Antarctic Flight: R.A.A.F. Operations in the Australian Antarctic Territory', unpub. ms, 1987, p.8; RAAF Historical Section.
37. Price, pp.128–31.
38. *Advertiser* (Adel), 25 Oct. 1947.
39. Price, pp.183, 185.
40. *Argus* (Melb), 1 May 1934.
41. Price, p.161.
42. Antarctic file: RAAF Museum.
43. Price, pp.168–71.
44. See *Aircraft*, Jan 1986, pp.46–8, Feb. 1986, pp.52–4; Antarctic file: RAAF Museum.
45. *Argus* (Melb), 3 Dec. 1935.
46. AA, CRS A705, item 153/1/786.
47. C.R. Taylor, op.cit., p.11.
48. Letter: WOff A.J. Skinner to author, Mar. 1989.
49. C.R. Taylor, op.cit.
50. Interview: SqnLdr G.L. Grendon, 26 Apr. 1989.
51. Wilson, op.cit., p.15.
52. Article by A. Argent (writing as 'Pelorus') in *Aircraft*, Feb. 1986, p.52.
53. Parnell & Boughton, p.55.
54. R.A. Swan, *Australia in the Antarctic: Interest, Activity and Endeavour*, Melb, 1961, p.219.
55. Copy of Douglas's log: RAAF Historical Section.
56. *CPP*, 1934–35–36–37, 3, pp.150, 152.
57. G.E. Douglas, 'The Search for Lincoln Ellsworth', unpub. account held by RAAF Historical Section; also Douglas' log, and report of flying operations during the rescue, written in Mar. 1936: RAAF Historical Section.
58. Douglas' log.
59. Antarctic file: RAAF Museum.
60. Wilson, p.18.
61. *CPP*, 1934–35–36–37, 3, p.157.
62. *Women's Weekly*, 29 Feb. 1936.
63. Wilson, p.18.
64. Douglas' log.
65. Williams to Secretary, Defence Department, 24 Mar. 1927: Williams papers.
66. *Argus* (Melb), 10 Nov. 1928.
67. Extract of file 165/1/539: RAAF Historical Section.
68. Charlesworth's report on his tour of duty in England and Somaliland: RAAF Historical Section.

69. Interview: Mr J.C. Miles, 15 March 1989.
70. Charlesworth's report on the Survey of Potential Oil Fields: RAAF Historical Section.
71. Charlesworth's report on the Photographic Survey of Potential Oilfields in Tasmania and in vicinity of Mt. Gambier: RAAF Historical Section.
72. *Argus*, 3 Jun. 1933.
73. See, for example, the passing reference to the RAAF reconnaissance in P.G. Purcell (ed.), *The Canning Basin, W.A.*, Perth, 1984, p.14.
74. Report of Air Survey IDC: RAAF Historical Section.
75. *CPP*, 1937, 5, p.2349.
76. ibid., 1934–35–36–37, 3, p.353.
77. *CPP*, 1934–35–36–37, 3, p.354.
78. Simms' report on Operations of the North Australia Survey Flight: RAAF Historical Section.
79. *CPP*, 1934–35–36–37, 3, p.409.
80. Simms' report, op.cit.
81. Jones' report of the tour of inspection: RAAF Historical Section.
82. G. Jones, p.55.
83. ibid., pp.55–7.
84. ibid., p.57, and interview with Jones, 21 Jan. 1988.
85. CPP, 1934–35–36–37, 3, p.437.
86. AA (Vic), MP187/4, item 159.
87. Lachal's report on Portland Roads Survey, 1936; RAAF Historical Section.
88. *CPP*, 1934–35–36–37, 3, p.439.
89. ibid., 1937, 5, p.2430.
90. ibid., 1937–38–39–40, 4, p.14.
91. Report on operations of the Western Air Detachment, 1937: RAAF Historical Section.
92. AA, CRS A705, item 153/2/142; Administrative Instruction No.12, RAAF Station, Laverton: RAAF Historical Section.
93. Hall, pp.33–40.
94. File on search for A3–2, May 1937: RAAF Historical Section; see also *CPP*, 1937–38–39–40, 4, pp.15, 110.
95. SqnLdr A.E. Moore, letter dated 15 Aug. 1937: copies in RAAF Historical Section.
96. ibid., 29 Aug. 1937.
97. ibid., 12, 20 Sep. 1937.
98. *Herald* (Melb), 28 Sep. 1937.
99. Letter by Moore dated 2 Oct. 1937.
100. *Herald*, 29 Sep. 1937.
101. *Argus*, 9 Jun. 1938; citation on Hely's personal file.
102. *CPP*, 1937–38–39–40, 4, p.130.
103. ibid., p.226.
104. 14 Sqn unit diary: RAAF Historical Section; telephone interview: SqnLdr G.A. Everingham, 27 Nov. 1989.
105. *CPP*, 1937–38–39–40, 4, pp.227, 237, 245.
106. Fowler's report dated 17 Dec. 1936, RAAF Historical Section; correspondence: Williams and Fowler, Jun. Jul. 1936, Williams papers.

107. File on Air Surveys of Pelagic Fish Movement in S.E. Australian Waters: RAAF Historical Section.
108. Fowler's report dated 12 Apr. 1937: RAAF Historical Section
109. Interview: ACdre J. Alexander, 28 Apr. 1988.
110. C.B. Schedvin, op.cit., p. 115.

Chapter 15

1. Parnell & Boughton, pp.131–32.
2. Gillison, p.47.
3. ibid., p.52; Parnell & Lynch, pp.51–2.
4. Interview: WgCdr R.C. Cresswell, 14 Sep. 1989.
5. Interview: AM Sir Charles Read, 4 Feb. 1988.
6. Taped recollections: GpCapt W.G. Compagnoni.
7. PRO, Air 8/221.
8. ibid.
9. AA, CRS A1196, item 35/501/14.
10. Bladin unpub.ms. and correspondence.
11. McCarthy, p.91.
12. Interview: AM Sir George Jones, 21 Jan. 1988.
13. Rayner, pp.110, 116.
14. University of Melbourne Calendar, 1938, Melb, 1937, p.1347.
15. Interview: AVM I.D. McLachlan, 5 Feb. 1988.
16. AA, CRS A705, items 56/1/213, 56/1/214
17. AA, CRS AA1969/100, item 108/2/P1.
18. AA, CRS A705, item 4/4/120.
19. *Herald* (Melb), 17 Apr. 1939.
20. Compagnoni tapes.
21. AA, CRS A705, item 9/18/28.
22. Compagnoni tapes.
23. Williams papers: RAAF Museum.
24. Compagnoni tapes.
25. Williams papers: RAAF Museum.
26. Telephone interview with Mr C. Eaton (son of GpCapt C. Eaton), 11 Jan. 1990.
27. AA, CRS AA1969/100, item 2806/8/W.
28. AA, CRS A705, item 56/1/294.
29. AA, CRS A1196, item 15/501/93; G. Jones, pp.60–1.
30. AA, CRS AA1969/100, item 2806/8/W.
31. AA, CRS A705, item 56/1/294.
32. PRO, Air 8/221.
33. ibid.
34. Air Board submission to Secretary, Defence Department, 3 May 1938: Williams papers, RAAF Museum.
35. ibid.
36. *Man and Aerial Machines*, no.4, May 1989, p.39.
37. File 'Defence Development Programme. Action Proposed in the event of an

Immediate Emergency', in papers of Secretary, Department of Air: RAAF Historical Section.
38. Parnell & Lynch, p.51.
39. File 'Defence Development Programme. Action Proposed in the event of an Immediate Emergency'.
40. AA, CRS A705, item 9/18/28.
41. Williams, pp.236–7.
42. Ellington Report, para.14.
43. Parnell & Boughton, pp.160–1, 168–9; D. Vincent, 'Quest for a bomber: the RAAF's supply problems, 1934–41', AWM *Journal*, no.9, Oct. 1986, pp.17–26.
44. Interview: WgCdr C.F. Gates, 1 Mar. 1988.
45. File 'Defence Development Programme. Action Proposed in the event of an Immediate Emergency'.
46. 1 Sqn unit diary.
47. Ross, p.126.
48. 1 AD unit diary.
49. Interview: WgCdr R.C. Cresswell, 14 Sep. 1989.
50. G. Long, *The Six Years War: A Concise History of Australia in the 1939–45 War*, Canb, 1973, p.15.
51. McCarthy, *A last call of empire*, p.5.
52. J.R. Robertson, in M. McKernan & M. Browne (eds.), *Australia Two Centuries of War & Peace*, Canb, 1988, pp.227; Williams, pp.248–9.
53. Gillison, p.58n.
54. Williams, pp.248, 250.
55. ibid., p.249.
56. Long, p.18.
57. McCarthy, *A last call of empire*, pp.13, 15; Gillison, p.79.
58. Williams, p. 251.
59. AA, CRS A1196, item 57/501/4.
60. ibid.
61. ibid.
62. Robertson in McKernan & Browne, p.229.
63. *Sun* (Melb), 21 Dec. 1939.
64. Interview: Cdre J.D. Goble, 4 Feb. 1988.
65. Goble to Street, 12 Oct. 1939, Goble papers in possession of Dr A.J. Goble.
66. Goble to Street (through Shedden), 2 Nov. 1939, ibid.
67. Goble to Newall, 14 Nov. 1939, ibid.
68. Goble to Shedden, 22 Dec. 1939, ibid.
69. Interview: ACdre U.E. Ewart, 7 Apr. 1988; correspondence with author, Jun.–Jul. 1988.
70. Whiskard to Anthony Eden, Secretary of State for the Dominions, 1 Jan. 1940, reproduced in Robertson & McCarthy, p.55.
71. ibid., pp.54–5.
72. ibid.
73. Letter: Dr A.J. Goble to author, 13 Jun. 1989.
74. *Argus* (Melb), 21 Dec. 1939.
75. Whiskard to Eden, 1 Jan. 1940.

76. Robertson & McCarthy, p.53.
77. Williams, p.253.
78. A.T. Cole, unpub.ms.
79. Robertson & McCarthy, p.53.
80. McCarthy, *A last call of empire*, pp.30–1.
81. ibid., p.30.
82. War report of the CAS, RAAF, 3 Sep. 1939–31 Dec. 1945, p.13: RAAF Museum.
83. F. Doak, *Royal Australian Air Force: A Brief History*, Canb, 1981.

Select bibliography

Unpublished official records

Australian Archives, ACT

Files in the following series have been consulted:

- CRS A1
- CRS A2
- CRS A289
- CRS A571
- CRS A664
- CRS A705
- CRS A1195
- CRS A1196
- CRS A1952
- CRS A1969/100
- CRS A2023
- CRS A2180
- CRS A6006
- CRS B197

Australian Archives, Vic

- CRS A705
- MP 124/6
- MP 153/20
- MP 187/3
- MP 187/4
- MP 341
- MP 367/1
- MP 1049/1
- MP 1118/2

Department of Defence

- Air 454/7/2
- DPR interviews (video and audio tape) with:
  - ACM Sir Frederick Scherger
  - AVM H.N. Wrigley

RAAF Historical Section, Air Office

A range of files are held here, as indicated in references; however, principal sources of significance are:

- Air Council minutes
- Air Board minutes
- Unit diaries

RAAF Museum, Point Cook

Holdings are similarly diverse, but main categories used were:

- AAC Routine Orders
- DROs No.1 FTS
- DROs No.1 Station
- Williams papers (see unpublished private papers)

Public Record Office, Kew, England

Material of value was examined in Air Ministry records held in the following series:
Air 2
Air 8
Air 9

Unpublished private papers and manuscripts

Banting, AVM G.G., papers, IWM, 74/95/1
Bladin, AVM F.M., unpub.ms in possession of Mrs J. Crock
Bostock, AVM W.D., papers (including 'Gwill' papers), AWM, PR 84/127
Cole, AVM A.T., papers (include unpub.ms), AWM, PR 88/154
Garrisson, ACdre A.D.J., letters, in possession of ACdre Garrisson
Goble, AVM S.J., papers, in possession of Dr A.J. Goble and Cdre J.D. Goble
Hartnell, AVM G.C., letters, in possession of Mrs J.M. Hartnell
Hughes, Rt.Hon. W.M., papers, NLA, MS 1538
Packer, GpCapt G., papers, NLA, MS 780
Reynolds, WgCdr J.E., letters, in possession of Mrs M. Bennett
Robertson, L.J., unpub.ms, held by RAAF Historical Section
Salmond, MRAF Sir John, papers, RAF Museum, Hendon, Eng, B2630
Symons, SqnLdr W.J., record of interview, held by RAAF Museum
Trenchard, MRAF Viscount, papers, RAF Museum, MFC 76/1/266
Wackett, Sir Lawrence, papers, NLA, MS 4858
White, General Sir Brudenell, papers, NLA, MS 5172
Williams, AM Sir Richard, papers, NLA, MS 6525, and at RAAF Museum
Wrigley, AVM H.N., ms.article, courtesy of FLt P.E. Hall

Interviews

The following persons were contacted by the author in person, or by telephone (indicated with *):
Alexander, ACdre J., of Wahroonga, NSW
Ambrose, Captain L.R., of Northbridge, NSW
Bennett, Mrs M. (daughter of WgCdr J.E. Reynolds), of Ballarat, Vic
Bostock, Mrs N. (widow of AVM W.D. Bostock), of Torquay, Vic
Bonwick, Mrs Y. (daughter of ACdre H.F. Delarue), of Beaumaris, Vic
Bradley, WgCdr W.A., of Heidelberg, Vic
Brookes, GpCapt Sir Wilfred, of Toorak, Vic
Candy, AVM C.D. (deceased) and Mrs E. Candy, of Kingston, ACT
Cleary, AVM C.G., of Deakin, ACT
Clerehan, Mrs S. (daughter of AVM A.T. Cole), of South Yarra, Vic
Creal, AVM D.A.J., of Curtin, ACT
Cresswell. WgCdr R.C., of Garran, ACT
Crock, Mrs J. (daughter of AVM F.M. Bladin), of Kew, Vic
Diprose, Mr L.M., of Strathmore, Vic
Eaton, AVM B.A., of Isaacs, ACT
Eaton, Mr C. (son of GpCapt C. Eaton), of Gooseberry Hill, WA (*)
Everingham, SqnLdr A.E., of Strathmore, Vic (*)
Ewart, ACdre U.E., of Balgowlah, NSW
French, WgCdr D.J., of Nagambie, Vic
Gardiner, Mr C.H.G., of Camberwell, Vic
Garing, ACdre W.H., of Turramurra, NSW
Garrisson, ACdre A.D.J., of Kingston, ACT
Gates, WgCdr C.F., of Ashmore, Qld

Goble, Dr A.J. (son of AVM S.J. Goble), of Heidelberg, Vic
Goble, Cdre J.D. (son of AVM S.J. Goble), of Terry Hills, NSW
Goble, Mrs K.D.L. (widow of AVM S.J. Goble), of Carnegie, Vic
Graham, GpCapt J.E., of Malvern, Vic
Gray, Mrs J. (daughter of ACM Sir Frederick Scherger), of South Yarra, Vic
Grendon, SqnLdr G.L., of Queanbeyan, NSW
Hakanson, FLt F.C.H., of Hurlstone Park, NSW
Heffernan, ACdre P.G., of South Yarra, Vic
Hey, AVM E., of Pearce, ACT
Hobler, AVM J., of Ashmore, Qld
James, Mr A.F.P., of Wahroonga, NSW
Jones, AM Sir George, of Beaumaris, Vic
Joyce, Miss G.E. (daughter of Mr A.C. Joyce), of Red Hill, ACT
Kingsland, Sir Richard, of Campbell, ACT
Lawrence, Mr T.F.C., of Campbell, ACT
Linnett, SqnLdr W.J., of South Oakleigh, Vic
McCauley, AM Sir John (deceased)
McCorriston, WgCdr J.H., of Bulleen, Vic
McLachlan, AVM I.D., of Darling Point, NSW
Miles, Mr J.C. (deceased)
Moore, SqnLdr A.E., of Kalorama, Vic
Podger, ACdre I.S., of Deakin, ACT
Read, AM Sir Charles, of Bayview, NSW
Read, Mr E.V., of Sorrento, Qld
Robey, Mr K.G.T., of Georges Hall, NSW (*)
Simpson, Mr C.H., of Kew, Vic (*)
Steege, ACdre G.H., of Palm Beach, NSW
Stewart, Mrs J. (daughter of AVM W.D. Bostock), of Frankston, Vic
Sunderland, SqnLdr E.W.F., of Beaumaris, Vic
Taylor, ACdre C.R., of Burwood, Vic
Walters, Mrs J. (widow of AVM A.L. Walters), of Croydon North, Vic
Warland, Mr I. (son of FOff H. Warland), of Manifold, Vic
Whetters, Mr W.A., of Coledale, NSW
Woodward, Hon. Sir Edward (son of Lt-Gen Sir Eric Woodward), of South Yarra, Vic

Correspondence

Austin, Mr R. (son of GpCapt H.A. Austin), of Berrigan, NSW
Chilton, AM Sir Edward, of Henley-on-Thames, Eng
Compagnoni, GpCapt W.G., of Wondai, Qld
Daish, GpCapt H.C., of Dibden Purlieu, near Southampton, Eng
Gilmour, Rev. W.F., of Point Lonsdale, Vic
Hancock, AM Sir Valston, of Dalkeith, WA
Lloyd, Mrs J. (daughter of GpCapt G. Packer), of Brighton, Vic
McAloney, GpCapt W.S., of Sandringham, Vic
McNichol, Maj-Gen R.R., of South Yarra, Vic
Nicholson, Mr A.W. (son of WOff W.H. Nicholson), of Bilinga, Qld
Oakes, Lt-Col R.F. (deceased)
Robertson, Mr L.J., of Staines, Eng
Scherger, ACM Sir Frederick (deceased)
Skinner, Mr A.J., of South Melbourne, Vic
Steel, Mr R.W., of Wamberal, NSW
Thomas, GpCapt Sir Frederick, of Barwon Heads, Vic
Truscott, Mr J.L., of Eaglehawk, Vic

Videon, Mr B.J., of East Bentleigh, Vic

Official printed sources

Commonwealth of Australia Gazette
Commonwealth Parliamentary Debates
Commonwealth Parliamentary Papers
Department of Air, *The Golden Years: Royal Australian Air Force 1921–1971*, Canb, 1971
Official Year-Books of the Commonwealth of Australia
Military Orders, held in AWM
RAAF Lists, held in AWM

Principal secondary sources referred to in notes

Bramson, A., *Master Airman: A Biography of Air Vice-Marshal Donald Bennett, CB, CBE, DSO*, Shrewsbury, Eng, 1985.
Brogden, S., *The History of Australian Aviation*, Melb, 1960.
Brook, W.H., *Demon to Vampire: The story of No.21 (City of Melbourne) Squadron*, Glen Waverly, Vic, 1986.
Brownell, R.J., *From Khaki to Blue: The Autobiography of Air Commodore R.J. Brownell, CBE, MC, MM*, Canb, 1978.
Clune, F., *D'Air Devel: The Story of 'Pard' Mustar*, Syd, 1940.
Copley, G.R., *Australians in the Air*, Adel, 1976.
Cutlack, F.M., *The Australian Flying Corps in the Western and Eastern Theatres of War 1914–1918*, Syd, 1940.
Doak, F. & Isaacs, J., *Australian Defence Heritage: The buildings, establishments and sites of our military history that have become part of the national estate*, Syd, 1988.
Ellison, N., *Flying Matilda: Early Days in Australian Aviation*, Syd, 1957.
Gillett, R., *Australia's Military Aircraft*, Syd, 1987.
——, *Wings Across the Sea*, Syd, 1988.
Gillison, D.N., *Royal Australian Air Force 1939–1942*, Canb, 1962.
Gunn, J., *The Defeat of Distance: QANTAS 1919–1939*, St. Lucia, Qld, 1985.
Hall, E.R., *A Saga of Achievement: A Story of the Men and Women who Maintained and Operated Radio and Radar Systems of the RAAF over 50 Years*, Box Hill, Vic, 1978.
Hamill, I., *The Strategic Illusion: The Singapore Strategy and the Defence of Australia and New Zealand*, Singapore, 1981.
Hewitt. J.E., *Adversity in Success*, South Yarra, Vic, 1980.
——, *The Black One*, South Yarra, Vic, 1984.
Horner, D.M., *High Command: Australia and Allied Strategy 1939–1945*, Canb, 1982.
Hudson, W.J., & North, J. (eds.), *My Dear PM: R.G. Casey's Letters to S.M. Bruce 1924–1929*, Canb, 1980.
Isaacs, K., *Military Aircraft of Australia 1909–1918*, Canb, 1971.
Johnston, W.W., *Ripcord Australia*, Syd, 1984.
Jones, G., *From Private to Air Marshal: The Autobiography of Air Marshal Sir George Jones*, Richmond, Vic, 1988.
Jones, R.M., *Seagulls, Cruisers and Catapults: Australian Naval Aviation, 1913–1944*, Hob, 1989.
Knight, F.F., *These Things Happened: Unrecorded History 1895–1946*, Melb, 1975.
McCarthy, J.M., *Australia and Imperial Defence 1918–39: A Study in Air and Sea Power*, St. Lucia, Qld, 1976.
——, *A Last Call of Empire: Australian aircrew, Britain and the Empire Air Training Scheme*, Canb, 1988.
Miller, H.C., *Early Birds*, Adel, 1968.

Odgers, G.J., *The RAAF: An Illustrated History*, Frenchs Forest, NSW, 1989.
Parnell, N.M., & Lynch, C.A., *Australian Air Force since 1911*, Syd, 1976.
Parnell, N.M., & Boughton, T., *Flypast: A Record of Aviation in Australia*, Canb, 1988.
Pentland, G., & Malone, P., *Aircraft of the RAAF 1921–78*, Dandenong, Vic, 1978.
Rayner, H.R., *Scherger: A Biography of Air Chief Marshal Sir Frederick Scherger, KBE, CB, DSO, AFC*, Canb, 1984.
Robertson, J.R., & McCarthy, J.M., *Australian War Strategy 1939–1945: A Documentary History*, St. Lucia, Qld, 1985.
Schaedel, C., *Australian Air Ace: The exploits of 'Jerry' Pentland, MC, DFC, AFC*, Adel, 1979.
Sinclair, J., *Wings of Gold: How the Aeroplane Developed New Guinea*, Bathurst, NSW, 1983.
Sutherland, L.W., *Aces and Kings*, Syd, 1935.
Wackett, L.J., *Aircraft Pioneer: An Autobiography*, Syd, 1972.
Walker, A.S., *Medical Services of the RAN and RAAF*, Canb, 1961.
Williams, R., *These are facts: The Autobiography of Air Marshal Sir Richard Williams, KBE, CB, DSO*, Canb, 1977.
Wixted, E.P., *The North-West Aerial Frontier 1919–1934*, Brisb, 1985.

Index of Persons

Ranks shown are those attained at the end of each individual's career, where known. Numerals in italics denote illustrations.

A1-22